William Hea

The Practice of Medi

Applied to the diseases and accidents in

Second Edition

William Heath Byford

The Practice of Medicine and Surgery
Applied to the diseases and accidents incident to women. Second Edition

ISBN/EAN: 9783337776077

Printed in Europe, USA, Canada, Australia, Japan

Cover: Foto ©berggeist007 / pixelio.de

More available books at **www.hansebooks.com**

THE PRACTICE

OF

MEDICINE AND SURGERY

APPLIED TO THE

DISEASES AND ACCIDENTS

INCIDENT TO WOMEN.

BY

WM. H. BYFORD, A.M., M.D.,

AUTHOR OF "A TREATISE ON THE CHRONIC INFLAMMATION AND DISPLACEMENTS OF THE
UNIMPREGNATED UTERUS," AND PROFESSOR OF OBSTETRICS AND DISEASES OF
WOMEN AND CHILDREN IN THE CHICAGO MEDICAL COLLEGE.

SECOND EDITION, ENLARGED.

PHILADELPHIA:
LINDSAY & BLAKISTON.
1867.

PHILADELPHIA:
CAXTON PRESS OF SHERMAN & CO.

PREFACE TO SECOND EDITION.

THE call for a new issue of this work, affords the author an opportunity of expressing his grateful acknowledgments to the medical press, and profession generally, for the kindness with which they welcomed his first edition. This kindness has stimulated him to renewed efforts, and he ventures to hope that this second edition will be more worthy of the patronage of his professional friends, than the first. By comparison, the reader will find a large addition of new matter, and a careful improvement in some of the former contents of the work. Notwithstanding this addition, the size of the volume is not materially increased.

CHICAGO, October, 1867.

PREFACE TO FIRST EDITION.

IN presenting the following work on the Medical and Surgical Treatment of the Diseases and Accidents Incident to Women, it has been the object of the author to furnish the student and junior members of the profession a concise, yet sufficiently complete, practical, and reliable treatise, to meet their wants in every-day practice. Whether this object has been accomplished will be determined by its success alone, and for the verdict thus indicated the author respectfully and hopefully dedicates it to his young friends.

It affords the author pleasure to acknowledge the valuable aid afforded by Dr. Quackenboss, of this city, in preparing this work for the press.

CHICAGO, September, 1865.

CONTENTS.

DISEASES AND ACCIDENTS

INCIDENT TO WOMEN.

CHAPTER I.

DISEASES AND ACCIDENTS OF THE LABIA AND PERINEUM.

ADHESION of the labia, and consequent occlusion of the vagina, sometimes occurs in infancy, or early childhood, as well as in adult life. The adhesions of infancy are so feeble and easily broken up, that they may be considered a trifling affair. Upon examining the parts, it will be found that there is no development of adhesive tissue, but the mucous membrane of the two sides is merely in strong coaptation. It probably is caused by the adhesive influence of dried mucus accumulating and drying between the parts, when in close contact, from want of cleanliness. The vaginal orifice is closed up to the urethra above, and down to the fourchette below. The treatment consists in separating the labia, by forcibly pressing each in opposite directions, until the adhesion gives way, washing and oiling them once every day, and afterwards to keep them from adhering again. Should we not be able to separate them in this way, the point of a silver catheter may be passed down so as to interrupt it. There will be no need of any other instruments in the case.

On one or two occasions I have seen firm tissual cohesions of the labia in childhood as the effect of ulcerative vulvar inflammation. In this form of adhesions it may be so firm as to require the use of the knife. They are, however, always superficial, and we may generally introduce behind the adhesions from above a bent probe or director. When this is the case, it is, I believe, the best plan

2

to separate them, by driving the bent director through the adherent part. The same care as in the infant will prevent them from adhering again.

The most grave sort of adhesions with which we meet is in the adult, as the effect of neglected inflammation of the vulva after childbirth. These adhesions are sufficient to entirely close the vaginal orifice by the coaptation and firm accretion of the entire inner surfaces of the labia. I have met with more than one instance in which the hairy margins of the labia were so nicely adjusted to each other, that you could not distinguish the point of original separation, from the perineum to the urethral orifice, and the finest probe would not enter the vagina anywhere. The depth of the adhesion may be very great, involving much of the vaginal cavity.

These cases are very embarrassing, and are seldom perfectly remedied. It is decidedly the best plan not to interfere with them, until the menstrual accumulation fills up all the vaginal cavity remaining open, and then our object should be to reach the accumulation with a small trocar as near the middle of the adherent parts as possible. Placing our patient in the lithotomy position, the catheter should be introduced into the urethra, the urine all drawn off, and the urethra held as near the symphysis pubis, or as far from the middle line of the vagina, as practicable. The catheter should be thus held by an assistant, while the forefinger of the left hand should be placed in the rectum. With this preparation, we may safely introduce the trocar into the collection of fluid as felt by the finger. The fluid being drawn off, the outer extremity of the perforation may be increased by the knife as far as may be desired, and as deeply as the surgeon may consider it safe. The opening may be increased as much as necessary by wax bougies, introduced and allowed to remain for twelve hours. The whole should be thoroughly cleansed by a syringe, with soap and water, as often as every twelve hours. The size of the bougies should be increased as often as once in twenty-four hours. If the opening is superficial, the treatment will not be protracted; but if it is deep, it will be tedious. It should be continued until all danger of closure is past, and it will be best to keep the patient under our supervision for some time after this appears to be the case.

Wounds.—The labia are sometimes wounded by accidents of

some kind extraneous to the patient, and they are sometimes
torn during labor. When the wound is deep enough to reach the
bulb of the clitoris, alarming and sometimes fatal hemorrhage is the
result. Professor Meigs gives an instance of great hemorrhage
from these parts in a woman who had fallen upon a chair so as to
cut through one of the labia. A case of fatal hemorrhage was
caused in this city about four years since, in the following manner,
as well as it could be learned from a legal investigation. A
drunken husband returned home late at night, and, as was his wont
under such circumstances, beat and kicked his wife, who was, probably, also inebriated. He kicked her with great violence in the
genitals, and the square-toed heavy boot, in penetrating the pelvis,
had cut off one labium and deeply wounded the other. In six or
eight hours after the occurrence, the woman was found dead, with
such copious effusion of blood from the wounds as, in the opinion
of the examining jury, to account for the fatal result. I saw a
case many years ago, where the patient was wounded by a knife
in one labium so as to cause very profuse hemorrhage.

The hemorrhage being the important effect of these wounds, our
efforts should be directed to its suppression; and this may, in most
cases, be easily done. The bleeding part should be pressed by
the hand firmly against the pubic ramus of the side upon which
it is situated until temporarily arrested, when an elastic air-bag,
or plug of oiled cotton or lint, may be introduced to fill up the
vagina, and a hard compress placed and held firmly by bandages,
so as to press the wounded part between the two. When wounds
of the labia are large and gaping, the hair should be removed,
and the wound treated according to ordinary rules for external
wounds. The rents occurring in labor do not, in the great majority
of cases, require any special treatment, cleanliness and quiet being all that is required.

Sanguineous Infiltration.—During labor, when the parts are
stretched to their utmost extent, some of the arterial twigs give
way and extravasate the blood in the loose structure of one
labium. The infiltration usually shows itself after the child has
been delivered, but sometimes, before the head has passed, the
swelling becomes very great, and proves an obstacle to the expulsion of the head. When this last is the case, the blood is
effused from a large branch of the pudic artery, and the forcible

injection into the tissues is so great as to urge the blood so far in every direction as to fill a large part of the space between the vagina and the pelvic walls. This is a very serious state of affairs, and calls for prompt and judicious interference. I once saw, in consultation, a case of this kind, so extensive as to arrest labor for several hours. These effusions, however, do not always call for surgical treatment, but when, as in the case here alluded to, the effusion is extensive, we must make a free incision in the inner surface of the labium, and allow the blood to escape; if it is coagulated, we should introduce the fingers and dislodge it. When the blood is thus evacuated, if hemorrhage continue, the bleeding artery must be compressed by the fingers until it ceases. The artery may be felt by the finger, where it crosses the plane of the ischium, just above its tuberosity, and as it runs along the ramus of the ischium and pubis. As it occurs after the expulsion of the foetus, the branch of the artery is smaller, and, as a general thing, the effusion not very extensive. Water-dressing, some evaporating lotion, or cooling discutient, will be sufficient, and absorption will be effected in from one to four weeks. Suppuration occasionally, I think not frequently, is excited by a small amount of effusion. This should be treated as an abscess. If the amount of blood is great, and the parts are tensely distended even after the child is expelled, it is better to liberate it by incision, for fear of sloughing or extensive suppuration and serious damage.

Œdema.—The distensible nature of the structure of the labia renders them liable to great œdematous infiltration in cases of general dropsy. Ordinarily, such distension is a matter of trifling importance, but the supervention of labor at a time when they are very largely swollen, is often a very embarrassing condition. They are sometimes so swollen as to occlude the vaginal entrance, and to yield only after protracted efforts, and even then, sometimes only after one of them has been more or less torn. When this excessive œdema is discovered before the head is pressing upon the external parts, or even when this is the case, no time should be lost before taking measures to lessen their size. This may be best done by everting first one and then the other, and making from ten to twenty small punctures through the mucous membrane only. A very sharp-pointed knife, taken between the thumb and finger of the right hand, so as to show only about the eighth of an inch, is the best instrument. Several quick, smart strokes with the instrument thus held suffice for the operation. The serum begins to exude from these punctures, and in half an hour the swelling is very much reduced.

Phlegmon.—Abscesses in the labia are apt to occur in three different forms. The first is common phlegmonous inflammation, occurring in the central part of one labium, very rarely in both. The heat, swelling, and pain, are very great, and the inflammation runs its course quite rapidly, generally suppurating and discharging in from six to eight days. This form of inflammation results from bruises, acrid discharges from the vagina, or the extension of inflammation from this cavity. It is located about the centre of the labium, and the swelling and tenderness are great from the beginning. The second form originates in over-distension of Duverney's gland, from a stoppage of its excretory duct. It is situated deeply at the lower or posterior end of the labium, and generally more slow in its progress. If the patient is intelligent, and has observed the case with care, she will tell us that there was a little tumor in the seat of disease for several days, and sometimes weeks, but slightly tender at first, gradually becoming more so until the inflammation becomes acute, when the abscess is fully formed. In this stage the labium is enlarged, tender, and hot, but there is not the acuteness of inflammation as is seen in the first variety. If the surgeon has an opportunity to examine the parts during the progress, he will perceive a well-defined tumor, pyriform in shape, with the small extremity directed to the vulva, while the larger passes beneath the ramus of the ischium. It will not seem to be, as it is not, in the central part of the labium, but beneath its under surface. It will bear handling somewhat freely, and by pressing against the ramus, and directing the pressure toward the vulvar end of it, the contents may sometimes be pressed out. The contents in the early stages are, for the most part, mucus. If examined later, the surrounding parts, the labium particularly, will be found in a state of phlegmonous inflammation, which, in ten days or two weeks, suppurates, and the pus is evacuated spontaneously. In this form of inflammation, if the duct of the gland can be opened before the inflammation becomes considerable, suppuration may be avoided. This may be done by pressing,

the fluid out, or introducing a very small probe into the canal of
the gland, thus opening it.   If these are both impracticable, it is
better to puncture it and squeeze the contents through the outlet
thus made.   If inflammation has begun, we may treat it like the
former variety, with leeches, purgatives, evaporating lotions, &c.,
in the earlier period, and afterwards, by poultices and anodynes,
until the suppuration is complete, when it should be evacuated by
puncturing it on the mucous surface of the labium.   The third
variety is characterized by a succession of small furunculi.   They
first show themselves as small points of induration immediately
below the mucous membrane or skin, which are very tender, and
in the course of a few days suppurate.   One scarcely passes
through these stages before it is succeeded by another, and so on
a continuation of them march along for weeks, and even months,
before they cease to return.   This condition has existed only in
such of my patients as were the subjects of some form of uterine
disease, attended with leucorrhœa.   They are generally anæmic,
have slow condition of the bowels, and poor digestion.   The radical treatment consists in curing the disease of the uterus, correcting the state of the bowels by mercurial and saline cathartics, and
reinvigorating the patient by the judicious employment of tonics.
We may palliate the sufferings of the patient by cleanliness;
as bathing the parts thoroughly several times a day with pure
cold water, and using cold water injections per vaginam, and
making such application to every hardened point as soon as it
shows itself as will arrest its progress.   I have used successfully
the strong tincture of iodine applied to the part, and the solid
nitrate of silver.   If either of these applications is used as soon as
the inflammation begins to come, sometimes it will be arrested, and
the patient escape for several days, or until another begins to
form.   Should we be unable to thus cut short the inflammation,
we must use poultices of bread mixed with a solution of acet.
plumbi, and anodynes, until suppuration is perfect.   These small
points of suppuration usually break themselves, and they will seldom be lanced.   Notwithstanding the fact that inflammation of
the labia is very painful, the patient will bear her distress until
suppuration is complete, or, at least, unavoidable, in almost all
cases, so that our treatment is confined generally to that appropriate to the suppurative stage.   The whole processes of inflam-

mation are rapid, so that this may be an additional reason why the first stage is not the subject of observation.

Rupture of the Labia and Perineum.—The perineum and labia majora are liable to be torn during severe labor. A number of causes may, under certain circumstances, lead to these accidents. A straight sacrum, by allowing the head to emerge from the pelvis farther back than usual, although not a frequent, is an occasional cause. Rigidity of the perineum, or undilatable state of the external organs, a condition frequently found in aged primipara and occasionally in other patients, is also a cause. A large and unusually ossified head, malposition of the head when the occiput emerges too much posteriorly, and a too narrow arch to the pubis, may also act as causes of rupture.

The perineum may be, and doubtless is, not unfrequently ruptured by the unskilful use of the forceps: 1st, by not making the proper spiral change in the position of the head, so as to bring the occiput under the arch of the symphysis; or, 2dly, not causing this part to keep close to the symphysis, by raising the handles at the proper time, and to a sufficient extent; or, 3dly, the forceps may be allowed to slip off the head under powerful traction. Mere slipping of the forceps when the points of the blades pass behind the head, and become detached entirely, and the convexity of them is not increased, will not generally produce this effect. When this is the manner of missing the hold of the instrument, the blades will be pressed close together, and pass through the parts without great distension. But if, instead of this mode, the blades spring so that the points are made to pass out over the largest part of the head, and thus widely separate the blades, the convexity becomes so great as to distend the parts enormously, and thus split through the fourchette first, and then the perineum, and finally, in some instances, the sphincters; 4thly, by elevating the handles of the forceps too much, the points of the blade may be brought in forcible contact with the perineum, and thus, added to the great distension, cause rupture.

The breach of substance, of course, differs very considerably. Mr. Brown divides the accident into slight and grave. He regards those as slight which are not ruptured through the sphincter, and considers them of but little practical importance; and believes that when the sphincter is violated, and then only, need much importance be attached to it.   The external sphincter is sometimes injured considerably, and the rupture stops short of its complete division, and at others both are torn through, and half an inch of the recto-vaginal septum also divided.   I saw one instance in which the two sphincters were torn through, while the larger part of the substance of the perineum in front of them was uninjured; the child having passed through the septum into the lower rectum, and through the anus, producing the above rupture.   This case did well without any operation.   The wound generally commences at the fourchette, and extends backward towards the anus, but occasionally it takes a direction to one side and passes outside the sphincter, leaving the anal opening untouched.   At other times, the rupture commencing at the fourchette, the rent is directed laterally outward, so as to separate to a greater or less extent one or both of the labia from the perineum.

Effects.—I think Mr. Brown underrates the importance of some of the slighter forms of this accident; for, reason as we may as to the means adapted to the support and maintenance of the uterus in its proper position, as the *floor* of the pelvis, the perineum serves an important part in sustaining that organ.   When the perineal support is lost, the positions of all the pelvic viscera are likely to be disturbed in their relations one to the other.   It is very rare to see, indeed I never have seen, the uterus, bladder, or rectum, protrude from the vaginal orifice when the perineum retained its perfect integrity.   On the contrary, one or all of them, when other causes co-operate, may be comparatively easily displaced, after the main portion of the perineal substance is lost. It will only be necessary to remember that the perineum being in the virgin triangular, the base at the skin, and the apex looking up into the cavity of the pelvis, and that the upper part, or apex, extends up at least an inch, and reaches obliquely forward above the tuberosities of the ischium, and that farther behind is quite a depression, into which the uterus, bladder, and rectum, in a state of distension, are lodged, gravitating there in a direction with the superior strait, to understand the great inconvenience of its loss. When the perineum anterior to the sphincter is split, this muscle will draw the anus farther back, and thus destroy this *pelvic pouch*, leaving its contents to settle still lower down.   I think that it is in this wise the most distressing protrusion of the vagina, bladder, rectum, and uterus, one or all of them, is permitted, if not caused. As an evidence that Mr. Brown does not differ from me so much as he seems, he has devised an operation to restore and even extend the perineum forward, for the cure of these displacements; and, as is usual with his surgery, it is a success. I do not wish to be understood as intending to say that in all cases of loss of the perineum, protrusion of these organs will necessarily occur, but that when extensive displacement of this kind is observed, it is almost always in connection with deficient perineal support. More serious and invariable are the consequences of the most extensive ruptures,—the loss of the functions of the perineum and sphincter both. Prolapse of the viscera and involuntary discharge of the contents of the rectum result. If the fæces are hard, the patient can generally manage to seek a proper place to perform defecation; but if fluid, there is no warning until they flow upon the person. The mucous membrane of the vagina is generally irritated and inflamed, while the skin is chapped and excoriated from frequent contact with the fæces.

Treatment.—Prevention, always the best treatment when available, will vary with the cause of the rupture. When, in labor, the perineum is very rigid, and relaxes with difficulty, the patient should be placed under the influence of chloroform, which induces relaxation with more certainty, perhaps than any other remedy. Minute nauseating doses of ant. et pot. tart., every half hour, is next in efficiency to chloroform. I would not consent to bleeding in such cases, unless there was evident approach to inflammation in the part, and in no case is tobacco to be thought of. In this condition of the perineum, the irritability of the structures ought not to be increased by attempts to support it. The perineum may be supported when greatly distended, and its integrity is threatened by too great inclination of the presenting part backward. The object of the support, in cases where it is deemed advisable, should be to keep the head as close to the pubic arch as possible, but not to retard its expulsion. Not much force is allowable for this purpose, or any other in relation to the perineum. It is needless, after what has been said as to the manner in which this accident occurs from the use of the forceps, to indulge in special admonitions as to their use. When a patient is the subject of this distressing circumstance and its consequences, she may derive some palliation from the perineal pad,—to prevent protrusion,—cleanliness, and astringent washes and ointments. The curative measures depend upon the extent of the rupture, and the consequences of it. If the perineum, with the sphincters, is separated in the central line, an operation to restore the integrity of the parts will be indispensable to a cure. If the sphincters are not torn through, and there is no prolapse, the inconvenience will be so slight as not to demand such aid. In cases where the sphincters are not ruptured, much may be done immediately after the accident to close or fill up the chasm without a surgical operation. The student may be reminded that a very easy way of ascertaining whether the rent extends through the sphincters, without inspection of the part, is to learn whether the patient can control the escape of gas from the rectum. If she can, that muscle is not materially damaged. A good way to arrive at the same object, is to introduce the finger into the anus. The resistance to the introduction is a measure of the quantity of muscular fibre left. For the first four or five days after confinement, the patient must be confined to her side, and it would be better, also, to surround the limbs at the knees with a roller, or bandage, to keep them constantly in contact. By lying on the side with the limbs close together, the parts are kept in almost perfect contact, and the lochial discharges flow out anterior to the wound. These two circumstances are essential to a cure. A diligent observance of the position on the side for a number of days, and a close proximity of the knees, is apt to result in adhesion of a part of the wound by the first intention, and much more of it by granulation. After the lapse of eight or ten days, the parts ought to be inspected, and a healthy state of granulations encouraged by cleanliness, good diet, and, if need be, by a stimulating application of nitrate of silver every four or five days. It will not be best to use suture or other surgical measures in recent cases of this kind. After the opportunity for treating such cases in their recent condition is past, and prolapse of the bladder, rectum, uterus, or vagina, renders interference necessary, the operative procedure is so similar to that necessary for the worst cases, that I will consider them in this respect together, and point out the difference as I proceed. For the most destructive form of rupture, the operation taught by Mr. I. Baker Brown is so perfect, and has been so entirely successful in my own hands, that I will not apologize for recommending and describing it without variation. A patient, to undergo this operation and be cured by it, must be in good general health. If she is not so, the operation ought to be delayed until proper means can be used to effect it. A firm, plastic state of the solids, without unusual tendency to suppuration, will be the most favorable condition; and I have observed that patients coming from the country will do better to have the operation performed at once, and that it is better, if practicable, to send our town patients into the country for a month or more. Thirty-six hours before the operation is to be performed, we must administer an efficient but not drastic laxative; castor oil or rhubarb will do very well. The patient should be placed in the lithotomy position before a strong light. If an anæsthetic is administered,—and it will very much facilitate the management of the patient,—it may be given at this stage of the proceeding. One assistant is placed at each side of the patient to steady the knees and hold the legs, while another assists in the use of instruments. The instruments necessary are a scalpel, a blunt-pointed bistoury, a pair of scissors, three large curved needles armed with double hemp cord ligatures, eighteen inches long,—quite strong, common hemp twine,—three or four small curved needles with silver wire in them, and two pieces of flexible catheter, three inches long, for quills. Sponges, warm and cold water, of course, must be at hand. The surgeon seats himself in front of the patient within easy reach. He commences by removing the hair from all the parts on which he is to operate. After which the edges of the cleft part are to be thoroughly denuded. The cicatricial tissue should be all removed, smoothly and evenly on both sides and up to the septum, the lower end of which should be trimmed in the same way. No part of the mucous membrane or superficial tissue of any kind should be left, as it will inevitably prevent union. An incision on each side of the central line posteriorly, so as to divide most of the fibres of the superficial sphincter, must be made by introducing the blunt-pointed bistoury, about one inch and a quarter into the rectum, and then carrying the handle of the knife obliquely outward, so as to make the incision extend outward, from the verge of the anus between the coccyx and tuberosity of the ischium, about one inch. This will pretty thoroughly divide the external sphincter.

After the bleeding ceases the rent is now ready to be closed. One of the large needles is made to enter the side of the wound, to the right of the operator, and at the upper angle, about an inch and a half from the edge of the cut surface, and dip down deep enough to go to the bottom of the torn portion, inserted into the corresponding part upon the opposing side, and come out as far removed from the edge to the left hand of the operator. Another ligature should be introduced at the central part of the cut, in the same manner as the first, but it will be necessary to penetrate to the depth of the septum, and it would be very proper to include it. It must be deep. A third, at the posterior part of the wound, will suffice to adjust the parts well. In introducing the ligatures we must be careful to place them so as to make the approximation of the surface equable and true. The quill placed on either side, the ligatures drawn and tied over them tight enough to bring the lips of the wound firmly and perfectly together, and from four to six silver sutures passed through the edges of the integuments and secured, and the operation is finished. It is advisable, I think, before, or immediately after the operation, to give the patient about two grains of opium, or its equivalent in some of its preparations, and continue it at intervals, to keep the bowels from moving, and allay irritability and pain. The patient is to be placed on her side, and have the limbs secured by a bandage at the knees. The position may be carefully changed from one side to the other, being always particular to keep the legs close together, and not to allow them to be used so as to contract the muscles at the pelvis. Every six hours, or oftener, the catheter is to be used to draw off the urine, lest it runs into the wound and vitiates the inflammation. The wound should be kept covered with pledgets of lint saturated with simple cerate or cold water. The deep ligatures commence ulcerating on the third day, and sometimes sooner, and if this does not progress too rapidly, they may be left in place four days, but if the ulceration is active, they may be removed on the third. It will only be necessary to cut the ends of the ligatures over one quill, when they may be removed by drawing upon the other. Soon as the suppuration begins we cannot be too careful about cleanliness. Plenty of clean tepid or cold water must be injected into the vagina and rectum two or three times a day, while the external parts are sponged and cleansed as

often.   The young operator need not be discouraged if, upon examination, the wound is not all closed by adhesive inflammation.
My experience is that this immediate and perfect closure does not
usually take place, but that much of the deep-seated portion is left

Fig. 1.

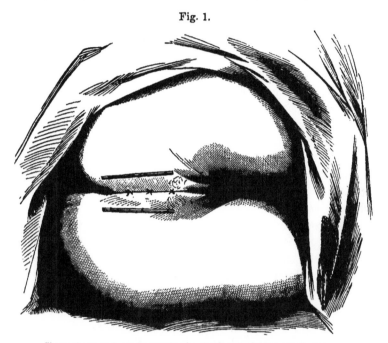

Shows the parts brought together by the deep and interrupted sutures.

to be filled by granulations, and it is sometimes several weeks before this is accomplished.   The skin and integuments generally
unite by the first intention, and when this is the case, there is not
much danger of failure, provided we keep up a granulating surface
all over the unhealed portion of the wound, and observe perfect
cleanliness.   The superficial silver sutures may remain for ten or
twelve days, as they produce no irritation whatever.   At the end
of twelve days some laxative will be necessary if the bowels have
not been moved.   The diet and medicine of the patient while in
bed, after the operation, cannot be uniform in all cases, and are
to be governed wholly by the state of the system; it will be better, I think, to err in favor of good supporting diet, stimulants and
tonics, rather than risk impairing the general health by abstemi-

ousness.   Adhesive inflammation is promoted by a high state of physical health, and suppuration by a low condition of it, and aside from imperfection of the operator's proceedings, we have most to fear from early, copious, and persistent suppuration.

When the rupture is recent, the parts may be restored by merely using the sutures, and the incisions through the sphincters, in the case of complete rupture, or without these when the muscles are not torn.

The operation is performed quite differently by Dr. Emmet, the skilful and ingenious surgeon of the Woman's Hospital, New York.   I subjoin a description as I find it in the *New York Journal of Medicine* for December, 1865, from the pen of Dr. Emmet:

"As early as 1855 Dr. Sims, in the Woman's Hospital, simplified this operation by bringing the scarified edges of the laceration together by means of deep interrupted silver sutures, and from this time the use of the quill suture, or a division of the sphincter ani, has been abandoned.   Further experience demonstrated a necessity for the use of a short rectal tube for some ten or twelve days after the operation, that a free escape of flatus might be unobstructed.   Where the laceration of the perineum has extended only to the sphincter, the rectal tube is not needed, and three interrupted sutures are generally sufficient ; if more extensive, so as to involve the muscle, two in addition are required.   The first suture passed should be the one nearest to the rectal mucous membrane, and should be made to follow the laceration entirely around, so as to bring together the sphincter.   The second should also include the sphincter, and be passed in the recto-vaginal septum, just beyond the first one.   The remaining sutures are introduced (as in the operation for a partial laceration of the perineum) through one labia about half an inch from the edge on one side, introduced from within outward into the other, and withdrawn at a point equally distant, so as to approximate perfectly opposite surfaces.   If the laceration has extended up the recto-vaginal septum for some distance beyond the sphincter ani, the edges should be brought together down to the sphincter by interrupted silver sutures, at a distance of about five sutures to the inch.   On introducing the first suture to clear the perineum, care must be taken that it is passed between the first and second sutures uniting the septum, and the next one in turn between the second and third. Without this precaution an opening into the vagina will be produced just behind the sphincter, from the fact that, as one set of sutures is passed at a right angle to the other, on twisting those of the perineum, tension would be exerted. This is a weak point, for if the tube is allowed to become obstructed, a small recto-vaginal opening will always result from the escape of flatus in this direction. I always scarify by means of scissors; it can be done rapidly, and with less hemorrhage. The knees should be kept tied together ten days after the operation, and the urine drawn with care, so that none is allowed to escape over the surfaces brought in apposition.

"The sutures of the perineum are usually removed about the sixth day; those within the vagina must remain for two weeks or longer, until the parts are strong enough to admit of the introduction of a speculum. The bowels are to be constipated for two weeks at least in all cases where the sphincter has been lacerated. When the bowels are acted on by either a purgative or a warm mucilaginous injection, the success of the operation will greatly depend on the dexterity of the nurse in properly supporting the parts.

"During the past ten years, in the hospital practice of Dr. Sims and my own, this operation has been uniformly successful. There has not been a single case of failure in uniting the sphincter and perineum by the first operation. I have, however, occasionally partially failed in private practice, where the laceration has been extensive, from a want of care in keeping the rectal tube unobstructed, or in not properly supporting the parts during the evacuation of the bowels, and necessitating a subsequent operation to close the rectal opening."

CHAPTER II.

DISEASES OF THE VULVA.

ERYTHEMATOUS, papular, vesicular, and pustular inflammations of the vulva are not unfrequently observed, as are, also, squamous diseases. They resemble the same form of disease in other muco-cutaneous cavities and the skin, and hence, will not here claim a separate description. A disease somewhat more distinctive, how-ever, and yet resembling a disease of the mouth, is known as purulent vulvitis. This affection is characterized by severe in-flammation of the mucous membrane of the vulva, attended with minute points of ulceration, numbering from one to two dozen. The ulcers are small, an eighth of an inch in diameter, slightly excavated, and almost always covered with pus. The vulva is intensely red, and bathed in pus and mucus. The inflammation sometimes extends into the vagina and causes a copious flow of pus and mucus from that cavity. Not unfrequently the labia are very much swollen, and occasionally the deeper tissues are in-volved in phlegmonous inflammation. This form of inflammation is not unfrequently, in its early stages, attended with considerable febrile excitement. To a superficial observer, it strongly resembles gonorrhœa, from the swollen labia, burning pain, copious muco-purulent discharge, and the difficult and painful micturition. Its occasional sudden and unexpected development adds to this simili-tude, and legal proceedings have been instituted against parties supposed to have been instrumental in imparting the disease to little girls. It occurs in children generally from two to ten or twelve years of age, and probably results from want of cleanliness, heat, and local irritants, accidentally applied. If allowed to pursue a course undisturbed by treatment, other than cleanliness, it will generally subside spontaneously in two or three weeks, or in the course of that time become very much subdued, and run into chronic inflammation without ulceration. This last is often ex-tended into adolescence, and as vaginitis, gives origin to the leucorrhœa of girlhood, and finally endometritis of the woman. It is sometimes attended with a debilitated and scrofulous constitution, indigestion, constipation, and ascarides; but it is not likely originated, though it is aggravated and fostered, by these attendant symptoms.

The treatment is general and local. In the beginning, where the inflammation is high, it should be antiphlogistic and soothing. We may administer a mercurial cathartic, and quicken its action by a saline laxative, and after the bowels have been thoroughly moved, nitrate of potassa may be given internally, every three or four hours, in doses to suit the age of the patient. The parts should be frequently bathed or fomented with a decoction of poppy-heads, or with the watery extract of opium. In the course of four or five days the acute symptoms will begin to subside, when, in addition to attention to the bowels, we may administer an acid solution of quinine internally, and begin the use of astringents locally. A solution of tannin, sulphate of zinc, acetas plumbi, or other such astringents, weak at first, and afterwards increased in strength, may be applied freely to the parts four or five times a day. These remedies will generally remove the inflammation in a reasonable time. The astringents should be increased in strength to a sufficient degree for the purpose. If those mentioned are not strong enough, the chloride of zinc, sulphate of copper, or even nitrate of silver, may be very properly resorted to. Should the inflammation extend into the vagina, the astringents may be injected into that cavity, by means of a small hard rubber syringe. We ought to be careful to use a very small syringe, and not to introduce it too far. The nurse should be carefully instructed in this kind of application. I feel impelled to insist upon the complete removal of the inflammation as early as it can reasonably be done, believing that if it continues until puberty, the inflammation extends into the body of the developing uterus, and entails a very distressing train of suffering upon the patient, that might have been avoided by an early and complete cure of the vaginitis. I am persuaded that too much importance cannot be attached to these views.

Corroding Ulcer.—I have met with a number of cases of corroding ulcer of the vulva in children, which have been the cause of great suffering and apprehension. It occurs most fre-

3

quently in children, but is occasionally met with in adults. There is in each case usually but one ulcer, and it is most commonly situated on the lesser labia at first, and spreads to surrounding parts. The ulcer is ragged and irregular, not much excavated, with a dark foul-smelling covering, and the discharge from it is sanious, fetid, and excoriating. It is not generally rapid in its progress, and sometimes lasts for months, creeping from one part to another, until the anatomical features of the vulva are almost entirely effaced. I have not met with this form of disease except in very debilitated, sallow, and badly nourished persons. The state of the system leading to this sort of ulceration, I have thought to be more the result of living in poorly ventilated houses, particularly; but coupled, also, with imperfect nourishment, or with nourishment of an improper character.

It is generally obstinate, and yields slowly to judicious treatment.

We should endeavor, as one of the main objects, to correct the constitutional condition as speedily as possible. To this end the circumstances of the patient should be changed to the most favorable sort. Good ventilation at home, frequent and prolonged exposure to the fresh air, nourishing diet, of which animal food should be a large ingredient, and comfortable clothing, kept thoroughly clean, are indispensable to success. The bowels should be kept in as correct a condition as possible by gentle laxatives. The digestion, which is always feeble, if not otherwise faulty, may be improved by the administration of infusion of cinchona, quassia, or colomba, with the mineral acids; the sulphuric being, perhaps the best. The chlorinated tincture of iron is also an excellent general remedy. The next thing to be accomplished is to convert the ataxic, half-sloughing, and corroding chronic ulcer, into an acute inflammatory one. This is done by profoundly stimulating it with the stronger caustics. The one which has seemed to me to be most successful is the caustic potassa. It should be applied to the whole surface, by passing a stick, not very rapidly, all over it. After this burning, we may dress the ulcer with calamine ointment twice a day. This will almost immediately improve the condition of the sore. Unless there is some considerable firmness around and beneath it, in thirty-six or forty-eight hours after the application, caused by the effusion of fibrine in the submucous substance, not much good will result from it, and it will be necessary to resort to it or some other in a few days. The strong nitric acid is also very useful. I have not tried the actual cautery, but should expect it to be very useful. We may often cure this ulcer by the weekly use of the solid nitrate of silver to it, dressing between times with lint saturated with black wash or calamine ointment. We ought not to be afraid of strong treatment, nor to continue it, or reiterate it in conjunction with a highly roborant general course of exercise, diet, and medication.

Gangrenous Vulvitis, or Noma.—This is a very severe and generally fatal affection of the genital organs, occurring almost, if not wholly, among children. It may attack one or both sides simultaneously. In the few cases I have seen, there appeared a bleb or blister on the inside of the mucous surface of the labium, which at the same time became enlarged, hard, tender, and painful. In a few hours the blister breaks, and from its side a not very abundant, but acrid serum is discharged. At this time a peculiar odor is emitted from the parts. All around the ash-colored surface, which represents the place where the blister was developed, the substance of the labium is very hard and much swollen. In two or four days the affected side is in a state of gangrene, the discharge is very much increased, the parts upon which it runs are excoriated and inflamed, and an intolerable stench is exhaled. I have not seen an instance in which the gangrenous parts were cast off; the patients having died beforehand. Generally, though not always, in the very beginning, the circulation and nervous system are very much disturbed. The pulse is quick and feeble, the patient nervously restless, or else stupid, the extremities cool, the body—particularly about the pelvis—hot, the tongue furred, generally brown, and the skin dingy and sallow. As the disease advances the pulse becomes still more rapid and weak, the extremities cold, the mind wandering, and the restlessness amounts to the frantic efforts of some sort of delusion. The tongue becomes dark-brown or black, the teeth are covered with sordes, and in the end, the patient sinks into profound collapse, and often coma, and dies.

The disease runs its course sometimes in forty-eight hours, and again, in milder forms, it may last five or six days. The causes, although unknown, must undoubtedly be of a depressing nature, overwhelming the organism very rapidly. It occurs sporadically, when it is comparatively mild, and epidemically when severe. In this last state it is very rapidly fatal.

The prognosis is very bad, as it is always, or pretty nearly always, fatal. The profession, so far as I am aware, has not decided whether the disease is a general one, and the affection of the genital organs an incident, or whether the local disease inaugurates the general symptoms. The former is most likely the truthful interpretation of the phenomena.

In such a disease there is but little prospect of a cure by treatment; we should, nevertheless, institute a course clearly indicated by the symptoms and signs. The general treatment should be strongly stimulant, tonic, and supporting; quinia, brandy, tincture of cantharides, and beef essence, as much as the patient can bear, should be administered. I do not think the strong caustic local treatment, generally advised, any better, if as good, as the charcoal and yeast poultices, chloride of lime, anodyne fomentations, and cleanliness. Much attention should be devoted to thorough ventilation, isolation of the patient, and the neutralization of the fetor by disinfectants.

Pruritus Pudendi.—A very annoying, and often obstinate affection of the genital organs, is an inordinate itching of the vulva. The itching returns by paroxysms. The patient will sometimes be free from it except when standing by a warm fire, or becoming heated by exercise, passion, &c. Or she may be affected only for a short time before, at the time, and after menstruation. Again, the paroxysms return without any apparent reason. The sensation sometimes is that of a burning glow, attended with an irresistible desire to rub or scratch the parts, a desire which the most delicate sense of propriety cannot always keep within due bounds. At other times the sensation is such as might be produced by the crawling of pediculi, and the patient is assured that thousands of these animalcula are moving upon her person, and will be convinced to the contrary only by inspection. This sense of formication, although very disagreeable, is a slight inconvenience compared to the sufferings of the other variety.

The former variety is almost always attended with inflammation of the mucous membrane of the vulva. The accompanying inflammation may be simply erythematous, papular, or vesicular.

Dr. Dewees describes a variety of vesicular inflammation resembling aphtha, attended with pruritus. I am sure that the papulæ or vesiculæ are neither of them always present in very distressing cases of this affection, although I have not seen it when the parts were not in some way inflamed. It may be observed that, in the formication variety of pruritus, the itching is generally most, if not wholly, confined to the cutaneous surface of the labia. It will be inferred that I consider pruritus but a symptom of several diseased conditions, generally of the genital organs, but sometimes it undoubtedly may be caused by the state of the intestinal tube, particularly the rectum, or some other remote condition. An intelligent scrutiny of the cases as they arise, will most frequently result in the discovery of the originating condition. It is often an obstinate affection, lasting for weeks, months, and even years in bad cases, but more frequently it is amenable to treatment, and a judicious course will be rewarded by success.

Treatment.—The first thing to be done is to remove the cause, when practicable. In order to do this, the abdominal organs will require attention. The sluggish secretions and bowels must be corrected by alteratives and laxatives. A mercurial, say five grains of massa hydrarg., may be given at night, to be followed in the morning by a saline laxative, sufficient to cause one or two stools. This may be repeated at intervals of from one to four days, until the object is gained. Meantime, if the stomach is weak and digestion imperfect, the bitter infusions, with alkalies or acids, as the condition may require, will be demanded; and should the patient be anæmic, iron may be given. As is sometimes the case, the patient will be plethoric, when the alteratives, with spare diet, will do better. With the above treatment, if the health be faulty, or without, if this is not the case, we will generally be obliged to resort to local remedies. And first of all is cleanliness. The parts, externally and internally, must be subjected to thorough and frequently repeated ablutions. The syringe may and should be brought into use for this purpose from three to a dozen times a day. The water used for ablutions may be impregnated with sal soda, very appropriately, or some fine toilet soap. I have found much advantage, when there was no eruptive accompaniment, from ӡij tinct. ferri chl. in a quart of water, three or four times a day. This is especially useful when there is leucorrhœa, and a

congested, dark appearance of the mucous membrane.   When there
is a vesicular eruption, the recommendation of Dr. Dewees, to
sprinkle the parts with pul. sodæ biboras, and keep them exposed as much as possible to the air, will be of great service.
Prof. Simpson uses chloroform in the forms of vapor, liniment, or
ointment, with good effect.   The infusion of tobacco, applied freely
two or three times a day, is recommended by the same author.
When the mucous membrane is much inflamed, a solution of hydrocyanic acid, ten drops to the ounce of water, often gives great
relief.   A strong solution of tannin and aqueous extract of opium,
is also applicable to this class of cases.   An excellent palliative
is pure glycerine.   It may be introduced into the vagina by
saturating a plug of cotton with it, and passing it up through a
glass speculum and allowing it to remain there for ten or twelve
hours.   We should take the precaution to attach a thread or cord
to the cotton so that it may be readily removed.   One of them
introduced every twelve or twenty-four hours is often enough
generally.   We should also apply it between the labia in the same
way.   As explained by Dr. Sims, who first recommended its use,
the glycerine induces copious serous depletion from the congested
mucous membrane, thus relieving it.

In cases of some duration, I have often been enabled to produce
a decidedly favorable change by applying the tinct. ferri chl. in
full strength with a brush once a day to all the mucous membrane
of the vulva, and as far in the ostium vaginæ as I could pass the
hair brush.   The first burning sensation is succeeded by great
amelioration of the sufferings, and finally, in many cases, by a
cure.   When this fails, we may sometimes succeed by making a
similar application of a solution of nitras argenti, in the strength
of ℥ss. to ℥j of water.   This last application should not be used
oftener than once in two days.   In the use of all these remedies,
we must not lose sight of the ablutions, nor fail to search for particular local causes, and try to remove them.   As has been very
judiciously remarked by Prof. Simpson, we will find great advantage in alternating the use of appropriate remedies, instead of
using the same kind all the time.   The obstinacy of this affection
will require great patience in many instances, as well as ingenuity
in using remedies.

Urethral Excrescences.—Caruncles of the urethra. Vascular tumor at the orifice of the urethra.

These names have been given to small tumors springing from the mucous membrane of the vulva, immediately round the urethral orifice, or from the lining of the urethra itself. They are generally solitary, but sometimes there are several. Sometimes they are sessile, and seem to be a hypertrophied fold of the mucous membrane of the orifice; at others, they are polypoid in their attachment. In size, they vary from a pin's head to a small nut. They also vary in their appearance. As before remarked, they sometimes resemble in color, consistence, and polish, the mucous membrane upon which they are planted; while in other cases they are quite red, almost scarlet, very soft, and easily broken. They differ in their anatomical properties quite as much as in appearance, seeming, in some instances, to have no more vessels and nerves than other portions of the neighboring tissue; while at others, they are formed mostly of capillary bloodvessels and loops of nerves. They are a morbid development of existing tissues, instead of a growth of abnormal substance. These tumors are often observed, particularly the more dense and light-colored varieties, without giving origin to any symptom that would lead to their detection; on the other hand, in many instances they often produce the most excruciating suffering. The kind of caruncle that has seemed to me to be the important one, is the blood-red tumor projecting from the mouth of the urethra, and attached by a small neck. A few weeks since, I met with one of these of crescentic shape, attached by a neck that arose from the concave margin, and had its other attachment inside the urethral orifice. It would not have weighed two grains, but it caused agonizing symptoms. It must not be supposed that all of the varieties will not occasionally cause great pain. The symptoms of their presence are almost always connected with the evacuation of the bladder, and attempts to handle the part. The passage of urine causes the most excruciating suffering from pain and tenesmus, the patient often straining for several minutes after the complete discharge of the urine. The slightest touch, also, is the cause of great pain. The diagnosis cannot be clear without an ocular examination. If the parts are exposed to a good strong light, and the labia separated, the excrescence will be at once discovered, unless it be quite inside the urethra. If any doubts exist, we should introduce the finger in the vagina, and press the urethra forward. It is difficult to say, with truthfulness, what are the causes of these caruncula. My cases have been in patients obviously deficient in cleanliness. This seems to have been the case in that which came under Dr. West's observation.

The treatment is simple, and consists in two main objects: 1st, the removal of them; and 2d, the production of a profound impression upon the point of origin. In fact, the tissues from which they spring should be destroyed to a slight depth. The first object may be most readily gained by snipping it off with scissors; and the second by holding caustic potassa, or the actual cautery, to the place until the virus is destroyed.

Vascular Urethra.—Analogous to the caruncle is the vascular urethra. It gives rise to the same train of symptoms, though not so intensely distressing, and is very persistent. It occurs more frequently in patients near the climacteric period, although I have seen it in much younger persons. When the labia are separated and the parts exposed to a good light, the urethra is seen to be patent, and the tissues around the orifice swollen and of deeper hue than usual. The mucous membrane of the urethra is of an intensely scarlet color, and, upon minute inspection, the vessels may be seen enlarged; it is very tender and sensitive to the touch, slight contact producing exquisite pain. There is great burning and sense of cutting when urine is voided, and all the symptoms, even the sympathetic nervous derangements, attendant upon caruncle. This condition is not incipient caruncle, for there is no elevation, no protrusion, and the condition lasts for years without material change of substance. The treatment I have found most effective, is strong nitric acid or caustic potassa applied cautiously to the membrane inside the urethra.

I have not tried the actual cautery, but believe it would be very effective. An application of the acid on a piece of lint moistened by it, to the whole membrane in sight every ten days, for two or three times, generally is sufficient; sometimes once only is required.

Hypertrophy of the Clitoris and Nympha.—It is very rare that we meet with hypertrophy of these organs without morbid change in the tissues. There is either cystic development in their substance, or degeneration of the membranous tissues. The two diseases that would seem to contribute most frequently to this enlargement, are syphilis and elephantiasis. In either case the degeneration is in the mucous membrane almost exclusively. The size is sometimes but slight, and sometimes very great, making a tumor from the size of a pigeon-egg to a volume long enough to reach to the knees. The surface is generally rough, uneven, nodulated, and sulcated. From exposure and friction they are almost always excoriated, and give origin to an ichorous irritative discharge, that produces inflammation in the skin round about. These tumors are always a great annoyance, and sometimes very materially impair the general health. There is but one eligible means of relief from these disgusting enlargements, and that is amputation. Fortunately it is an easy and safe operation. The parts are included in the coil of the ecraseur, and slowly crushed off. Where the enlargement extends to all the parts, the clitoris may be operated on first, and then each of the nympha separately amputated. This instrument is much better for these operations than the knife, as there need be no apprehension of hemorrhage. If the hypertrophy is simple there will be the same necessity for amputation, as the like inconveniences will almost surely arise as in the morbid variety.

CHAPTER III.

STONE IN THE BLADDER.

VESICAL calculus in the female is of very rare occurrence, absolutely and relatively. Of all the cases of calculus vesical, only about one in twenty is met with in the female sex. This may be accounted for from the size, straight form and dilatability of the urethra, and consequent direct escape of small sanguineous and mucous accumulations, and even sandy concretions. Indeed, quite large stones are expelled through the urethral canal, making their way out, in some instances, in a few moments with acute suffering, while in others they are many hours in forcing a passage. It would seem that these hard substances are evacuated more readily during the state of pregnancy, than at any other time; doubtless, because of the urethra partaking in the general increased dilatability of the genital organs which precede labor.

Symptoms.—There are probably no symptoms attendant upon stone in the bladder in woman, but is produced more frequently from other causes, hence they are quite unreliable, and can be taken only as suspicious instead of diagnostic evidence of its presence. They are great and persistent irritability of the bladder, severe pain after voiding the urine, sudden cessation of the flow while there is yet a desire to urinate, and evidently some fluid in the organ, enlargement or relaxation of the urethra, and incontinence of urine. The urine is also charged with mucus, pus or blood, or all three of these in greater or less quantities. The symptoms will be more strongly marked if the calculus is rough and jagged in shape, and less so if the surface is smooth and even. All these symptoms are not present in any given case, but some of them are certain to be prominent and very distressing.

Diagnosis.—The only way to positively determine the diagnosis, is by physical examination of the cavity of the bladder. This is done by means of the fingers and the sound. If two fingers be passed deeply into the vagina, as far as the cervix uteri, the most dependent part of the bladder may be pressed strongly up against the internal face of the pelvis, or lower portion of the anterior abdominal wall. If this latter be pressed well down into the pelvis with the other hand, while the fingers are still in the vagina, careful manipulation will scarcely fail to distinguish a calculus of moderate size. When the bladder is full of water, if the calculus is large, it may be raised and its presence pretty conclusively determined by *ballottement*. The stone is felt, however, more distinctly through the urethra by the sound, the same as it is used in the male. The operation may be facilitated by the fingers in the vagina moving the stone around. The same difficulties in preventing or making difficult and perfect diagnosis are met with, as in the male, if the stone be encysted or adherent to the upper or anterior wall of the bladder; but if the instrument is sufficiently curved and moved about in various directions it will be detected; its position and size ascertained with more precision and certainty than in the male.

Treatment.—The only means of relief available is the entire removal of the calculus. This may be done by dilating the urethra, and extracting through it, by lithotomy or lithotrity. All these operations are less hazardous in the female than in the male, in fact, we scarcely take the subject of danger to life into consideration in operating for stone on a woman, but one very great inconvenience likely to follow dilatation of the urethra and lithotomy is incontinence of urine, and the attention of recent operators is turned mainly to the matter of avoiding this most distressing sequel. The preference is given by some surgeons to lithotomy, because they think this evil less frequent after it, while for the same reason others resort to dilatation of, and extraction through, the urethra. Very few now practice lithotrity in the female, and this operation is looked upon as attended with more hazard than either of the others. It is astonishing with what facility the female urethra may be largely and rapidly dilated. I have seen it stretched so as to admit the index finger in ten minutes without violence to its integrity. Where the stone is not very large—over an inch in diameter—we may expect to succeed by dilatation without much damage, if the proper caution and gentleness are used. When the stone is much larger, and especially if it is rough, we should cut. The operation of dilatation is simple. It may be performed by compressed sponge, Weiss dilators, or bougies. The

sponge is very rough and almost intolerable, and slower in its action than the instrument. If we use the sponge it should be inserted as much as six to ten hours before the time to operate. The
patient should have a full dose of opium, and be under its influence
before the sponge is used, and we will succeed much better if she
is rendered insensible by chloroform. The sponge when compressed ought to be a little larger than a female catheter, three
inches long, and slightly conical in shape. It should have a perforation reaching two-thirds or three-fourths the length of it, from
the base toward the pointed end. After the patient is prepared, a
strong sound or staff is passed as far in the perforation of the
sponge as it will go. With the sponge thus mounted upon the
sound it should be passed slowly but firmly through the urethra
until one end is in the bladder. If this can be endured for six
hours, the sponge may be removed, and the urethra will be large
enough generally to admit the finger. The sponge, I think, is less
likely to rupture the fibrous structure than any other means of
dilatation now in use.

A less painful mode, and one next in safeness, is performed
with graduated bougies. After the patient is under the influence
of the anæsthetic, a bougie large enough to require some force to
introduce should be passed into the bladder, and allowed to remain
in the urethra for from two to five minutes. This should be followed by one a size larger, and so on until the canal will admit
the finger. One advantage of this mode is that the distension is
uniform all around the urethra. If the dilator is used the blades
are passed folded together, and then separated so slowly as not to
tear the tissues. Whatever means we use to dilate with, they
should be laid aside so soon as the finger will enter the canal, and
further dilatation effected by this means until there is room for
the introduction beside it of a small lithotomy forceps, with which
to apprehend the stone. Soon as the finger can be made to enter
freely the bladder, other fingers should be passed into the vagina,
and caused to press the stone forward so that its size, shape,
consistence and the character of the surface be ascertained. If
there is a long diameter, the end must be directed to the urethral
opening, and retained with as much security as may be, until the
forceps are introduced and the stone seized. Traction should be
made in the direction of the urethra with the instrument, while

with the fingers in the vagina the efforts may be governed so as to keep up the right direction and steadiness, and also to push the stone into the urethra. Swaying the instrument in different directions, and performing slight rotation, the force used should be very gently applied and slowly increased, giving the parts time to stretch, and no more exerted than is just sufficient to accomplish the extraction. We should not be in a hurry, but take plenty of time; more damage is done by too great hurry than too great dilatation, I think. The parts are torn instead of being stretched. If the stone is too large to be removed in this way, which will be very seldom indeed, we may perform lithotomy, or the mixed operation of cutting and dilating. There are several methods of performing lithotomy in the female.

Dr. Sims has proposed and performed lithotomy through the vesico-vaginal septum. He exposes the parts as for operation for vesico-vaginal fistula, introduces a curved director through the urethra, and cuts into the bladder upon it until the opening is large enough to permit the stone to pass. The finger is then passed through the artificial opening by which the forceps is guided, the stone seized and extracted through it. The wound is then closed with silver sutures, and the patient otherwise treated as for fistula. In experienced hands this must be a prétty and complete operation, but the young operator will hesitate very properly to substitute this for dilatation. The operation for lithotomy and dilatation combined is more simple, and easily effected by the inexperienced. A grooved director is passed into the urethra until the end of it enters the bladder, with the groove directed up to the arch of the symphysis pubis. A straight, long, narrow, blunt-pointed bistoury is placed with the back of the blade in the groove, the blade touching and parallel to the director in its whole length, and then urged forward until it enters the bladder. This will be known when the sensation of dividing tissue ceases to be experienced. This should be done slowly and no more division made than will be caused by simply passing the blade along the groove. When this is accomplished the instruments are withdrawn, and the canal dilated to any necessary extent with the finger. The rest of the operation is performed as in the case of dilatation and extraction. A few days' rest on the bed, and as much opium as is necessary to keep the pain from being excessive, will suffice to restore the patient to health.

CHAPTER IV.

VAGINITIS.

ACUTE VAGINITIS begins generally in the lower part of the vagina, with swelling, intense redness, and dryness of the mucous surfaces of the labia, vulva, and vagina. There is great heat in the parts, and the patient complains of burning pain in them. Difficult, painful micturition, pain in passing the fæces, sense of weight in the pelvis, and tenesmus, are generally present also. Not unfrequently there is backache and pain, radiating down the thighs, into the hips, up the spine, and into the head. Sometimes the symptoms are so acute as to produce general febrile disturbance. When this is the case, there is chilliness alternating with heat, more or less increase in the frequency of the pulse, furred tongue, pain in the limbs, &c. In the course of thirty-six hours, the pain, redness, and swelling have spread to the whole of the vaginal cavity, and soon there is a profuse secretion of mucus, which, after two or three days, or even sooner, is mixed with pus-globules in greater or less abundance. When this last is the case, the discharge is either green or yellowish in color, and less tenacious. This state of things lasts for from ten to twenty days, when the inflammation gradually subsides, the discharge becomes less in quantity and lighter in color, until in four or five weeks the disease is entirely gone, or it merges into the chronic form. The inflammation usually involves the urethra, and sometimes the bladder, and its greatest intensity is almost always in the lower third of the vaginal canal. The inflammation sometimes spreads to the rectum. Sometimes it attacks the mucous membrane of the cervix uteri, and even invades the cavity of the corpus uteri, remaining longer in these localities than in the vaginal cavity.

Diagnosis.—The diagnosis of acute vaginitis is not difficult, as the parts may be easily seen and touched.

Prognosis.—As has been heretofore intimated, it subsides spontaneously, and leaves the parts free from disease, or in a state of chronic inflammation. The prognosis, therefore, is favorable.

Cause.—It is caused by contagion more frequently, perhaps, than anything else, but does doubtless arise from abuses, injuries, and want of cleanliness, and probably other causes. I have seen the non-contagious form in children very much more frequently than adults, spreading usually from the vulva upwards. Simple acute vaginitis is not a very common affection. At first it involves the mucous membrane and submucous tissue, but before many days it is confined to the membrane alone.

Treatment.—This at first should be slightly antiphlogistic. A few grains of calomel, followed in ten or twelve hours with a saline cathartic, should be the first step. This may be succeeded by nauseating doses of ant. et pot. tart., until the dryness and swelling have subsided. In the meantime, perfect quietude in the recumbent position should be enjoined, the parts bathed every hour or two thoroughly with tepid water, and the patient should abstain from stimulating or nutritious ingesta. So soon as the discharge has become copious, and yellowish or green, and the swelling of the parts has entirely subsided, the treatment should be changed for astringents, specifics, laxatives, and baths. We may give half a drachm of balsam copaiba in emulsion or capsules every six or eight hours, and have the vagina syringed copiously with a saturated solution of alum, or acetate of lead, two or three times in twenty-four hours. Every third day a few ounces of solution of argent. nit., in the strength of ten grains to the ounce, may be advantageously used. The bowels should be kept open, and the patient should abstain from stimulants at all times during the treatment. The astringent injection ought to be changed every five or six days, using alum, sugar of lead, and sul. zinc, alternately. Perseverance in this treatment will very materially shorten the course of the disease.

Chronic Vaginitis.—This is a more frequent form of disease than the acute, and its importance will be understood from this consideration. It is in many instances a very distressing affection, and often mistaken for diseases of the uterus, bladder, or rectum.

Symptoms.—There is generally pain in the back, more frequently in the sacrum and coccyx, but not seldom higher up; pain in the groin, weight and sense of bearing down in the perineum, dragging in the hips and pelvis. A burning sensation in the vagina, extending all over the lower part of the person, very distressing and depressing, is sometimes the chief symptom complained of by the patient. In married patients it is the cause of distress during the act of coition, to such a degree, sometimes, as to entirely preclude such indulgence. I am now treating a patient who assures me that although she has been married fifteen years, she does not remember a single instance of sexual intercourse that did not give her discomfort; generally it is the cause of decided pain, and sometimes entirely intolerable to her. Leucorrhœa is a common, but not invariable symptom; it may be yellow or white in color, but when the case is not complicated with cervical inflammation it is always thin. In chronic vaginitis there is generally a long train of sympathetic symptoms not unlike those observed in diseases of the uterus. The nervous centres are disordered in their functions, and we have nervous symptoms of almost every description. The mind is sometimes affected by it to irascibility, despondency, suspiciousness, peevishness, and purposeless instability. In other, or, perhaps, the same cases, there is palpitation of the heart and large vessels to such a degree as to cause alarm for the life of the patient. Headache should be mentioned as quite common; it is more commonly located in the occipital region, but may be in the top, forehead, temples, or all over the head. The eyes are generally weak. The stomach is frequently deranged to a considerable extent, and in various ways; and there is generally a constipated state of the bowels, though diarrhœa is an occasional symptom. There often is pain, too, in urinating, and in passing the fæces through the rectum. The uterus is almost always affected, also, and through it the symptoms may become greatly diversified and increased. We should expect this complication.

Diagnosis.—Upon examining the vagina, the introduction of the finger will give some pain, sometimes a good deal, and the speculum causes a great amount of suffering. There is general redness of the mucous membrane; it is usually bathed in its own mucous secretion; sometimes it is so raw as to bleed upon the use of instruments in making the examination. The sensitiveness, redness, and exaggerated secretion, are conclusive and diagnostic symptoms when they are permanent.

Causes.—Chronic vaginitis is often the result of an acute attack. The inflammation only partially subsides at the time, and is continued indefinitely. Some of the most obstinate cases I have met with have thus resulted from gonorrhœa. Another set of cases are seen in patients whose husbands were the subjects of syphilis in early life, but to all appearance have been cured. I am inclined to the opinion that chronic vaginitis is not an uncommon occurrence in women thus situated. ·It is more likely to follow recent cases of syphilis, and is sometimes subacute in grade. Another form is apparently produced by abortions, colds, and other causes, with, at the same time, inflammation of the cervix· uteri. Constipation, causing sluggishness of the vaginal circulation, or any other causes producing this vascular condition, as the pressure from pelvic tumors, phlegmonous effusion, &c., contribute to the causes of chronic vaginitis. There is no doubt but that certain constitutional taints, as scrofula, rheumatism, and, as before intimated, syphilis, are efficient co-operating causes.

Prognosis.—Chronic vaginitis, in its simpler forms, is apt to be obstinate, and resist judicious treatment for years. It is more particularly so when originating in constitutional diseases. When connected with incurable tumors, it will, of course, resist all sorts of treatment.

Treatment.—The constitutional treatment of chronic vaginitis is sometimes of the first importance, while at other times it is unnecessary, or nearly so. The variety which seems to be connected with the syphilitic condition, requires the alterative remedies which are found beneficial in this affection under other circumstances,—the preparations of mercury, iodine, and the vegetable alteratives, for instance. When associated with scrofula, the vegetable tonics, with alterative treatment, plenty of outdoor exercise, cod-liver oil, and cold bathing, sea bathing, &c., will be appropriate measures to be employed. As it is not unfrequently complicated with rheumatism, or this diathesis, it may be necessary to prescribe for it with such consideration.

But in more simple cases, where there is no such taints or complications, conditions exist that require a judicious course of general treatment for their removal, before we can be successful in our main object. Such is a torpid state of the bowels and portal circle, with scanty secretions. Mercurial and saline laxatives,

4

vegetable tonics, as the bitters, also alkalies, will, when judiciously used, assist us very much. We should be particularly careful to avoid a loaded or impacted state of the rectum, as this is the cause of much vaginal congestion. An injection once or twice a day, when necessary, will suffice for this.

In all forms, in addition to the general treatment, when that is necessary, we will be under the necessity of resorting to local measures. Much benefit will be derived from a sitz bath twice a day. The bath should be tepid, as a general thing, as being more likely to agree with the largest number of patients. When it is more agreeable, the bath may be cooler. It should be large enough to cover the hips, and the patient should remain in it for an hour at least, and often it is better to use it for a greater length of time. Of more importance are injections. Simple water in large quantities is sometimes sufficient, but more frequently astringent substances will be found essential. The injections should be administered through a perpetual syringe, and the quantity should be large, say from one quart to a gallon of water at each time. The common astringents, as alum, sul. zinc, acet. lead, of the strength of one drachm to the quart of water, will generally suffice. We find cases, however, in which none of these substances can be used, because they disagree with the patient, producing dryness of the parts or increasing the inflammation. In such cases we must carefully search for the right local remedy. We may find it in tannin, tr. iron, perchl. of iron, astringent decoctions, nitrate silver in solution, &c. The last, used once in four or five days, with a glass syringe, and the other astringents between, often proves to be the best course.

An excellent and very convenient mode of applying medicinal substances to vaginal surfaces is to make small sacs of domestic or linen and fill them with the substance intended for use, and introduce them into the vagina. A sac the size of a small glove finger, with a piece of thread attached to it, will hold an abundance of almost any remedy we desire to use. Tannin in powder or ointment, gall ointment, belladonna ointment, and other articles are used in this way. A mixture I have used very commonly consists of two drops of creasote, half drachm of tannin, and one grain of belladonna extract, introduced at bedtime each night. The little bag may be removed in the morning by traction on the string. There are, I think, some advantages in the use of these little bags over the other sorts of medicated pessaries used. I not unfrequently inclose copavia capsules in these little sacs, and think it an admirable mode of making balsamic applications to the vaginal mucous membrane. Where the astringents or other remedies are thus used they will not replace the injections wholly. Indeed the vagina should be well washed out before the introduction, and at the time of the removal of them. Patients of course can manage these application without aid.

Perseverance and time are important items in the treatment. If we can remove this chronic inflammation in three or even six months, we ought to be satisfied. And we ought not to be surprised to have it return one or more times after it is apparently cured. It is well, also, to teach our patient patience in this respect.

Puerperal Vaginitis.—It might not seem necessary to consider the vaginitis occurring after labor as a separate affection, but there is so much difference,—in the causes, nature, symptoms, and termination,—between ordinary vaginitis and this form, that I think it may be profitable to do so. In some cases of labor, circumstances occur that induce a severe form of inflammation of the vagina. The one most potent is long detention of the fœtal head in the pelvis. The pressure thus exercised upon the vaginal walls interrupts the circulation more or less completely; and if continued for a number of hours, violent reaction in the parts result when the pressure is removed. This pressure does not affect the mucous membrane of the vagina so deleteriously as the deeper-seated tissues. The fibro-cellular part of the vaginal walls is the seat of the inflammation. I do not think the use of instruments, however awkward, does so much damage as the long-continued pressure. It must not be denied, however, that instruments do give origin to this form of inflammation. When they do so, the inflammation is more circumscribed; it does not extend to all parts of the vagina, as is apt to be the case when pressure by the child's head has been the cause. On account of the nature of the causes, this form of vaginitis runs its course rapidly, and is most sure to end in structural lesions. It is in intense forms of this sort of vaginal inflammation that sloughs and deep ulcerations are met with, which open the bladder and cause vesico-vaginal fistula, recto-vaginal fistula, and cicatrices, which result in contractions and even occlusions of the vagina. It is astonishing how much destruction sometimes

is effected by intense post-partum inflammation. I remember being
called to a case in consultation where the child's head had been
pressing down sufficient to bulge the perineum and labia for sixty
hours without any motion. I delivered her with the short forceps
in a few moments, without any violence to the parts. The patient
was then unavoidably left in the hands of the same careless practitioner that had so outrageously neglected her before the delivery.
I saw her three months afterwards, and found the whole septum
between the bladder and vagina gone, the urethra terminating
abruptly, as though it had been cut straight across, in a great
irregular cavity, that was bounded by the pubis before and the
uterus behind, and without any defined sides to it. In still a
worse case, where shoulder presentation had prevented the passage
of the child, the woman was in the second stage of labor six days.
The woman arose from her bed with a large undefinable cavity,—
without any bladder, apparently, but the very top portion,—and
the loss of two inches of rectum, into which the urine and fæces
were poured involuntarily. In more than one instance I have seen
the whole vagina sealed up, from the fourchette to the urethra, and
as far up, as well as I could judge, to the os uteri, as the effect of
intense and neglected puerperal vaginitis, arising from unaided
difficult labor. Every practitioner must meet with cases in which
the cavity of the vagina is misshaped, and partially closed, from
the cicatrices resulting from it. Now, much of these direful effects
may be averted by the rational management of inflammation after
it has been initiated.

Symptoms.—When injurious pressure has awakened inflammation in the vagina, the labia and walls become swollen, hot, and
very tender. The patient does not generally complain of much
severe pain, but there is a sense of soreness and heat. There is
almost always fever, chilliness, and other evidences of disturbance
to the circulation; the tongue is coated, ordinarily white, sometimes yellow, or even brown, from the beginning. As the disease
advances two or three days from the beginning, the discharge from
the vagina becomes more than ordinarly fetid, the labia excoriated, while the heat of the vagina is still very great, and there is
much mucus and some pus issuing from it; and later, shreds of
decomposed substances, and sometimes considerable sloughs, are
mingled with the discharge, increasing the fetor. The pulse is

more accelerated, and sometimes becomes quite rapid; the patient is much prostrated; the tongue brown and dry, and the teeth foul with a dark clammy mucus, while the skin is bathed in a copious perspiration. In from two to six or eight days, to these symptoms are added an evacuation of urine through the vagina, at first in small quantities, and afterwards more considerable, until, in a short time, the contents of the bladder are passed through this way; the parts around are excoriated by the urine and other acrid discharges, and a slow, uncertain convalescence succeeds, with a permanent vesico-vaginal fistula. Occasionally, though not so frequently, the fæces pass through the vagina after a few days from the beginning of the inflammation, and we have a recto-vaginal fistula. If neither of these evils occur, there is extensive ulceration, not so deep, but extending over a large surface of the vagina; thus pus and acrid ichor are poured out in copious quantities, for a long time, gradually decreasing as the surface heals. As these ulcerations heal up, the tissue becomes condensed and contracted, until such strictures or occlusions result as are above mentioned. The practitioner should be wide awake to this frequent course of post-partum vaginitis.

Treatment.—As most damage from this form of vaginitis usually accrues to the bladder and rectum, our first and most solicitous care should be bestowed upon them. The bladder should be frequently emptied with the catheter; at least every four hours the urine must be drawn off. To appreciate this direction, we have but to remember that this organ may be considerably distended in that time, and as the septum between the vagina and bladder is in a state of intense inflammation, it is softened, and therefore is easily ruptured. My impression is that fifty per cent. of the vesico-vaginal fistulæ which now occur might be avoided by following this rule. Its importance cannot be over-estimated. In very bad cases the catheter might be used even more frequently, or kept in the urethra. The rectum should be kept free from any accumulation of fæces by frequent injections of tepid water. In addition to this prevention against fistula, the utmost cleanliness must be observed. The vagina should be washed out with soapsoads or other bland detergent fluid, from four to six times a day. For the first four or five days the parts may be kept lubricated thoroughly by the injection, after the water, of very bland sweet oil, or almond oil. When the slough begins to be thrown off, or pus and sanies become copious, an injection of half a pint of tepid water, containing six or eight drops of creasote, twice a day, will serve to cleanse and stimulate the parts better than soap and water alone, which should be used between times. After the lapse of a week or ten days, if the ulceration is not healing, an injection of ten grains nitrate of silver to the ounce of water may be used quite advantageously. This solution should be injected from a hard rubber or glass syringe, directed to the ulcerated part by the finger. As the case still further advances, a solution of tannin, alum, sulphate of zinc, or other astringents, with the detergents, may be used. As the parts begin to contract by the advanced healing of the ulceration, the closure, partial or entire, should be anticipated by the introduction, daily or oftener, of wax, rubber, or other sort of bougies. It is well, when this last expedient is necessary, to smear them with ointment that may exert a healing influence on the ulceration. The physician cannot be too attentive to these cases. He should see to it personally that his directions are carried out, and feel himself responsible for any serious permanent injury that can result from want of diligence. Women or their nurses cannot understand, and it is feared physicians do not properly appreciate, these means of averting the awful accidents which result from the sloughing and ulceration in these cases.

Vaginismus.—Dr. Sims described this affection first to the Obstetrical Society of London, December, 1861, and has since given it to us in his "Clinical Notes on Uterine Surgery." It is an "hyperæsthesia of the vulva and hymen attended with involuntary contraction of the sphincter vagina." The parts are so very sensitive that the slightest touch with the finger causes great pain, and, in some instances, coition is entirely impracticable. In all the cases I have ever examined, there was very decided redness and increase of the secretion of the parts exposed by separating the labia. Dr. Sims thinks that the sensitiveness is confined to the vulva and hymen, but I apprehend that more extended observation will convince him that the whole vagina is involved. In one of my cases now under treatment, the sensitiveness of the vulva has almost entirely disappeared; the finger may be introduced into the vagina, but the upper part of this cavity is so exquisitely tender that the patient screams with pain as it approaches the cervix uteri.

The general symptoms are grave according to the chronicity of the case. It generally shatters the constitutional energies of the patient, rendering her a wreck, according to the expression of Dr. Sims. Dr. Sims says it is independent of inflammation. Mr. I. B. Brown agrees with him. It is according to them mere hyperæsthesia. In my cases the parts were always in a state of inflammation; but I cannot think the hyperæsthesia was wholly of inflammatory origin. Of course I am not prepared to say that inflammation is even a general attendant. The observation of the profession will soon determine that point, as the disease is now fairly set before it, and, from its distressing symptoms, will attract much attention. My patients have all been barren, and were apparently not aware of their condition until married. The intensity is not always sufficient to prevent coition, and sometimes is much greater than others. The sensitiveness is greater near the menstrual epoch, occasionally in a very marked degree.

Diagnosis.—The sensitiveness is characteristic, and hence there is no need of much labor in forming a diagnosis. The least touch of mucous membrane of the vulva, with a feather, soft brush, or fingers, gives the patient great suffering, and sometimes agony unlike anything else.

Prognosis.—Judging from all I have seen and read upon the subject, there is very little, if any, tendency to spontaneous subsidence. Its duration, therefore, is perplexingly long. But all agree as to its curability.

Treatment.—Dr. Sims has succeeded in curing *all his cases* by dividing the sphincter vagina deeply on either side of the vaginal orifice. He makes the division sufficiently deep to permit of free dilatation, and then keeps the vagina open with large bougies until the wound cicatrizes. The results of this operation are all that might be expected from it. The hyperæsthesia disappears, and the obstacles to coition are removed, but there is necessarily great mutilation. Mr. I. Baker Brown, in his Surgical Diseases of Females, condemns this procedure as severe and needless, and gives two cases where the sensitiveness was cured by the relief of fissure of the rectum. He thinks the hyperæsthesia is a symptom of some disease of the rectum, generally fissure; and that by inci-

sion of the fissures it will disappear.   Dr. Braun, of Vienna, according to Mr. Brown, has cured one case by removing the clitoris.   A case of some severity is reported in the London Lancet,
American reprint for March, 1867, in the care of Dr. G. C. P.
Murray, in which the hyperæsthesia appeared to depend upon inflammation of the cervix uteri and vagina.   It was cured by
making a free application of the solid nitrate of silver over the
inflamed cervix, and a solution to the vaginal surface.   These applications were repeated in a fortnight, and were succeeded by the
tinct. iodine.   While there can be no doubt that Dr. Sims's plan
is efficacious, I cannot think it necessary, and the success of other
means by different practitioners, bears me out in this opinion.
We almost always find the patients in a state of health unsatisfactory, and, according to my observation, evident local disease
besides that of sensitiveness ; and, from what we have learned
from Mr. Brown and Dr. Murray, more than one kind of local
disease.   As in the treatment of all other diseases, therefore, we
should carefully and diligently search for the cause of the hyperæsthesia.   If it is fissure of the rectum, this should receive our
first attention ; if inflammation of the vagina, uterus, or vulva, we
ought to cure this.

In all the cases I have seen, and I now have three under treatment, nothing I have tried has been of so much advantage as remedies directed against inflammation of the vagina and vulva.   The
course I usually pursue, is to apply the solid nitrate of silver to
the vulva every ten or fourteen days, and in the interval use glycerine and tannin.   The first application reduces the sensitiveness
very decidedly, and it becomes less after each successive touch,
until finally cured.   We should bear in mind that the hyperæsthesia does extend into the vagina and near to the uterus, and
that there is equal necessity to treat the vaginal cavity as the
vulva.   I have been in the habit, in fact, of managing it as I would
vaginitis.   The strong astringents, glycerine and narcotics, applied by means of medicated pessaries and injections, are valuable
adjuncts.   With the local treatment, rational general treatment
is very beneficial.   Attention to the bowels, the condition of the
stomach, and the secretions generally ; tonics, exercise, change of
air, bathing, attention to clothing, and all the regimenal circumstances calculated to benefit the general condition of the patient.

Vesico-vaginal Fistula.—Although generally resulting from puerperal vaginitis, it is sometimes produced by other causes. Extensive ulcerations from pessaries sometimes penetrate the septum
between the vagina and bladder. Stone or other foreign bodies in
the bladder may act as causes of ulcerative processes of sufficient
gravity, to do the same. Malignant diseases, as cancer of the
uterus, vagina, or bladder, not unfrequently lay open these cavities. And, in some rare instances, perforations by the unskilful
use of instruments have been observed.

The fistula, when established, is oftener than otherwise associated
with other effects of the disease from which it is produced. Cicatrices and contractions of the vaginal walls are very common accompaniments. These embarrass examinations and operations
very much when extensive. They also change the size, shape,
and direction of the vaginal cavity. Fistulæ of the bladder are
located near the urethra, involving this tube ; at the cervix uteri,
involving that organ ; and again, in the middle, between these two
points. But these fistulæ are also occasionally urethro-vaginal or
vesico-uterine, without affecting any other parts. Their size varies from a mere pin-hole to several inches in diameter. They are
situated lengthwise, crosswise, or diagonally, with reference to the
direction of the vagina.

The constant flow of urine through the vagina, instead of the
urethra, is a sufficient symptom to decide the existence of fistula,
but we meet with cases where the flow of urine is not constant ;
the patient being able to retain for some time, and then discharge
her urine naturally. This circumstance is due to the plugging of
a small opening by mucus, or the prolapse of some part of the bladder into the fistula. In all instances it is proper and necessary
to make a clear diagnosis of its existence, size, shape, position, and
complications. This is done by the fingers and probe. The patient should lie on her back with the hips near the edge of the bed,
and her legs flexed so that we may have free use of both hands.
The fingers will readily pass through a large fistula into the bladder,
and, by moderate care, be made to thoroughly survey it and the
surrounding parts. But the fistula may be so small, or situated
so as to entirely escape detection by the finger. We will be aided
in such cases by introducing a probe, slightly bent, through the
urethra with one hand, while the fingers of the other are in the

vagina.   The bent extremity of the probe is turned toward the
septum, pressed gently upon and passed over every part of it, until it is made to pass through the opening, when it may be recognized by the finger in the vagina.   Having found the fistula, ascertained its size, position, shape, direction, &c., we ought to survey the vagina, to find strictures or other deformity, and ascertain the distensibility of this tube.   We do this in part to determine the prognosis of the case.   Can the fistula be cured? is a
pertinent and important question, which will be decided by this
kind of examination.   Fortunately, *now*, thanks to Dr. Sims, almost anything short of loss of the whole septum may be cured.
If the fistula consists of a defined opening, it matters little how
large, we are justified in expecting success.   If, as is sometimes
the case, there are no sides, edges, or ends to it, but the vagina
and bladder are one cavity, smooth and continuous, we cannot
reasonably undertake an operation, unless it be to close the vulva,
as has been suggested, if not practised, by Dr. Sims.   Some circumstances, independent of the character and size of the fistula,
are necessary to insure success.   The vagina should be healthy.
If the walls of this cavity are in a state of inflammation or congestion, the prospects of a cure are more remote.   Great nervous susceptibility is sometimes difficult to overcome, and should be a reason to defer the operation.

The general health of the patient is also a matter of the first
importance.   A highly nervous condition of the system, with an
abundance of lithates in the urine, is a condition in which there
are many chances for failure.   Some surgeons lay stress upon a
preparatory process, for a number of days, or even weeks, to avoid
an aplastic condition.   They administer the tincture of iron, wine,
and nourishing diet, to render the diathesis plastic, or favorable to
adhesive inflammation.   The patient should be in the best possible
health, and if she is not, whatever is the deviation, it is best to
take time to correct it.   In many instances, as before remarked,
the inflammation which causes the fistula also produces cicatrices,
that narrow, distort, or tie down the vaginal walls.   These difficulties should be corrected before we undertake to operate.   The
strictures should be divided, and the vagina kept dilated until its
normal condition is restored as nearly as possible.   After the ef-

fects of these preparatory operations are over, we may consider the patient ready for the main operation.

The only other step will be the administration of a cathartic to evacuate the bowels. The catharsis ought to be entirely over at least twelve hours before the operation. With these preliminaries accomplished, we should have a large window on the sunny side of the house, a sunshining day, four assistants, a table of convenient height, five feet long, and two wide, and the necessary instruments. The table, covered with one or two quilts, is to be placed with the end toward the window, from four to six feet distant. The patient lies on her left side, the limbs drawn up, the right rather most, with the left arm behind her, so that she rests full on the front of the chest. One of the four assistants uses the chloroform, another the speculum, a third the sponges, and the fourth attends to the instruments. On a tray within easy reach of the operator, the instruments should be placed. They are the speculum, tenaculum, scissors, two long sponge-holders, forceps for carrying the needles, one wire adjuster, blunt hook, forceps to twist the wire, half a dozen needles, slightly curved, about one inch long, armed with small silver wires about twelve inches long, an elastic male catheter, or one of Sims's S-shaped instruments, with an India-rubber tube, a little larger than the catheter, to carry the urine clear of the bed. The surgeon takes his seat at the end of the table next the window, near the breech of the patient, introduces the

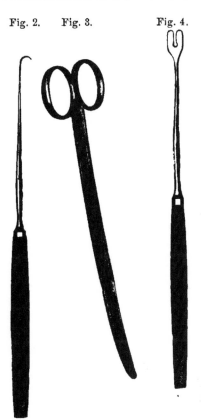

Fig. 2. Fig. 3. Fig. 4.

Fig. 2, tenaculum, with which to hold the edge of fistula while being pared; 3, curved scissors, for paring edge of fistula; 4, wire adjuster.

speculum, dilates the vagina, and thus brings the parts thoroughly in view, and then gives the instrument to the assistant to keep in

Fig. 5.

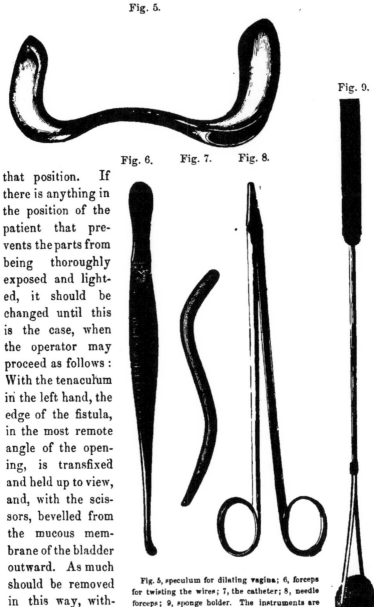

Fig. 6.   Fig. 7.   Fig. 8.

that position.   If there is anything in the position of the patient that prevents the parts from being thoroughly exposed and lighted, it should be changed until this is the case, when the operator may proceed as follows : With the tenaculum in the left hand, the edge of the fistula, in the most remote angle of the opening, is transfixed and held up to view, and, with the scissors, bevelled from the mucous membrane of the bladder outward.   As much should be removed in this way, without changing the

Fig. 9.

Fig. 5, speculum for dilating vagina; 6, forceps for twisting the wires; 7, the catheter; 8, needle forceps; 9, sponge holder. The instruments are represented half size.

place of the tenaculum, as practicable. Another place on the edge of the fistula is seized, and trimmed in the same manner, and so on, until the whole circle is denuded completely of the cicatricial tissue. As this part of the operation is being accomplished, the assistance of the sponge will be called into use on account of the bleeding. I do not see the necessity of removing as much substance from the edge of the fistula as is directed by some authors. The main object, I think, is to have the edges evenly and thoroughly denuded of the mucous membrane. This much should be done with a completeness that admits of no doubt, and if we have a good light, there need be no doubt, as we can see and examine the part sufficiently well to be positive. After the bleeding has ceased, we may insert the sutures. We commence at the angle of the wound most remote and difficult to reach. The needle is to be introduced first into the lip of the wound nearest to the operator, by starting it in about half an inch from the freshened edge, dipping it down, so as to make the point come out in the denuded portion, just at the junction of it and the vesical mucous membrane. The needle being brought through at this point, is again inserted in the opposite edge, corresponding as nearly as possible with that part whence it emerged, and carried forward far enough to emerge half an inch beyond the edge of the wound, and drawn through. The ends of the wire thus fixed are held by the assistant engaged with the speculum. The next suture is to correspond with and be placed within two lines of the first. They are thus placed in sufficient numbers to close the opening completely. (See Figure 10.) Having all the sutures introduced, the one nearest the operator must be isolated and twisted by the forceps made for that purpose, until the angle of the wound is evenly coapted. The next, and so on of the remainder, is to be managed in the same way. Great care must be taken to see, as the closure is effected, that the lips of the wound are drawn evenly and smoothly together. (See Figure 11.) If we are not particular, the edge of one side or the other rolls slightly in, and unfreshened mucous membrane is brought up to denuded surface. This, I think, is a circumstance that is very liable to occur in the hands of an inexperienced operator. Both the insertion of the sutures and bringing together the edges may be facilitated by the skilful use of the tenaculum and the adjuster. The tenaculum will enable us to disengage and straighten the edges, in adjusting them, and in inserting the needles keep them firm.   The adjuster will place the twist of the wire in any position with reference to

Fig. 10.

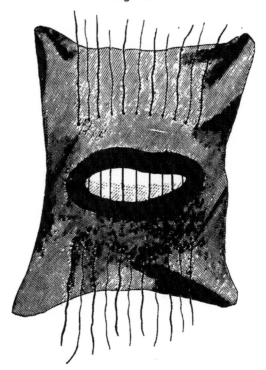

The fistula with edge pared and the sutures placed.

the junction of the wound we may desire.   In twisting the wire there are two things to be avoided,—one is tightening it too much, and the other leaving it too lax.   Experience will fix these items after a few operations, but I think that the operator may venture to tighten the twist of the wire until it fixes but does not strangulate the part included in the stitch.   After the twist is completed, we ought to be able to pass an ordinary probe through the circle of the stitch without much force, and yet, upon its removal, there should be no apparent space.   If the stitch is drawn too tightly, the parts will be strangulated and cut through early by ulcera-tion; if too loose, the urine will pass through as the bladder

becomes filled, and prevent adhesion. The last thing in the operation is to cut the wires about half an inch from their insertion,
and bend them down in the direction of the wound, leaving the
ends lying centrally upon it. The operation finished, the catheter
may be inserted, the patient placed carefully in bed, on either
side, and a grain of opium administered. The catheter will sometimes become foul with deposits, and require cleaning every twelve

Fig. 11.

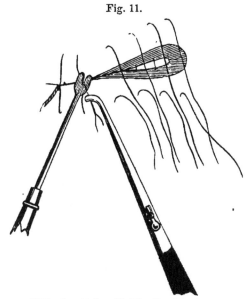

Closing the wounds and twisting the wire sutures.

or eighteen hours, but as a rule, while the urine is running freely,
it may remain in place. Great watchfulness will alone prevent this
instrument from being misplaced. The great desiderata of the after-
treatment, are to prevent an accumulation of urine in the bladder,
and the bowels from being evacuated. The former can be certainly accomplished in no other way than by having a competent
assistant by the patient, or very near her all the time, and when
the catheter does not deliver the water freely, to have it removed
and replaced by a clean one, however frequently that may be
required. We may keep the bowels quiet by administering a
grain of opium twice or three times a day. If the patient is very
restless, we ought to give as much more as is necessary to quiet

this. The only other important item of treatment, as a general thing, is cleanliness, and for this purpose, vaginal injections of tepid water with fine toilet soap, twice or three times a day, will suffice. The vagina will thus be kept clean with much certainty.

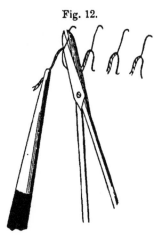

Fig. 12.

Removing the sutures.

The diet should not be too sparing. The ordinary diet of the patient, in half or two-thirds of the quantity, I am convinced is better than any considerable change in quality. The patient must remain quiet as practicable for nine or ten days, when, prior to taking out the sutures, we should try the capacity of the bladder, by leaving the catheter out for an hour, and then drawing off the urine. There will be no good in leaving the sutures in place longer than ten days, perhaps, but there can be no harm result from their presence longer. The removal of them is easily accomplished, by passing one blade of the scissors within the circle of the stitch, and dividing it, when the wire may be withdrawn by the forceps. For the first three or four days the patient should wear the catheter at night, and pass her urine every two hours in the day, gradually lengthening out the time.

I have given this description of operation, because it is the one most nearly approaching, if not identical, with that now performed by Dr. Sims, after all his great experience,—a sufficient commendation without any other,—the simplest and easiest of comprehension and performance, as well as, in my humble judgment, most uniformly successful. Dr. Sims does not now use chloroform in this operation, but I have been obliged to resort to it in most of my cases. When a patient can be induced to do without it, we will get along better than if she is anæsthetized.

Recto-vaginal Fistula.—This accident does not so frequently as vesico-vaginal fistula result from puerperal vaginitis. Stricture of the rectum, abscess of the recto-vaginal septum, rupturing into both cavities, and accidents with instruments, as often cause it, perhaps. It is not so common or frequent as vesico-vaginal fistula, nor so distressing. The passage of the fæces, if proper cleanliness is observed, although disgusting, is not so productive of inflammation and excoriation as urine, and their discharge may be controlled by appropriate fixtures. A cure is also more easily accomplished; indeed, it is often spontaneous. As the contents of the bowels pass intermittingly, and, when in contact with the raw surface, do not irritate it considerably, the ulcer has time to contract, and healthy granulations, in a good state of the general health, is the result.

The symptoms and diagnosis of this fistula are so obvious, that I need not dwell upon them; but we sometimes meet with cases where the opening is so small and tortuous, that great patience in the use of the probe will be required to satisfy ourselves as to its position and direction. The injection of water into the rectum while the parts are under inspection will generally clear up all doubts.

Treatment.—If we are associated with these cases during the ulcerative condition, we may conduct them to a cure with great certainty, and, perhaps, more readily than after the edges of the opening have cicatrized. The important items of treatment at such times are, 1st. Proper attention to the bowels; 2d. Great cleanliness; and, 3d. Maintenance of healthy granulations until the contraction obliterates the opening. The bowels should be kept quiet as much of the time as possible. To accomplish this, the diet should be concentrated and nourishing in character; beef essence, milk, eggs, crackers, coffee or tea, and, if necessary on account of debility, wine, or medicinal tonics; and if the bowels have a tendency to move, opium in such quantities as will restrain them. Every four or five days a gentle alterative, say three grains of massa hyd., followed by a saline cathartic; after the bowels have moved from this, the opium may be given to restrain them for four or five days again, and so on until the opening is closed. During this treatment, there should be frequent injections in the vagina of water. The part should be examined with the speculum every day, to see that the edges remain raw. When there is any tendency to cicatrize, the edges may be freely touched with pure nitric acid. If the cure is protracted, the acid should give place to the actual cautery. Toward the last, as the opening becomes small, especially if it is tortuous, a piece of twine, or what is, per-

5

haps, better, a silver or iron wire, may be passed through it, and the ends brought out through the anus and vagina. If the case is chronic, and the opening small, the application of the acid may be made every day until the edges are denuded, and then the same course followed as above directed. Of course, these applications must be made through the vagina with a speculum that completely exposes the part touched. If the place is large and chronic, we will very much shorten the process of cure by an operation similar to that for vesico-vaginal fistula. After having thoroughly evacuated the bowels, the patient may be placed in the lithotomy position, and exposing the parts to a strong light, the perineum may be retracted by the rectangular speculum blade of Sims, while the vulva is held open by assistants. The edges are then to be pared thoroughly, and the aperture closed with silver sutures. The bowels will require the use of from two to four grains of opium daily to keep them quiet. They should not be allowed to move for ten days, when a saline cathartic should be given, and after it has operated well, the stitches removed. During the time between the operation and the removal of the stitches, the patient is to remain quiet in bed, and have injections, per vaginam, of tepid water with soap, twice a day. If by this operation there is imperfect closure of any part, the treatment recommended for recent cases will suffice to complete the cure. Even these larger sized fistulæ are sometimes cured by the caustic acids, the actual cautery, or tinct. litta; but it takes a longer time, and is attended with more pain and annoyance. The operations on these fistulæ will be greatly facilitated by having the breech of the patient projecting somewhat over the end of the table.

CHAPTER V.

SEVERAL conditions are necesssary to the healthy performance of the functions of menstruation.

1st. The ovaria must be present, and sufficiently healthy to produce ova.

2dly. The uterus must be sufficiently perfect, anatomically and physiologically, to be the medium of this elimination.

3dly. A certain, but not as yet very well-defined, state of the blood and nervous system.

I do not think that these are all the conditions necessary to perfect menstruation; but they are the obvious and undoubted ones.

The physiological chain of circumstances that give rise to menstruation may be given thus: the organs concerned being fully developed, the blood and nervous system matured to a certain degree, an ovum is produced, and during the time it is being matured and cast off from the ovary, all the organs of generation are intensely congested by the increased energy of the capillary circulation; the congestion and stress of blood upon the delicate capillaries of the mucous membrane of the uterus become so great, that the walls of some of these vessels are ruptured, and an effusion of blood takes place in the cavity of the uterus, which, finding its way out of the vagina, is called menstruation. If ovulation does not take place, the congestion does not occur, and in the absence of the congestion there is no effusion. All these processes which accompany menstruation should be painless, and only of a certain grade of activity to be healthy. When absent, deficient, painful, or in excess, the function is deranged.

This very brief summary of the general physiological principles of healthy menstruation, will do for a starting-point, in the consideration of the disorders of them. And as I proceed, I shall have occasion to mention pathological states, which are superadded to a deviation from the above conditions, as also some that induce derangements of menstruation, notwithstanding the presence of them in a healthy degree.

I shall consider the disorders of menstruation under four different divisions :

1st. Amenorrhœa.

2dly. Menorrhagia.

3dly. Dysmenorrhœa.

4thly. Misplaced menstruation (Metatithmenia).

Under these four heads may be included all the deviations met with in ordinary practice.   It is usual with authors to make only three distinct divisions.   My fourth division is spoken of by those who have described it as uterine hæmatocele, hæmatoma, &c.; but I shall give what I consider good reasons for classing it under the general head of menstrual disorders.

AMENORRHŒA.—Amenorrhœa simply means the absence of menstruation, and may appear under several different circumstances.

1st. Menstruation may never make its appearance in the individual.   2dly. After having occurred it may cease, or, as the term is, be "suppressed;" and this suppression again may be suddenly brought about, and attended with acute symptoms, and hence properly denominated acute suppression; or it may not be attended with acute symptoms, and last long enough to be called chronic. 3dly. I should think it right to consider deficient menstruation as suppression, although but partial.   This partial suppression assumes two forms, viz., infrequency, when the intervals are uncommonly long; and scantiness, the return being regular, but the quantity of the discharge much less than it should be.   Or there may be both scantiness and infrequency.   4thly. The menses may be retained in the cavities of the uterus or vagina, or both, after having been effused.   This retention is very different in many respects from the suppression; giving rise to quite a different set of symptoms, and requiring a separate sort of treatment, agreeing with it only in the non-appearance of the blood externally.

The pathological states upon which the symptom amenorrhœa is based are very numerous, and sometimes inscrutable.   The more obvious are the following: congenital absence of the uterus or ovaria, or both; congenital non-development, or acquired atrophy of these organs; disease of the uterus and ovaria, acute or chronic.

These are the ordinary local causing conditions. The general are, anœmia, cachexia (as tuberculous, scrofulous, &c.), pregnancy, and nursing. Serious diseases of any of the vital organs may, under certain circumstances, give rise to amenorrhœa.

The local symptoms which attend the absence of the menses will differ according to the conditions which give rise to it. In acute amenorrhœa, we shall have signs of great congestion, or inflammation of the uterus. The patient, after commencing to menstruate, being subjected to the causes necessary to suppression, such as the partial or general application of cold, is seized with pain in the back, hypogastric region, and hips, attended with a sense of chilliness more or less intense. These symptoms are usually succeeded by febrile reaction, headache, pain in the limbs, general languor, white tongue, and a persistent pain of varied severity in the region of the uterus. There is, in this state of things, as there seems to be, inflammation of the uterus. The symptoms may subside, and generally do in a very few days, leaving more or less local discomfort in the pelvis and neighborhood. At the next menstrual period, if the uterus is not much inflamed, and the system not greatly deranged, the blood is effused, but seldom with the same naturalness in quantity, quality, and painlessness, as before; but there is more or less pain, which is again manifested henceforth at each successive period.

At other times the discharge fails to show itself after having been thus suppressed, and the case becomes chronic, and lasts an uncertain length of time. When this is the case, the non-appearance is likely to be attended by chronic inflammation of the uterus, or ovaria, as the result of the acute attack; or the morbid effects brought about by uterine sympathies derange the stomach, bowels, liver, in fact, all the chylopoietic organs, to such a degree, as to render chymification or chylification imperfect. Sanguification will be thus vitiated, anæmia or cachexia results, and the patient is broken down and "miserable;" or we may have a combination of these two states and the symptoms to which they give rise. I cannot but see in this catenation of circumstances, the complicated effects resulting from inflammation of the uterus. Should the amenorrhœa be primary,—by this I mean to say, should the menses never have made their appearance,—the girl, if old enough and sufficiently developed, will suffer differently. And there is very nearly, if not quite, the same set of symptoms present in cases where they have made their appearance imperfectly, in quantity and quality, or for a few times, and then ceased.   The patient suffers under the symptoms of imperfect sanguification : inability to exercise, palpitation of the heart, shortness of breath, torpid liver and bowels, want of apppetite, or depraved appetite ; desiring improper food at improper times, despondency, great apathy, and timidity.   The surface is pale, and either white and translucent, or, what is much more common, of a greenish hue.   This condition of the system was formerly styled chlorosis.   The sufferings are often very great and protracted, and not unfrequently merge into those of tuberculosis, insanity, or other serious organic diseases. It is not unusual, even in cases where menstruation has never been perfectly established, to find the patient afflicted, also, with symptoms of inflammation of the uterus.

The general symptoms accompanying scanty menstruation, when the scantiness is the result of imperfect establishment, are very much of the above character, viz., those connected with anæmia, &c. But, as is frequently the case, the scantiness and infrequency, as also the entire suppression of menstruation, depend upon organic changes in the uterus gradually brought about by chronic inflammation.   What these are we cannot always determine.   Sometimes, however, we find the fibrous structure condensed until the bulk of the organ is smaller and harder than natural ; at other times it is greatly enlarged, as I have verified by examination.   The most common, I think, is condensation and atrophy.   In such instances, there will, of course, be quite a different set of symptoms, in fact, many, if not all, the symptoms found described in connection with chronic inflammation of the substance of the cervix and body of the uterus.   I need not enumerate them here, but refer the reader to the article in which the general symptoms of these conditions are given.   Chronic amenorrhœa, scanty or infrequent menstruation, are in this way associated with the most miserable states of the general health.   I do not believe, however, that the mere absence of the menses is the cause of such terrible nervous suffering as we often see associated with it, but that much of it is caused by the condition of the uterus and other organs upon which the irregularity depends.   The non-appearance of the menses on account of the absence of the uterus, is not usually attended with the chronic suffering I have here alluded to, ordinarily, and indeed in all the cases—which are three—of this kind to which my attention has been called, the patients appeared to be perfectly well. One of these patients was thirty-three years of age, another twenty-seven, and the third twenty-two, and although they all ought to have menstruated several years before I saw them, they were in perfectly good health. This is an argument, I think, in favor of the opinion just expressed, that the serious and annoying symptoms arise from the pathological condition of the uterus, or general conditions giving rise to it. The only symptoms these patients complained of at any time that seemed to be attributable to amenorrhœa, were the backache, weight about the hips, &c., which denote the presence of the menstrual molimen in well-formed persons. In the cases where amenorrhœa exists before the organs are sufficiently developed to assume the function of menstruation, we often observe a good state of health, even after the person has attained to an age when they are expected. I have had occasion to see, examine, and watch for several years, two cases of chronic amenorrhœa from deficient development of the uterus and, perhaps, ovaria. They were both married. One of them is twenty-eight years of age, has been married nine years, has never menstruated, has no sexual desires, but lives happily with her husband, and desires to be like other women merely to have a child for him. There are no distressing symptoms in her case. Her breasts and uterus are developed to about the size of a little girl of thirteen years of age. There is hair upon the pubes, the.mons is well developed, as is also the clitoris. The other has been married three years, is twenty-five years old, and resembles the first completely.

Tuberculosis when causing amenorrhœa is usually well manifested before the suppression occurs, but sometimes this symptom shows itself so early in the case that it is regarded as the cause of the consumption instead of the effect.

From what is said above, the reader will see that amenorrhœa is a symptom of the absence, imperfection, or disease of some of the organs of generation, or some grave deterioration of the blood or nervous energies, and that we are to look into all the circumstances which attend upon it, with a view to learn the causing conditions. We are not always fortunate enough to be able to do

this, and we will then be under the necessity of contenting. ourselves with conjecture, and the necessary uncertainty of treatment
we adopt.

Retention of the menses, as I have remarked, is quite a different thing from a suppression of them. It is always the result of
congenital or acquired occlusion of the genital canal. The os or
cervix uteri or the vagina may be occluded as a malformation.
The ostium vagina may be occluded by a hymen, in such way as
to entirely prevent the discharge of the menses. Adhesions occurring in infancy, childhood, or adult age, may give rise to the
same occlusion. The most common cause of adhesion is inflammation, from accidents or negligence in childbearing, or want of
cleanliness in childhood.

After the menses have been retained for a few months in the
vagina or uterus, the abdomen begins to swell more or less rapidly,
according to the quantity produced each month, until the distended
uterus may be felt above the symphysis, rising higher each month.

The distress in these cases seems to be caused by distension, and
not from any other cause. There is weight about the loins, on
the perineum, and finally, a sense of disagreeable tightness in the
abdomen. We should always be suspicious of the existence of obstructions to the flow of the blood when the amenorrhœa is complete, with but slight constitutional symptoms, and has withstood
judicious treatment for a length of time.

Diagnosis.—It is not usually difficult to determine positively
when there is amenorrhœa, and yet, sometimes, there may be good
reason to doubt in some instances. It is not necessary that there
should be an effusion of blood to constitute menstruation, for there
are periodical discharges from the genitals which indicate the process of ovulation, and, under certain conditions of the system, are
more appropriate than an effusion of blood. I allude to a periodical discharge of mucus or sero-mucus. The uterine congestion is
not sufficient in quantity or force to give rise to hemorrhage, but
to cause effusion of the thinner portions of that fluid. The point
of diagnosis in which we are most interested, is to definitely ascertain the causing condition. We are often obliged to treat patients
for a time without having more than their statements as a basis
of our diagnosis, and fortunately, in most cases, this is sufficient.
We are not justified, however, in continuing the care of an obsti-

nate case for any great length of time without making an effort to verify or ascertain the fallacy of the grounds for our opinion. And, if need be, we must resort to physical examination. The fact of our patient being a virgin should cause deference, but not forbid an examination indispensable to a correct understanding of the cause of a condition that is destroying her life. I need only mention that suppression attended with acute inflammation of the uterus and ovaries, will be attended with marked and almost invariably unmistakable symptoms. The pain, fever, tenderness, and sympathetic symptoms will leave no room for doubt. Anæmia, cachexia, nursing, &c., are obvious conditions, and will be easily made out by very little attention.

It should be borne in mind, in this connection, that we cannot always determine the causing condition of amenorrhœa, and after most ingenious and searching investigation, we are unable to do more than conjecture why the patient does not menstruate. Correctness in diagnosis may be attained with great certainty when there is physical defect in the genital organs, by proper direct examinations of them, and they should be instituted when other means fail to satisfy us. The presence or absence of the uterus, in most instances, can be pretty satisfactorily determined by introducing the finger into the rectum and a catheter into the bladder, and approximating them. If it is present, its thickness interposed between the two will prevent the finger from defining the shape of the instrument; if it is absent, they may be made to touch with the intervention of the walls of the rectum and bladder. The catheter, in this examination, should be introduced deep into the bladder, and the finger as far up the rectum as possible. With this precaution, there can hardly be a mistake. I have met with three instances of congenital absence of the uterus, and in all the vaginæ were absent, but they presented all the external evidence of womanhood, such as perfect mammæ, mons veneris, labia majora, clitoris, and hair. The patients had the demeanor of women, and assured me that their desire for the society of men was as great as usual, and that they experienced strong sexual feeling. One of them had married, and was defending herself in a suit for divorce, upon the ground of her entire ignorance of any anatomical defect in organization; and the other was about twenty-two years of age, and submitted to an examination with the hope of having a correction of the physical defect, preparatory to entering matrimony. It is possible that the vagina may be absent while the uterus is perfect in formation,—the same examination will furnish us with proof,—or the vagina may be occluded from defect of formation. This can be determined in the manner I shall presently describe. Absence of ovaria can perhaps not be determined by physical examination, but there is always such a complete absence of the signs of womanhood that we cannot long hesitate. There are no prominent mammæ, the manners peculiar to the sex, desire for society of males, and sexual propensity, are absent. There is no hair on the pudenda, and the whole external organs are not developed. The signs are the same at any age. The patient at mature age presents no more evidence of sexuality than the little girl.

I have very recently met with an instance of *congenital atrophy*, apparently, of the uterus. The patient, although now twenty-eight years of age, has not menstruated, unless, as she doubtfully said, twice when about seventeen years of age, very scantily. She is rather above medium size, and possesses all the characteristic appearances of womanhood. She has enjoyed fair health until the last twelve months. For the past year she has suffered from distressing palpitation of the heart, which almost incapacitates her for business. She has been married nine years, during which time she has enjoyed sexual intercourse indifferently. She has no monthly pains, the signs of menstrual congestions, and nothing by which to know when to expect that function. Her mammæ are developed to a degree about equal to that of a girl of thirteen or fourteen years. They are very small in circumference, the diameter being about two inches and a half, with a thickness at nipple of about three-quarters of an inch. The nipples are very small. The orifice to the vagina is small, the labia very thin and undeveloped, and very slightly formed mons veneris. The uterus could be felt to occupy its usual position, only rather higher up in the pelvis, but was very light and small. When the fingers were placed under it in the vagina, and it was pressed down from above, it gave the sensation of diminutiveness, apparently not exceeding half its natural size. The ordinary uterine sound would not enter it more than half an inch. A probe, with an extremity about the twelfth of an inch in diameter, freely passed up one inch and a half. From all this, it was plain that the uterus was in a state of atrophy; and I infer that the ovaria were so, from the absence of the nervous signs of menstruation.

The size of the organs, as measured by the means and on the plan above indicated, determines, together with the history of the case, that it is congenital atrophy. Acquired atrophy is confined generally to the uterus, while congenital atrophy generally involves all the genital organs, including the breasts and nipples.

I have met with a number of instances of acquired atrophy, and in carefully tracing their history, could attribute it to some early miscarriage, which it seemed to follow. And this atrophied condition, doubtless, was hyper-involution of the organ after abortion. I believe I know of no case that did not thus originate. In looking over the menstrual history of these sufferers, there was a time when they menstruated normally, and the function was disturbed after having been thus established; and upon making a physical examination, the uterus is found much smaller than it ought to be. When amenorrhœa is attended by chronic inflammation of the uterus, a not unfrequent occurrence, the speculum and probe will reveal the condition beyond the probability of making a mistake. I have seen the worst forms of indigestion, and very great emaciation, attend this condition; in fact, I have seen no other benign disease of the uterus produce so much emaciation as this. The patient is sometimes bedridden for months. In two instances recently cured by local treatment and proper dietetics and hygienic regulations, the patients were reduced to two-thirds their ordinary weight.

Occlusion with retention may be diagnosed by the history of the case, tumidity of the uterus, protrusion of the distended part into the vagina or external to it, a fluctuating insensible tumor felt through the anterior walls of the rectum, by introducing the finger into that bowel and pressing forward. A thorough inspection will disclose the point of occlusion. This condition is so obvious, that there is scarcely any likelihood of being mistaken, if we are at all careful and thorough. It is not always an easy matter to determine the extent of these adhesions. I saw an instance of the effect of inflammation succeeding to a disastrous parturition, where the vaginal orifice was completely closed, commencing with the labia minora; I was unable to determine how far the adhesion extended, but believe that the whole vaginal cavity was obliterated. This did not, at the time I examined the patient, make much difference, as there was no distress from distension; indeed, I could not find that the uterus was at all enlarged from distension, although it was two years after the inflammation had subsided that I saw the case.

Prognosis.—The curability of amenorrhœa will depend on the causing conditions. When occlusion of some portion of the genital canal prevents the discharge of the menses, we can usually, by surgical means, evacuate it, and establish an outlet for the future. Although simple and easy of accomplishment, the evacuation of a long-retained and considerable accumulation is always attended with hazard. In the first place, inflammation may foil our efforts to establish a permanent viaduct for the blood which may be discharged from the uterine vessels; and in the second, this process may be so great and extend to the peritoneum in sufficient intensity as to cause the death of the patient. Amenorrhœa from anæmia may be pretty surely cured; in fact, it is the curable variety compared to those connected with other causing varieties. When arising from inflammation, it will also generally yield to appropriate treatment, as the cure wholly depends upon the removal of the causing conditions. The cachexia which may produce amenorrhœa is often entirely incurable, and, therefore, must our prognosis be unfavorable when they are associated.

In all cases of absence of the ovaria or uterus, we could not expect to do good by any treatment. Where there is only atrophy of the organs, we may hope that some of the ingenious contrivances to increase their development which our profession of the present day affords (they have almost all emanated from, or been perfected by, the fertile genius of Prof. Simpson, of Edinburgh), may enable us to succeed. It cannot be concealed, however, that these causing conditions will often resist every means within our reach. To sum up, then, according to my observation, when amenorrhœa arises from any other causing condition than general anæmia, or inflammation of the uterus or ovaria, the prognosis is not very promising, and we should be cautious in promising a permanent and speedy cure. Failure in the function of menstruation is pretty sure to be accompanied with an inability for conception; imperfection of it is, likewise, very frequently an evidence of barrenness. This is particularly the case with scantiness. When there is infrequent menstruation, but the function is otherwise perfect, the patient is often prolific. I have known a woman for several years, who does not menstruate more than three times in a year, and then not at regular intervals, and yet in the last six years she has had two children, conception following immediately after one of these irregular menstrual discharges.

Treatment.—We should always bear in mind the fact that amenorrhœa is but a symptom, and endeavor to amend the condition or disease upon which it depends. This rational mode of procedure, however, is not always practicable, for unfortunately, as has been more than once stated, we cannot in every instance ascertain precisely such condition. In such cases we make use of remedies, or plans of treatment, which, from the success that has occasionally followed their use, have gained the title of emmenagogues. This term signifies promoter of menstruation. Are there any direct emmenagogues? I think, in the nature of things, there cannot be. To cause a flow of the menses proper, which depends upon ovulation for its existence, they must produce or promote the evolution of ova. That there are remedies and plans of treatment which indirectly promote the menstrual discharge, I think there is very little doubt. In a general way, we ought to consider this class of remedies as producing their effects in two different modes : one is to cause the growth and discharge of ova, and the other the discharge of blood as a hemorrhage. It would be better then to say, that they were oviferous in their nature in the first case, and hemorrhagic in the second. To the first order belongs the preparations of iron and some other mineral and vegetable tonics, nutritious diet, exercise in the open air, diversion of mind, travel, sea-bathing, and, in fact, almost everything which is promotive of healthy functional action generally, by correcting the derangement of vital organs, and generating good blood and plenty of it. To the second, such as aloes, savin, cantharides, and any such hygienic measures, as foot, hip, and leg baths, sinapisms to the feet or legs, &c., such as determine blood to the pelvic organs.

They may very properly be combined in some instances. I do not consider it necessary to go any farther in attempts to explain the *modus operandi* of remedies used for these purposes in this place; they will come up in a different way as I proceed. I shall now take up the recognized, and, for the most part, palpable conditions presenting themselves for consideration as causes of amenorrhœa.

When amenorrhœa results from cold applied to the surface or lower extremities, or other cause suddenly acting to suppress the flow, the uterus and ovaria are bordering on if not in a state of acute inflammation, and the remedies for it should be directed to the relief of the diseased organ or organs. The question very naturally arises, can we, or ought we, to do anything to cause the return of the flow immediately upon its suppression? and if so, what? Experience teaches us that if the flow can be reproduced in a very few hours after its suppression, the turgid and phlogosed condition of the sexual apparatus may subside into a condition of health, and that this may sometimes be done by judiciously managed stimulation; but if the flow is not re-established in a few hours, that we need not expect it to recur until the next period, if then, and that it is injudicious to continue stimulation beyond a very short period. Then what is the proper course of stimulation? If our attention is called to the case within twelve hours, and there is not much febrile reaction, we may very properly direct a hot bath to the whole person of the patient below the waist, for half an hour. The patient should then be put to bed, and two large mustard plasters placed upon the inner portion of the thighs and hypogastrium, and allowed to remain until a strong rubefacient effect is produced, when they may be removed, and the whole replaced by a hot linseed-meal poultice. While these measures are being accomplished, we should administer copious draughts of some warm tea. I cannot approve of the gin-slings or toddies, given so freely under such circumstances; they often do harm by their excessive stimulation, rendering the inflammation a fixed evil.

Should the flux not return in twenty-four hours from the time of suppression, it would be unreasonable to expect and injudicious to continue treatment to cause it to do so. It then remains for us, if possible, to remove the phlogosed condition of the organs, so that they may be in a state to resume their functions at the return of the next ensuing menstrual period.

It will be found, I think, that for the first month, in case of an acute suppression, especially in plethoric patients, the most successful course of treatment will consist in a moderate antiphlogistic and alterative means, kept up steadily. The one I have ordinarily followed is to use counter-irritatives to the hypogastric region, the hip-bath twice a day, of tepid water, six grains of blue mass every third night, to be followed in the morning by a seidlitz powder, and abstinence from all stimulants and highly seasoned food. If, however, the suppression continue beyond the second period after the suppression, it may be attended with chronic inflammation, with or without general anæmia, &c., and will come under some of the conditions hereafter to be considered.

Amenorrhœa connected with chronic inflammation of the uterus or ovaria may be treated as I have directed those affections to be managed elsewhere. I think that it is not very common for suppression, in the chronic form, to depend upon inflammation alone. More frequently the condition of these organs following inflammation, and sometimes remaining after its subsidence, are the causes of amenorrhœa, such as the condensation of the fibrous tissue without diminution in size, or the dense condition with atrophy. The same treatment, with little variation, is applicable to both. I shall have occasion to detail this treatment in speaking of atrophy and want of development.

Another condition which succeeds inflammation of the uterus and ovaria, after an acute suppression, is anæmia. For there certainly are cases in which the poverty of the blood succeeds an acute suppression, and in turn prevents the establishment of the flow. A tonic, roborant treatment, applicable to anæmia arising from other causes, may be instituted, if need be, even before the inflammatory condition of these organs has entirely subsided. Perhaps a little more attention to alteratives, in connection with the tonics, is necessary in this class of cases. When anæmia is the primary condition upon which amenorrhœa depends, it will almost always be found to be a sequence to some preceding affection. Indigestion, connected with a slow or depraved state of the secretions of the alimentary canal, often, by preventing the introduction of nutritious elements into the blood, induces anæmia. These conditions arise, for the most part, in two different ways: either the nervous energy necessary to the sustenance of these functions is diverted to other objects, as mental training in the school-girl, or the circulation in the abdominal organs is too sluggish on account of sedentary habits, as with the sewing-girl. Sometimes want of exercise and too great a tax upon the brain from studies, anxiety, &c., co-operate in the same individual. Anæmia may be produced by a great variety of causes besides those above-mentioned, but, according to my experience, these are greatly the most frequent. I would not have the reader to believe that because I have given the school-girl and the sewing-girl as instances of amenorrhœa that they are the only persons in whom the same character of causes operate in the same way. Too many fashionable young ladies, who might enjoy the blessings of relaxed, diverted, or healthily employed minds, and appropriate and enlivening exercise, become anæmic from sheer laziness, and the nervous anxiety connected with envy.

Bearing in mind, then, the habits which induce indigestion and anorexia, we must, first of all, thoroughly revolutionize the habits and circumstances of the patient, making plenty of outdoor exercise one of the main conditions. Riding in a carriage is not outdoor exercise for these patients, but they must ride a horse, or, what is very well, walk, run, and romp. An excellent sort of diversion for the mind is occupation in domestic duties; making beds, sweeping, caring for and attending children, cooking, washing, &c. The mind and body are both employed in a varied and diverse manner in these household duties; and it will be found that exercise of body and mind both is most profitable as it is most diverse and varied. While it is true that some kinds of exercise, as walking or riding, may be made to call into play a great many muscles, yet the whole routine of duties presenting themselves in the business of housekeeping, by personally doing the work, is more beneficial than all others devised. All this lesson is taught by the contrast between the young mistress and her servant.

In addition to the adoption of a more rational course of habits for the patient, much may be done by the judicious use of medicines. Almost invariably the tonics must be preceded by, or accompanied with, alteratives and laxatives. The stomach will no more recognize and respond properly to any tonic that is introduced into it until prepared, by correcting the secretions, quickening the gastric circulation, and unloading the bowels, than it will digest food under similar circumstances. The alteratives suitable, generally, are mercury in some form, taraxacum, and turpentine. When the bowels are torpid, the stools dry and of unnatural color, particularly if the color is light, from three to six grains of blue mass, given every third night, and followed next morning by a seidlitz powder, or sufficient sulphate of magnesia to cause one or two evacuations, is an admirable alterative. Ten grains of good extract of taraxacum, with a minute quantity, say the twentieth of a grain, of bichloride or biniodide of mercury, three times a day for two or three days, generally does very well. The mercury should not be given with the taraxacum longer than three days, and then intermitted for a week, but the taraxacum may be given for weeks steadily forward. An excellent alterative for the stomach is Venice turpentine. Ten grains, three times a day, after eating, on some sugar, alternated, or given with some of the mercurial preparations, proves often of great service. I cannot but mention the compound confection of black pepper, made in imitation of Ward's paste, as having frequently an excellent laxative and corrective effect on a weak state of the stomach attended with constipation. I have known it to cure some of the most obstinate cases of constipation attended with anæmia.

If there is not scantiness of secretions, but slowness of peristaltic movement, we ought to depend on rhubarb and aloes. The compound aloetic pill is a good preparation. In the selection of tonics, we should bear in mind the difference between the stomachic and blood tonics. Iron is, perhaps, the only, or almost the only, blood tonic directly, while there are a great many articles that act as stomachics. Almost all the bitter vegetables ranged under that head in the books are useful under certain circumstances. The stomach tonics, by improving digestion, are indirectly blood tonics, so that they are sometimes all that are necessary. In many instances, too, the stomach must be prompted by the bitters, or other stomachics, before it will absorb or assimilate iron. The bitter may be mixed with the iron, precede it, or administered simultaneously with it. It is sometimes convenient and profitable to combine the alterative and stomach tonic. A mixture of this kind, often used, is the compound tinct. cinchona, with bichl. mercury dissolved in it. The tinct. gentian, or colomba, answers very well compounded with mercury. Ext. gentian, and Quevenne's iron compounded, in a pill, operate well on the weak state of an anæmic patient. If we understand the principle that governs the

7

treatment in such cases, we may readily find the means to accomplish our ends,—alterative, stomach tonic, and blood tonic.

The cachexiæ, several of which interfere with the regularity of the function of menstruation, must be treated as if the menses were present in their normal quantity, and in these cases the amenorrhœal complication is of no importance; hence, special efforts to restore the flow are injudicious, and, in most cases, injurious.

In patients well developed in most respects, whose genital system is deficient, the menses cannot be produced unless these organs grow and become more active. Anything that will stimulate these organs will occasionally bring this result about. Wedlock is a remedy in these cases sometimes. The indulgence in society, and the recreations of it, in company with men, sometimes, through the moral faculties, stimulate the genital organs towards development. The stimulus thus afforded by society is one of the beneficial effects resulting from the change of habits required in young girls who go to boarding-schools until sexually dwarfed by confinement to the uninteresting society of their own sex.

Prof. Simpson has recommended an instrument which he calls an "intra-uterine pessary," to bring about this development. It is equally applicable to cases of atrophy of the uterus arising after the menses have been established. I have had occasion to use it, and am now employing it in the interesting case to which I have alluded above. It is theoretically better, I am afraid, than it will be found practically; yet there is no doubt much good may be done by it. The object of the intra-uterine pessary is, 1st. By being introduced into the cavity of the organ, and allowed permanently to remain there, to cause a development analogous to the development around a growing polypus or fœtus. If this can be effected, the uterus becomes more vascular, and hence is more likely to bleed under the stimulus of ovulation; and, 2dly. To be the medium, or generator, rather, of galvanism, to stimulate the nerves of the uterus. Both of these objects are promotive of uterine hemorrhage, if not of correct menstruation. They are necessary to the development of an atrophied uterus, whether congenital or acquired. But this instrument is recommended and used in obstinate cases of amenorrhœa, where there is no apparent deficiency in the size and development of the organs concerned. It is in this class of cases that most may be effected by it, and yet it sometimes entirely fails to produce any effect. To do good in the cases of atrophy and want of development, it should be used continuously. Where the development is good, I am inclined to think that it will do more good by using it intermittingly. In these last we may introduce the instrument, and allow it to remain one week before the time of the expected period, and then, after the time is passed, remove it, and again introduce it at the proper time. We should remember that we cannot use an instrument of the same size in all cases. In the uterus that is much atrophied, it would be violence to use an instrument that is applicable in a case of a fully developed organ. In the former, we must have an instrument that will pass into it easily, and in a couple of months use one larger; and after the lapse of a similar time, make another one still larger, &c., until development is complete. The instrument is made of copper and zinc, and consists of a stem and bulb. The bulb is hollow, in order to be light as possible, flattened, and oval in shape, one inch long, three-quarters of an inch wide, and half an inch thick. It should be perforated through its thinnest diameter by a hole two-twelfths of an inch in diameter. Into this perforation the stem is to be inserted. The stem should be two inches long for a uterus not atrophied, and as much less as is necessary in the judgment of the attendant, when atrophy has taken place. It should be hollow and light like the bulb. The bulb, and one inch of the stem next the bulb, is made of copper, the extremity of the stem of zinc. This completes the instrument as made and used by Prof. Simpson. I find, in some instances, great difficulty, if not an entire impracticability, of wearing it, on account of a tendency to fall out. Sometimes, too, the gavlanic stimulus is not sufficient. In either of these cases I have made an addition to it, which, I think, adds to its efficiency as well as security of position. This consists of a zinc ball, about an inch in diameter, attached to a copper rod four inches long. The ball is introduced into the vagina after the intra-uterine pessary has been introduced, while the stem is attached to a framework outside the pelvis, to keep the whole in position. As will be seen by a study of this apparatus, we have quite a galvanic battery,—the copper rod reaching from the framework of zinc outside to the zinc ball inside, this last lying in con-

tact with the copper bulb of the pessary, &c. If we do not desire any galvanism in the case, the whole apparatus can be made
of copper. Made in this way, the instrument is quite efficient.
The young physician or student may be embarrassed in his attempts to introduce the pessary without a little consideration.
The plan I have found most convenient is, to expose the os uteri
by means of the quadrivalve speculum ; secondly, to secure the
pessary by inserting a piece of whalebone, properly shaped, in the
perforation in the bulb ; thirdly, thus mounted, to insert the stem,
and with great gentleness, urge it forward to its full length, or
until it is arrested by the contracted internal os uteri, or the end
touching the fundus. If this arrest occurs, the instrument is
either too large, or too long, and must be replaced by one more
suitable in this respect. After the pessary is inserted, we may
withdraw the speculum, and, if necessary, apply the ball and external framework above described to keep it in position. All this
direction does not include a fact which should ever be borne in
mind by the student, viz., that the instrument is sometimes utterly
intolerable; and, at others, a good deal of address and patience
is required to habituate the parts to it. The patient should be
forewarned that pain and inflammation are the possible effects,
and that she must inform us should they be considerable. There
is always some pain, sometimes a great deal. When the irritation
is too severe, the instrument must be removed, quietude observed,
and, if necessary, anodynes, and even antiphlogistic treatment, to
remove the symptoms. After all these have subsided, it may be
again introduced. A little perseverance and care will render most
cases tolerative of its presence. During the time the instrument
is used, the vagina must be thoroughly cleansed, at least twice a
day, with tepid, warm, or cold water, and some fine soap, used as
injections.

Obstructions to the flow of the menstrual blood must be removed by the finger, when practicable ; when not, by the knife.
In recent, and sometimes in cases of long standing, the adhesions
may be torn through by pushing the index finger forcibly against
them. If the obstruction is caused by the hymen, this may generally be done. This failing, we should introduce first an exploring trocar into the containing sac, and guided by this, freely in-
cise, by means of a tenotomy knife, the most dependent part. The

incision must be made in the lateral direction, being careful to avoid the bladder and rectum. These two organs ought to be emptied previous to operating. The opening ought to be free, the blood all evacuated, and the cavity in which it was contained thoroughly washed out with tepid water. The opening will have a tendency to close, and partially or wholly restore the obstruction. We can avoid this best by introducing the finger every day for two or three weeks. The injections into the uterus must be repeated twice a day for the first ten days or two weeks. As the cavity decreases in size, we should be careful to adapt the instrument to this circumstance. From the time the operation is performed, the patient should be carefully watched, as there is often a strong tendency to peritoneal inflammation soon afterward, which not unfrequently proves fatal. When there is much pain, a full dose of opium, administered and repeated afterward as often as necessary, will go very far towards averting such evil. Toxæmia is another danger to which the patient is subjected, and is to be avoided only by the thorough cleansing of the cavity I have described. When either of the sequences do occur, their treatment will, of course, be governed by the same principles as if they originated under other circumstances.

CHAPTER VI.

MENORRHAGIA.

HEMORRHAGE occurring at the time of menstruation beyond the usual quantity for the patient, is menorrhagia. Hemorrhages occurring at other times do not belong to this denomination. Often there are both hemorrhage and menorrhagia in the same individual, which depend upon the same conditions of the system or reproductive organs, and are alike symptomatic of some local or general disease. It ought never to be forgotten that menorrhagia is a symptom and not a disease, and that we must look beyond the discharge for indications of treatment. Unfortunately, we cannot always discover the conditions in which it originates, and consequently are often left to conjecture.

In considering the pathology of menorrhagia we must make two general divisions. 1st. Those cases in which the reproductive organs are at fault; and 2d. Cases in which, with a healthy condition of these, there is general disease, found in the blood for the most part. These two are sometimes combined. The former is by far the more frequent. Local causes operating upon the uterus and ovaria are more commonly efficient than the general conditions which sometimes give rise to it.

Chronic inflammation of the mucous membrane lining the cavity of the corpus or cervix uteri, or both, is, in my observation, a condition attended with menorrhagia in a great many instances of this affection. In young persons this is particularly the case.

There are cases also of general inflammation of the uterus attended with menorrhagia, which, when the inflammation is cured, the patients are no longer the subjects of this symptom. Temporary congestion of the uterus from long standing, too much heat about the pelvis from heavy clothing, excitement from any cause at the time of menstrual congestion, too great indulgence in sexual intercourse, sometimes excite temporary menorrhagia. The mucous membrane of the uterus is sometimes covered more or less completely with little vascular growths, not large enough to be called polypi, that bleed at the time of menstruation. Sometimes one or more minute, hardly distinguishable fibrous growths, intruding into the mucous membrane of the cavity, or embedded in the walls, render the organ more than naturally vascular, and consequently liable to bleed more at the time of menstruation. Larger polypi of a soft or hard nature, existing in the cavity or attached to the cervix, are first suspected on account of increased menstrual flux.

Common ulceration, by inducing permanent congestion, is often attended with menorrhagia, but almost always the malignant ulcerations, as cancer, &c., give rise to increased discharge of blood at each menstrual period.

Tumors situated near the uterus, so as to press upon the veins which carry the blood away from this organ, induce hemorrhage and menorrhagia. Cancer, chronic inflammation, or other organic diseases in the ovaria, not unfrequently give rise to this symptom. In fact, any local disease in the pelvis that acts as a focus of congestion, may do the same thing; therefore, hemorrhoids, prolapse of the rectum, chronic inflammation, polypi, common or cancerous ulceration, are sometimes attended with this effect.

I can readily see how displacement, particularly if attended with inflammation, may act as a causing condition; but I think its effect in this respect is overrated by some authors.

The general conditions attended by menorrhagia are not very numerous, but some of them are undoubtedly efficient operating causes. Plethora from high living and indolent habits, from hereditary predispositions in persons beyond thirty years of age, generally from intemperance and other irregular habits, &c., may be enumerated among them. I think I have known two patients affected with menorrhagia from a possession of the hemorrhagic diathesis, so well established by the observations of surgeons as being the cause of hemorrhage of a different nature. These general conditions seem to act by causing a rich and abundant state of the blood, which gives rise to rupture of the capillaries from the stress of undue distension. The opposite states of the circulating fluid may also be attended with menorrhagia; hence we have this symptom from anæmia, or watery condition of the blood, if this is a correct expression. My idea of anæmia is exemplified by the condition of the blood when deprived (from whatever cause) of a large portion of its globules; plethora, when there is a relative increase in the amount of globules. Either of these conditions may be attended with a relative increase or decrease of any of the other ingredients of the blood, except, perhaps, water. They may also be complicated by the retention or introduction of poisonous ingredients,—septicemia. In anæmia there is probably, at least in many instances, no great reduction in the amount of fluid in the vessels; and hence, so far as the fulness of the vascular system is concerned, we may have plethora. I think the hemorrhages and menorrhagia that result from the watery condition of the blood, are often, if not generally, brought about by undue stress upon the capillaries of the organ from which the flow takes place, and not because of the thinness of the blood, which, as is supposed by some, enables it to permeate the walls of these minute vessels.

Anæmia resulting from indigestion, confinement, depressing passions, deprivation and toil, damp low situations, miasm, and degeneration of the kidneys, all induce menorrhagia occasionally. More occult causes of anæmia, such, indeed, as are inscrutable in their action, may be expected to exist.

The peculiar condition of the blood—which, if not anæmia, resembles it more nearly than the opposite condition of the circulating fluid—brought about by albuminuria, should be suspected when any of the symptoms indicating it present themselves in females who have suffered from eclampsia in their first confinement, or in patients approaching the climacteric period of life. Scurvy has seemed to me, on several occasions, to give rise to menorrhagia. During a prevalence of this affection in my circle of acquaintance one year, among the German population more particularly, in consequence of their inability to procure proper vegetable diet, menorrhagia was a very troublesome symptom.

The above are most of the obvious conditions of menorrhagia, but there are instances which cannot be traced to any of them, or any other state of the system or particular organs. It is believed, however, that they are the exceptions, and that, a careful survey of each individual case will enable us to get a clue to conditions so obvious as to warrant us in using them as a basis for treatment. Constipation, whether it be regarded as a general or local affection, always acts mischievously on the uterus, and when there is a tendency to menorrhagia it will be aggravated by this condition of the bowels. And perhaps in some instances costiveness may of itself be sufficient to induce menorrhagic congestion by obstructing the circulation.

Diagnosis.—This is not generally difficult, so far as making up our judgment merely as to its existence is concerned, but, as already intimated, it is often exceedingly difficult to get precisely at the conditions which give rise to it. Individuals must be judged by their own standard as to the quantity of menstruation. What would be too much for one would be an ordinary quantity for another. If a woman who has been in the habit of losing twenty ounces of blood at each menstrual period from the beginning of her menstrual life, is in the enjoyment of good health, and suffers no inconvenience from it, that may be regarded as her normal quantity. If, however, for several years six or eight ounces was her ordinary limit, and her health is affected injuriously by the increased flow, twenty ounces must be considered menorrhagia, and treated accordingly. The general standard must be very carefully erected. Probably an average quantity of loss in health is not far from eight ounces. Many lose much more, some much less, and yet all may be equally well and healthy in the discharge of this function.

It may be well to instruct the student how to measure the quantity lost. As women generally wear napkins over the genital fissure at such times to protect their clothes, we may judge correctly enough for practical purposes, by inquiring the number saturated each time. One napkin will require an ounce of blood to thoroughly saturate it. Counting each one soiled as an ounce of blood lost, we can easily ascertain the quantity. Having determined that menorrhagia exists, whence does it originate?

The symptoms of general suffering when menorrhagia has become an inveterate symptom, are often grave, but not peculiar. There is much of a sameness in suffering from any form of uterine disorder or disease. Nervous pains in the head, back, groins, perineum, hypogastrium, lowness of spirits, indigestion, and legions of disagreeable symptoms, which no man can enumerate, but which all experienced physicians who have treated diseases of females must have witnessed, and set down as uterine, will be found to attend upon them. The obligation to determine the causing condition of menorrhagia is imperative, as without this knowledge our treatment is entirely empirical and uncertain.

First of all we should try to fix upon the distinction between the general and local conditions. I need not enter minutely into the consideration of the circumstances and symptoms of the general conditions. I will, however, caution the reader not to mistake the cause for the effect; for, as will readily suggest itself, anæmia is one of the effects of menorrhagia, and if we do not get an intelligent history of our patient, we may fall into this error. With proper caution we may satisfy ourselves which was first in the chain of circumstances, the anæmia or the bleeding. When we suspect albuminuria, a test of the urine will enable us to decide as to its existence; when we have decided that a given case is not of general origin, we are to investigate the condition of the pelvic organs.

I need not dwell upon the mode of examining them, as full directions for this purpose are given elsewhere. It should be thorough when there is obscurity, and conducted so as to reach all the organs in the pelvis, but especially the uterus, which is the organ much the most frequently at fault. The finger, speculum, probe, and, if need be, the dilating influence of the compressed sponge, should be brought into requisition. In obstinate and otherwise inscrutable cases, any doubt may be cleared away by dilating the cervix so as to feel, and, if need be, see the cavity of it and the corpus uteri. Any excrescences, vascularities, or other abnormal condition, would thus be discovered, and be placed in a condition to be removed or corrected.

Prognosis.—Immediately, there is not much danger in menorrhagia, though there are instances on record of its proving fatal by reason of its immediate effects, the patients literally bleeding to death. Dr. West gives one instance of this kind, and speaks of another that came to his knowledge. There were no signs of pathological lesion, in either case, discoverable by dissection, that could account for their disastrous termination. This event is, however, undoubtedly very rare.

A much larger number of cases prove disastrous indirectly, by inducing other diseases. The loss of a large quantity of blood impairs the functions of all the important organs. Nervous and cerebral affections arise thus. We may trace to this circumstance

sometimes obstinate forms of indigestion, hepatic and splenic congestions, &c., &c. In various ways, indirectly, these losses affect
the health injuriously, and, acting conjointly with the causing
condition, it would be surprising if grave cases of menorrhagia did
not do so much mischief to the general organization as occasionally to prove fatal. Another point in the prognosis, of very great
interest, is its curability. Can it be cured by judicious scientific
treatment ? This question can be answered in each case only by a
scrutinizing examination of the individual circumstances connected
with it ; and will depend for its solution upon the character of the
causing conditions, the length of time it has been in existence,
and the effects already produced by it upon the general health of
the patient.

The general statement, that very great benefit can be derived in
almost all cases, and a cure very generally be made, may properly
be ventured upon.

In coming to a prognosis, we must bear in mind the influence
which a long-established habit has in keeping up the discharge
after the causing condition has been removed. The physician and
patient should, therefore, not be discouraged at the slowness with
which the curative effect of the treatment is produced.

Treatment.—The treatment is palliative and curative. The
palliative has for its object the relief of the hemorrhage; to prevent it from being excessive until the period passes by, when it
will spontaneously subside. The curative is intended to remove
or cure the causing condition. In conducting the palliative treatment, we may very properly commence before the time for the appearance of the menses. We ought especially to rectify and remove the circumstances that aggravate it. We will often find
constipation an aggravating accompaniment. In such cases, it is
well to give ten grains of blue pill, the fourth and second night
before the expected attack, and follow it in the morning each time
with a saline cathartic. This removes the abdominal plethora for
a time, and prepares us for an antiphlogistic regimen, which
should be strictly observed where there is general plethora, for
two or three days before, and during the time of the flow. The
diet should be light, and taken cold; and the patient may take
the twentieth of a grain of tartarized antimony four or five times
a day, or, what I have thought sometimes better, ten grains of

nitrate of potassa as frequently. The patient should also assume the recumbent posture for two or three days beforehand, and be, as far as possible, free from all causes of mental excitement. All these precautions are intended, of course, only for cases of seriousness and gravity. To some extent they may, with propriety, be resorted to in most cases. The cathartic, rest from fatigue, cool diet, &c., will always be comfortable as well as profitable.

It is a difficult matter always to command the co-operation of our patient in such measures as are necessary to moderate the hemorrhage after it has begun. There is a popular fear of the consequences of using measures to stop the menstrual discharge; and I have known medical men who always allowed a certain amount of blood to escape, and time to elapse, before they would put in requisition such measures as they considered most effective. There can be no question about the propriety of beginning with, and even anticipating, the time, by the use of the most energetic means to arrest the discharge. The danger is that they will fail, and not that they will be too prompt in their action. So far as we can, we ought to be governed by the general condition of the system and the uterus in the choice of our remedies. If the phlogistic pulse and general habit of the body are obvious, and steadily maintained through the first days of the attack, our remedies must be mainly directed against the condition thus indicated. Antimony, or nitrate of potassa, have always been favorite remedies in such cases with me. Now, sometimes, when I can anticipate the attack in this class of patients, I am in the habit, in addition to the nitrate, to bring about the influence of veratrum viride, and keep the patient moderately under it during the whole time. It is in this condition that the digitalis, so strongly recommended by Dr. West, will be found most beneficial. He recommends us to give $\bar{3}$iv of the infusion every four hours. It is but fair to state that Dr. West does not advise it in this class of cases only, but says, "The existence of a very considerable degree of exhaustion, and the necessity for the administration of wine, do not contraindicate a trial of digitalis." He gives it in these cases, however, in only $\bar{3}$ii doses to begin with. The digitalis should be continued until some dizziness or sense of faintness is produced, when it should be stopped.

In cases where the hemorrhage is more passive, and unattended with arterial reaction, the general treatment must be different. Very frequently the uterus is enlarged and lax of fibre, and the remedies must be such as will condense its tissues, as ergot, borate of soda, alum, and strychnia. Probably turpentine should be mentioned second to ergot in this connection; it is certainly very effective in many instances. The ergot should be given in large doses. My mode of administering it is the same as in cases of labor. Twenty grains of the powder in water that has been poured upon it boiling hot, and allowed to stand until cold. This should be repeated as often as once an hour for three or four times. Whatever the effect, we should then wait for the lapse of twelve or eighteen hours before having recourse to it in the same way. In the course of the paroxysm, it may be profitably used four or five times, if necessary.

In this way ergotism may be produced, and after wearing off, which it is apt to do in less than twenty-four hours, reproduced. Turpentine should also be given in large doses; 3i every four hours, in an emulsion, is very effective. In threatening cases it may be given even in larger doses, 3ii, for instance, every five hours. Its effects are ordinarily more marked than those of any other remedy, not excepting ergot. Acet. lead, and the vegetable astringents may all be tried, but I have not had much good result from them at such times. Positive quietude is indispensable to the success of every general course of treatment. In cases attended with great pain and suffering in the region of the uterus, full doses of opium sometimes produce admirable results. In addition to this general treatment, and ice-cold drinks, there is a certain kind of local treatment, applicable alike to almost all sorts of severe cases. We may always use cold to the hypogastric region. Ice in a bladder, or piece of oil silk, is the best application of this sort; or, if in winter, frozen wet cloths make an excellent application. Small pieces of ice, introduced into the vagina as near the os uteri as can be, ought to be used, as affording valuable aid. Copious ice-water injections is sometimes more effectual than any other local remedy of this kind I have tried. I have not seen any bad effects from this free use of cold internally.

In cases of peril or great exhaustion, the tampon is a valuable means of relief. It is not best to wait long enough for the patient to become greatly exhausted, in cases where our *former observation* teaches us the flow will continue, or where we have tried the above treatment without moderating the discharge, but we should at once resort to the tampon.

For several years I have not used any other form of tampon than the gum-elastic air-bag. It may be found in almost any of the shops at present, and if its tube is properly supplied with a stop-cock, it makes the handiest and, I think, the best in use. It can be introduced up to the mouth of the uterus while empty, and then by means of the inflater, which goes with it, distended to any desirable dimensions. In cases of this kind, it ought to be inflated until it produces slight sense of pain in the vagina from distension. This pain will soon pass off, and with it the vagina may slightly dilate; when that is the case the distension may be increased. It should not be allowed to remain in the vagina longer than twenty-four hours, without being removed and cleansed. The vagina should also be syringed with cold or even ice-water. After these things have been done, it may be again introduced, and so on as often as necessary. The physician may use cold water in a gum-elastic bag, in place of air, if he choose, but it is not so convenient, and probably not much, if any better; as, by distension, the instrument may be pretty certainly made to check the hemorrhage.

By employing such judicious and sufficiently energetic means as we may command, we will very greatly relieve the extreme cases; and sometimes the repeated interruption of the discharge proves curative. This will be the case most likely when the hemorrhage outlasts the condition which originally produced it, which is not very common. In most instances, it will be necessary to place our patient under proper treatment during the intervals. The object of the curative treatment is to remove the causing condition, which should be carefully ascertained when at all practicable.

We have seen that anæmia is occasionally an efficient causing condition. When this condition is uncomplicated, it is almost always produced, more directly, by a failure on the part of the digestive apparatus. I do not think it would be proper here for me to do more than indicate a few general principles by which to be governed in the management of anæmia. Almost always we need medicines and management, to give muscular tone to the stomach and bowels, a healthy state of the secretions in the whole of the alimentary canal, and more directly to add iron to the blood. We must effect perfect chymification, chylification, sanguification, and defecation; and if any of these processes is imperfectly performed, the result will not be attained. We cannot, in most cases, simply give iron with the hope of causing better sanguification. The first two steps of digestion must be well performed before good rich blood can be made, however much iron we may pass into the stomach or blood, were that possible. Then defecation, to keep separate the effete materials of the ingesta, is just as necessary to the formation of pure blood. We, to be sure, occasionally meet with cases that may be cured by any process that will correct one of these derangements, but this certainly is not common. By examining carefully each case as it comes up, we generally ascertain which one of these processes is most at fault. But we must not only furnish the vessels with well-digested material, but, to perfect sanguification, we must introduce oxygen in sufficient quantities, or we meet with failure still. In more special terms, our anæmic patients will need the bitter tonics, and, generally, some stimulants given at meal-time, so as to stimulate the stomach at the time of digestion, get mixed with the chyme, and pass with it through the duodenum and jejunum, and not be absorbed wholly until taken up with the chyle. Iron may be superadded to these, more directly to enter the vessels; and probably two hours after eating is the best time to administer it. For the purpose of promoting proper secretions, a few grains of blue mass, administered every second or third night, to be followed in the morning after with a saline laxative, large enough merely to move the bowels gently, has seemed to me to suit very well generally. Then there are some substances that may be added, if thought best, more directly to act upon the muscular portion of the alimentary canal; as strychnia and ergotine, both of which may be employed in certain cases, to great advantage. Or, if we prefer, we may use five grains of pulverized nux vomica, or powdered ergot, in place of their alkaloid principles, three times a day.

By keeping in view these leading ideas of the pathological causes of anæmia, and the modes of relieving them, the treatment will be comparatively simple and efficacious. Healthful muscular exercise in the open air, at least one-fourth of the twenty-four hours, and proper diet, will be indispensable addenda to these.

But we should not forget that, generally, underlying all these things, there may be a *cause* that it will be necessary to remove.

In women, the uterus may be diseased, or the genital organs may be abused by masturbation. We should examine carefully, in order to ascertain the originating difficulty.

Uræmic anæmia, for I think this is the condition albuminuria usually indicates, is, happily, of more rare occurrence; and it is unfortunately, also, much more difficult to manage. Iron, ergot, and the terebinthinates,—the balsam copaiba particularly,—are the most useful means in such cases. The tincture of iron in pretty large doses, say forty drops three times a day after eating, is extremely useful. It should be administered during the intervals; and a few days before the attack, it may be omitted for five grains of the powdered ergot three times a day. One capsule of balsam copaiba, containing half a drachm, may be given in place of the iron, followed by the ergot. Such use of these remedies, although generally incapable of restoring the function of the kidneys, often improves the symptoms very much.

General plethora may accompany menorrhagia, but I think very seldom. Plethora, or rather congestion of the abdominal organs, is quite common; and the judicious management of it will afford great relief, if it does not cure the symptoms. I need not dwell at any great length on the subject in this place, as I have already spoken of it in the treatment of the paroxysm; but I must refer to the great benefit I have seen result, where this condition was prominent, in the use of minute doses of bichloride of mercury, and large doses of taraxacum, administered for a long time.

In one of the first cases I ever treated,—and one in which the most marked benefit resulted,—I put ℥ii of the extract of tarax. in ℥xvi cinnamon-water, and then dissolved gr. i of corrosive chloride of mercury in it. The patient took a large tablespoonful of this mixture three times a day. Slight ptyalism occurred in three weeks from the time she began to take it; the paroxysm came on in three days after this occurrence; the discharge was very greatly reduced. The same process was repeated the next two months, when the patient was entirely cured.

When the hemorrhage depends upon endometritis, which, I think, is very frequently the case, the disease is to be treated as directed in another part of this work. (See Endometritis.)

When inflammation of the body or cervix uteri is present, our treatment will be ineffective, as a general thing, until it is removed. And we should look and expect their presence in a great many instances. Subinvolution and the congestive or hyperæmic condition, which often accompany it, are not unfrequently the obstinate causes of menorrhagia. These conditions may be removed with a good deal of certainty by patient and persevering treatment. The treatment I have found most effectual is the application of nitrate of silver to the inner surface of the organ pretty thoroughly every seven or eight days. This stimulates the uterus to contraction and condensation, and in the course of four or six months the difficulty may be entirely removed. We should be sure to persevere long enough. On theoretic grounds, we might expect ergot and strychnia, administered for a length of time, to lead to the condensation of the uterus tissues; but I have so generally cured this class of cases by the local application of the nitrate of silver to the inner mucous membrane of the uterus, that I confess I have but little observation with any other treatment. These strong astringent applications to the inner surface of the uterus, made by the flexible caustic-holder, or the whalebone with cotton or lint upon the end of it, will generally cure those cases that are kept up by too great vascularity of the mucous membrane. We may thus introduce the nitrate of silver in substance, and apply it thoroughly to the inner surface, by freely moving the caustic-holder about in it, or if we fail in making a strong impression, a small piece of the nitrate may be left in the cavity to dissolve and diffuse itself. I have done this a number of times, and with the exception of trifling pain, lasting only for a few hours, and some sanguineous discharge for two or three days, have had no bad result.

The application of the nitrate to the inner surface may be done once a week with the caustic-holder, and in general will succeed well. The acid nitrate of mercury may be used with the whalebone probang once in two weeks, when the nitrate of silver is not used. Creasote, muriate tinct. of iron, and, in fact, all the remedies I have recommended for endometritis can be used in the way recommended under that head. But it sometimes happens that the menorrhagia continues in spite of this treatment, and I think generally on account of the imperfect applications to the inner surface of the uterus. It then becomes a question whether the os and cervix ought to be dilated. I think that when the bleeding resists all other judicious means for its arrest, this measure is justifiable, and, if the loss is great, demanded. We may have two objects in view in resorting to dilatation : 1st, to ascertain the precise condition of the cavity of the uterus when practicable in this way, and 2d, to be able to remove any small growths that may be found there, or to make a thorough application to the mucous membrane. In one instance, not long since, I cured the menorrhagia by dilating the os and cervix so that I could introduce my finger to the fundus, and then with a cotton swab cauterizing the membrane thoroughly with acid nitrate of mercury. I operated this way twice, at intervals of two months. There was but little inconvenience from its use, and the case was entirely cured. We should be careful to cure, if possible, all local complications of menorrhagia, as having a very important bearing upon our success. Hemorrhoids should be particularly attended to, as keeping up and aggravating the discharge. Chronic vaginitis, urethritis, cystitis, &c., or any disease in the pelvis that keeps up constant congestion of the parts, should be remembered as quite worthy of our notice in this connection. If ovaritis be complicating the case, it should be treated by the means which would be used were it an independent affection. Ovarian inflammation is supposed to have an important relation to this and other menstrual aberrations in two ways, 1st, by its directly stimulating the uterus by unusual manifestation of ovarian orgasm, and 2d, by acting, from proximity, thus keeping up a large flow of blood to the whole pelvic viscera.

CHAPTER VII.

DYSMENORRHŒA.

DYSMENORRHŒA means painful menstruation, or difficult menstruation with pain. Menstruation may be attended with severe suffering, when the flow is not otherwise attended with perceptible improprieties. It may be performed with difficulty, as when the flow is prevented by obstruction, or when a membrane is expelled with great pain. Whether there is a class of cases in which the flow is established slowly and imperfectly at first, or membranous productions are expelled without pain, I am not informed ; but I have certainly seen instances in which both these conditions were present with a very trifling amount of pain. To be dysmenorrhea, the pain must occur at the time of the menstrual epoch, or immediately before or after it ; and it must emanate from, if it is not confined to, the pelvic region. It is taken for granted, also, that it is necessarily connected with this process. The pain is not always seated in the uterus ; it may be in the ovaria, in the plexus of nerves supplying the uterus, or the lower extremities ; or it may be in the bladder, rectum, or urethra, one or more of them. And in some cases it seems to occupy all these organs and plexuses of nerves.

A question has often occurred to me in this connection, which I think of great interest to decide in each case. Is the pain always dependent upon a morbid condition of the uterus or ovaria, or is the pain dependent, in some instances, on the morbid condition of the organ in which it is located, the rectum or bladder, for instance, influenced by the normal congestion of all the pelvic viscera at the time of menstruation ? I think I have met with cases of this last kind. From this mode of viewing the subject, we might expect that any pelvic organ, in a state of chronic inflammation, may become very painful by having the inflammation exaggerated, as a consequence of the menstrual molimen ; or any pelvic organ, with strong neuralgic proclivities, may become af-

fected severely by the hyperæsthesia of that period.   Thus arises
the sciatica, and other neuralgia of menstruation ; the disease not
being in the uterus, or ovaria, but in the nervous plexuses, and
the suffering rendered paroxysmal by normal periodical congestion
of them as the effect of the menstrual molimen and hyperæsthesia.

These cases are not painful or difficult menstruation, although
the menstruation is attended with pain.

The pain of dysmenorrhœa is not only differently located, occupying different organs and tissues, but, as might be readily inferred,
different in character.   Sometimes it is neuralgic, sometimes inflammatory, sometimes tenesmic, and again, it may combine all
these qualities ; terrible lancinating or shooting pain, accompanied
with distressing aching, burning, and bearing down.   The pain
may also be of a cramping or spasmodic character.

I make these remarks to show the student that the study of dysmenorrhœa is not a very simple one, and that if his views of the
subject are not somewhat comprehensive, his inferences and practice will fall short of the requirements of the case, and that an
examination which does not embrace the whole assemblage of pelvic organs and tissues will be imperfect.   Nor can the nature of
dysmenorrhœa be understood by confining our investigations to
the pelvis ; but the blood, the nervous system, and abdominal organs, will sometimes be at fault, and demand attention in a searching investigation of the subject.

While the above statements are true in regard to the general
subject of dysmenorrhœa, and should be always present in our
minds, the majority of cases are more simple, and not difficult of
comprehension, and confined in their nature to some appreciable
morbid state of the uterus or ovaria.   Hence the propriety of the
divisions entertained by most authors, and which are supposed to
render the subject more lucid and comprehensible.

The divisions which I think sufficiently comprehensive to embrace almost, if not all varieties, are, 1st, inflammatory ; 2d, neuralgic ; and 3d, membranous.   Some authors enumerate many
more, such as congestive, obstructive, &c.   I need not mention all
of them, as they will readily recur to the reader.   It is hardly necessary to point out how a uterus, in a state of inflammation, may
become the seat of very great pain at the time its vessels are rendered excessively turgid by the menstrual congestion.   The in-

flammation and, consequently, the pain, may be in the ovary, instead of the uterus, when the same obvious reasoning would be applicable. Hence another subdivision of dysmenorrhœa,—ovarian.

The congestive variety of authors properly falls under the inflammatory form. Congestion recurring at every menstrual period, so far beyond its ordinary bounds of moderation as to give rise to excessive suffering, without some permanent morbid state of the parts, is certainly very improbable; and, judging from my own observation, which, of course, is not so extensive as many observers, does not take place. In other words, I have not seen a simple case of congestive dysmenorrhœa, but I have seen numerous instances in which the inflammation was very moderate but persistent, and, aggravated by the recurring congestion, and seemed to me to be the cause of the pain. This has been proven, too, by the removal of the inflammation, and contemporaneous subsidence of the dysmenorrhœa. To add to the difficulty, and even perplexity of the subject, however, I must acknowledge that painful menstruation is not always present when inflammation exists in the cervix. It is with some timidity I state my conviction, that the obstructive dysmenorrhœa of authors should be arranged under this head.

From the time of Dr. McIntosh's writing on this subject, his doctrines, in this respect, seem to have been received as true, universally. Notwithstanding this universal agreement of able and experienced authors, I must be permitted to express the opinion that the idea does not have its foundation in fact. If obstruction to the flow of the menstrual fluid is sufficient cause for the excessive suffering of dysmenorrhœa, why not have these terrible pains when, from adhesion of the cervical canal, or external os, the flow is entirely prevented? I am sure, also, that these symptoms do not frequently coincide with a very small opening; that the internal and external ores are often both very much contracted, without giving rise to any pain during menstruation. During the time that my attention has been specially directed to this subject, I have examined many patients in whom the external os was so small as to scarcely admit one of Anel's probes, and yet they menstruated freely and easily through it. Indeed, I have not seen an instance of dysmennorrhœa from obstruction, while I have seen many cases in which the conditions were present, and they ought to have existed, according to the doctrine of obstruction, but did not.

Dr. Bennett (page 229, of his work on the Uterus) speaks of what he calls a marked illustration of this class of cases. After attempts at dilatation, not sufficient to allow the probe to enter the uterus through the os internum, the patient was entirely relieved.

I must respectfully decline to believe the correctness of his inference in this case. It may be asked how I would account for the relief afforded in these cases by dilatation? The bougies, sponges, or other foreign bodies, frequently introduced to "dilate" the os and cervix uteri, cure or relieve the inflammation. The treatment, in fact, is one of the best kind to speedily modify the painful character of the inflammation. Dr. Bennett lays great stress upon the entire immunity from suffering between menstrual periods, and yet, in another place, says that this is possible with the existence of inflammation.

Whatever may be the difference in theory, however, from the testimony of Sir James Y. Simpson and Dr. Sims, there are many cases very much benefited by operations upon the cervix, and I think I can do no better than to give a synopsis of the conditions upon which Dr. J. Marion Sims thinks dysmenorrhœa depends. In his recent most excellent work on Uterine Surgery, published by William Wood & Co., he says (page 142), " It (dysmenorrhœa) is only a symptom of disease, which may be caused by inflammation of the cervical mucous membrane, retroflexion, anteflexion, fibroid tumor in one wall of the uterus or the other, contraction of the os externum, flexures of the canal of the cervix either acute or greatly curved, either at the os internum, at the insertion of the vagina, or extending throughout the whole length of the canal, all of which are but so many mechanical causes of obstruction which must be recognized and remedied if we expect to cure the dysmenorrhœa."

The following table is on page 132 :

Of 100 cases of painful menstruation,	os was normal in but	6
	os was contracted in	90
	cervix was flexed in	61
	congested in	7
	there were polypi in	2
Of 29 cases of excessively painful menstruation,	os was normal in	0
	os was contracted in	26
	cervix was flexed in	23
	had polypi in	2
	was congested in	1

This tabular testimony of 129 cases is a strong argument in favor of Dr. Sims's theory that dysmenorrhœa is almost always caused by obstruction.

The inflammatory variety is attended with every grade of suffering, from a very moderate amount of pain to the intensest agony. The pain may, and, I think, usually does, begin before the beginning of the flow, and subsides after the discharge becomes well established ; or it may begin at the time of the beginning, and last during the whole flow ; or it may not begin until after the discharge becomes free, and then cease with it; or, again, the pain may come on with the cessation of the flow. I have seen instances in which the blood made its appearance, and continued for two or three days, and then ceased without pain ; and in a day or two the flow was re-established with very great suffering. In inflammatory cases the pain is not uniform in character; sometimes it is constant and aching in the region of the uterus, or ovaria, or both; sometimes cramping, and compared to colic, with intervals of ease, and again of a terrible tenesmic character. The cramping and tenesmic pains would indicate obstruction, if any kind of pain would ; but in many of these cases there is an enlarged, instead of a contracted, state of the cervical canal, and upper and lower os uteri. It is often attended with chilliness, followed by reaction. I think, also, there is no foundation for the opinion that simple displacements cause dysmenorrhœa.

The neuralgic dysmenorrhœa described by authors, but of which I have seen so little that I cannot speak authoritatively, occurs in patients subject to neuralgic attacks elsewhere, as in the head, face, &c. Instances that seemed to me to be more nearly like it than others I have found to be attended with some endocervicitis, or endometritis, and to yield to local treatment alone. I do not remember ever to have seen the attempt made to cure these cases by local management fail. It is true that this trial has been made either under my own immediate attention or by my instruction.

And I wish to be specific in my statement that there were many of them just the kind of cases described by Dr. West, in his admirable work on females, as neuralgic. My own observation drives me to the conclusion, therefore, that many of the cases described as neuralgic dysmenorrhœa are at least attended by inflammation of the mucous membrane of the cervix or body of the uterus, and that the cure of this inflammation cures the dysmenorrhœa. But I am also thoroughly convinced that a neuralgic diathesis exists in many of the patients, thus aggravating the suffering in particular instances. If neuralgia was the essential nature of these cases, my treatment ought not to have cured them. I am inclined to think that the uterine inflammation causes the neuralgic diathesis in these cases. I am sure that although I have tried such remedies as were suggested by my reading and observation in neuralgic diseases, yet I have invariably failed with every kind of treatment, besides that of local applications to the diseased mucous membrane. It may be that I am wrong, but I regard the treatment as the proper diagnostic means when made use of in this way. I wish here to say that I hope no one will underestimate my respect for the fathers or masters in the profession. I have followed them with becoming veneration, and shall continue to do so; but I cannot always arrive at the same conclusions they do, and I must respectfully ask to enjoy the independence I am forced to assume by my own observation, and request correction from my compeers after they have reviewed the facts upon which my opinions are based. Although I have, in my division of the subject, enumerated neuralgia as a variety of dysmenorrhœa, out of veneration for authority, I believe that the nervous is the unimportant element instead of the essential one.

The membranous variety of dysmenorrhœa is more rare than the others. Whether inflammation has anything to do with the process of formation of this membrane, I am unable to decide. There is one thing I can say, however, and that is, that this condition is often associated with it. I am not disposed to discuss the question as to the formation of a delicate membrane during each act of menstruation, or the sameness of this dysmenorrhœal production with that of the deciduous membrane of pregnancy. It undoubtedly is the result of disease in the cavity of the corpus uteri. In all the cases I have had an opportunity of thoroughly examining, there was very considerable disease of the submucous or fibrous tissues of the uterus. I do not wish to be understood as arguing that inflammation is the cause of this membrane, for I am not prepared to do so. Indeed, I think the subject has not been sufficiently investigated by anybody to decide. The structure of the membrane appears identical with that of an early decidua. If it is so, it is interesting to inquire how this identity with pregnancy can be produced. The symptoms in this sort of dysmenorrhœa are generally quite intense, and the patients suffer longer than usual. If there is any difference in the character of the pain in this class of cases as distinguishing it from others, it is more expulsive, and resembling more nearly the throes of labor.

The diagnosis between the different varieties of dysmenorrhœa cannot otherwise be complete but by examining the discharges for the membrane, and making a thorough physical éxamination of all the pelvic organs. I need not stop to give the different steps of such examination. When there is obscurity, the rectum, bladder, ovaria, and urethra, should be carefully scrutinized. In most cases there will be found evidence of some form of disease of some of these organs. The detection of the membrane will of course determine to which division it belongs in a very clear manner. The only danger of mistake is between the membrane of early abortion and that of dysmenorrhœa. Abortion at the menstrual period only occurs occasionally, and the intervening "periods" are free from pain or the discharge of the membrane. In dysmenorrhœa the recurrence is regular at each menstrual time. The appearances of the membrane are not such as to warrant a decision based upon them alone, in all cases at least.

Prognosis.—The inflammatory form may be cured with a good deal of certainty by local treatment for that condition of the uterus. The membranous form is very difficult of cure, and can seldom be accomplished with the ordinary mode of treatment.

Treatment.—The urgency of suffering in many cases of dysmenorrhœa calls for very energetic palliative treatment. We can, therefore, with great propriety divide the treatment into palliative and radical, or curative. The former, intended merely to palliate the sufferings during and between the paroxysms; the latter, to prevent the return of them, by removing the conditions upon which they depend. When there is a strong element of inflammation in the case, the palliative measures must be derivative and antiphlogistic as well as anæsthetic. A vapor bath taken at the beginning of the paroxysm, sufficient to induce copious perspiration, is often attended with great relief. A brisk saline cathartic a few hours before the supervention of the great agony, will generally modify its asperity. Cupping to the loins, hot hip-baths, and general hot

baths, are all sometimes very beneficial.   I have been in the habit
of treating the cases attended with endometritis, with first a dose
of Rochelle salts the day before the paroxysm, dry cupping to the
loins upon the supervention of the first symptoms, to be immediately followed by from ten to twenty grains of Dover's powder,
and a vapor bath.   Soon as copious diaphoresis is fairly brought
about, the patient should be wrapped up in blankets and put to bed.
If the pain becomes intense, notwithstanding these means, the patient may be kept under the influence of sul. ether until quiet sleep
is induced.   If there is no obvious inflammation, the opiate of Dover's powder or morphia, followed by ether or chloroform, will do,
without the bath or cupping.   Dr. Simpson uses the chloroform
in the beginning of the paroxysm with great confidence, but does
not expect this anæsthetic to interrupt the paroxysm unless given
at the very beginning.   I think, given at the time of the administration of opium, it renders the operation of this drug more prompt,
while the opium prolongs the effect of the chloroform.   Usually,
they certainly act very pleasantly together.   Various local anæsthetics are recommended and used.   The carbonic acid gas, carried from a generating apparatus through a tube into the vagina
in a full stream a considerable time; obtunds the sensibilities of
the parts, and thus relieves the pain.   Chloroform can also be used
in the same way.   Opium, belladonna, hyoscyamus, aconite, and
other remedies of this class, used as suppositories, may very properly be resorted to under certain circumstances.   From the observation of the fact of the simultaneous irruption of the menses
and subsidence of the pain, the patients are apt to resort to the
use of strong stimulants, which have the reputation among them
of promoting the flow.   This practice should be discouraged, as
having, in the end, generally an injurious effect.   Antiphlogistics,
anodynes, revellents, and anæsthetics, are more rational and almost always more beneficial.   The paroxysm of the membranous
variety calls, in addition to some or all the above-mentioned remedies, for another important item of treatment.   The more readily
the membrane can be expelled the sooner the suffering is ended.
I have now used on several occasions the ergot, with, as it has
seemed to me, very great advantage, shortening the paroxysm by
many hours, without increasing the painfulness of the paroxysm at
all.   Under the influence of large and frequently repeated doses,

as grs. xx, every half hour for two or three hours, I have many times known the membrane expelled apparently very much sooner than it otherwise would have been done, judging from the course of former paroxysms. And what seemed curious, the pains were much more tolerable while the patient was ergotized. Sometimes ergot fails to have any effect, as it does under other circumstances.

The radical treatment of dysmenorrhœa is attempted by general remedial measures, or by local treatment of the uterus or other organs diseased, for the purpose of removing their morbid condition. As has been already remarked, dysmenorrhœa is sometimes attended, if not produced, by certain diathetic conditions. Rheumatism, gout and scrofula are all blamed for it in some instances, where the patient is evidently laboring under any one of these general states. And it is very correct to make use of general means to correct them, with a view to the cure of the dysmenorrhœa. Tinct. guaiac. ammon., given as directed by Prof. Dewees, in ℥j doses, three times a day for a long time, is sometimes curative and often palliative in its effects. Colchicum does much good in rheumatic patients, and the terebinthinates in various forms are popular remedies. When the patient's health indicates a marked scrofulous condition, cod-liver oil, and iodide of potassium, iodide of iron, and much outdoor exercise, with generous meat diet, warm clothing, and regulated bowels, are essential. In all cases, we should carefully study the condition of the health, and diligently endeavor to correct any deviation we may discover. I have seen several remarkable cures by thus attending, for a reasonable length of time (several months), to the general health alone. Sometimes a gentle salivation, connected with a course of tonics, does a great deal of good. I have sometimes administered ℥ss. of compound tinct. cinch. with the twentieth of a grain of the bichl. hyd. three times a day after eating, for this purpose, with great benefit.

Very frequently a large majority of the cases are connected with inflammation in the pelvic organs, or some one of them. When this is the case, our treatment may be general or local, or both combined, according to the organ affected. If there is evident inflammation of the ovaria, leeching, cupping, counter-irritation, and the alterative effects of mercury, very gradually induced, followed or accompanied with tonics, is the advisable course to be pursued. Leeches or cupping just before the menstrual period is the best time for their use, applied over the ovary that seems to be affected, or persistent counter-irritation by means of a seton, kept open for months. Of all the organs in a state of inflammation causing dysmenorrhœa, however, the uterus is very much the most frequently thus affected. And the form is generally that of endometritis, or endocervicitis.

I wish to discuss here the effects of some remedies that have cured this affection when used for other purposes,—to dilate a stricture, for instance. ·Since I have made these diseases somewhat a special study, I have had occasion to try the effects of dilatation by means of bougies, stem pessaries, and prepared sponge, and have found much good to result from all of them. But there is one thing I have learned of the use of these instruments that has not been mentioned by those who have heretofore recommended them, and that is, that they will do just as much good in many cases when there is no stricture as where there is. I have used the bougie as recommended by Dr. McIntosh, and the stem pessary according to Dr. Simpson, when the cervical canal was open more than naturally, with as much good effect as when the canal was very narrow. The explanation of all this is, that these means modify at first and finally cure the endocervicitis, upon which the dysmenorrhœa depends, and in this wise relieve the patient. The dilatation by the uterotome of Dr. Simpson, making free incisions, cures the cases, not by merely giving freedom of discharge to the pent-up blood, but by their effect upon the inflammation. It is rather emptying the capillaries of the part inflamed, than the cavity of the uterus, that effects the cure.

It is my conviction that mucous or submucous inflammation, or both, is generally the cause of the narrowing of the cervical canal, and the sufferings of the patient.

I do not consider the above-mentioned modes of treatment the best in such cases, although they do sometimes effect cures, and would refer the reader to the treatment of inflammation of the cervix in another part of this work. A few applications of the solid caustic to the mucous membrane so modifies the disease, that in many instances the first paroxysm after the commencement of them is attended with very much less suffering. Generally the improvement is gradual, however, but progressive as the inflammation subsides.

Whatever the condition of the lining membrane of the uterus may be, in the membranous variety of dysmenorrhœa, the object of treatment is to change it from the morbid state. I am not sure that there is ordinarily a state of inflammation. The morbid state may be removed sometimes by a long course of alterative treatment of a general character. But I cannot now recall a case in which I have been fortunate enough to succeed thus. These cases are very difficult to manage, and obstinate under the best treatment. The successful plan, in my practice, consists of applications made directly to the mucous membrane of the uterus, by introducing lunar caustic, acid nitrate of mercury, or tinct. ferri. chl. into the cavity. I have on several occasions—seven in all—thus cured it. In three of the cases it became necessary to dilate the cervix with a sponge, so as to make application freely. I dilated the cervix a week or ten days before the expected paroxysm, so that the finger could have been introduced, and through the dilated cervical cavity, by means of a piece of cotton on whalebone, made a thorough application of the acid nitrate of mercury to the mucous membrane of the cavity of the uterus. I have seen no disagreeable results from this treatment, and shall feel justified in again resorting to it when the urgency of the case demands it. Three of such applications as this were all that was required. In the other four instances the nitrate of silver was used by means of the flexible caustic-holder, alternated with the tinct. of iron, with the cotton swab. It required six months in one, and in the others from eight to eighteen months. The general treatment must be governed by the general condition of the patient to a great degree. If she is weak and debilitated, tonics, a generous diet, exercise in the open air, good company, and ease of mind, are indispensable in connection with alterative medicines. Continued use of sitz baths for months together will relieve the irritability very much sometimes. But I would urge efficient applications of the medicines above-mentioned to the uterine mucous membrane, as the most likely to cure this very obstinate and painful affection. Make the applications thorough if possible through the cervical canal in its natural condition, but if too narrow for this purpose, dilate it, and after dilatation, make the application thorough. And I think for this purpose the acid nitrate of mercury the best. These applications must be repeated until the cure is accomplished. The acid nitrate of mercury may be used once a month, or the nitrate silver once a week.

As I have given the opinion of Dr. Sims as to the causes of dysmenorrhœa, I cannot complete this article without giving the reader an idea of the mode of treatment found most successful by him, viz., that of dilating and strengthening the canal of the cervix. He exposes the mouth of the uterus by placing the patient in the same position, and using the same instrument as for vesico-vaginal fistula. With a tenaculum he seizes and firmly holds the cervix and draws it into the most convenient position. If the cervix is not flexed but merely narrow, he introduces one blade of the scissors into the canal of the cervix far enough to divide it on one side up to the junction with the vagina, and then closes them. The other side of the cervix is divided to the same extent in like manner, then, by means of the knife represented in figure, he divides the cervix up as high as the internal os.

Fig. 13.

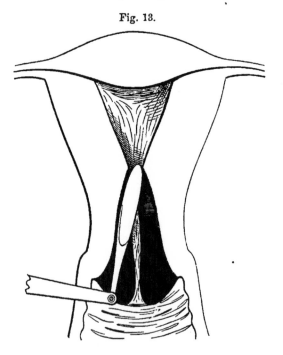

Fig. 13 represents the operation for dividing the straight cervix when too narrow. The dark part the portion cut. On one side the knife is shown in the act of dividing the tissues. This is Dr. Sims's plan.

Fig. 14.

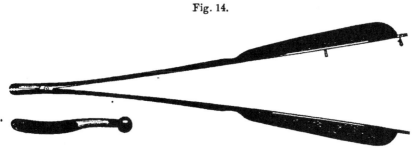

Emmet's Knife for dividing cervix. From a cut in the June Number, 1864, New York Journal of Medicine.

If the cervix is flexed, the lip of the uterus on the convex side is divided to the same height, and then the cervix opened with the knife. In this way the cervical canal is rendered rectilinear.

This is represented by Fig. 15, taken from page 169 of Dr. Sims's new work on Uterine Surgery. It shows the posterior lip already divided by the scissors, the tenaculum fastened into the anterior lip, and the knife being inserted as high as necessary.

Fig. 15.

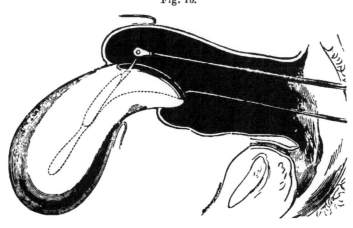

"The representation in the cut is taken from the perfected instrument made by Wade & Ford, of New York city. To their ingenuity is due the application of the principle. The representation is half the size of the instrument, but the blade at full size is out of proportion, as it should be represented both longer and narrower."

After having thus completed the operation, Dr. Sims places in the wound of the lip of the cervix some cotton saturated with glycerine, and then proceeds to fill the vagina with cotton to guard against hemorrhage, which he regards as always imminent. If there be but slight bleeding, it is not necessary to use more cotton than will keep the dressing in place. The patient should keep the recumbent posture for several days. The cotton in the vagina may be removed in twenty-four hours after the operation; that in the wound remains from two to three days. Dr. Emmet recommends that the sound be passed through the cervix every other day until the discharge ceases to prevent the parts from adhering. The sound need not be used for this purpose until the tampon is dispensed with.

In the American Journal of Medical Science for January, 1867, I find the following summary, which I present to the reader without apology:

"*Comparative Merits of Incision and Dilatation of the Mouth of the Womb in cases of Dysmenorrhœa*—Professor D. Humphreys Storer read, in August last, a highly interesting paper on this subject before the Boston Society of Medical Improvement. The large experience and sound judgment of Professor Storer not only entitle his opinions to a respectful consideration, but his conclusions to entire confidence. He says: 'From a somewhat extensive employment of sponge-tents during the ten past years for the treatment of dysmenorrhœa and sterility, I have formed conclusions different from those of the gentlemen of whom I have spoken (Drs. Barnes, Baker Brown, Greenhalgh, and Sims). I have not unfrequently been disappointed in the result hoped for. The local obstruction has almost always been overcome by the long-continued, persevering employment of the dilator; but the opened canal does not always remove the condition thought to depend upon its closure,—dysmenorrhœa and sterility still remain. I have, however, never seen the ill effects spoken of from the employment of tents. I cannot recall a single instance where more than a few hours' inconvenience has been produced; and in such cases the expanded sponge, when removed, has proved to have been originally much larger than it was supposed to be—showing that he who employs these tents should be acquainted with their uncompressed dimensions. My experience has taught me, then, that these contractions, however firm they may be, may almost invariably be overcome. The physician need not feel that the part is undilatable because the application of three, or five, or half a dozen tents does not overcome it; in a case occurring in my practice, about a year since, *eighteen* sponge-tents were introduced at intervals of two and three days before the canal was opened. My perseverance was rewarded by the perfect relief of the patient. I could point, were it necessary, to several cases where, after years of sterility, the sufferer has been relieved and borne children, and in the intervals of their childbearing have suffered no dysmenorrhœa. I have repeatedly seen cases of dysmenorrhœa remain relieved for years, and known no return. In a word, I have relied upon dilatation to relieve these affections, and whatever opinions may be advanced by others, so long as I feel we have a remedy from which we can confidently expect relief, and very rarely observe any injurious effects, I shall feel it my duty to employ it.

"That cases do occur where the difficulty *cannot* be removed by dilatation, there can be no question; but 'that incision is the only efficient and permanent remedy (in most cases) for dysmenorrhœa,' I unhesitatingly deny.

"Let us for a moment look at the method proposed. Those who advocate it should of course be satisfied that it has superior claims over the means now employed. I have thought the ill effects produced by *distension* might be occasioned by want of care; but those arising from incision *may* follow the operation of the most skilful surgeon who advises it, when the metrotome cuts through the walls of the *inner os*; and Dr. Barnes states, to employ his own language, 'there is no doubt that the surgeon has actually cut through the substance of the uterus, and wounded the plexus of vessels outside; hence severe and dangerous hemorrhage has ensued, and inflammation of the peri-uterine tissues.' And even supposing the operation should be successfully performed, it is acknowledged by Dr. Routh, one of its advocates, 'that such an amount of contraction frequently exists as to render it necessary to have a dilating substance worn for a considerable length of time to prevent its perfect occlusion;' and Dr. Williams observes that 'oftentimes no relief is afforded. He had seen a patient whose cervix uteri had been slit up on both sides, forming two large protruding lips, without affording any relief to the sufferer.' Where

8

the external os has been almost cartilaginous to the feel, I have overcome the obstruction with the hysterotome; but I have never attempted to divide the internal os. I cannot, however, recall the instance where it was required."—*Boston Medical and Surgical Journal*, September 2, 1866.

CHAPTER VIII.

METATITHMENIA (Μετατιθήμι μην); OR, MISPLACED MENSTRUATION.

THE accident to which I apply the above term is called retro-uterine hæmatoma, peri-uterine hæmatoma, and uterine hæmato-cele. This nomenclature has reference to the effusion of blood near the uterus, without reference to the time or mode of occur-rence; while the appellation I employ takes into consideration at once the cause and the fact of sanguineous effusion, which occurs at the time, and, generally, in consequence of menstrual congestion.

As will be gathered from what I have already said, it is an effu-sion of blood in tissues around and above the uterus; the effusion being sometimes very extensive, at others limited to a small space. The effusion may take place in the vaginal wall, between the vagina and rectum, tearing up their connecting tissue; or the posterior wall of the uterus, beneath the peritoneum; between the perito-neal layers of the broad ligament beside the uterus, or in the peritoneal cavity. The mode of the accident varies somewhat, owing to the locality in which this blood is found. The blood is effused in interspaces beneath the peritoneum and elsewhere, as the effect of a rupture of some vessel; but while effusion may be, and, perhaps, generally is, attended with a ruptured vessel in the peri-toneal cavity, the blood sometimes also arrives in that cavity from the uterus through the Fallopian tubes. We are not yet able to decide which of these circumstances is the more common.

The time of menstruation, or shortly before or after, is the time when this accident happens; and it may fairly be considered as a result of menstrual congestion, or effusion. As an accompaniment of menstrual congestion, the bloodvessels of the whole genital organs are greatly distended, and in certain cases this turgidity becomes too great for their capacity, and a rupture is caused at some particular place; or, the cavity of the uterus being filled with a profuse flow into it, the blood regurgitates through the tubes into the peritoneum. It is not likely, however, that any considerable effusions are thus caused; so that the sudden and copious collections sometimes observed, must be accounted for upon the supposition that a small arterial twig has given way in the ruptured ovisac at the time of the escape of the ovum, and poured the fluid rapidly into the sac formed behind the uterus by the descent of the peritoneum. I think the instances I have witnessed were more frequently connected with cases of disordered menstruation; but I have also seen it take place in patients whose menses were normal, so far as I could determine.

Dysmenorrhœa may be regarded as the most common deviation accompanying misplaced menstruation.

There can be no doubt but that effusions of blood, in every respect similar to misplaced menstruation, are caused by the condition of the uterus and appendages in abortion, and even after labor; but when so, the *modus in quo* is precisely the same, the congestion being caused, not by the menstrual molimen, but by the congestion of pregnancy and morbid excitement which sometimes attend these two states,—rupture of a small vessel, or regurgitation, being the immediate condition.

Sanguineous collections arising in this way may be minute in size, but sometimes the quantity of blood is dangerously, and even fatally, large. The small collections are forced into places where distension is most difficult, as in the cellular tissue, while the large effusions are met with in the peritoneal cavity. Immediately after the blood is extravasated, changes begin to take place in it, and the tissues occupied by it. Inflammation, to a greater or less degree, almost always is the result. In a mild grade, the inflammation produces an effusion of serum which augments the bulk of the accumulation, and makes the appearance of much blood, when in reality there is but a small quantity. When this is the state of things, the disappearance of the tumor, by absorption, may be expected in a comparatively short time. And we often see it removed by absorption in a very few weeks.

The intensity of the inflammation is frequently much greater, proceeding through the stage of serous effusion to the production of fibrinous deposit. A hard tumor is the result. This again may remain for a longer or shorter time permanent, and then very slowly disappear, or only be partially taken away, leaving a permanent hardness; or, what is not unfrequently the case, proceed to suppuration and discharge in some way or other.

I have seen as many as two cases terminate fatally by the exhaustion of suppurative fever, without the discharge of the contents of the tumor. When suppuration is fairly established by the inflammation thus arising, exulceration and evacuation follow as a general rule. The vagina is more frequently perforated by the ulcerative process, but the rectum, bladder, or uterus, may serve as the conduit of discharge. If the inflammation is of an acute character, and the steps in the process of evacuation rapidly succeed each other, the character of the discharge will partake largely of a bloody quality; but should the time required by exulceration be considerable, pus will prevail in the composition. In any case, however, the discharge is a mixture of pus and changed blood. This last is sometimes very greatly changed, but in others but slightly. In rare instances, the peritoneum is inundated by rupture into its cavity with this mixture of pus and blood, and overwhelmed with a general inflammation, soon resulting in death. I have seen two cases of this kind, which were verified by postmortem appearances.

After absorption in the cases attended with the milder grade of inflammation, very slight traces of it, if any, can be found by examination of the patient. When effusion of fibrine takes place, displacements, permanent adhesions of the uterus and other parts, and deformity, will be left behind, considerable or slight, as the amount of deposit was small or great. These changes will, of course, be greater after the process of suppuration and discharge has been reached by the inflammation. Fistulous and tortuous openings may also embarrass the convalescence of the patient, or even by their long continuance exhaust her.

Symptoms.—The symptoms of metatithmenia vary in different instances. The attack is generally sudden and well marked. At the time, just before or after the menses are flowing, the patient is seized with severe pain in the hypogastrium or one of the iliac regions. Frequently there is also a sense of faintness, sometimes slight, but often it amounts to complete syncope. In place of the faintness there are sometimes coldness and tremors. The pain becomes persistent, and, perhaps, less severe, but not unfrequently it increases for a considerable time and then gradually diminishes.

After the inception the pain usually spreads over the abdomen to the back and hips, and sometimes down the thigh and leg. As the pain becomes greater or extends over a greater space, febrile reaction is developed, generally moderate in grade, but occasionally the reaction is excessive; the pulse becomes rapid, the heat considerable, and the patient complains of great depression and thirst. The abdomen increases in size and becomes tympanitic, while there may be a distinct tumidity and hardness felt in one of the iliac regions; sometimes the hardness extends over the hypogastric to the other ilium. This hardness and swelling may scarcely rise above the pelvic brim, but it not unfrequently is perceived extending as high as the umbilicus. It is not much, if at all, tender to the touch. It is irregular in its outline also. In very rare instances the effusion takes place slowly, the symptoms are developed quite gradually, and the time of the beginning not so definite, but the subsequent course of them are apt to be the same.

After the symptoms are fully manifested they pursue a course corresponding to the grade of inflammation which is awakened by the effusion. In some cases the inflammation around the effusion is active and intense, and continues with severity until suppuration and exulceration end the process.

Of course the fever is corresponding in grade and persistence, passing through the high grade to hectic, attended with all its exhausting discharges. If the inflammation is less acute, the fever may be persistent for weeks, and sometimes for months, but of more moderate grade, until it gradually subsides, or slowly ends in suppuration and discharge. Fortunately, in the large majority of cases, the amount of the effusion is small, the grade of inflammation slight, and the duration but a few days or weeks.

Subsidence of an attack of metatithmenia and recovery from it are not generally followed by a series of them. But there are two ways in which individuals are rendered miserable by its frequent recurrence. One is, when all the symptoms subside entirely for months, and then return. The tumor entirely disappears, the inflammation is wholly gone, and the patient feels that she has fully recovered her health, when, suddenly, during a menstrual flow, she is again seized with pain, swelling, fever, &c., which again subsides to be repeated more or less frequently. I have a patient who has suffered attacks of this sort perhaps twenty times in the last six or seven years, in whom the tumor has at different times been mistaken for ovarian or uterine tumors. In the other way, the subsidence is only partial: there is all the time some tumidity, some inflammation, and more or less sympathetic suffering, with occasional severe returns. More blood is effused, the tumor is increased in size, and the inflammation intensified, and all subside to a partial extent and again return. When the tumor is much inflamed and suppurates, it may suddenly discharge through the vagina; all the urgent symptoms readily subside, and the patient becomes convalescent. Again, the discharge is sometimes slow and difficult, the relief is imperfect, and a protracted convalescence the result. But sometimes, after a course corresponding to the above description, sudden and general peritonitis is lighted up by extension of inflammation from the sac, or a discharge of some of its contents into the peritoneal cavity. The discharge is generally fetid and highly irritating, consisting of partially decomposed blood, pus, and ichor. It is always offensive compared to discharges from an ordinary abscess. I have seen one or two instances in which the general symptoms were not manifested at all, nor did the pain amount to anything more than an inconvenience, not very difficult to bear. It is interesting to observe the effects of this misplaced menstruation upon the flow *per vias naturales.* Occasionally no effect seems to be produced, the flow being natural in quantity and duration; in fact, it is just at the time of the cessation of the discharge that effusion into the tissues takes place, but at other times there continues for many weeks a constant stillicidium of blood. Or, occasionally,—when they occur during the course of the symptoms,—the amount of menstrual discharge is very much increased. I knew one patient that had a constant slight sanguineous discharge from the vagina for six months, and at the regular menstrual periods copious hemorrhages. In other cases they are more scanty than usual.

Diagnosis.—There are several conditions with which it may be confounded, if some caution is not observed. Inflammation of pelvic cellular tissue, or pelvic abscess, is the one most likely to be mistaken for metatithmenia, or this last for the first. And as I have already shown, the abscess is sometimes the result of misplaced menstruation, the effusion in the tissues exciting intense inflammation, which proceeds to the stage of suppuration. It is not unlikely, indeed, that a larger number of these abscesses have a hemorrhagic nidus than we at present suspect. However this may be, we will often have reason to hesitate in our diagnosis. The inflammation in cellulitis is not ordinarily ushered in by the same suddenly occurring acute pain and faintness. Chilliness and fever are more marked from the beginning, the pain usually commencing after the fever has begun, or, at least, increasing after the fever is established. The tumor above the linea ilio-pectinea is not perceptible for many hours, oftener one or two days; it is extremely tender, and even in its outline. The tumor in metatithmenia is observed in a few hours, and is not so very tender to the touch. It may be handled and pressed upon much more freely than the tumor of simple inflammatory origin. If examined per vaginam the inflammatory hardness and swelling is very firm. It is usually lower down and more to one side. The tumor from sanguineous effusion is quite elastic at first, and presents an edge-like projection down behind the uterus, entirely below the os and cervix. The finger may be pushed up between the cervix and the tumor, and the thick convex edge of the latter reminds one of a thick cake. There is very little tenderness, and vessels may almost always be felt pulsating over this projection. I need not say that this is never the case in the early stages of cellulitis. The vessels in this last are obliterated by fibrinous and serous effusion.

If inflammation of a high grade speedily follows the effusion of blood in the tissues, the symptoms of the two may be so intimately blended as to make it doubtful how the tumor began, and, in fact, to convert it into pelvic abscess. Tumors of the uterus, under certain circumstances, may be confounded with the tumor of sanguineous effusion; but their firmness, the want of conformity to the shape usually assumed by this last, the enlargement of the uterine cavity, our ability to isolate them by the fingers and probe, their gradual, unperceived growth, and their mobility, will almost always suffice to make the distinction manifest. From ovarian tumors they may be distinguished by the more regular outline, fluctuation on percussion, less grave symptoms, gradual development, absence of the projecting edge behind the uterus, the want of the beating vessels, &c., in ovarian growths. Displacements of the uterus may always be made out with great certainty by introducing the probe into its cavity to ascertain the direction of the fundus, and correcting its deviations. Hence the diagnosis need not be long embarrassed by any question in reference to them. Retroversion of the impregnated uterus is constantly attended with great urinary distress, while metatithmenia seldom is. Extra-uterine pregnancy, perhaps, in some instances, more nearly resembles it than any other, but the enlarged and flaccid cervix, open os, dark color, and enlarged cavity, in this sort of pregnancy, and their absence in the accident we are considering, will suffice to distinguish between them.

Causes.—Anything that induces intense and prolonged congestion of the vessels of the ovary and uterus may, under certain circumstances, produce effusion, by rupture of some of the minute ramifications. And there is reason also to believe, as shown by Dr. West, that when the os uteri "does not permit of easy flow from it, that the blood, after accumulating in the cavity during the menstrual effusion, may be passed out through the Fallopian tubes and escape into the peritoneal cavity." In what proportion of cases this takes place is not known. Both of these circumstances being present, congestion of the vessels of the uterus and ovaria, and effusion of blood in the cavity of the uterus at the time of menstruation, will explain why this last is considered the almost invariable causing condition. The whole etiology is not given, however, in the above explanation, and the balance finds its expression in "a peculiar susceptibility," or, in other words, it is not known why it should occur when it does, and why it does not take place at other times.

Prognosis.—The dangers to be apprehended in metatithmenia arise, 1st, from the shock of the effusion in the peritoneal cavity, which, however, is not generally considerable; 2d, fatal exhaustion from the amount of effusion in the abdominal cavity; and, 3d, inflammation and its effects. From inflammation we may fear death, permanent damage to the organs about the pelvis, and great suffering. Very few patients escape without protracted suffering, often weeks and sometimes months, from the inflammation lighted up by the effusion.

*Damage to a greater or less degree frequently follows the displacement, adhesions, perforations, and thickening of the uterus, vagina, rectum, and bladder. The exhaustion of protracted febrile excitement, perspiration, diarrhœa, and vigils, not very seldom

wears out the vital resistance of the patient, often of a very delicate constitution ; or sudden and violent inflammation of the peritoneum overwhelms and destroys her. The prognosis in any given
case will be governed by the intensity of the symptoms, and the
comparative strength of the patient. If the amount of the effusion
be never so large, and there be but little inflammation, the prognosis is more favorable than if the effusion be small and the inflammation great. In fact, I think we may with great propriety
form our prognosis by the amount and intensity of the inflammation alone, as it is almost the only source of danger.

The proportion of deaths compared to recoveries perhaps will
range not far from ten per cent. This at least is the result of my
observation and research up to the present time.

As before observed, a cause of death, though not frequent,
should nevertheless be mentioned as influencing the general subject of prognosis in misplaced menstruation, viz., a fatal amount
of extravasation of blood in the peritoneal cavity. More than one
case is recorded in which there was fatal prostration, coming on
and pursuing its course in a few hours, which, when examined, revealed as the source of an extensive and copious hemorrhage, a
ruptured twig of artery on the ovaria. Of the many cases that
come within our observation, the number that thus prove fatal are
extremely few.

Treatment.—The three great facts of this accident,—hemorrhage, pain, and inflammation,—afford us indications for treatment sufficiently plain to guide it correctly. It is very seldom
that we are sent for, or in any way see these cases, until after the
hemorrhage has exhausted itself or been stopped by backward
pressure, after filling up the space into which the bleeding takes
place. Should we, however, meet with an instance during the hemorrhagic stage, it would be very proper to make use of ice to the
pelvic region, perfect quiet, and astringents internally, until the
effusion ceased ; but, as I said before, such opportunities seldom
offer themselves. The cases as we ordinarily see them, have proceeded through this stage ; the effusion, in fact, is generally accomplished in a few moments, or at most, very few hours. When we
see the patient, she is either suffering with pain and prostration or
coldness, the primary effects of the hemorrhage ; or pain, fever,
and inflammation, and our treatment will be conducted according

to the conditions in these respects. Our resources in the first condition will lie in the use of opium or other anodyne, to relieve the pain as much as may be necessary, and if the prostration or chilliness is considerable, to stimulate sufficiently to establish equilibrium in the circulation, but not febrile reaction. In very many cases it will be sufficient to keep our patient quiet, and place her upon moderate anodyne treatment, good nourishing diet, and perhaps after the first week or two, tonics, and she will slowly rally from the first shock, absorption of the blood result, and she soon will recover her health. In these moderate cases we cannot be too careful not to overdo the treatment. The patients will generally recover spontaneously in a few days or weeks.

But another class of cases occur, as I have already said, in which inflammation very soon succeeds the sanguineous effusion. A knowledge of the mischief which this inflammation brings about should make us prompt in meeting it with appropriate remedies. If the inflammation runs high, adequate antiphlogistic measures will be indispensable to a favorable course. An active cathartic of calomel and jalap or some other alterative cathartic should begin at once, while at the same time, if deemed advisable on account of the force of reaction, we may apply a dozen or twenty leeches. These may be followed by the tinct. of veratrum viride, in doses of two drops every hour, until the pulse is brought down to its natural frequency and volume, if not below these conditions, and then continue its use in less doses, or less frequently repeated, for some time. According to my observations, the most of adults will be affected to this point by taking as much as one drop an hour; some will require more and some less. The energy of this antiphlogistic course must be graduated by the force of inflammation; but few cases will require as much as is described here. Should the inflammation advance to suppuration, the remedies required will be supporting; at first, sul. acid and quinine, and afterwards, these with wine or other stimulants, nourishing diet, &c. These cases are often so protracted, the patients are so much prostrated, and suffer so much pain, that great skill will be called for to adapt the anodynes, tonics, and nutrients to the various conditions of the patient for so long a time.

A question associated with the progress of inflammation, and one of great importance, is the propriety of evacuating the fluid.

To evacuate the blood soon after its extravasation would seem to remove the cause of inflammation, and thus avoid it. To say that an early evacuation of the effusion would never be proper, is perhaps to assume an extreme position, and there may be cases where such evacuation is advisable, but I think the number requiring it must be very few, and I do not know any circumstance which would induce me to do it. Indeed, I should fear inflammation, from the sudden discharge of a large amount of blood from the cavity, almost as much as if it were allowed to remain in it. There is another condition in which an operation for discharge of the contents of the tumescence is less a question of doubt, viz., when pus has become mixed with the blood, on account of inflammation. It is very important in some instances to puncture and discharge the fluid. When the patient is being worn out by the protracted course of the disease, and the perspirations and diarrhœa which so often attend it, we must interfere surgically for her relief. And again, when the fluid is increasing, and the tumor rising in the abdominal cavity, without showing any disposition to "point" in the pelvis, or any other place where it is desirable to have it do so, there is danger of the discharge of the pus and blood in the peritoneal cavity, by rupturing the sac above, and we must anticipate it by choosing the place and mode. When we have determined to relieve the distension by puncture, we ought to use an exploring needle or trocar to ascertain the contents before evacuating them. After being satisfied by this corroboration of our diagnosis, we may plunge a large trocar or even a knife into the most dependent part of the tumor. This point will almost invariably be immediately behind the uterus, but occasionally it will be at the side of the pelvis.

After free puncture, either with the trocar or knife, the discharge readily takes place, and the patient immediately experiences great relief. If the puncture is made to remove the blood before inflammation has begun, the evacuation may be more difficult, as it is often coagulated; in that case the opening must be made large with a knife, and if the blood does not easily flow, the finger may be introduced to break up the clots and facilitate their expulsion. After the contents are thus expelled as near as can be, they sometimes reaccumulate and are again discharged, and repetitions of these processes lead to still more chronic suffering, until the patient becomes a permanent invalid or dies from the exhaustion of such long standing. We may, with a good deal of certainty, cause contraction, granulation, and obliteration of the cavity, by injecting it with iodine, wine, or other irritant. The best way to secure efficiency in injections is to introduce through the fistulous opening, or one made for the purpose, a small flexible catheter, so as to reach the bottom of the cavity, and throw the fluid through this tube. We thus place the fluid used in full strength in contact with the walls of the cavity, while the injection thrown out of a common syringe will mix it up with the contents of the sac, and thus dilute it.

CHAPTER IX.

ACUTE INFLAMMATION OF THE UNIMPREGNATED UTERUS.

ACUTE inflammation, not arising from specific causes, generally affects the fibrous portion or substance of the walls of the uterus. It almost, if not quite always, pervades the whole of the organ, the fundus, body, and cervix. The whole organ is inflamed and tender. Exposure to cold is the most frequent cause. The cold may be applied to the general surface when the uterus is in a state of turgescence from menstrual congestion, sexual excitement, or incomplete involution after labor or abortion., The same agent acting upon a portion of the surface, the feet and legs, under similar condition of the organ, may give rise to the same disease. It is not likely that cold, however applied, would be a sufficient cause, but for the predisponent condition I have mentioned. The excitement of excessive sexual indulgence may be carried so far as to cause a moderately acute inflammation of the substance of the uterus, as also blows upon the abdomen, &c.

It is not a very frequent disease, and yet I do not think it can be regarded as an unfrequent affection.

Symptoms.—In speaking of the symptoms of the disease, I wish the reader to bear in mind that their intensity will vary from a mildness that will scarcely confine the patient to her couch to a very severe and grave disease, almost overwhelming the nervous system, with delirium and convulsions, and calling the stomach into excruciating sympathy with it. In considering the subject, I wish to be understood as attaching more importance to the suddenness than the intensity of the attack in determining the nomenclature.

It is somewhat owing to the exciting cause, as to the symptom which is likely to usher in the attack. If the cause is a moderate one, as excessive sexual indulgence, pain will generally begin some time before the general symptoms. If the cause is cold suddenly and extensively applied to a menstruating patient, chills and rigors may precede the pain. However that may be, when the case is fairly developed there is fever, aching in the back, pain in the head and extremities, flushed face, and furred tongue. In addition to these general manifestations there is local pain, indicating the organ affected. This pain may be confined almost entirely to the sacrum and lumbar region if the inflammation is moderate, but generally there is pain in the pelvis behind the pubis, or in one or both iliac regions. Sometimes the pain radiates in several directions up the abdomen, down the thighs, and around the body. The pain is usually of a dull aching, but sometimes of a sharp character. In addition to these symptoms indicating inflammation in some of the pelvic organs, the nervous system is often affected with hysterical symptoms, convulsions, coma, laughing, crying, or unreasonableness of some kind. I should have mentioned among the local symptoms, dysuria, and difficult and painful defecation. Should the peritoneal covering become involved there is swelling, and greater or less tenderness of the abdomen. Nausea and even vomiting are not unfrequent symptoms.

After a week or more of this kind of suffering the symptoms gradually subside, and the patient slowly recovers her usual health; or sometimes the subsidence of the pains is not complete, and she continues to suffer with a chronic form of inflammation. The termination is almost always in resolution or the chronic form of the disease. Possibly, in some exceedingly rare instances, the force of inflammation is spent in some circumscribed locality, and it proceeds to suppuration. I have lately seen an instance of this kind where the suppuration was in the anterior lip of the cervix.

Prognosis.—The termination is so frequently in resolution or a moderate form of chronic inflammation, that we may almost always expect complete or partial recovery. Death probably never results in uncomplicated cases of acute metritis, but unfortunately we occasionally meet with grave and even fatal peritonitis, apparently resulting from extension of the disease from the uterus. It has been my misfortune to have lately met, in consultation, with two instances of this sort. Although the prognosis is favorable as a general rule so far as the recovery of the patient from the attack is concerned, it is not so favorable for the complete re-establishment of health, as the patient is likely to be affected with chronic inflammation in the body or cervix. Not unfrequently we trace chronic inflammation back to a moderate attack of the acute.

Diagnosis.—Inflammation of the cellular tissue beside the uterus, metatithmenia, rectitis, or cystitis, cause symptoms which may be mistaken for metritis. When doubt exists it may be easily and certainly solved by a digital examination. If the bladder is the seat of disease, the tenderness complained of by pressing it between fingers in the vagina and others above the pubis will be sufficient proof; pressure may be made upon the rectum by including it between the introduced fingers and the sacrum. The inflammation at the side of the uterus, or cellulitis, causes tenderness and hardness close to the iliac bones on the side, and the hardness seems to be continuous with the bones. The greatest tenderness is, therefore, close to the side of the pelvis. In all these cases the uterus may be touched, provided it is not moved so as to press upon the inflamed part or organ without causing pain. If it is the seat of inflammation the tenderness will be confined to that organ, while all the rest are free from it, and may be handled freely. We should not forget that all these organs may be implicated in one great mass of acute inflammation, and all the pelvic contents be intolerably tender to the touch. In an examination to diagnosticate inflammation of the uterus, I need hardly say that a resort to instruments is unnecessary.

Treatment.—The intensity of the inflammation will govern us in the activity of treatment. If it is not attended with great pain or febrile reaction, although our remedies must be the same, there is no need of using them with the same energy. We should, however, bear in mind the great likelihood of leaving the chronic form behind, and be diligent in our medicinal and hygienic appliances, until every vestige is gone, when practicable. If the attack is moderate, it may sometimes be interrupted in the beginning, by measures to induce a copious perspiration, more particularly if caused by an exposure to cold. Even a smart attack may sometimes be relieved by a large dose of opium and a steam bath, used within a few hours after the commencement of the symptoms. After the symptoms have become fairly established and have lasted for twenty-four hours, we must not expect to find immediate relief, and should begin the systematic use of antiphlogistic treatment. In the subacute form, a brisk cathartic, foot-bath, and fomentations over the uterus, should be followed by tart. antimony, muriate of ammonia, and calomel. ℞. Ammon. hydrochloras, ℈ij; ant. et pot. tart. gr. j; ext. glycyrrhiza, ℥ss; aqua bull., ℥viij. Mix and dissolve. This is a good old formula, and may be taken in tablespoonful doses every four or six hours. Perfect quietude should be enjoined also, and rest at night may be insured by giving one grain of calomel, with twice the amount of opium, in a pill at bedtime. Continued for five or six days this will generally induce slight mercurial effect, when the pain and other symptoms will pretty surely subside. If it does not do so, a blister over or a little above the pubis will aid in banishing them. If the attack is severe, we ought to add to the above remedies the more immediately depressing. The patient may be bled from the arm until a decided impression upon the pulse is produced, or we may apply from ten to twenty leeches to the vulva and groins, as the depletant measure. In the country, where leeches cannot be had, scarification and cupping can be profitably substituted for them. Should arterial excitement be high after the depletory measure, the tinct. of verat. viride, in doses from four to six drops every four hours, with the ammon. mixture, will be an efficient adjunct to our remedial measures. The calomel should be withheld as soon as its specific effects are produced.

I should not discharge the obligation I feel to the student, in the treatment of this disease, were I not again to caution him against an imperfect cure of it. Very often it becomes chronic, and renders the patient miserable for years. We should try to avoid this consequence. Too early a resumption of duties and active exercise should be especially prevented. When practicable, a continuation of treatment and avoidance of the causes which produced the inflammation, are of equal importance. As a means of perfecting the cure which the more active treatment has brought about, the sedative effect of water affords us valuable aid. The sitz bath and vaginal injections are the modes of using it. The sitz bath ought to be used as much as the time and patience of the patient can be made to allow. An hour is short enough time, and two hours is better, twice or thrice in twenty-four hours. The injections should be copious, and may be used in the bath and of the same water. From two to four gallons of water ought to be passed

9

through the vagina in this way each time the bath is used, by means of the perpetual rubber syringe.

Acute Inflammation of Mucous Membrane of the Uterus.—I do not know that I have ever met with an uncomplicated case of acute inflammation of the mucous membrane of the uterus. Cases that I have seen have been connected with inflammation of the vagina, and have arisen as the effect of some poison directly applied to the membrane. Most of them were gonorrhœal, but in some I have been puzzled to determine whether the poison of this affection was the cause or not. Probably this poison gets into families, where and in ways it ought not, and thus deceives us. However this may be, I think one of the worst features of gonorrhœal inflammation is the frequency with which it invades the mucous membrane of the uterus, and the difficulty of completely eradicating it. It is very apt to lurk in the uterus after the acute symptoms are removed and the inflammation gone entirely from the vagina, and thus require treatment as chronic endometritis.

If I am not deceived by my observation, acute endometritis, of a non-specific character, is a very rare affection, and as I have not seen it and doubt its existence, I do not feel justified in compiling a description of it.

CHAPTER X.

CHRONIC INFLAMMATION OF THE UTERUS AND CERVIX.

It would not be necessary to go into even a short detail of the sympathetic accompaniments of disease of the uterus, were I not convinced that they are often considered independent affections, and their origin not suspected by very many, if not the great majority of practitioners. It is my wish to impress the conviction that these diseases are overlooked, misunderstood, and neglected; and that an immense amount of suffering that is now borne as a necessity by women, might be relieved, if we would investigate and study their ailments with as much patience as, and with no more reserve than, we approach and investigate lung diseases or throat affections.

Dr. Scanzoni* says: "The sympathetic phenomena which very distant organs so often present during the course of uterine diseases are of the highest scientific importance." They are the more important, because our attention is more frequently called to them than to their original exciting cause. The secondary or sympathetic diseases distress patients most, and the fact of their mentioning no other troubles may, without inquiry, mislead us into the opinion that they are independent affections.

Sympathy of the Stomach.—The stomach is apt to be disturbed as early and as frequently as any other organ by chronic uterine disease. This is no more than we would expect, considering how often and intensely it is influenced by pregnancy, and its great readiness of complication in most affections of other parts of the system. Simple anorexia is, perhaps, the most common of the sympathies of the stomach, as is also its contrary, voracity; but some unbecoming and even disgusting depravity of appetite is also not uncommon. Inappetency sometimes proceeds to the extent of loathing of food, and to longing for inappropriate articles of diet.

* Diseases of Females.

Nausea, with loathing of food and disgust at the smell of it, is another feature of stomachic trouble; also, frequent vomiting when the stomach is full; an absence of discomfort when it is empty, and the vomiting is sometimes worse when there are no ingesta, and nothing is expelled but some of its secretions, which are usually acid, but sometimes bilious. Gastralgia occurs when the stomach is empty; or during digestion, or immediately after swallowing food. The capacity of the stomach to digest food of any kind is often impaired, but more frequently some particular sort of food disagrees with the stomach and embarrasses digestion; in short, almost every form of disordered stomach may be looked for as the result of the sympathetic influence of diseases of the uterus upon that organ. The grade of functional disturbance may vary from the slightest inconvenience to that complete arrest of digestion which rapidly induces inanition and death. Extreme cases of indigestion, however, are not of frequent occurrence, and the disturbances are rather those of great annoyance than such as result in very serious impairment of nutrition; and many patients who constantly complain of suffering very seriously from sensitiveness connected with digestion, attain to a state of apparent robust embonpoint.

Sympathetic Disease of the Bowels.—The bowels probably sympathize in diseases of the uterus next in frequency to the stomach, and their functional derangements are multitudinous. Constipation is very common. The bowels, in many instances, have apparently no natural tendency to move. I have one patient, who assures me that she often has been fourteen days without any fecal discharge whatever, and that she dare not try how long she could go without it, but that she always used some means to promote the alvine evacuations. In other cases, constipation terminates with diarrhœa; and an alternation of diarrhœa and costiveness, which lasts from two to six days, is a constant and habitual state with the patient. In cases of constipation resulting from this cause, the constipation seems to depend upon a want of muscular tone in the intestines; peristaltic action is deficient, and the appearance of the evacuations is in all respects natural, and their consistence proper. In other cases the secretions are deficient, and the stools are dry, hard, and deficient in quantity. But constant diarrhœa and irritable bowels are also frequent accompaniments of uterine disease. The passages may be profuse, watery, and exhausting, or profuse and fecal. A peculiar kind of discharge in cases of diarrhœa in uterine disease presents a muco-fibrinous cast of the intestines. These casts are sometimes quite tenacious, and of variable length, from two to ten inches, and are often complete casts of the intestinal tube; at other times there are shreds of this false membrane of irregular shape and size. The discharge of these substances is usually attended with some dysenteric symptoms. The diarrhœa sometimes seems to be excited or aggravated by certain articles of food; at other times one kind of ingesta seems to agree as well as another; and again the bowels may be quite regular, except at or near the period of menstruation. The irregularity is often entirely confined to that time. With or without diarrhœa there may be tumultuous gaseous commotion in the bowels; they may be more or less distended, or without distension there may be annoying borborygmus and motion from the gas passing from one part of the bowels to another, inducing the opinion that pregnancy exists. The gaseous distension of the abdomen is sometimes so extensive and permanent as to induce the over-willing patient to believe that it is caused by gestation, and being connected with hysterical craftiness, she may impose the same belief on a careless practitioner.

Sympathetic Affection of the Liver.—Closely connected with, and of course very much influencing the condition of the alimentary canal, is the condition of the liver. Sometimes the bile is poured out in such copious quantities as to induce full and free discharges of it from the stomach by vomiting, and to stimulate the intestines to copious bilious diarrhœa when they are not irritable, but subject to the ordinary stimulation of ingesta. This overflow of bile comes in paroxysms, and produces a sort of cholèra morbus. When it occurs only once a month, it is apt to be near the time of menstruation, but it may return several times between the monthly periods. But there is often a persistent absence of secretion for a time, or this condition may alternate with the other; or the bile, instead of finding its way into the alimentary canal, may pass into the circulation and give the skin a jaundiced hue. When the functions of the liver are seriously disturbed, there is apt to be at one time a deficiency of bile, and at another a great redundancy. I have not seen this organ congested to any great extent, as observed by Dr. Bennett. But I have seen an enlargement of the spleen in such instances, though I have not supposed it to be the result of the influence of uterine disease. When these copious effusions of bile take place somewhat suddenly, all the pain and spasmodic action observed in bilious colic are likely to present themselves.

Sympathetic Affection of the Nervous System.—Much more distressing, if not more serious suffering, is experienced in the nervous system than in the digestive apparatus. Aches, pains, and complaints of evident nervous ailments, are the peculiar province of uterine disease. There is hardly a disagreeable or even excruciating sensation that these patients do not experience; and too often this real suffering is mistaken by the friends for imaginary, and the woman's complaints are treated with unreasonable impatience and rudeness by persons from whom she ought to receive kindness and sympathy, because her appearance does not correspond with her morbid sensations, as we are apt to observe them in other examples of disease. It is remarkable, too, and a fact that often impeaches them with insincerity in their complaints, when the uninitiated are the judges, that these patients will pass from a state of excruciating suffering and loud complaints, under a little excitement, to one of actual enjoyment and hilarity, or conversely. The transition from the excitement of private company, or a public party, gives way in a few minutes to a doleful condition of suffering and unappeasable complaints. The inconsistency of the complaints and enjoyments, the incapacities and the performances of these patients, are almost characteristic—at least in their sudden alternation—and are inexplicable in any other way than by supposing that the pains in the different organs to which they are referred, are more dependent upon the general nervous susceptibility than upon the organic disease of even trivial character. They are strictly neuralgic in their nature, and confined to the nerve-matter or tissue of the parts. A great number of the disagreeable sensations and pains appear more frequently in particular parts, and hence may be distinctly referred to in this description.

Cephalalgia.—Cephalalgia, in some form, either partial or general, is a very common attendant upon the nervous susceptibility of uterine patients. Often general, the whole head seeming to pulsate and thrill with terrible pain, rendering the patient almost frantic with the intolerable aching. In a few hours general cephalalgia subsides, leaving the nervous energies prostrate for a short time, but otherwise the patient is free from all pain. This subsidence would not be complete if the cephalalgia were anything but nervous pain in the head. These general cephalalgiæ are often, but not necessarily, attended by nausea and vomiting, or other stomachic, hepatic, or intestinal disorder; and may be relieved, when that is the case, by emesis or an alterative cathartic. This is what is commonly called sick headache. The most frequent forms of pain in the head, however, are partial, and confined to some particular part; as hemicrania, confined to the whole of one side, or a lancinating pain in the temple, brow, or eye. All these are very common pains in uterine disease; but persistent or frequently recurring pain in the occipital region, or on the summit of the head, is nearly pathognomonic of uterine disease. It is almost invariably the case that a woman has chronic uterine disease if she complain of persistent pain in either of these regions. The occipital pain I have observed in this connection much oftener than the pain on the top of the head. It is ordinarily a dull aching, that completely unnerves the patient, and renders her unfit for her duties for some days; it is usually very persistent, in some patients being almost constantly present, but in other cases only occurring once a month, ordinarily at the menstrual period. The pain on the top of the head is described generally as a burning pain; patients complain that they have all the time a hot place on the top of their heads. This pain is probably more constant in patients that have it, than is any other of the pains about the head. I think I have observed that when patients suffer greatly from pain in the head they complain less of suffering which is more directly referable to the uterus, than when any other symptom seems to exceed. Indeed, I have met with patients who were martyrs to these excruciating headaches, that did not complain of anything which pointed directly to the uterus as the origin of their suffering, and yet upon examination that organ was found ulcerated and inflamed; and when these conditions were cured by appropriate treatment, the headache ceased to annoy them. A remarkable instance of this kind occurred to me some years ago. The patient came to town to consult me about what she called neuralgia. The pain was located in the occiput; it lasted one week in every four (her menstrual week), and when very severe she had hysterical convulsions. Convulsions occurred at almost every recurrence of the headache. She had no backache at any time; her menses were natural in every respect, as far as I could gather from her history, on which I placed the more reliance, from the general intelligence of the patient. She could walk long distances without inconvenience, had no pains in the hips, groins, or legs: in short, she made no complaint from which I could infer the origin of the nervous suffering to be in the uterus, except that the headache was sure to come on at the time of menstruation. Her uterus was ulcerated and inflamed; and after appropriate treatment, was cured, when the suffering vanished, and she has since enjoyed complete immunity from them. This woman was about thirty years old, and in the midst of her child-bearing period, and it might hence be supposed that the uterus exercises more sympathy than at any other time of life; but, as the following case will show, this is not the fact: Mrs. ——, 49 years of age, had ceased to menstruate three years before I saw her, but was subject to the most excruciating headache every six or seven days, each attack of which would so prostrate her that she would scarcely recover from one before the next would appear. She had some backache and inconvenience in walking, but these symptoms scarcely attracted her attention amid the terrible sufferings caused by her headaches. Six months' treatment addressed to the uterus alone sufficed to remove all this great trouble, and render the woman comfortable and capable of her duties in life. The overwhelming influence of this terrible cephalalgia on the nervous system seems to occupy so completely the capacities of it, that less pain is unheeded by it, and no cognizance is taken of the sufferings of the less sensitive but inflamed and mischief-making uterus.

Affections of the Spinal Cord.—The spinal cord seems to partake very much of this sensitiveness of the nervous system, probably more so than the brain. Pain in some portion of the cord or cauda equina is almost universally present in uterine disease. The most common parts in which the pain is situated are the sacral and lumbar. Pain is so general in those regions, that it has come to be regarded as necessary in the estimation of very many persons to establish the probable existence of this affection. The pain is fixed and almost constant; but aggravated by anything that excites the uterine vascular system; as standing or walking for a long time, lifting or jumping, or sudden emotions. Fright, anxiety, or anger, as the patient says, "flies to the back," and aggravates this pain. It is especially apt to be worse during the menstrual congestion. Sometimes walking increases the pain so much as to incapacitate the subject of it from this kind of exercise. An expression often made use of to signify sensitiveness of the back, is "weak back." Women will say, I have not exactly pain in my back, but it is so weak that I cannot move on account of it, or can hardly stand, or cannot arise from a stooping posture. The pain may be fixed in any part of the spine. I have a patient, whose backache is at the junction of the dorsal and lumbar region. In connection with these pains there is often tenderness in the same region, so that pressure causes great complaint. The pain is not only increased in the part pressed upon, but it sometimes darts along the nerves around the body.

Sympathetic Pains in the Pelvic Region.—A number of painful localities is generally found about the pelvis; in the inguinal or internal iliac region exceedingly common. Immediately above one of the groins a constant and fixed aching may be found, which is aggravated by all the circumstances that increase the pain in the back. Most generally there is some tenderness or soreness in the part in which there is pain, which is increased by pressure. This pain sometimes extends to the hip and side of the pelvis. It is much more frequent in the left side, but is often confined exclusively to the right, and less frequently it is in both right and left side alike. In more rare instances, the pain is centrally situated behind the symphysis pubis.

Extension of Inflammation to Bladder and Rectum.—The patient will often say she has pain in the bladder, or pain in the rectum, and believes that these organs are affected. These two last pains when complained of, are generally very appropriately stated to be in the bladder and rectum, and are indicative, for the most part, of an extension of inflammation to these two organs. When this is the case, pain accompanies or rather is increased by micturition, or immediately after it the pain may occur. The same remarks are applicable to the alvine discharge; at that time the pain is increased, or then only occurs. These pains are not, strictly speaking, sympathetic, but occur as consequences to the extension of the inflammation. The pains indicate correctly the locality of the inflammation. The pain in the iliac region sometimes extends up the side as far as the mammary region; or there may be pain in this latter region not connected with the former. The pain may likewise be situated between these regions and be independent of any pain in them.

Affections of the Sciatic and Anterior Crural Nerves.—Pain in the course of the sciatic, obturator, or anterior crural nerve is very common in uterine affections of an inflammatory nature. These pains are often so severe and aggravated by any exertion as to incapacitate the patient from walking. Particular motions cause pain, according to the nerve affected. When the sciatic is the seat of pain, sitting down, especially on a hard chair, increases it, so that the patient resorts to cushions for defence against pressure. Pain in the course of one or more of these nerves is often the most distressing circumstance connected with the case, and it is often treated as neuralgia seated in these nerves, while the cause is not even suspected. The pain may occupy the whole length of the nerve, or it may be confined to its upper or lower parts, or to an intermediate portion of variable length. The part of the limb traversed by the nerve may be tender or not; most frequently there is no tenderness. The pain may be fixed, or darting and transitory. It may be constant or paroxysmal; the patient may enjoy immunity for hours and days, or even weeks, or she may be a constant sufferer from them. They are apt, as other pains are, to be greater during menstrual congestion than at any other time. The pains emanating from the pelvis are not sympathetic, nor are they probably reflex; but they are caused very likely by pressure of the uterus on the nerves, or they may be produced by an extension of the inflammation to the nerve sheaths.

Hyperæsthesia.—Akin to pains in various parts is hyperæsthesia without inflammation; great sensitiveness of particular parts. Tenderness of the scalp is often complained of. The whole surface of the head is so tender as to require great care in dressing it, and no pressure can be tolerated without an effort. Of a similar nature is tenderness along the spine. The different spinous processes in some sections of the column cannot be touched without giving the patient great suffering. Pressure upon these tender vertebræ sometimes causes pain to shoot along the spinal nerves, passing out of the intervertebral foramina in the neighborhood. There is occasionally also general tenderness of the abdomen.

Anæsthesia.—Much less frequently there is anæsthesia of some particular parts. The patient complains of a want of the ordinary sensitiveness of the parts; or there is a feeling of numbness which lasts for some days, and which recurs so often as to obtain the distinction of a symptom of the case. The muscular system, through the nervous, is in many cases very seriously affected. Cramps and spasmodic action are very frequent in particular cases, and they are confined almost constantly to certain limbs. They occur more frequently in the lower than in the upper extremities.

Spasms.—A worse state of things, however, exists when there are general spasms of the limbs and abdominal walls, and hysterical convulsions. They are apparently induced by fatigue, or occur at the time of menstruation. The patient, after complaining of severe pain in the stomach, falls into a state of general convulsions, which lasts from thirty seconds to some hours, and subsequently sinks into a state of quietude, but not of insensibility. These attacks are usually repeated several times and then subside, leaving the patient in the possession of her usual physical condition, which is one of nervous misery.

Accompanying Manifestations of Moral and Intellectual Perverseness.—During the spasmodic action which, in the majority of cases, have to a critical observer the appearance of being partly voluntary, there is apt to be a singular perverseness of moral and intellectual manifestations, which was on a certain occasion very graphically expressed by a clerical friend in speaking of a patient, that she "seemed to be actuated by an evil spirit." In the midst of great suffering they not unfrequently try to bite and otherwise wound those who endeavor to restrain their violent agitation; they attempt to throw the covering from them, with the apparent object of exposing their person, or say some very perverse things. At other times they attempt to imitate the symptoms of some grave organic affection. One patient, by heaving up the lower part of the chest spasmodically at rapidly succeeding intervals, induced her friends to think that she was the subject of violent palpitations of the heart, and therefore that she must be the subject of cardiac disease; she also imitated throbbing of the temples by spasmodic contractions of the temporal muscle. When this throbbing of the temples was very violent, I requested her to hold her mouth open so as to relax those fibres, but she looked up and said some very perverse things, and became contemptuously calm. A request to hold her breath when the palpitations were violent, induced her to act in the same way, and caused an instantaneous cessation of them. The great peculiarity in these spasms has always seemed to me to be a guarded cunning, a deceitful and perverted consciousness. To a close observer this is always easily detected. By using the foregoing epithets descriptive of the peculiarity of this kind of hysterical phenomena, I do not wish to be understood as saying that deceit, cunning, &c., are indications of freedom from disease on the part of patients who are thus affected. I think this is not usually the case, but that these indications are the result of the morbid state of the mind and body. The spasmodic action of the muscles is not contemporaneous in the corresponding extremities as in epileptiform hysteria or epilepsy; but is so irregular as to move the body in many different directions, instead of giving to it frequently repeated similar motions. I do not know that I have seen the epileptiform hysterical convulsions arise from uterine disease.

Syncopal Convulsions.—But there is a singular variety of semi-convulsions or syncopal convulsions, which I have noticed in a few cases, that I do not remember to have observed in any other connection. They occur very frequently after they have once seized the patient, as often as three or even six or eight times during the twenty-four hours. They take place in the daytime or at night, during the sleeping or waking condition, and do not seem to result from any particular excitement at the time. If the patient is sitting and talking, or is engaged in work, she suddenly ceases, and slowly sinks down to the floor; she turns her head to one side, almost ceases to breathe, becomes pale and trembles, sometimes very gently, sometimes violently. This state lasts only for a few seconds; she arouses, looks about confusedly, and although she knows she has had a fit, as her friends call it, she does not remember distinctly anything which passed during the time. As these attacks become chronic, they may be attended with very slight convulsive movements, frothing at the mouth, and sequential somnolence; but, ordinarily, this is not the case. If the patient is attacked in the night while asleep, unless some person observes the attack, it will not be known to have occurred, the patient being unconscious of it. There is generally, however, movement enough to awaken anybody who may be in the same bed with the patient. In all cases of this kind, I have noticed great impairment of memory, particularly of recent circumstances, followed after a while with impairment of other mental faculties, and finally dementia. There is not usually any severe pain in the head or spinal centres; there is, in fact, no prominent painful circumstance apparently connected with the case. Patients having such paroxysms are generally worse at or near the time of menstruating; but sometimes they are quite exempt from them at this time, and have them not long after menstrual congestion is over.

Muscular Weakness.—Extreme muscular weakness—I do not mean that which results from general debility, but of some particular set of muscles—is often present as an accompaniment of uterine disease. This is most frequent in the back and lower extremities, not often in the upper extremities. It is probably imperfect innervation of the part, or it may be some affection of the muscles themselves. I have been inclined to look upon it as partial paralysis, resulting from reflex irritation. More or less numbness of the parts exists in connection with the weakness of the muscles.

Circulatory System.—The circulation and its organs are very often deranged to a distressing degree. Palpitation of the heart is often troublesome, and patients are apt to think themselves the subject of disease of the heart. We are often consulted solely with reference to this symptom, it having absorbed the attention and awakened the apprehension of the sufferer to such a degree that her other inconveniences were forgotten or overlooked. These palpitations are sometimes attended with pain in the region of the heart, which occasionally shoots up to the left shoulder and down the left arm to a greater or less distance, the distress being so great as to amount almost to angina. The palpitation is worse during nervous excitement. It occurs generally in paroxysms. We meet with instances in which it oftener occurs after lying down at night than at any other time. Sometimes it seems to be increased during digestion. The sensation of palpitation does not seem to be at all commensurate with the increased excitement of that organ, and *vice versa*. I have observed instances in which the patient complained of violent palpitation, while the pulse and heart, as far as I could judge, were not at all disturbed. In such cases we might say that the sensitiveness of the heart was increased until its ordinary motions were perceived by the patient. Indeed, the pains and increased irritability of the organs supplied with the great sympathetic nerves, seem to result from increased suscepti- bility or sensitiveness instead of organic changes. There is also sometimes a sensation of throbbing, as though the blood was pass- ing through the arteries in increased quantities, and with increased force in some parts of the system; this occurs mostly about the head, sometimes in the hands and feet, and occasionally inside the head, apparently in the brain; also about the genital organs. Great irregularity of distribution of the blood is often observable, the hands and feet being uncomfortably cold, and continuing in that state for twenty-four hours at a time. In connection with cold extremities, the head is apt to be hot, or warmer than nat- ural; this heat of the head may also be present when the feet and hands are of the common temperature. This heat about the head and face is sometimes almost constantly present in certain patients, and is the source of great annoyance to them. It is apt to be caused by anything that excites the person. The heat is greatest and frequently exclusively located on the top of the head. I do not think that this sensation of heat arises from any other cause as frequently as from uterine disease, and I am sure it is one of the most common symptoms in such disease. There is great heat complained of in the back of the head also, in many instances, and sometimes it extends along the spine, affecting the whole or only sections of it. Burning in the sacrum and loins is very common. Flashes of heat and flashes of color in the face and head, and even in other parts of the body, are very common and annoying occurrences. The power or nervous energy of the heart may be impaired to such an extent as to render the patient liable to faintness on the application of very slight causes,—anger, fear, surprise, or even the more tender emotions, overcoming the patient very readily.

Respiration.—The respiratory apparatus is not so frequently or so severely affected as some of the rest of the organization ; and yet we often meet with some very curious and considerable deviations from the natural condition of its functions. The constriction about the throat, or the feeling as if a ball rose up to the throat and obstructed respiration, and the feeling as if smoke or dust was in the air which the patients breathe, are complaints we hear almost every day. All these sensations, or any one of them, may be aggravated to an agonizing degree, inducing the fear that the paroxysm may be fatal, and causing the patient to suffer for some moments, and sometimes for hours, the horrible sensations of impending suffocation. The breathing may be spasmodic from painful and unnatural contractions of the respiratory muscles. There may also be pleurodynic pains during each ordinary effort of respiration. Imperfect respiration, or partial inflation of one lung, or of parts of the lungs, occasionally occurs. The modification of the respiratory murmur arising from this imperfect inflation of one of the lungs I have observed on several occasions, and not without serious apprehension of the result ; but in all cases where this was the only modification of physical sounds, the patients have done well, and the inflation improved as the returning nervous energy of the rest of the system was established. The respiration is not often apt to be hurried as a constant circumstance, but occurs as the effect of excitement from mental or moral emotions. In some cases, amid the tumult of nervous excitement during a paroxysm, I have seen the respiratory efforts increased to sixty in a minute ; and, generally, these nervous patients constantly have increased frequency of respiration. There are cases in which cough is a very constant symptom ; it is a peculiar, nervous cough, as a general thing, and is excited or made worse by anything that renders the patient more nervous. Sometimes it is difficult to distinguish it from the coughs which arise from insidious affections of the lungs. It is possible that the coughs arising from slight lung difficulties may be aggravated by the nervousness consequent upon uterine disease. I once saw a patient affected with a peculiar nervous cough as the effect of uterine disease, which sounded like the barking of a small dog, and the sound was made at every expiration during the waking condition of the patient, except when the mind was intensely occupied. She was an intelligent young married woman, about twenty years of age. While her whole attention was absorbed, she forgot to cough, but as soon as her attention was relaxed, she habitually produced the same sound. This had lasted when I saw her six months or more. When she was embarrassed by a conversation which related to her case, the sounds became much louder and persistent, appearing in perfect synchronism with every respiratory effort. It must be added, that I did not have an opportunity to treat this patient, nor have I heard from her, so that I cannot give her subsequent history; but the rest of the symptoms plainly indicated uterine suffering, and an examination established the fact that she had inflammation and ulceration of the neck of the uterus. She had never borne children, or miscarried.

Sympathy of the Excretory Organs.—The excretory organs also sympathize with the uterus, particularly the kidneys. It has been a long time observed that female patients, in a state of nervous excitement, secrete a large quantity of urine, which is usually limpid, almost odorless and insipid. These qualities are most likely dependent upon the amount of water being so much greater proportionately than the salts: these last scarcely seem to be present at all. It is extremely dilute urine. Uterine patients are very prone to these large discharges of limpid urine. This kind of alteration in the functions of the kidneys is, doubtless, indirect, and does not occur except in connection with a greatly excited condition of the nervous system as the medium between the kidneys and the uterus. More considerable deviations, however, are apt to take place; the salts are likely to be increased in quantity compared to the amount of water; or one sort of the salts may be greatly over or under the proper proportions in relation to the others. The urine may be decidedly morbid in its composition. It is probable, too, that this deviation is secondary to derangements of the stomach and liver, but nevertheless, it is often present. The urine may be highly alkaline, or highly acid in reaction, showing the production, to an unusual degree, of salts having such chemical qualities. The presence of these salts in excess, whether of the one kind or the other, is pretty sure to produce painful micturition and other disagreeable sensations, as burning and smarting in the urethra and bladder. There is no doubt, however, that these painful and disagreeable symptoms may arise as the more direct effect of inflammation of the uterus when the urine is correct in composition; hence the examination of the urine will be necessary to determine the cause of the symptoms. But the urine is often secreted in very diminished quantities in cases of uterine disease; and that, too, without apparent general febrile excitement. Patients frequently complain of this symptom. Whether there is an increase in the excretory functions of the skin at such time, I am unable to say. The skin is probably not very much affected in its excretory capacity as a general thing, but some very curious deviations have been observed.

Mammary Bodies.—More direct are the effects upon the mammary bodies. They are often highly excited by uterine disease; this is no more than would have been expected from the close sympathetic relations between these organs. Congestion is the most common sympathetic condition. The mammæ increase in size, become hot and painful as a general thing, but sometimes there is no change in their sensible or sensitive conditions. The appearances are natural, but the patient complains of a peculiar and painful condition, not unlike the sensations perceived during the suppurative stage of inflammation; but there is neither tenderness, nor swelling, nor heat, nor other deviation, than the unnatural sensation. Sometimes the breasts are really inflamed. The lymphatic glands in the axilla, and from the axilla to the border of the mammæ, in some instances, become affected at the same time; in other instances, however, they do not partake in the sympathies of the mammæ. They also become tender in some cases when the mammæ do not seem to be excited.

Moral and Mental Derangement.—No more constant derangements, perhaps, occur, than are observed in the mental and moral qualities of the patient. The patient loses the complete control which she has been in the habit of exercising over her emotions, and finds herself becoming despondent, fretful, suspicious, and unsteady in her purpose; whimsical, having desires not before experienced, indulging in thoughts and feelings toward her friends which in her former days she did not entertain. She will often call herself a changed woman. If the source of irritation is not discovered and removed, she loses her strength of will entirely; and, instead of her moral feelings being guided by her will under the influence of a sound judgment, she exhibits indecision, and wavers in matters about which she heretofore had no difficulty in making decisions. She finds herself giving way to peevishness to a frightful degree; nobody can please her. In place of her usual satisfaction in the attentions of her friends, she finds fault with their efforts to make her comfortable. Sourness, moroseness, jealousy, carelessness, timidity, and some peculiar perverseness, change her nature entirely. Sometimes one class of ideas will seize her whole faculties, and she will scarcely think or talk of anything else. She has no patience with anybody who will not listen to her, and believes everybody to be her enemy who cannot sympathize with her in her imaginary troubles. The different phases of mental and moral troubles under which the patient labors are almost innumerable. As will be seen, this state of things is closely bordering on insanity, and there is no doubt that insanity is often the result of uterine irritation in patients who are hereditarily predisposed to it. I think I have seen cases of insanity that were excited into activity by the great nervous irritation connected with uterine disease. But, in place of this steady deviation from her natural mental condition, the patient may generally be sane, and show an abnormal state of mind only when circumstances occur which are likely to excite her, when she loses all control and indulges in excessive anger. Sometimes, in a fit of despondency or melancholy, she contemplates or even attempts suicide. Or, if her sense of wrongs weighs heavily upon her, and no means of redress shows itself, she thinks seriously of fleeing from what she fancies is the cause of them. Still another sort of paroxysm exhibits the acts of a depraved and indecent nature ; so disgusting as to shock the witnesses of them, and in her recollection of them to mortify her exceedingly. The common hysterical paroxysm of crying without a sufficient cause, the indulgence in unbecoming and unseemly levity, rapid transitions of despondency and hope, need hardly be mentioned, from their familiarity to every observer. . When, in reference to such unbecoming exhibitions, patients are kindly remonstrated with, they will, in general, acknowledge the impropriety of them, but will end with saying, "I cannot help it;" which is the unanswerable and doubtless truthful exposition of their mental condition. Neglect of duty in all the relations of life is one of the phases of their mental state. Sometimes a wilful selfishness, caring for nothing but what they fancy will make them happy, or conduce in some way to their interests, absorbs their whole mind and governs all their actions. At times there is an intelligent appreciation of the impropriety of their actions.

I have dwelt so long on these general symptoms, and have made so much of uterine sympathies, that I am forced to recall an expression made use of in a notice of Professor Hodge's work on "Diseases of Women," that "if all this is true, it is almost a pity that a woman has a womb;" but I conscientiously believe I have fallen very far short of mentioning all the sympathetic evils resulting from chronic diseases of the uterus, and I only design this as an outline view of a subject that will fill itself up in painfully warm colors in the observation of those who devote themselves to a close study of the diseases of women. While this is my conviction, I do not wish to be understood as saying that nearly all of the above symptoms will show themselves even in a majority of cases: some of them will be prominent in some cases, others in other cases; and in rare instances we meet with nearly all of them in some sufferer, and in nearly all chronic cases we shall find enough to move us to commiseration for the ruined health of women thus affected. I know there are thousands of my peers in the profession, who do not see in the foregoing array of symptoms any indication of disease of the uterus; and when uterine diseases are obviously coexistent, they are apt to be arranged in the order of sequency. This does not shake my faith in the facts I have observed for myself, nor disturb my judgment formed from an observation of a very large number of cases carefully watched through all stages of progress to their termination. That all the above symptoms may occasionally be present in cases in which the uterus is healthy, I have often observed; but that they are also frequently present as the proximate and remote effects of uterine disease, I am well satisfied. Another well-established fact, according to my judgment, is, that the direct symptoms referable to the uterus may be feebly pronounced, while some or even a large number of the sympathetic disturbances are very prominent; and judging from the freedom of pain and other inconveniences experienced in the uterine region, there are even cases in which the uterus does not seem to suffer at all. These cases are well calculated to mislead us, and to induce the opinion that the womb difficulty is of minor importance, and need not be the object of solicitude until we get rid of the more troublesome and prominent symptoms. We cannot be too careful in our consideration and management of this class of cases, and I insist, that while we adopt judicious remedial means for the removal of the more afflicting symptoms, that we be sure to address ourselves to the disease of the uterus, however slight it may appear to be. I have seen too much good result from the observance of this direction not to dwell with emphasis upon its importance. The cure of the uterine disease will be a valuable diagnostic measure in such cases. Not only may there be a great difference or want of correspondence in the severity of the local and general symptoms, but in many cases in which the general symptoms have almost made a wreck of the health and happiness of the patient, the local inflammation and ulceration will be found upon examination to be trifling in amount and degree. The inflammation may be very slight and the patient suffer very greatly from it, either generally or locally, or both; or the ulceration may be extensive and the inflammation very considerable, and yet the patient hardly be sensible of any inconvenience whatever from its presence. This statement will be confirmed by careful observers in this field of research. It, however, will prove a stumbling-block to those who entertain the opinion that uterine disease is of small importance in the consideration of woman's ailments. They seem to think that there is of necessity an exact and invariable correspondence between the magnitude of cause and effect, and they point to these cases and say,—the symptoms were present, but very trifling, if any, uterine disease showed itself upon examination; or they will say, there was great ulceration, but the patient did not suffer from its presence, at least not in proportion to the amount of local disease. I need not particularize instances in which other diseases are comparatively latent, or cases in which the symptoms are unduly severe compared to the amount of actual disease, as they will suggest themselves to every intelligent practitioner. But recurring to the sympathies of the uterus, we find that while some patients are not affected at all by pregnancy and others are favorably affected, their health being better then than at any other time, that some absolutely perish on account of the same functional derangements inaugurated by pregnancy; and, as is shown on a former page, organic diseases are not unfrequently lighted up. We will probably always be at a loss to understand precisely this difference; but there can be no doubt that it is more on account of constitutional differences than local ones. The concatenation of sympathetic influences may be caused by the greater susceptibility of the organs secondarily affected. In fact, the only mode of accounting for it is by supposing this increased susceptibility. I am convinced that this great but inexplicable diversity of sympathetic effects is as likely to result from uterine disease as from pregnancy. We must expect a very great range of difference in the extent of sympathetic derangement from uterine disease. It is interesting to observe the rise and development of the sequences to diseases of the uterus. How far can the uterus produce a direct effect in creating this large amount of sympathetic disorder? Are most of the symptoms produced by the direct sympathetic relation of the uterus to other organs, or does the diseased uterus first affect some other more influential organ detrimentally, and then this last the organism generally? I am inclined to think, from a large observation, that the uterus has direct sympathy with only a few organs, and no one probably is so powerfully affected by it as the stomach. It is the first organ affected by it in pregnancy, being brought into a morbid condition in a very few weeks. The well-known, powerful, and almost universal sympathetic influence exerted by the stomach upon other viscera is sufficient, when it is diseased, to account for the great variety of subsequent symptoms. The stomach is the great centre from which radiate abdominal, thoracic, cerebral, and spinal disturbances almost *ad infinitum*. And there can be no reasonable doubt that it is an active agent in originating the disturbances of the great vital organs. The subject of the sympathetic influence of the uterus then becomes the more interesting and important, from the fact that very slight deviations from its ordinary condition arouses the most influential of all the organs to a state of disease, which depresses the functional energies and increases the susceptibilities of almost all the rest of the organism. In addition to the chain of sympathetic susceptibilities produced by this state of the stomach, too frequently the digestive powers of that organ are impaired or perverted, so as to supply the chyme in deficient quantities or in deteriorated quality, and in this way injuriously affect the composition of the blood, inducing anæmia or oligæmia. Imperfect nutrition will follow, as a matter of course, in the one case, and perverted nutrition in the other, so that emaciation or obesity will be ordinarily present. Another organ probably in direct sympathy with the uterus is the cerebellum, as it seems to me to be as frequently affected as the stomach. The mammæ are, of course, in direct sympathetic relation with the uterus, and yet they are not uniformly affected in many cases when the uterus is very seriously diseased. I do not believe that we are able to say, at present, whether there are other organs that come directly under uterine influence. A proof of the powerful and very ready effect upon other organs of irritation of the uterus, may be found in the fact, that very often when the patient is in a condition of comfort, so far as her general suffering is concerned, an application of nitrate of silver to a morbid os uteri will give her excruciating pain in the head, render her exceedingly despondent and irritable, and very much aggravate the symptoms with which she is affected. This I have so often observed to be the case, that I cannot but regard it as one of our diagnostic means. After such an application, the patient will generally complain of an aggravation of the general symptoms, whatever they may have been, and say that all the pains are made worse by the application of the caustic. When an organ has been the subject of irritation or functional derangement for a long time, in consequence of sympathy with the uterus, it may become the subject of organic disease, and then continue as an independent affection of perhaps a dangerous character; or if organic has not succeeded to functional disease, the power of habit which is so frequently thus engendered will perpetuate morbid action for an indefinite period after the cause of it has been removed.

CHAPTER XI.

CHRONIC INFLAMMATION OF THE UTERUS AND CERVIX.

LOCAL SYMPTOMS.—The symptoms which more directly indicate inflammation and ulceration of the cervix uteri, should be dwelt upon with some minuteness. Some of the symptoms have been already mentioned, and the rest are reserved for a separate notice.

Pain in the Sacral or Lumbar Region.—Pain in the sacral or lumbar region is one of the most constant, and when persistent, indicates, with a good deal of certainty, disease of some kind in the pelvis. The pain in these regions which is caused by the uterus is ordinarily central, being in the middle of the sacrum at its lower extremity. It is sometimes at its upper extremity, or it extends the whole length of this bone. Not unfrequently a painful spot may be found on one side, over the sacro-iliac junction. Some patients describe the pain as if a bundle of nerves were pulled upon from the inside of the sacrum, and others describe it as an aching or burning pain. Accompanying the pain in the sacrum is often a sense of soreness upon pressure, an inability to sit with comfort, on account of the tenderness of the lower part of the sacrum.

Pain in the Loins.—Pain in the loins is probably not so common as that in the sacrum, but is quite as various in its nature. Very frequently there is great weakness in the loins, to so great an extent sometimes as to prevent the continuance of the erect posture for any length of time.

Inability to Stand.—I have had a number of patients who were unable to stand upon their feet long enough to dress their hair, on account of the weak back. It is remarkable that patients often feel this weak back more when standing than walking; and they are sometimes able to walk a distance, without any great inconvenience, but as soon as they stop, the weakness is apparent to a distressing degree.

Inability to Walk.—Ordinarily, the weakness disables the patient from walking. The pain in the back is almost always increased by walking or standing, and on this account the patients avoid being on their feet, although the back is strong enough. There are many patients who have severe inflammation of the cervix, who do not experience any of the inconveniences in the sacrum and loins already described; but some of them are very generally present.

Great pain in the back, closely resembling pain arising from diseased uterus, is also caused by hæmorrhoids, prolapse, or inflammation of the rectum. The pain caused by diseases of the rectum, I think is much more frequent on the left side of the sacrum, and in the left nates or hip, than centrally situated; in fact, I have come to regard pain confined to the left nates and hip, as indicating, with considerable probability, rectal disease, and I always inquire into the functions of that organ, when that pain is present. It differs in position from the pain in the iliac region, so common as the result of uterine disease. It is situated near the sacrum, and more in the side of the pelvis than the latter.

Pain in the Iliac Region.—Pain in the iliac region is very common. In frequency, it is next to pain in the back. The pain is commonly situated near the anterior superior spinous process of the ilium, and below the level of it. It is not referred to the iliac bone, or fossa, but to a place a little above the groin. We often meet with it on both sides, but much more frequently on one only; on the left side much oftener than on the right. Dr. Dewees considered this pain in the left groin, or a little above it, as almost diagnostic of prolapse of the uterus. It is certainly very frequently indicative of inflammation of the uterine cervix.

Soreness in the Iliac Region.—This pain is generally accompanied with soreness upon pressure, and sometimes there is soreness upon pressure when there is no constant pain. Walking, standing, or riding generally increases it. A severe shock or strain, from lifting, will sometimes cause pain suddenly to appear in this region, when it had not before been observed.

Pain in the Side, above the Ilium.—Instead of the pain situated, as here described, there is often pain higher up in the side, or in the iliac fossa, or along the crest of the ilium, and even midway between the crest and ribs of the side. These pains are not in the ovaria, although they seem to point to the ovaria more directly than to the uterus, and are by some regarded as a symptom arising from ovarian inflammation. Dr. Bennett admits that it may be a sympathetic painful condition of the ovary. It is not material whether this is true or not; it is certain that it is very frequently present in uterine disease, and almost invariably cured by remedies addressed to the uterus, instead of to the ovaria.

Weight, or Bearing-down Pain, or Uterine Tenesmus.—Another indication of uterine disease, of less frequent occurrence, is a sense of weight in the loins or pelvis. This sense of weight is experienced in the loins and iliac regions more frequently than elsewhere; but it is often felt at the pelvis, and oftener in the perineal and anal regions. Patients express themselves as feeling a heavy weight dragging upon the back and hips, and others feel as though the insides were dropping through the vagina. Occasionally we meet with such urgent uterine tenesmus, that the patient is obliged to keep the recumbent posture, in order to enjoy any comfort. In these cases, the patient in the erect position cannot resist a constant desire to "bear down," resembling the tenesmus of dysentery. This sensation is sometimes more distressing than any other symptom, and obliges the patient to desist from walking.

Leucorrhœa.—Leucorrhœa is one of the symptoms usually relied upon as an evidence of disease of the uterus. In the healthy condition of the uterus and vagina, there ought to be no discharge; the vaginal canal is merely moist, and no mucus should make its appearance externally. When the mucous membrane is temporarily excited, there is more than ordinary secretion; but it ceases as soon as the cause of excitement passes.

Effects of Inflammatory Excitement.—We should *a priori* expect increased vaginal discharge to be accompanied with some form of disease, especially when it continues for more than a few days. Our knowledge of the discharge from mucous membranes lining the cavities elsewhere, will afford us enough data to confirm these views. We do not expect to see a constant flow, however moderate it may be, from the male urethra, when it is perfectly healthy; and we take gleet as an evidence of chronic urethritis, and it is generally the sequence of an acute attack of that disease. A constant discharge from the nose is an evidence also of more or less disease. It is just so with the vagina. The indications from leucorrhœa are derived from the color or consistence of the discharge, or both. The discharge from the vagina, resulting from mere excitement of the mucous crypts, is thin, glairy, and not very tenacious. It is ordinarily acid in reaction. There is no color, and but little consistence to it. When a moderate excitement of the internal mucous membrane of the neck of the uterus produces a discharge of mucus, sufficient to appear at the orifice of the vagina, the discharge is white, not unlike milk, and when examined closely, will be found to consist of minute coagula swimming in a little clear fluid. When the mucus flows from the mouth of the uterus, it is thick, and resembles very closely the albumen of an egg, and is alkaline in reaction. When it passes into the vaginal canal, it meets with the acidity of the vagina, and is coagulated, and the whole changed from a colorless translucency to an opaque white. The reason that the coagula are small and so numerous, may probably be found in the fact that mucus arrives in the vagina in such small quantities; each coagulum represents a minute drop of mucus, changed in quality. As, however, the mucous membrane of the vagina furnishes only a small quantity of acidity, when this alkaline discharge from the cervix is copious it soon neutralizes the vaginal acid, and passing through this cavity unchanged, appears at the external parts possessing its characteristic qualities. We then hear the patient complain of a tenacious albuminous leucorrhœa; she will nearly always compare it to the white of an egg, but state that it is more tenacious. Unless the quantity is considerable, the mucus from the internal cervical membrane does not appear at the external orifice unchanged, but passes into this curdled condition. There is often a considerable quantity of this milk-like leucorrhœa in the whole length of the vagina, and hence it has been supposed by many that this is the vaginal mucus, in its natural condition, and they have called it vaginal leucorrhœa. I am satisfied that it is changed cervical mucus.

Character of the Mucus in the Vagina.—The vaginal mucus does not undergo any kind of coagulation, but appears at the vaginal orifice as a clear, thin, almost watery fluid, which moistens but does not stain the linen. This colorless mucous secretion is indicative of increased vascular and glandular excitement, without detriment to the integrity of the membrane whence it is derived, and the excitement may be due merely to temporary congestion, in which case it will disappear, or it may be inflammatory, when it will become more persistent, and possibly permanent.

Amount of Leucorrhœa not always Proportioned to Extent of Disease.—The abundance of this discharge is no criterion by which to judge of the amount of inflammation, or its intensity, but it will scarcely remain colorless after the integrity of the membrane is invaded. When the albuminous fluid appears at the orifice of the vagina, there is persistent cervical disease, almost of a certainty.

Yellow Leucorrhœa, when there is Inflammatory Ulceration.— The thick, white-of-egg-like albumen will be mixed, when there is ulceration in the cervix, to a greater or less extent, with pus, so that it will be stained yellow. If the quantity of ulceration is considerable, and its surface is producing pus, the yellow will preponderate in the color, and sometimes the whole of the production becomes yellow. The yellow color may be in streaks through it, or intimately mixed with it, so as to stain it uniformly; or the pus may be mixed with the white creamy secretion found in the vagina. Pus may be mixed with any of the varieties of leucorrhœa, and impart to it its tint, more or less completely.

I am in the habit of considering the pus-colored leucorrhœa as indicative, with great certainty, of destructive or ulcerative inflammation of the mucous membrane of the genital canal. This inflammation may be situated in the vagina, in the cervix of the uterus, or in the uterine cavity; and I can hardly conceive of the production of pus by a mucous membrane with a whole epithelium.

Ulceration sometimes exists without Leucorrhœa.—While I am almost confident of the existence of ulcerative inflammation somewhere, when this purulent leucorrhœa shows itself, and persists for some time, I am not, on the other hand, at all confident that ulceration does not exist, if the yellow leucorrhœa is not present. Indeed, I do not regard leucorrhœa necessary to establish the existence of ulcerative inflammation. There are many cases in which it is quite evident that mischievous inflammation is plainly the cause of the invalid condition of the patient, and yet neither mucus nor pus ever shows itself at the vaginal orifice. This may probably be attributed to two circumstances: in the first place, all ulcerated surfaces do not discharge pus, or if pus is discharged, it may be in very small quantities; in the second place, the absorbents of the vagina may be so active as to take it up before it arrives at the external parts. Cases have been observed in which a large secretion is caused by a small amount of ulcerative disease. Notwithstanding the fact that in uterine disease leucorrhœa is a common and significant symptom, it will not do to base an absolute opinion on its absence, in any given case.

How is the Pain Produced?—How are the local painful symptoms produced? Is the pain in the groin or ilium caused by prolapsus and traction on the broad or round ligaments? I think not. Pain and sensitiveness in the ilium are so frequently present, when I cannot detect any kind of displacement, and so generally disappear when the inflammation is cured, that I am convinced displacement is not necessary for their production. They are of that character of pains which range themselves in the category of that vague, yet indispensable term, sympathetic; or, of the not less fashionable, yet equally indefinite term, reflex.

Bearing Down not caused by Displacement.—The sense of weight or bearing down in the pelvis is one as to which there would, from its nature, seem to be no doubt as to its origin being in displacement. It gives the patient the idea that the womb is bearing with unusual weight on unusual places, viz., the perineum, the rectum, or the bladder; and yet, in a great many instances, we will fail, I think, to detect any deviation from the natural position of that organ, and as soon as the inflammation is cured the symptom vanishes, without any treatment being directed with reference to displacement. How can we account for this symptom? I think its explanation may be found in the fact that the pelvic organs, on account of the inflamed condition of the uterus, and the general pelvic vascular turgescence, are unusually sensitive, and receive painful impressions from contact, which, in the absence of these conditions, would have no effect in causing inconvenience of any kind. I also think that moderate prolapse, retroversion, or other displacement, when unattended by congestion or inflammation, may exist for a long time, without giving rise to any disagreeable sensation whatever. When the uterus is slightly displaced, with considerable pain and sense of weight accompanying this condition, the displacement is commonly considered to be the cause of the distress. When, however, the uterus occupies a normal position, and a sense of weight and pain still exist, they are regarded by most practitioners as the results of an "irritable uterus." That the uterus is sensitive, "irritable," if the term suits better, there is no doubt; but that it is ever so, without congestion or inflammation, I do not believe.

Severity of Suffering not commensurate with Amount of Disease.—The great error in the estimate of the importance of uterine inflammation, is in endeavoring to measure the amount of inflammation by the severity of suffering; in assuming, that because the woman suffers a great deal there must necessarily be extensive inflammation or ulceration. I believe I have seen more nervous prostration, more keen suffering, and have heard louder complaints from a small amount of endocervicitis, than from extensive and obvious external ulceration. Pelvic congestion and increased sensitiveness of the viscera contained in the pelvic cavity, caused by a small amount of persistent inflammation in the neck of the uterus, calls into action, in an exaggerated and intensified form, all the sympathies which are excited by the uterus in its physiologically congested condition, and its persistence wears the more upon the general organism, on account of the increased sensitiveness produced from day to day, by virtue of its chronicity alone. It is anticipating what I shall say in the chapter on Prognosis, to state that endocervicitis is not only more difficult to cure, but more destructive to the health and happiness of the patient, than inflammation and ulceration external to the os. Indeed, we often find cases of extensive ulceration very apparent through the speculum, and consequently entirely unmistakable to the most careless observer, which produces less inconvenience than an amount of endocervicitis so small as to escape the attention of any but an experienced gynecologist. The fact is perplexing, but the knowledge of it will cause a proper appreciation of what is apparently a trifling matter.

Effects on the Functions of the Uterus.—Having given the foregoing sketch of the general and local symptoms of ulceration and inflammation of the neck of the uterus, I purpose to glance at the effects produced on the functional action of that organ. The first function assumed by the uterus and the last to continue is menstruation. It becomes a matter of interest to the physician to ascertain the cause of deviations in a function so persistent, so general, and so important to the health of woman. As inflammation is the cause of injurious and even destructive tissual changes, and of functional aberration in the vital organs much more frequently than any other pathological condition, so I think that the functional aberrations of the uterus depend much more frequently upon inflammation than on any other one cause.

Pain during Menstruation.—Pain during menstruation is not necessarily attended by deviation from the normal monthly flow. That there are varieties of dysmenorrhœa or painful menstruation, with unusual quantities and extraordinary kinds of discharge, is true; but, in many instances, the discharge, though accompanied with pain, is right as to its character and quantity.

Kind of Pain attendant upon Uterine Inflammation.—The kind of pain attendant upon uterine inflammation is, for the most part, the same in quality but varying in intensity. It is a continuous sore pain, with heat in the parts, so slight as to give the patient very little inconvenience, and it varies from this to pain of considerable severity. The pain is at times sufficient to cause the patient to keep her bed for several days, and sometimes for the whole period of the menstrual flow; occasionally it amounts to agony, prostrating the·patient by a paroxysm of pain, which may last a few hours or even several days.

Cramping Pain.—Instead of this continuous sort of pain of varying intensity and duration, there are less frequently painful throes, "coming and going," like labor-pains or after-pains. This kind of pain is often mistaken for colic. They are often very severe, and may last a few hours or several days. They may depend on some substance contained in the uterus, as shreds or membranes of fibrous exudation, and cease at their expulsion. But more often no such cause can be discovered in the evacuations; nothing can be found but fluid blood or coagula evidently formed in the vagina. In other cases, the os internum uteri is small, and does not readily admit the passage of the uterine sound.

Effects of Partial Closure of Os Uteri on Menstruation.—Many practitioners believe that this condition of the os internum, by preventing the ready flow of the blood, causes it to accumulate until the quantity is sufficient to arouse expulsive efforts for its extrusion. In a large majority of cases of painful menstruation I have had the opportunity of observing there was no coaptation; and in several of the worst cases I have met with, the os internum allowed the sound to pass with so much freedom that I could not distinguish its locality. It is also true that in many cases in which the os externum was not larger than a small pin-hole, the patients menstruated without any pain whatever. I do not wish to assert that painful menstruation does not occur as the effect of any other cause than inflammation, though my conviction is, that inflammation is its most frequent cause. The pain may occur at any time during the menstrual flow, and before and after it. Not unfrequently a paroxysm of severe pain, lasting several hours or a day, warns the patient of the approach of the discharge; and subsides suddenly and completely, or gradually and incompletely, as soon as the discharge is fairly established. Frequently the pain continues during the whole time of menstruation, beginning shortly before or synchronous with the discharge, and subsiding with it, though in occasional cases it continues after it. We sometimes meet with patients who begin to menstruate without any suffering, but who have pain during the flow, or after its discontinuance. I think that a majority of patients affected with uterine disease have some pain during menstruation; but there are some who have none whatever, and pass through their period with little or no suffering. (See Dysmenorrhœa.)

Manner of the Flow modified by Inflammation.—The manner of the flow is often modified. Instead of the continuous flow commencing moderately, gradually increasing, and then as gradually declining, every manner of deviation almost may exist. With some the discharge begins naturally, increases very rapidly, until at the end of twenty-four or thirty-six hours an average amount is lost; and then the discharge suddenly declines and ceases, or continues in very moderate quantity for a time longer, and gradually or suddenly stops. With others the flow may begin and proceed naturally for a day or two, cease for one or two days, and then reappear and flow freely for a sufficient time. When menstruation proceeds in this way, it is generally attended with pain. These two varieties are more frequent than any other.

Duration of the Flow.—The duration of the flow may not be affected by it. The flow may continue three weeks or the whole month. This, however, is not frequent. It does not much affect the periodicity of return, of menstrual congestion, and of effort; but it is not unusually the case that we cannot distinguish the discharge which attends ovulation from the hemorrhage which proceeds from an ulcerated surface, as hemorrhagic congestion is so constantly present. We often meet with patients who are so confused by the frequent irregular returns of uterine hemorrhage, that they lose all reckoning as to the time for the menstrual return. Occasionally, continuous hemorrhage is present. The most frequent deviation from regularity in periodicity of the menstruation consists in a slight anticipation of the time of its return.

Menorrhagia.—Menorrhagia or hemorrhage at the menstrual period is not an unusual functional deviation. The hemorrhage is often very considerable, and continues after the usual period has passed by. The flooding is usually greater while the patient is in an erect posture, and it is greatly moderated by recumbency. Occasionally, however, it is not moderated by this means. It would seem probable, *à priori*, that menorrhagia would be the rule with patients affected with uterine inflammation, but such is not the case. I am not sure that even a majority of patients have it.

Menorrhagia frequent in Cervicitis.—I have observed that menorrhagia occurs much more frequently in patients in whom the inflammation occupies the cavity of the neck; this also is the case with painful menstruation. All cases in which there has been either great pain or hemorrhage, or both, for they are frequently coexistent, have been, in my observation, cases in which endocervicitis is the principal disease. Menorrhagia is not always the result of inflammation of the uterus, though inflammation is its most frequent cause; and in such cases it cannot be cured without first curing the inflammation. (See Menorrhagia.)

Amenorrhœa sometimes Results.—Amenorrhœa is the least frequent of menstrual deviations as the effect of inflammation in the cervix uteri; but this inflammation is frequently the cause of scanty menstruation. It is curious to notice the manner in which this scantiness occurs. It seems to come on after the inflammation has lasted for a considerable time, and is almost always associated with sterility. In cases I have watched for some time, I have been induced to believe that the organ was atrophied and rendered less vascular and erectile; probably on account of a deposition of fibrine throughout the general structures of the uterus. I have not been able to verify my opinion in any case by dissection. The scantiness is sometimes attended with irregularity, which consists in postponement or lengthened intervals. I treated one patient for endocervicitis in whom the uterus did not appear to be, as far as I could measure it per vaginam, more than one inch and a half in length, and correspondingly small in the other dimensions. This patient would menstruate sometimes only a day every month, and discharge but half an ounce of blood each time, and occasionally the discharge would not return for five, six, and even nine months. In early life her menses had been regular in quantity, quality, and times, and unattended with pain. She was barren, having never conceived, as far as she was aware. She dated the beginning of her disease from vaginitis during an attack of fever, which occurred two or three months after marriage. (See Amenorrhœa.)

Function of Generation affected by it.—The great function for which the uterus was formed, that of generation, seems very frequently to be disturbed by inflammation of the neck of the uterus. Some practitioners think, because a woman bears children with frequency, the uterus cannot be much diseased. This is unquestionably a mistake. I have known many women with extensive ulceration bear children very frequently. Conception may be entirely prevented by inflammation, or gestation may be arrested by miscarriage, or labor may be rendered difficult by it. It has already been stated that many women will bear children, having at the same time very considerable disease of the uterus, but there is always great liability to embarrassment of the function in such cases. There is no doubt that many cases of sterility depend wholly upon inflammatory action about the neck.

Sterility.—Sterility is attended by different circumstances. Some women are sterile their whole lifetime; others, after having borne children to the full period and giving birth to them, become sterile for years, or for the whole of their subsequent life; others again become pregnant soon after marriage, miscarry at an early period, and never again conceive. In most cases of sterility which I have had the opportunity of examining, I have invariably found evidence of inflammation in the cervical cavity. Very often the inflammation is confined to this cavity. The history of these cases showed that ulceration and inflammation had existed from the time of menstruation; these were cases in which conception had never taken place. In cases of sterility in which the women have become sterile after having once borne children, ulceration is usually situated around the os, extending upwards into the cavity of the neck. This is almost certain to be the case if the woman has borne several children. When the patient has miscarried but once, there is not likely to be external inflammation to any great extent; but if there have been several abortions, the ulceration is apt to creep out and manifest itself upon the labia uteri, and sometimes becomes very extensive. Although the foregoing statements with reference to the position and extent of ulceration in sterility will generally be found to correspond with the appearances, yet we must not be surprised to find pretty extensive ulceration external to the os uteri in the originally sterile patient; and in those who have borne children and become sterile afterwards, we shall sometimes find no external ulceration. The result of my observation is, that when sterility originates in uterine inflammation it is in that form of it known as endocervicitis. Sterility is oftener associated with the condition and quality of the leucorrhœal production than on any apparent incapacity of the uterus. In many of these cases the secretions from the vagina are very abundant and intensely acid, so as to produce irritation of the external organs. Although the semen is diluted and defended from the influence of acid vaginal secretions by mucus of alkaline reaction, yet when these vaginal secretions are abundant and possess strong chemical qualities, they may destroy the vitalizing influence of the seminal fluid, and thus prevent fructification. Or the very thick, tenacious, albuminous fluid which sometimes plugs up the os uteri and whole cervical cavity, may prevent the ingress of the spermatozoa, which, by their independent motion, according to present belief, penetrate the uterus, meet the ovum somewhere on its passage to the os uteri, and produce their fructifying influence upon it; and thus is precluded the possibility of effective insemination.

Abortion.—But conception may readily occur and pregnancy be complete, and after gestation has continued for a certain time abortion may take place. Abortion is a very frequent effect of inflammation and ulceration of the os and cervix uteri. The seat of inflammation or ulceration which most frequently induces it, is inside the cervical cavity. We find some patients who have aborted very frequently and never had a full term child; others, who have had one or more children, but who miscarry every pregnancy afterward; and again, others who miscarry frequently and occasionally go to full term. It is not strange that miscarriages should result from this cause; à priori, miscarriage might be regarded as its necessary effect. Many patients bear children at term who labor under severe ulceration, and who are prostrated by the constitutional sympathies accompanying pregnancy.

Conditions of the Uterus in Abortion.—Two general conditions of the uterus exist as the effect of the cervical inflammation, and are probably the proximate causes of abortion, viz., congestion or arterial injection of sufficient strength to cause hemorrhage; and, perhaps, by means of insinuation of the clots, separation of the placenta, or irritability of such a nature that contraction and expulsion follow conception; or, perhaps, increased sensitiveness of the mucous membrane may increase its excito-reflex influence so as to arouse uterine contraction, and thus cause the foetus and membranes to be expelled. When abortion is caused by congestion, it is apt to be ushered in by hemorrhage. The hemorrhage, after continuing for a varied length of time, from a few hours to several days, is followed by uterine contractions. When abortion is the result of increased irritability, the first symptom is contraction, with the paroxysmal pains attendant upon it. This continues for a time, when hemorrhage and expulsion succeed. When abortion occurs once, it is very likely to recur in every subsequent pregnancy about the same time until the disease is cured upon which the abortion depends. While abortion is very likely to recur in the congestive or hemorrhagic variety, it is generally not so exact in the time of recurrence. This variety, however, takes place more frequently at the time when the monthly congestion is present, while the other is independent of such influence. The probability is, that in the congestive variety the foetus perishes before expulsive efforts arise; while in the other the foetus is not affected until the contractions have continued long enough to partially separate the placental attachments. Whatever doubt, however, may be cast upon all this, there can be no question as to the injurious effect produced upon gestation by ulceration or inflammation of the cervix uteri. Mr. Whitehead, of Manchester, England, has written a book, full of information, almost solely to illustrate this consequence of uterine inflammation.

Effect upon Labor.—The effect which inflammation of the os and cervix uteri exerts upon labor is not so apparent as upon the progress of gestation. Although I have watched patients whom I knew to be laboring under inflammation of the neck of the uterus in parturition, I have not been able to perceive any increase in suffering or tediousness.

Even when induration and hypertrophy were both of several years' standing, no ill effects from them, so far as I could see, attended labor either at full term or prematurely. I have observed cases of abortion occurring in such patients quite as readily and without more troublesome symptoms than in one whose uterus was healthy. The general tissual changes going on in the uterus would lead us to expect this in advanced pregnancy, but I confess to some astonishment at having seen kindly, rapid, and complete dilatation in abortion at the early periods. It is equally singular to see the return of the induration after the involution of the uterus is fairly completed. One would suppose that the softening accompanying pregnancy would be permanent, and this is the case with indurations of recent date. I have not observed in such cases that the abortions were attended with more hemorrhage, or were more tedious or painful than when they occur as the result of some transient cause.

Effects upon 'the Post-partum Condition.—Of its effects upon the childbed, or post-partum condition, a favorable opinion cannot be given from my observation. A good getting up is not to be expected with much confidence in patients affected with uterine disease. The most common effects in childbed is retardation of the processes of involution. The congestion consequent upon labor is protracted, the uterus remains larger and more sensitive than is usual, so that instead of the organ recurring to its primitive dimensions and susceptibility in one month, two or more may be required. The lochia, instead of subsiding in fourteen or twenty-one days, continues for weeks, or even months, after it should have subsided; and when it goes off, it is apt to merge imperceptibly into leucorrhœa, which becomes pesistent. Inability to walk or stand without great distress is the effect of the size and sensitiveness of the organ. A sense of bearing down, or of weight in the pelvis, pain in the sacrum, down the sciatic nerve or in the hip, harass the patient greatly, and these symptoms pass off so slowly that she is kept in bed an unusual length of time. Acute metritis not unfrequently supervenes, or acute inflammation of the cellular tissue at the side of the uterus. Phlebitis, pyæmia, and phlegmasia dolens are more likely to arise in patients who have chronic inflammation of the cervix.

On the other hand, it is a fact that these subsequent acute inflammations sometimes operate very favorably upon the cervical inflammations. Instances are not uncommon of patients being entirely cured of ulceration by the effects of gestation and labor upon the tissue of the neck and its mucous membrane. We are to hope for this favorable result only as a remote probability, because, as already stated, the condition of the parts is generally left in *statu quo*, or, if any difference is perceptible, it consists in an aggravation of the disease, and the patients get up from childbed rather worse than better.

CHAPTER XII.

CHRONIC INFLAMMATION OF THE UTERUS AND CERVIX.

ETIOLOGY.

Sexual Indulgence.—The unnatural social habits of woman, and the circumstances which surround her, render her extremely susceptible to uterine disease. Coition, indulged in by the lower animals only for the purposes of generation, and periodically, at long intervals, is resorted to by man as the most common indulgence of his lower nature. The continued and extreme excitement in the sexual system ruins many of both sexes, but it produces the most disastrous effects upon women for obvious physiological reasons.

Improper Reading.—Less powerful but still efficient sexual excitement is found in the influence of lascivious books, so generally read by the young, as well as in the nature of the associations connected with most of the amusements of society.

All this is aided by heated rooms, stimulating diet, improper clothing, &c. At or near the periods of menstrual congestion these excitements operate with much more efficiency than at any other time.

Cold.—During the menstrual congestion, the application of cold to a large portion of the surface is also a fruitful source of uterine inflammation in very young girls.

Constipation.—Chronic and obstinate constipation keeps up a predisposing uterine congestion, and I have long since been led to regard continued constipation as a condition the most deleterious to female health.

Standing.—Constant standing also produces much evil; it is much worse than walking, or even than going up and down stairs.

Abdominal Supporters, Pessaries, &c.—Pressure upon the abdomen by miscalled uterine supporters, the improper use of pessaries, sponges, &c., may be enumerated as causes in certain cases of uterine inflammation. There can be no doubt, also, that prolapse of the uterus, when considerable, and other displacements, are sometimes the cause of inflammation of that organ. This, however, is a rare occurrence, as I think displacements are much more frequently the effect than the cause. Circumstances occur which may mislead us, if we are not careful, as to the proper relation between displacement and inflammation.

Severe Exertion, Jolts, &c.—We not unfrequently meet with patients who tell us that they were "perfectly well" up to the time of some severe exertion, jolt, or lift, when suddenly they felt something give way in the lower part of the abdomen succeeded by pain in the back, hips, loins, groin, accompanied by a sense of prolapse and weight upon the perineum. Soreness and great permanent inconvenience persist, thereafter, until the case becomes chronic. In such cases, the patient dates the beginning, not only of her trouble but her disease, from the strain or jolt, and believes it to be the whole cause of her disease. A critical inquiry into the history of the case will convince us that inflammation had preceded the accident, and that the uterus was probably rendered susceptible of the sudden depression by its increased size and weight. However this may be, the inflammation is greatly aggravated, if not originated by the circumstance.

Hemorrhoids.—The turgidity of the pelvic vessels, kept up by hemorrhoids, prolapse of the rectum, vagina, or bladder, or inflammation of any of these organs, must contribute largely to swell the number of uterine cases.

Pregnancy.—Although pregnancy is a physiological condition, and, in the nature of things, ought not to even predispose to disease of the uterus, recent investigation seems to indicate it as a prolific cause of ulceration. Dr. Cazeaux and other French obstetricians have examined a large number of cases of pregnancy, with a view to determine the frequency of ulceration in this condition; and having found ulceration almost always present they have determined that the leucorrhœa of pregnancy caused ulceration of the uterine mucous membrane. As well might we expect to see ulceration of the bladder in consequence of diabetes mellitus, or ulceration of the skin in diaphoresis. Inflammation, undoubtedly, has the effect in this case, as in all others, of giving rise to the profuse and perverse secretion of mucus as well as the ulceration. There is no doubt but that, in consequence of the dependent position of the uterus in its relations to the vascular system, it is more liable to congestions of both a transient and persistent character than any other viscus, not even excepting the rectum. These congestions are the predisposing conditions of inflammations generally, and if they are persistent and long continued, excite as well as predispose to inflammations. Constipation, standing on the feet for a long time, tight dressing, &c., act by impeding the upward tendency of the blood, causing it to leave the pelvis tardily, and thus keep hyperæmia in the uterine vessels until organic disease occurs. I cannot but believe that anything which will keep up these congestions for a sufficient time will bring about inflammation of the uterus in some part.

Abortions.—Abortions are both the cause and effect of inflammations of the uterus. It is unnecessary to point out the deleterious effects of abortions produced by violence.

All the circumstances exist that are required. The violence is nearly always sufficient of itself at once to give rise to more or less acute disease. In cases occurring from other causes than intentional or accidental violence, there are many efficient causes of congestion and inflammation. Probably one cause not usually thought of, is the too early assumption of the erect posture.

Bad Management after.—Being nothing but an abortion at an early period, it is not considered important by the physician that the patient keep the horizontal position; the patient sits up, walks about, &c., and the congestion existing continues sufficiently long to produce inflammation. Now, I think it is quite as necessary for the patient to remain quiet in bed, until involution is well advanced, in cases of abortion, as in labor at full term. Many of the conditions inducing inflammation in cases of abortion are the same as arise in parturition. I shall, therefore, speak of them under that head, and the intelligent reader will at once perceive them as they are brought forward.

Labor.—The uterus, at the time of labor, is predisposed to vascular disease, on account of its extremely vascular condition; when labor comes on the excitement is tumultuously increased, the nervous susceptibilities are enhanced, while the forcible contraction of the muscular part of its composition greatly adds to its excitation. When we remember the powerful compression of the os and neck by the child's head in passing through them, and even the frequent lacerations to which the mouth is subjected, it is astonishing that nature is competent, under the circumstances, to so completely restore so many parturient women to their former condition of perfect uterine health.

Decomposition of Productions of Labor.—In addition to all these, however, there is generally more or less decomposition of organic matter in the vagina, near the os and neck, giving rise to irritant products which, without proper cleanliness, might remain long enough in contact with the highly sensitive parts to cause inflammation. I know that nature should, and may, in most instances, safely be trusted to repair all the damage done in these ways when other circumstances are favorable; but these favorable circumstances are often wanting. The erect posture is too early assumed in many women, on account of their necessitous condition, or thoughtlessness and ignorance. This prolongs congestion of the dependent uterus, arrests or retards involution, and excites the uterus to inflammation; this inflammation is often prolonged by the continuance of the same cause until it becomes a fixed condition. The number of circumstances which cause and increase inflammation, in cases of parturition particularly, will be seen and understood without dwelling further upon them. We should remember them, and give our best care to patients passing through the conditions of the lying-in month, and thus avoid much suffering.

Vaginitis.—Inflammation originating in the vagina often spreads to the neck of the uterus, and occupies its mucous membrane externally, passes into the cavity of the cervix, and often, I think, to the cavity of the body of the uterus.

Gonorrhœa.—Gonorrhœal vaginitis is very prone to do so, and if not arrested while yet in the vagina, and that soon after its commencement, the neck of the uterus is seldom left without permanent damage; and after gonorrhœal vaginitis is cured, it is frequently the case that the cavity of the neck is left inflamed. This may, and I think generally does, become chronic, unless removed by appropriate applications made directly to the membrane.

There is reason, too, for believing that the vaginal inflammation, in which profuse leucorrhœal discharges originate, arising from other than contagious causes, may pursue the same upward course, and leave behind the same grave chronic difficulties. It is well known that vaginal discharges are sometimes the result of general conditions, such as the scrofulous, for instance, so that we may have scrofulous vaginitis, and this may spread to the mucous membrane of the genital canal. Vaginitis may also arise from immoderate coition, masturbation, or the introduction of foreign bodies from perverse habits. What I have said above of the effect of vaginitis in causing cervical inflammation of the uterus, was intended to apply more particularly to this disease occurring in adults; but there is another condition under which it occurs, that I think has escaped the attention of medical men, or at least has not attracted sufficient notice, viz., the vaginitis of children.

Vaginitis of Children.—I think I have observed several instances in which, before the appearance of the menses, the cavity of the cervix must have been affected with inflammation extending from the vagina. Indeed, if the history of patients who very early commence to complain of signs of inflammation of the cervix be properly traced, it will be nearly, if not always, found that they were to some extent the subjects of leucorrhœal discharge during their childhood. The kind and locality of the disease arising from infantile vaginitis is almost peculiar. It is situated inside the cavity of the neck, and if the os uteri is examined with the speculum, when the disease is not great there will be found but little, if any, unnatural appearance, save the issuing of muco-pus from it. The os is often contracted in size; it is very seldom enlarged.

These young patients do not generally complain much of suffering from their vaginal inflammation until the commencement of their menstrual visitation, when they have severe pain at each time. The suffering ordinarily increases as the functional activity of the uterus increases, until the patient is a confirmed sufferer with dysmenorrhœa or menorrhagia. At other times, instead of having much direct uterine suffering, the general nervous system is most affected, or the vascular or nutritive systems become seriously deranged at the period when the menses should appear. It is not the usual opinion, but I am, nevertheless, inclined to the belief that chlorosis and chorea are sometimes the effect of derangements thus produced.

The above short and imperfect sketch of the causes of inflammation of the mucous membrane of the uterus will give but an inadequate idea of the vast number of causes which produce the inflammation in question. There is no mucous cavity in the body that is subject to so many causes of intense excitement, arising from the nature of its functions, from its accidents and abuses, as is that of the cavity of the female genital canal. Hence it is not wonderful that this cavity is very much more frequently the seat of disease than any other mucous cavity in the human body.

CHAPTER XIII.

CHRONIC INFLAMMATION OF THE UTERUS AND CERVIX.

PROGNOSIS.

A JUST estimate of the chances of making a cure, or of spontaneous recovery; or, in other words, correct notions of the prognosis of a disease in any given case, has necessarily great influence upon our treatment; and a truthful prediction of the progress of a case, or of its ultimate result, has an important relation to our reputation and to the confidence of our patient. It is especially important to be able to give a reliable prognosis in cases in which the profession as well as the patients are not perfectly satisfied about the pathology and therapeutics in reference to them. Too unfavorable an opinion discourages our patient, and precludes us from having a fair opportunity of exercising our efforts; too favorable an opinion, one not justifiable by the result, brings disappointment to the patient, injures the reputation of the practitioner and the profession, and is also apt to influence improperly the inexperienced medical man against the treatment adopted. The general principle that should govern our prognosis is temperance. We should temperately encourage our patient, if we can conscientiously do so, and if our judgment will not allow us to do this, we should express, temperately and cautiously, an unfavorable prognosis; and hope should never be extinguished until a patient is moribund. Too many good reasons will suggest themselves for the last course to require any argument in support of it. What I have said of a guarded prognosis, and the necessity of not giving a sweeping and absolute opinion, seems to me peculiarly applicable to the disease of which I am now treating. Physicians have not all been convinced of the propriety of treating uterine diseases with the speculum; a large number are entirely, and conscientiously, opposed to it. They are made so, undoubtedly, by the failure of local treatment to fulfil the hope originated by its most ardent advocates. It does not do what they are told it will do; it certainly does not in all cases. The only grave error I think committed by that benefactor of womankind, Dr. Bennett, in his work on the Unimpregnated Uterus, is that his book leads his readers to believe that he scarcely, if ever, fails to cure his cases. This is the impression made upon most physicians who read his book. However true it may be, with reference to the practice of so able a master, I think it would be an unjustifiable expectation on the part of the profession at large. From what I have heard and read of the opposition of medical men to local treatment in uterine disease, I think this unrealized expectation of success from local treatment, is one of the main causes of it. Upon trial, medical practitioners become disappointed with the results as they were led to expect them, and abandon the plan as a failure. While I cannot coincide with Dr. Bennett as to the almost universal success of local treatment for uterine inflammation, I am of the opinion that it is greatly superior to any other with which I am acquainted. Prognosis must depend for its reliability, to some extent at least, upon a correct and complete diagnosis of the whole condition of the patient.

Uncomplicated Case Favorable.—The probability of recovery of health will depend upon the absence of any important general diseases in conjunction with the local. We should remember that the patient aims at recovery of health, instead of merely the cure of any one part of the ailments. An important matter is to determine the pelvic complications, if any exist, and how far they are curable, before we pronounce a prognosis.

Prognosis without Treatment.—What is likely to be the progress and result of the disease when allowed to go on without interference? Generally, it will go on from bad to worse. This is particularly the case with the childbearing woman; it is almost equally true of the menstruating. unmarried woman. In the latter, however, if she avoids the causes which aggravate it, she may not get worse; but if her situation, or her inclination, subjects her to the aggravating causes, she will also become worse. Not unfrequently. the patient recovers after the "change of life" takes place. The cessation of the menstrual congestions, if other things are favorable, seems to determine a gradual recovery. This I fear, however, is far from being as frequently the case as we might suppose from reading, and judging physiologically of the matter: Indeed, some of the most obstinate cases I have met with were patients in whom the disease had outlasted the change of life.

Not often directly Fatal.—Notwithstanding the tendency of the disease to get worse during the whole menstrual life of the patient, and to subside only with the subsidence of uterine activity, it seldom proves fatal directly. Nor do the most common and immediate effects of it proceed to a fatal extent. The debility, the imperfect or perverted hæmatosis or the nervous energy, seldom becomes so great as to be the immediate cause of death. This, however, sometimes does occur, and we should indulge a false security to suppose that our patient could not thus die. I think I have seen more than one instance of death thus resulting. The nervous and muscular centres very rarely become so influenced by perverted innervation and hæmatosis as to assume dangerous or even fatal complicating conditions.

Indirectly Fatal.—As very correctly stated by Dr. Bennett, such an unnatural condition of the nervous system and blood is engendered by the disease as to destroy the capacity of the patient to resist or ward off the attacks of the acute diseases to which she may be exposed, or the chronic ones for which she may inherit a strong predisposition. It is difficult, also, to resist the belief, although I have not verified it by observation, that puerperal fevers and post-partum affections are more likely to occur and assume a dangerous or fatal state in patients affected with chronic uterine disease. I need hardly mention the increased hazard to married women from abortions, and the diseases intercurrent with them.

Prognosis in different Varieties.—There is some difference, other things being equal, in the gravity or seriousness of different varieties of inflammation. Some produce much worse effects upon the constitution, are more obstinate and protracted in their duration, and even resist treatment with greater persistence than others. When the inflammation is confined to the mucous membrane outside the os uteri, the prognosis is most favorable; if the inflammation exist in the mucous membrane of the cavity of the cervical canal, it will be more obstinate and difficult to eradicate; and some of these cases are exceedingly so. In the class of cases in which the inflammation has extended to the submucous tissue, the prognosis, so far as a perfect cure is concerned, is unfavorable; it becomes especially unfavorable when the inflammation has lasted so long as to materially alter the shape, size, and consistency, by deposition of fibrine, of the neck of the uterus. In these cases the inflammation is not all that has to be encountered, but the organic alteration must be corrected. This cannot always be perfectly done. If the neck of the uterus is indurated, enlarged, and nodulated, we can only partially restore the organ to its original softness, evenness, and size; and to do this requires a long time, and patient and judicious management.

Prognosis under Treatment.—In cases of uterine inflammation and ulceration in general, what is the prospect under properly conducted treatment? The prospect of cure is comparatively favorable. I mean by this statement that, compared with other diseases which produce as much suffering, the prognosis, under proper treatment, is quite favorable. What the per cent. of cures would be if summed up, I could not say; but it is large. A more circumstantial consideration of the prognosis I think would be profitable. With reference simply to recovery or death, the prognosis is favorable, because, even when a cure is not effected, as we have seen, it is not usually fatal.

Can the Inflammation be always removed, and if removed, will the local symptoms always subside?—The local inflammation can nearly always be removed; but with its removal, the local symptoms do not always leave the patient.

Will the several Symptoms always subside?—The inflammation, so far as we can see, may generally be removed; but many of the symptoms, as the pain in the back, groin, or elsewhere, may persist, to the great discomfort of the patient. I have endeavored to show that many of the symptoms depend upon the congestion kept up in the whole of the pelvic organs; and that these congestions are not unlike those produced by the menstrual molimen, and that this persistent congestion depends upon the presence of the inflammation in the cervix and os uteri. This congestion sometimes outlasts the inflammation, and thus keeps up some of the local symptoms. But by far the most frequent reason why the local symptoms do not subside, is the persistence of inflammation to some extent. This may be out of sight, and consequently undiscovered; but if it is mucous inflammation, we may know that it is not cured while there is a superabundance of mucous secretions or vitiated mucus or pus in view. If it is submucous inflammation which still exists, there will be unnatural tenderness when touched by the finger or instruments. This tenderness being unnatural, would indicate still some inflammation.

How long will it take to cure the Inflammation?—In what length of time can we reasonably expect a cure of the local inflammation? No certain answer can be given from any mere observation of the case in the beginning. From three to twelve months should be the latitude given in most instances. A shorter time than three months is uncommon; and we might in many instances not reasonably expect a cure in twelve. In order to fulfil the expectations of the patient and of ourselves, we should take plenty of time, and we should not lead our patients positively to expect a removal of all the symptoms when the treatment has terminated; for they sometimes subside so slowly that they continue many months after the treatment has ceased. The general or sympathetic symptoms sometimes become a sort of habit, and continue after the disease which called them into existence has been cured. This is particularly the case with the great degree of general nervousness which renders some patients miserable; but in the great majority of cases they do subside very readily as the ulceration is cured. When they do not, judicious treatment directed to them will do more for them, after the ulceration is removed, and will almost invariably relieve the system of them. For the removal of these general symptoms, time is an item of the utmost importance, and we do not do justice to our own reputation or to the patient, by fixing the time too positively in which relief may be expected.

Prognosis influenced by Age of Patient.—The age of the patient, I have thought, had a good deal to do with the readiness and completeness of recovery from all the troubles of uterine disease. Young women will recover quicker than the old; the naturally robust and active woman sooner than the delicate and inactive. If there is an hereditary predisposition to insanity, or any general nervous disease, there is great likelihood of its being excited into activity by uterine irritation; and when once started, they are apt to assume a permanent and durable form. We should not, therefore, promise too much in patients whose general health has been long seriously affected, as it is impossible to predict the measure of benefit to be derived from a removal of the cause of the general affection in the cure of the local.

How and when does Relief come in Favorable Cases?—But, in cases in which relief from the general and local symptoms readily succeeds the treatment, there is considerable difference as to the mode in which the relief comes. In very many cases the patient experiences benefit from the beginning. In the first month she feels the cessation or a great amelioration of some of her symptoms, generally of the local pains, and she continues to improve until entirely cured. In other instances, the symptoms are aggravated for several weeks, and there is no improvement until after the local treatment is discontinued; again, relief does not follow for some months, and yet by judicious general management it is secured. In a great majority of cases, we may very plainly see the beneficial effect of our treatment, if not before, certainly by the time we have procured the complete resolution of the local disease. As I have before intimated, the general sympathetic effects are sometimes kept up by local complications, and will subside only when they are removed.

Will the Functions be Restored?—An important part of prognosis, one in which our patient often feels a deep interest, is the determination of the prospect of restoring the functional derangements of the uterus. As it has been before stated, inflammation and ulceration of the cervix uteri often cause sterility. This condition occurs under two different sets of circumstances; in one, the patient never conceives after marriage, and may remain sterile during her whole lifetime; in the other, she conceives and miscarries, or even goes to full term of pregnancy for one or more times; and then, as the inflammation and ulceration become established, she ceases to become pregnant. Where the patient has been married for several years, and does not become pregnant, the cure of the disease is not generally followed by productiveness; and when it is, it is usually after the lapse of a long time, sometimes amounting to several years. Although this is the most common condition of the functions, sometimes, after treatment, fertility is the immediate effect of a cure. I have noticed that patients who remain sterile in this way, usually have a very scanty menstruation; and

12

we cannot reasonably hope that our patient will cease to be sterile when she is cured, if this very scanty menstruation is not, or cannot be corrected.   I am inclined to think that in cases of this kind the ovaries are in some way, probably by chronic inflammation, also rendered unfit for their duties.   Those who are restored to fertility by the cure of the inflammation, always, or nearly always, have a normal condition of the menstrual flow.

Patients who have conceived and miscarried, or borne children, but become sterile, are usually cured of their sterility with the cure of the disease of the uterus.   Yet repeated instances have come under my observation, where a miscarriage soon after marriage has resulted in permanent and incurable sterility.   Almost all these cases were abortions induced by forcible means.   The inflammation seemed sufficiently intense to destroy the capacity of the uterus for lodgment of the fœtus; or, at any rate, to render that organ in some manner unfit for the discharge of its part of the function of generation.   If a woman has had several miscarriages, or borne a number of children, and then becomes sterile, there is great reason to hope that she will at once become fruitful as soon as the inflammation is cured.   This result will be the more likely if menstruation retains its natural characteristics.   The habit of miscarrying is generally quite effectually broken up by the cure of the disease on which it depended, so that we may pretty confidently assure our patient that as soon as the inflammation is cured, pregnancy will go on uninterruptedly to the full term.   We should, however, promise this only of future pregnancies; as, according to my observation, a cure undertaken during the existence of this condition is not very promising, although we have good authority for making the attempt.   I am not satisfied that the attempt is always best to be made, and generally wait until pregnancy is over, and even stop the treatment if I have begun.   I could cite many cases corroborative of the statement that habitual abortion is cured by the relief of the inflammation.   It will not be amiss to state the result of my observation as to restoration from the menstrual deviations which attend, and for the most part depend upon the diseased uterus.   It may be stated that, generally, this restoration takes place, but it certainly does not always.   I think different sorts of menstrual trouble are differently influenced by the cure of the inflammation.   Scanty menstruation often remains permanent after the cure of the diseased cervix, and much more frequently resists treatment than any other derangement. Where the menses have been wholly suppressed, we may hope for better results from judicious management. In fact, the stimulation of the uterus generally restores this function when absent on account of chronic inflammation, unless, as is sometimes the case, so much organic alteration has been brought about as to destroy, to some extent, the texture of the organ. Menorrhagia often continues with considerable obstinacy after all the disease of the cervix is removed; but it is nearly always much moderated, and quite frequently entirely cured, and ceases to trouble the patient before the inflammation has wholly disappeared. Where it is obstinate, it will nearly always be found to be the case that after the lapse of a few months it begins to improve, and after a while the menstrual discharge will not exceed the natural quantity. I think we may pretty confidently hope that by the exercise of a little patience we will cure this functional disorder of the uterus, where it has depended upon inflammation of the cervix.

Dysmenorrhœa, when dependent upon this cause, disappears often very readily under the influence of treatment directed to the cervix, but we should be careful to distinguish between it and that which depends upon other causes. Very commonly one of the first good effects of local treatment is to ameliorate the suffering during the menstrual discharge. This is often remarkably the case, the first menstrual effort being so much better as to astonish the patient and her friends. It would hardly be justifiable, however, to promise, generally, such ready relief; for sometimes this feature of the case remains quite obstinate, and causes the patient a great deal of suffering after the inflammation is entirely cured.

Complicated with Phthisis.—In the course of my practice, it has occurred to me to have cases complicated with tuberculous disease of the lungs, and some of these patients have seemed to run down more rapidly after their recovery from the uterine disease than before, on account of their softening and discharge. I have not had an opportunity to observe a sufficient number of such cases with that scrutiny so necessary to arrive at a correct conclusion. It might be supposed that, on account of the derivative influence of the uterine disease, the consumption was kept in abeyance by its continued existence. On the other hand, the debilitating effects upon the system at large, which it undoubtedly exerts, might with equal propriety be expected to co-operate in the general prostration.

Throat Disease.—The frequent complication and the effects exerted by the one upon the other of throat affections,—pharyngitis and laryngitis,—and uterine affections, makes it a matter of interest to determine what, if any, is the effect upon the diseases of the throat, of curing ulceration of the cervix uteri. This may be regarded by most of my readers, and probably is, an irrelevant question in this connection, but I think careful attention to it will lead to a different way of thinking about it. I am persuaded that some, at least, of the chronic sore throats of this climate can be much more easily cured after the uterine complication is removed. Women often believe that there is an intimate connection between them, and hope that the cure of inflammation of the uterus will relieve the throat; and I have seen cases in which I was almost ready to believe there was some encouragement for the opinion.

Skin Diseases.—Psoriasis, lepra, and some other of the chronic forms of scaly eruptions, coexisting with inflammation of the cervix uteri, have been aggravated or ameliorated as the uterus grew better or worse; when the uterus is better the eruption is worse, and the converse. I have noticed several cases in which this seemed to be unequivocally true; and it is remarkably the case in two patients now under my observation. Without my speaking of it, they have both remarked it. If this observation should prove true on a large scale, it would indicate the extension, in a modified form, of this chronic inflammation to the mucous membranes, and afford us a valuable hint for the appropriate mode of managing a class of very obstinate cases. The above facts have an important and direct bearing upon our prognosis; for, according to my experience, the cases attended with these chronic skin diseases are very obstinate and protracted.

Cure remains Permanent.—When cervical uterine inflammations are once cured, are they likely to return? It is a popular belief that these uterine diseases cannot be permanently cured; that they will keep returning. This belief is, no doubt, supported by the fact that many of our patients are constantly laboring under the causes that originally produced the affection; and, therefore, are likely to have them reproduced. Of course, patients thus situated will have a return of the diseases, but where the causes can be avoided and the cure completed, there is no reason why they will not remain cured with as much certainty as any other disease susceptible of perfect removal. I cannot refrain from here expressing the opinion, however, that a large majority of the cases that thus thwart our hopes never are entirely cured, and I believe great discouragement arises from want of the experience necessary to decide when the disease is entirely removed. I have met a number of instances in which the practitioner supposed he had removed the inflammation, but the symptoms remained; where an examination revealed a discharge of muco-pus from the mouth, showing inflammation still remaining inside the neck, and discoverable only by the discharge.

CHAPTER XIV.

COMPLICATIONS OF INFLAMMATION OF CERVIX.

Complications.—Various and troublesome intra-pelvic complications are often observed in connection with uterine disease. These complications for the most part arise during the existence, and generally as the effect, of the disease of the uterus; they may, of course, also arise as independent affections. Notwithstanding the frequent secondary origin of these complications, after they have continued for a considerable length of time, some of them become permanent, and after the originating disease has subsided, they go on indefinitely if not cured.

Vaginitis.—Probably the most common of them is vaginitis, in some form; ordinarily in that of erythematous inflammation of the mucous membrane, which is indicated by an increased mucous discharge, some tenderness and heat. Instead of the inflammation being thus moderate, there may be copious muco-purulent discharge, great irritation, and so much tenderness as to render an instrumental examination very painful, and often unsatisfactory. Such severity of inflammation is apt to be of short duration, and dependent upon some superadded cause of the inflammation. The inflammation is usually more moderate and persistent, continuing more or less for weeks or months together. Another form of complicating vaginitis is eruptive, and, although not usual, it yet sometimes accompanies the simple variety. The eruption in the milder form is vesicular. Small vesicles appear somewhat thickly studding the inner surface of the labia, on the nympha, the membrane of the vestibule, and sometimes the cutaneous surface on the edges of the labia majora and the anterior edge of the perineum. This eruption is attended with great heat, or a burning sensation, and not unfrequently with intolerable itching. The vesicles are not very thickly set upon the surface, but the latter is of a fiery red color. A greater or less amount of serous discharge keeps the parts wet and sticky. Almost always this mild eruptive variety is paroxysmal, and generally appears simultaneously with the commencement of the menstrual discharge, and has seemed to me to be dependent upon the acrid discharge accompanying it, and the congestion present at such times. The eruptive variety of vaginitis is sometimes much more severe in grade, and the vesicles are changed to pustules, and the accompanying inflammation much greater. Fortunately this is not nearly so common as the first two forms. Vaginitis sometimes has its origin, I have no doubt, in an extension of the mucous inflammation from the neck; but frequently, I think, the inflammation is caused by the acrid irritating nature of the perverted secretions from the mucous membrane of the cervix, and by want of proper cleanliness. The vaginal inflammation, although exceedingly annoying to the patient, is otherwise of much less importance than some of the other complications.

Urethral and Cystic Inflammation.—Urethral and cystic inflammation also not unfrequently result from or accompany cervical inflammation. It is not necessary to give their symptoms in detail. The main fact to which I desire to give expression is, that when there are symptoms of cystitis or urethritis, we should be watchful for the probable occurrence of inflammation of the bladder and urethra, and be aware of the importance of giving attention to them as complicating diseases. For I think I have seen indubitable instances of cystitis and urethritis which could be traced to this cause, continuing after the uterine disease was cured. When not properly attended to, they may induce nephritis. The inflammation of the neck no doubt directly induces inflammation of the bladder, by reason of its immediate apposition to its walls; and while this inflammation ordinarily is of short duration, yet it sometimes becomes very persistent, and even permanent. The attacks, when acute in grade, as they sometimes are, become extremely distressing, and absorb the whole attention of the patient, and demand the prompt interference of the medical attendant. More commonly the grade of inflammation is mild, and confined to the mucous membrane of the organ. The scalding micturition, indicative of urethritis, is often distressing to a great degree, and is not unfrequently very persistent. This urethritis and cystitis I think are caused by migrating inflammation from the vagina in some cases, and the inflammation probably goes on through the ureters to the pelvis of the kidneys. When cystic inflammation is persistent and somewhat severe, it often passes for the disease. The symptoms of cervical inflammation of the uterus being overwhelmed and obscured by the more urgent and distressing vesical affections, it is not thought to be the origin of the trouble. Although the vesical symptoms, as before stated, may become urgent, and the inflammation assume an important prominence in the case, usually this complicating affection is slight, and manifested by very mild and transient symptoms. In this form, cystitis and urethritis are very common indeed.

Cellulitis.—A more formidable, troublesome, and perplexing complication, however, is a chronic or subacute form of cellulitis, as it has been named by Prof. Simpson. It consists of inflammation and suppuration of the cellular tissue contained in the duplication of the peritoneum, at the side of the uterus. I think this is a frequent complication, and more frequent, according to my observation, than we are led to believe by any description I have ever met with. When it is present, it embarrasses our diagnosis, and should very materially modify our prognosis. I have met with instances in which it remained unnoticed, and exercised a very embarrassing effect upon the treatment and the progress of the case for a long time. This complication is important for two main reasons at least, viz., 1st, the great obstinacy with which in the chronic form it resists treatment; and 2dly, from the fact that the pelvic or uterine symptoms do not subside while it lasts, even when the uterine disease is removed. It is likely to occur in two forms, differing considerably in intensity and duration,—the acute and the chronic.

Acute Cellulitis.—In the acute form the symptoms are violent, and run their course somewhat rapidly. The patient, after some exposure, or more than ordinary exertion, experiences a great increase of pain in the pelvis; it usually occurs on one side, and rigors supervene, which are succeeded by febrile reaction of high grade. The pain is constant, and often excruciatingly severe, of a tense and aching character. It is sometimes attended by paroxysmal exacerbation, but it is generally free from it. The fever, pain, and great soreness, continue from six to twenty days, or even longer. The fever gradually becomes more remittent, and finally intermittent, being terminated, or nearly so, every night by copious perspirations. The pain continues, however, until it is relieved by a discharge of pus per vaginam, rectum, or urethrem.

Suppuration in Cellular Tissue.—If the discharge is free and copious, immediate and almost complete relief follows; if, as is much more frequently the case, the discharge is small, the relief is only partial, and the patient lingers in a state of great suffering for weeks, and even months, before the discharge is completely effected, and the cavity of the abscess filled up. During the existence of these acute symptoms, if we examine per vaginam with the finger, we will find the mucous membrane hot and exceedingly tender to the touch.

Diagnosis of Cellulitis.—In seeking to ascertain the relation of the organs, the uterus generally will be discovered situated near one side of the pelvis, and fixed in its position, so that it cannot be easily moved in any direction; or it may be in the middle of the pelvis, and a little lower down upon the perineum. When it is to one side, we may feel on each side of it solid tumefaction, filling up to a considerable extent, if not completely, the lateral and anterior portions of the pelvis; and if we press upon this hard and tumefied part, we shall cause great complaint of tenderness. The patient will cry out with the pain produced by it. If the uterus is central in its position, the hardness, pain, and swelling will occupy one side of the pelvis, and while it will give the patient great pain to carry the finger up the side of the uterus, where this tumefaction is situated, on the other side there will be no tenderness.

Extent of Cellulitis.—These inflammations invade the cellular tissue in the pelvis to a greater or less extent in different cases, and sometimes the infiltration is so great as almost wholly to fill up the cavity of the pelvis. In other cases there is only a very small amount of induration, not larger than the thumb. Now, attacks of the kind above described cannot deceive a careful practitioner; but the milder and less pronounced variety may go unnoticed, unless we are watching for it.

Chronic Cellulitis.—The patient in the milder form is seized with some increase of pain in the back or groin, or elsewhere about the pelvis, which lasts for three, four, or five days; and after a discharge of very little pus, it subsides, and leaves the patient in her usual condition. This mild form may, and indeed often does, occur as an original condition, but much more frequently it follows at some distance of time an acute attack, such as I have above described. However this may be, it nearly always represents a small nidus of chronic inflammation by the side of the neck of the uterus. The chronically inflamed cellular tissue in this region is not so great in amount as to cause any febrile excitement, and in fact attracts but little attention, except when it is aggravated into the suppurative process from time to time. I have met with instances in which suppuration and discharge of pus from a small chronically inflamed point of cellular tissue had recurred every few weeks for twenty or more years. And it is often the case that patients having this inflammation will experience exacerbations every month, before or after menstruation, and thus these attacks may pass for cases of dysmenorrhœa. The frequent discharges of pus with these slightly painful exacerbations should cause us to make an examination, when we may generally find a point of induration and tenderness. The results of the examination will be most satisfactory at the time of the exacerbation, as the parts will be more tender and the swelling greater.

It is not necessary for me here to go any further in the description of this intra-pelvic abscess, as I only wish to call attention to the fact that it is not an unfrequent complication of uterine disease; that the symptoms attending it very much resemble inflammation of the neck of the uterus; and that when it continues after the inflammation and ulceration of the neck are cured, the uterine symptoms do not subside as readily as when these last are not thus complicated. To the inexperienced practitioner it is a troublesome and perplexing complication, and if not particularly cautious, he is betrayed into an unjustifiable prognosis, if nothing worse. Intra-pelvic inflammations of this kind, although occasionally independent and uncomplicated, I think are much more frequently associated with chronic inflammation of the uterus. And I cannot but determine, as the result of my own observation, that they are secondary to the uterine inflammation in a large majority of cases, and caused by an extension of it.

Cause of Cellulitis.—Dr. Bennett thinks increase of inflammation of the uterine tissue, produced by strong cauterization, occasionally the immediate cause of cellulitis. Although this is doubtless true, yet a great many cases occur in which no local treatment has ever been resorted to. I do not remember to have met with but one case in which this could have been the cause, and in that case it did not manifest itself until four weeks after the caustic potassa had been applied, for cervical induration and tumefaction. It is reasonable to suppose, however, that any circumstance which would excite the vessels as this does, might, and most likely would, enhance the probabilities of cellulitis.

Rectilis as a Complication.—The rectum is very often diseased in uterine cases; in fact, it is not often that inflammation of the uterus lasts for many months without affecting the rectum to a greater or less degree. Chronic inflammation of the rectum is quite a common complication with certain kinds of uterine diseases. The inflammation is evinced by the tenesmus, frequent discharges, and the increased secretion from the mucous surface. The symptoms are those usually present when the rectum is inflamed from any other cause. The degree of inflammation will cause quite a difference in the intensity of the symptoms. In very many instances there is moderate tenesmus, causing five or six stools in the twenty-four hours. These are partly fæcal, but thinner than natural, and loaded with mucus; or there may be more tenesmus, with more frequent efforts at stool, less discharge, which consists mostly of mucus, streaked with blood. The discharges from the rectum in bad cases may be more or less purulent in character, or may consist exclusively of blood.

Diagnosis of Rectitis.—Where there is rectitis, it is usually tolerably high in this organ, being two, three, or four inches above the anus; and in our examinations, if we press upon the rectum from the vagina, it is found to be quite tender to the touch, and always empty. It is too irritable to retain fæces for any length of time. So that when we find a mass of hardened fæces occupying the rectum, very perceptible through the posterior wall of the vagina (and we will often find such), we may be pretty sure that the rectum is not much affected.

Stricture of the Rectum.—Another condition of the rectum which is apt to be associated with rectitis in uterine disease, is stricture of this organ; the stricture varying, of course, as to the time it has lasted and the severity of the cause.

Fistula in Ano.—They may both be succeeded and accompanied with fistula in ano. These complications have their own symptoms, and must be investigated and treated as though they were independent affections, while we attempt to remove the cause.

Causes of the Rectal Diseases.—How these three different forms of rectal disease are produced by the disease of the uterus, although not very plain, may, I think, be generally explained. The inflammation doubtless extends from one tissue to another in rare instances, but more frequently, I think, it is caused by the pressure of the uterus upon the rectum. The rectum lies on the sacrum; and the uterus often becomes so heavy that its supports are not sufficient to keep it in place; it settles upon the rectum and presses it against the hard surface of the sacrum, thus irritating it very much, bringing about congestion and inflammation first, spasmodic and then organic stricture, and subsequently ulceration and perforation of the mucous membrane of the rectum. The lumps of fæces or other substances burrow through the rectum in this ulceration, when suppuration and exulceration establish a fistulous opening.

Prolapse of the Rectum.—But without much inflammation the rectum is sometimes prolapsed so that it protrudes from the body, either through the anus or the ostium vaginæ. In long-standing cases of uterine disease, great relaxation of the mucous membrane of the rectum is a frequent occurrence; and then, in every effort at stool, it falls in large folds through the anus, often entangling the fæces in them, so the patient is under the necessity of picking them out before the evacuation can be completed. Or, what is less frequently the case, as the tenesmus of defecation attempts the expulsion of the contents of the rectum, this organ is forced forward into the vaginal cavity, and externally between the labia, so as to form a tumor external to them with the fæces contained in it. The evacuation of fæces from the rectum is very difficult in this complication, and the patient will tell us that she is obliged to introduce her fingers into the vagina, pressing the whole mass backward and downward toward the opening of the intestine. It is astonishing to what extent such displacements of the rectum are carried. Its folds often protrude sufficiently to cause a tumor below the anus or external to the vagina large as a man's fist.

Hemorrhoids.—Hemorrhoids form another disease of the rectum and anus, which complicates diseases of the uterus. They of course will not require a distinct description; their frequent occurrence renders everybody familiar with them. The pain resulting from inflamed hemorrhoids often masks or simulates inflammation of the neck of the uterus; and when they are associated, the cure of either will not remove the symptoms, so that we need not be surprised at their greater obstinacy when they coexist. The prolapse of the rectum and hemorrhoids are the unquestionable results of uterine pressure. The continued congestion kept up in the rectal vessels by the constant pressure of the uterus upon the rectum, hypertrophies the mucous membrane, and causes varicosity of the extremities of the veins, and in this way induces both results. They are, therefore, the *indirect* results of inflammation of the uterus, this last bringing about a change in the position of the uterus, so that in some portions of it it presses the rectum against the sacrum so firmly as to embarrass its circulation and cause the changes above described.

Hypertrophy of the Rectal Mucous Membrane.—The rectum is not only prolapsed, but the mucous membrane is hypertrophied quite largely, before it can appear externally; and in conjunction with this hypertrophy there is also great relaxation of the fibres of the rectum and sphincter ani, or of the fibres of the vaginal walls, to allow the escape of the parts to the enormous extent which sometimes takes place.

Displacements of the Uterus.—The most common displacement of the uterus, where these two last rectal complications are present, is the subsidence of it in the axis of the superior strait. This brings the neck of the uterus straight down upon the rectum, and the whole weight of the uterus rests upon it. This brings me to the consideration of the most frequent of all complicating circumstances connected with chronic inflammation of the uterus, viz., uterine displacements. So frequent are these displacements in this relation, that, as I have before stated, they are regarded as the causes of all the associated difficulties. While I cannot assent to this view of the subject, I believe them to be frequently, if not almost invariably, the effects of inflammation, and am confident they are most important and mischievous complications, and probably give rise to more suffering than any complicating condition whatever. As I have already stated, it is most frequently the displacements that cause stricture, hemorrhoids, and prolapse of the rectum. By the uterus being crowded down upon the rectum, these affections may be produced. It will not be expected that I shall dwell with any great degree of minuteness upon the different degrees or characters of displacements, or give a full description of them here, as I only wish now to speak of them as a complication of chronic inflammation of the uterus.

Subsidence of the Uterus.—The most common displacement I meet with is a subsidence, or lapse, of the organ, while its vertical axis remains what it was before the change of position. This does not bring the uterus, or any part of it, nearer the vaginal orifice; the lower end of it settles down upon the lower bone of the sacrum, while the fundus points upward toward the umbilicus. In examining per vaginam, instead of finding the os uteri upon, or nearly upon, a level with the inferior border of the symphysis pubis, and touched by introducing the finger almost directly backward, it is necessary to bend the finger over the upper edge of the perineum, and carry it back and downward to the lower end of the sacrum. This displacement is very frequent, according to my observation, and does more injury by pressing upon the rectum, and gives more distress than almost any other displacement. It almost always obstructs the passage of the fæces through the rectum, and makes the patient feel as though the bowel was constricted at the point of pressure. After long continuance, it induces, in many instances, organic diseases of the rectum, inflammation attended by tenesmus, mucous and even bloody discharges, hemorrhoids, &c. All these rectal complications above mentioned may arise in this way.

Anteversion.—The inflamed uterus is also anteverted, more or less, in many instances, so that the fundus presses heavily upon the bladder, while the os, higher up than in the first-named displacement, presses the rectum against the sacrum. But as most of the weight of the uterus is upon the bladder and anterior wall of the vagina, the rectum is not so distressed.

Pressure upon the Bladder.—The greatest inconvenience is felt on account of its pressure upon the bladder. Frequent micturition, sense of weight behind the pubis, &c., are its symptoms.

Retroversion.—Retroversion is also not unfrequent as a troublesome complication. As the fundus presses upon the lower part of the rectum and perineum, while the neck and os press upon the urethra and bladder, there is dysuria and rectal tenesmus of greater or less intensity. The symptoms will be modified by the greater or less degree of malposition.

Prolapse.—Common prolapse, with the mouth following the axis of the vagina, is the least frequent of these displacements, as I have observed them. It sometimes occurs, however, to a very great extent, and produces a great deal of distress. Compared with the other forms of displacement, it produces less inconvenience when present in the same degree. It certainly does not interrupt the function of the other pelvic viscera so much as subsidence, retroversion, or anteversion. Where excessive, it gives a sense of dragging and perineal tenesmus that are very disagreeable, but it does not cripple the patient, and render her unable to walk or stand, as is the case with the other displacements. While displacements aggravate the sufferings connected with diseases of the uterus, they render the treatment more difficult, and often imperfect, on account of the difficulty of exposing the os, and bringing the axis of the uterus to correspond with the direction of the speculum.

Theory of Displacements.—I cannot now enter into the theory of displacements as complications of inflammation. I believe they are one of the effects of the pre-existing inflammation; that they are brought about by the inflammation increasing the size and weight of the uterus, and thus causing it to settle down by virtue of its weight in spite of its supports; that the suffering caused by the displacement results from its pressure on morbidly susceptible organs, made so, perhaps, by a long continuance of the pressure, and by the sense of soreness in the inflamed uterus itself, and also in part by traction upon the lateral and round ligaments. Still, I have no question that in very rare instances the displacement results from other causes than inflammation, and then I can easily comprehend how it may produce inflammation in the uterus. The circulation must be embarrassed, congestions will readily occur on account of pressure and forcible flexion of the veins and arteries, and inflammation is very apt to follow long-continued congestion, &c. (See Displacements.)

CHAPTER XV.

Submucous or Fibro-Cellular Inflammation.—Chronic inflammation may originally attack any portion of the uterus, from the fundus to the lower extremity of the neck, and be seated in either the fibro-cellular or mucous tissues. The part of the organ most likely to be attacked, however, is the cervix, and of this the mucous tissue is nearly always the seat of disease. When the inflammation originates or invades the fibro-cellular tissue of the uterus, it is soon followed by enlargement of the portion inflamed. If the posterior wall is thus affected, this is on account of an effusion of serum to some extent; the part is thickened, and as more and more fibrinous coagulation takes place within the cellular tissue, it becomes hard as well as enlarged, and then we have a hard, tender tumefaction in that part of the uterus. When the substance of the cervix is chronically inflamed, with or without coexistent mucous inflammation, it is enlarged, or, as Dr. Bennett has it, hypertrophied, at first not very hard, but if the inflammation continues, there is hardness; hence we have hypertrophy, induration, and enlargement. Hypertrophy is not the word for this condition of things; the part does not enlarge by an increase of existing tissue or a development of more of the same kind, but it is enlarged by an effusion of fibrine, which assumes an imperfect arrangement. It is increased in size in this way and also indurated. This kind of enlargement should be distinguished from the enlargement of congestion,—a condition in which the uterus is injected with an unusual quantity of blood, and its substance distended by it. This is the case every month, but it becomes more permanent by the continuance of some point of irritation which keeps up an afflux of blood, and yet the irritation is so moderate as not to induce that stress of circulation necessary to an effusion in the tissues. We can, therefore, have chronic enlargement of the neck, and even the body of the uterus, without induration or actual structural changes. This is often the case where the inflammation is confined to the mucous membrane. Enlargement is no evidence, therefore, of fibro-cellular inflammation; induration must be superadded to make the whole of the changes necessary to constitute a case of it. When, therefore, we meet with an enlarged and *indurated* uterus, or cervix, we may with safety conclude that it is suffering under chronic inflammation of the fibro-cellular tissue, with certain provisions that I shall have occasion to mention in future. When the uterus is hypertrophied, as in pregnancy, or in consequence of a growth or other substance which causes a development of tissue, the fibro-cellular structure is softer than natural.

Hypertrophy.—The hypertrophy from growth is general, including the neck, body, and fundus; the enlargement from congestion is not always, though most frequently, general; the enlargement accompanied with induration, and indicative of chronic inflammation, is apt to be partial; confined to the cervix, sometimes to one lip of the os uteri, or some part of the body near the neck. When the whole cervix is chronically inflamed, it enlarges in every direction; the thickness is increased from the size of the end of a man's thumb to half the size of his fist, or even larger than this, and it is hard and tender to the touch. The cervical canal is decreased in calibre in most instances, and somewhat lengthened. The induration is not always of the same intensity; its hardness is often very great, at other times but little more than natural. As the induration and enlargement may be quite partial, the shape as well as size of the neck or portion of the body attacked, will seldom present its natural contour. The *proportions* of the different parts do not correspond in shape or size as they do in the healthy condition. Then we have in chronic inflammation of the differents parts of the uterus increase in size, hardness, and disproportion of corresponding parts, and hence alteration in shape, to which is almost always added tenderness upon pressure or touch, particularly with instruments.

Hardness with Atrophy.—Although these statements will be found to correspond with facts so frequently as to constitute the rule with regard to the subject, yet there are important exceptions. I have observed quite a number of instances, in which long-standing inflammation of the body of the uterus seemed to have brought about a shrunken condition of the organ. So that, notwithstanding the presence of all the symptoms, the uterus was very much diminished in size. It appears in such cases also to be indurated as well as decreased in its dimensions. It is barely possible these uteri were congenitally smaller than usual, and what seemed to be atrophy was natural. In two instances I had assurance that in the early part of married life there had been pregnancy and abortions. If this was true in these two cases, they must have been pathological. It has been supposed that the fibro-cellular form of uterine inflammation always precedes inflammation of the mucous membrane for a greater or less length of time. This is certainly not always the case; for we meet with inflammation of the mucous membrane entirely unconnected with the submucous tissue, as a simple affection. They are, however, much more frequently combined than separated from each other.

Mucous Inflammation.—As a simple affection, that of inflammation of the mucous tissue is much the most frequent. Where they coexist, we have the increase of size, hardness, and irregularity of shape, indicating inflammation of the submucous substance combined with the evidence of mucous disease.

Seat of Mucous Inflammation.—The inflammation of the mucous membrane may extend to the whole of it, from the fundus through the cavities of the body and neck to the os, and then cover the whole of the vaginal portion of the uterus. This extent of inflammation is not very frequent, and when it occurs it almost immediately succeeds parturition or abortion, or is produced by gonorrhœal inflammation. I have seen it under these circumstances oftener than any other. It almost always causes a great deal of distress and suffering.

Cavity of the Cervix.—Probably the most common extent of inflammation is to the mucous membrane of the cavity of the cervix, and a portion or the whole of the membrane covering the intralabial portion of the os. By far the greater number of instances that come under observation in practice are inflammation of the membrane around the os and inside the cavity of the cervix. I fear that this statement represents a fact that has not been generally apprehended by practitioners. I am disposed to believe that too many practitioners have failed of success in curing their cases, because they have not followed up the inflammation sufficiently above the os, in the cervix, being satisfied with curing that which was visible only, and, in consequence, leaving really the most important part of the affection untouched.

Cavity of the Body of the Uterus.—Sometimes the inflammation is limited to the cavity of the body, to the cavity of the cervix, or to the membrane in and external to the os uteri. Inflammation limited to the cavity of the body of the uterus is not common, but I am quite sure that I have met with at least two instances. One of these had been treated for inflammation of the os and cervix, and cured of this, but the inflammation in the cavity of the body was left. The other had not had any treatment, as far as I could learn, for uterine disease. She had habitual leucorrhœal discharge of rusty-colored mucus, very much like the brickdust sputa of pneumonia ; the os externum was very small, and the os internum uteri large, as was also the cavity of the body. This patient did not menstruate, and had not for a number of years, and although married, did not become a mother ; the disease was caused by miscarriage in early life. She was thirty-four years of age.

Endocervicitis.—Endocervicitis alone, or inflammation limited or confined to the cavity of the cervix, is, on the other hand, an extremely common form of the disease. Not unfrequently this form of inflammation exists without any appearance of it in the os or external to it. When inflammation of the mucous membrane of the cavity of the cervix alone exists, it has certain effects upon the shape and other properties of the neck that are apt to attract our attention. Dr. Bennett describes the os as patent and the cavity of the neck enlarged, so as to admit the finger and permit the opening of it by a speculum to some extent, so that we may see the inside. Now, while this is very generally the case, it certainly is not always so. This condition of the os and cervix is more frequently met with near the menstrual periods than at any other time, and is probably always owing to the congestion of the vascular tissue of the cervix and about the os.

Endocervicitis with Diminished Size.—I have undoubtedly seen many cases of this endocervicitis, in which neither the os nor cervical cavity were in the least enlarged, and others, in which the os uteri was contracted much below its natural size. The secretions of the mucous membrane are always modified ; generally they are very much increased, and often changed in character.

They may become purulent or sanguineous, owing to the grade of the inflammation and the degree of congestion. The inflammation situated external to the os on the end of the uterus, between the labia or their external surface, is very common, but it is not often limited to this part. It is almost always combined with endocervicitis.

Certain forms of these mucous inflammations are found more frequently in certain sorts of patients.

Endocervicitis in Virgins.—Virgin patients seldom have inflammation external to the os uteri; their disease is endocervicitis almost always; very rarely there is a little rim of inflammation around the os upon the end of the uterus.

Endocervicitis in Aged Women.—Again, in senile patients, women who have passed the climacteric period, and ceased to menstruate for some years, we find the inflammation in the cavity of the cervix.

External Inflammation combined with Internal in Child-bearing Women.—In the married, child-bearing women, we find the external inflammation combined with the internal uterine inflammation of the mucous membrane. They are the kind of patients in whom most frequently the enlargements, indurations, and fibro-cellular inflammations are observed. The form of disease in persons who have been married, but never been pregnant, partake to some extent of the character of both the virgin and the child-bearing woman. They often have external, combined with internal, mucous inflammation, but not often fibro-cellular. Now, what I mean by these statements is, that these kinds of patients are likely to have the forms of disease which I have ascribed to them, but there certainly are exceptions to all of them.

CHAPTER XVI.

PROGRESS AND TERMINATIONS.

THE intensity, terminations, and effects of inflammation upon the parts immediately implicated, of course will vary very greatly indeed. There can be no doubt that suppuration takes place in the fibro-cellular tissue of the uterus, especially the neck ; but that such an occurrence is very rare is also true.

Progress in Submucous Tissue.—The inflammation of the submucous tissue seldom proceeds any further in pathological changes than an effusion of fibrine, and its more or less complete solidification. When once arrived at this stage, it is likely to continue indefinitely, unless interrupted by some artificial or naturally intercurrent circumstance. The tendency of inflammation of this tissue is not to stop short of fibrinous effusion, and remain stationary for any length of time; it is either resolved before or soon attains to it. Whether inflammation commences in the deeper tissues, and affects the mucous membrane secondarily, is a subject which cannot be very often demonstrated. The probability is that this is occasionally the case ; but what occurs more frequently, I think, is the transition of the inflammation from the mucous membrane to the submucous tissue, particularly in the neck and about the os. Hence it will be found that a case, as I have verified more than once, which this year presents only a tolerably bad form of mucous inflammation, without any tumefaction or hardness of the neck, in twelve months will present the tumefaction and induration characteristic of fibro-cellular inflammation.

Intensity of Mucous Inflammation.—In the mucous membrane the inflammation continues for an uncertain length of time without complicating the other tissues, and there is a very great difference in its intensity and effects. We often meet with inflammation of the mucous membrane of sufficiently mild grade to merely cause a slight increase in the color, heat, and secretion, without producing tissual changes. One thing which ought to be remembered, and I shall not apologize for the reiteration, is, that a permanent increase of secretion in a mucous membrane is, and should always be, regarded as an evidence of inflammation in it. Another not less important fact is that discolored mucus, either yellow, red, or otherwise, is not produced by a mucous membrane which retains its tissual integrity. Blood cannot get through the capillaries of a sound membrane, and pus is not produced by a mucous membrane while the epithelium retains its perfect integrity.

Progress of Mucous Inflammation.—After the inflammation has lasted for a time, if its intensity is increasing, the epithelium gives way more or less completely. The destruction or rather the want of reproduction of epithelial scales, is generally observed in patches. At the point where the inflammation attains to the greatest intensity, the epithelium is not maintained. However small this point may be, the redness is increased; and if we look at it we see that the place is scarlet-colored instead of a pale rose-color, as when the epithelium is entire. Inasmuch as this is loss of substance, although slight, it is ulceration, or abrasion, the beginning of ulceration; as yet the secretion is merely increased in quantity, or at most very slightly discolored with pus-globules, and rendered a little thinner by the exudation of a small amount of serum. This absence of epithelium is generally observed, where it occurs, in one continuous patch of greater or less dimensions, and indicates a not very intense degree of inflammation. When this effect of inflammation is first observed, it is apt to be situated around the os uteri; but I have occasionally seen it over the whole intra-vaginal portion of the neck. The cases in which I have observed this state of extensive abrasion were in persons who had passed the climacteric period of life, and they were the subjects of copious, watery, very irritating, and slightly yellow leucorrhœa; and upon examining them I was forcibly reminded of the chafed condition of the thighs in fleshy persons, so red and fiery was the appearance presented by it. They were obstinate, and it required great care in the use of remedies not to aggravate the inflammation.

Forms of Ulceration.—This epithelial denudation is the simplest and the most common form of ulceration met with in practice. Of course, in this form of ulceration the red portion is not depressed; it retains its level with the adjoining surface, and consequently the term ulceration is not considered applicable to it by those who do not believe in uterine ulceration at all. After this description, as faithful a one as I can give, the reader will form his own judgment. I hope I may be allowed to consider it a breach of continuity of the mucous membrane, while anatomists persist in describing the epithelium as a part of that membrane. After the epithelium is lost for some time, there is a gradual increase in the size of the papillary structure of the membrane covering the neck of the uterus; and if the membrane is now examined, instead of the smooth redness there is something of a velvety or plushy appearance. The intensely red surface is covered by, or rather seems to be formed of, an infinite number of extremely minute projections, so closely apposed, that there is hardly any space between them. Scarlet velvet is a very good representation of its appearance. The papillary projections do not seem larger than the minute silk fibres of velvet, as short and as thickly set. This surface is almost always covered with mucus and pus in different proportions of admixture. There is always pus, however, when this complete absence of the epithelium is observed. Still, the evenness of the mucous surface is not disturbed. There is no excavation at least. If there is any change in this respect, the red patch is very slightly elevated above the surrounding surface. As the inflammation advances, the papillary development is greater, but also somewhat different; some of the papillæ increase faster than others, crowd upon the smaller ones, cause them to disappear, and usurp the space occupied by their oppressed neighbors. If the membrane is now examined, there will be seen, instead of the numberless minute, closely set papillæ, a greater or less number of larger ones, varying from the size of a small sewing-needle to a large pin's head, thickly studding the red surface. The redness now, as a general thing, is not so intensely scarlet. The ends of these papillæ, which rise from half a line to a whole line above the level of the surface upon which they stand, are darker red, inclining to lividity. The papillæ thus increase in size, and decrease in number, by strangling each other, until some of them attain the size of small shot, and look like warts. The larger they are, the greater is their lividity of color. As will be inferred, the diseased surface is more and more elevated and irregular, until very considerably raised above the surrounding level. In such cases, pus is generally poured out, in considerable quantities, from the spaces between the papillæ, and the whole surface is thickly covered with tenacious mucus, colored with pus, or with nothing but pus. Sometimes, however, such surface produces no pus or mucus, and seems preternaturally dry.

Complication of Mucous with Submucous Inflammation.—This sort of mucous inflammation is seldom observed without being accompanied by submucous inflammation as a complication. There is nearly always considerable enlargement and induration of the whole cervix where these greatly enlarged papillæ present themselves. In such cases as these, I think we may safely conclude that the inflammation commenced in the mucous membrane, and passed from it to the deeper structures. But there is another kind of enlarged, hardened neck, which with equal certainty begins in the fibro-cellular tissue, viz., when in connection with great hardness and enlargement, the surface deprived of its epithelium is extensive, and is uneven, or nodulations of moderate elevation, but greater extent of superficies than the papillæ, exist, reminding one of the rough surface of very coarse sacking or sea-grass carpeting.

Ulceration and Enlargement.—This kind of a surface is always seen upon a greatly enlarged cervix, which also is very much indurated. It is a very obstinate and very discouraging state of the disease, but will usually yield to sufficiently energetic and long-continued treatment. The boldness in the use of caustics necessary to the cure of such cases as these, requires strong nerves to institute and thoroughly execute. In the varieties I have here noticed, the surface is more or less elevated. But instead of papillary development after the destruction of the epithelium, the integrity of the mucous membrane is further invaded; the surface becomes somewhat depressed, with the edge of the red portion well defined; in short, ulceration, as it is usually understood, becomes quite evident. I should have stated before that in many cases, where the epithelium only is destroyed, the red patch shades off into the healthy rose-color imperceptibly. In this last kind of ulceration the termination of the two is more abrupt.

Aphthous Inflammation.—Other sorts of ulceration occur less frequently on the neck of the uterus,—isolated, small ulcerations, several of them set upon a red surface, not unlike what we see upon the lips and inside of the cheek; also there are occasionally aphthous, or curdy spots, elevated somewhat, but soon degenerating into little yellow ulcers. I have, on one or two occasions, seen such ulcers in patients who were the subjects of nursing sore mouth, and I always regard these minute isolated ulcers as the effect of constitutional disease ; or they at least receive their peculiarity from the condition of the system, and indicate a general unhealthy state of the mucous membrane. It would hardly be proper for me to stop here to describe all the particular sorts of ulceration that occur ; in addition to those resulting from inflammation, there are some which are the effect of specific diseases. These specific ulcers do not assume any peculiarity, nor are they particularly modified by their location upon the neck of the uterus. A chancre possesses its characteristic, when planted upon the neck of the uterus, as distinctly as when seated upon the glans penis. There is no difference between the peculiar, ragged, insensible, foul ulcer of scirrhus on the neck of the uterus and the mammary gland. The phagedenic ulcer of the uterus is the same as when observed to dissolve down so rapidly the tissues of organs elsewhere, and it would not be proper for me to draw their diagnostic characters here. A very little experience, care, and reflection, will save anybody from error of diagnosis or treatment, when the question of difference between common ulcer of inflammation and specific ulcer presents itself.

CHAPTER XVII.

DIAGNOSIS.

FORTUNATELY for suffering woman, we may arrive at demonstrative knowledge of the extent, nature, and locality of diseases of the uterus; and, as a consequence, treat her diseases with the certainty which a positive diagnosis always insures. The evident advantages of a physical diagnosis will render it quite unnecessary for me to use any argument in favor of it, or to induce medical men to resort to it. A physical examination, however, of the genital apparatus of females, is quite a different matter from a physical examination of the chest, eye, ear, or any other organ of the body; and hence the necessity of approaching and conducting it under conditions rendered imperative on account of the circumstances connected with it. The education and natural sense of modesty, so appropriate to female character, and which always command the respect of gentlemen, make such examinations disgusting and disagreeable above almost all others demanded by the necessities of woman's circumstances. With a view to this fact, it is our duty, by our conduct toward our patient, and the management of the examination, to divest it as nearly as possible of every disagreeable feature. Medical men generally I think are, as they should be, actuated by the above considerations, and I fear that they are often so influenced by their own sense of delicacy as too frequently to abstain from the enforcement of essential investigations. This is an error we should always bear in mind, and I think we shall less frequently regret a thorough, although somewhat indelicate examination, when dictated by an honest and intelligent conviction of its necessity, than regret a neglect of such examination from too great a deference to the mere shame of our patients. We should not be, in important cases, constrained to take things for granted that we are not sure exist. Our bearing to a female patient should be deferential, candid, and modest. She should be convinced by our demeanor that everything we do and say is strictly necessary and relevant to her case, and has its foundation in our solicitude for her welfare alone. Nothing therefore should be said or done but what is called for and obviously proper. This sort of treatment from her medical adviser will always command the confidence and earnest co-operation of an intelligent female patient. There should be a full and explicit understanding between the physician and the patient as to the necessity of a physical examination, in what it consists, and how it is to be conducted. The good sense of the practitioner will enable him to judge whether he should commit the detail of explanation to the husband or some other appropriate second party, or whether he impart it directly to the patient; all the circumstances of the case will enable him to determine this matter without much difficulty. After the preliminaries are disposed of, I would insist upon conducting the examination without exposure. It is needless in ordinary uterine examinations, and should be permitted only when the disease is upon the external parts. One position and kind of preparation, so far as the patient is concerned, will suffice for most cases, whether we wish to make a manual or instrumental examination. There is no necessity for the patient to unclothe herself.

Position of Patient.—She should lie down on her back across a bed, so that the breech will be very near the edge; draw up her limbs by flexing the thighs and knees, and place her feet, separated about twelve inches, upon the side of the bed very near the nates. In this position a sheet should be thrown over her, so as to completely cover her person, and hang down several inches below her feet, over the edge of the bed. If we wish to use the speculum, or our eyes in any way, the bed should be placed immediately before a large window, in which the light is not obstructed by blinds or curtains. Thus placed, by kneeling down before her, we can have free use of both hands, a matter of much importance in delicate manipulations. Let the patient be very near the edge of the bed, lest by reaching too far, our examinations may be difficult, if not imperfect. When we wish to make a manual examination, we have need of nothing further than a little oil. Our objects in making a manual or digital examination are to ascertain the position, size, consistence, and sensitiveness of *all* the organs in the pelvis; the presence or absence of anything that does not belong there; and if anything unusual is there, what are its properties, connections, and nature. Upon making examinations for the first time, the whole of this investigation should always be attended to.

Digital Examination.—The mode of examining the pelvis with the fingers is of the utmost importance. After oiling the fore and middle fingers the index should be very gently introduced, and the examination conducted as far as possible with it; then the two should be introduced, with which nearly all the cavity of the pelvis can be reached. The index finger will not reach as far, by one and a half to two inches, as the two together. As the finger is introduced, it naturally and easily comes in contact with the rectum, which may contain fæces, and consequently will appear as a round, full ridge along the middle line of the posterior wall of the pelvis, or a mere soft fibrous cord, hardly perceptible to the touch. The full rectum is generally a healthy one, as the fæces cannot remain long in a rectum rendered irritable by disease. By pressing upon the rectum with the finger, we may ascertain the presence of inflammation by the increased sensitiveness; the organ is absolutely insensible to moderate pressure when in a state of health. We should seek for internal hemorrhoids, which are small tumors in the bowels, or the induration and contraction indicative of stricture; and, in short, examine it as completely as possible in this way. Next we should turn our finger forward, pass it up behind the symphysis pubis, and along the front wall of the vagina, and as well as practicable ascertain the condition of the bladder. It may contain a calculus or other foreign substance, or, what is very much more common, be inflamed. In the first case, the foreign body may be felt by the finger. The examination is more complete if the fingers of the left hand are used to press into the pelvis from just above the symphysis pubis. The substance can thus be grasped by the fingers of the opposing hand. With the fingers of one hand above the bladder, and the other in the vagina below it, we press it and thus ascertain its sensitiveness. With the two fingers of the right hand pressing up by the side of the uterus, between it and the walls of the sides of the pelvis, first on one side and then the other, while the fingers of the left hand press downward toward them from above, so as nearly as possible to meet them, the cavity may be pretty thoroughly explored, and any unnatural substance or uncommonly sensitive tissue be easily discovered. All these manipulations should be performed with the utmost gentleness, remembering that rudeness may deceive us as to the sensitiveness of organs, as well as give the patient unnecessary suffering. While we are gentle, we should be as thorough as possible. The main object, however, for which we institute these examinations, is to ascertain the condition of the uterus with respect to position, size, shape, consistence, sensitiveness, &c., &c. Where is or ought to be the os uteri and cervix, and how shall we find them? In the virgin, the os uteri ought to be in the middle of the pelvis, upon or a little below the level of the arch of the symphysis pubis, and within easy reach of the index finger, two inches and a half from the entrance of the vagina. We may know when we feel the neck of the uterus by its consistence, shape, size, &c. It has more consistence than any part with which our finger comes in contact, as we push it backward into the vagina. In passing through the vaginal canal, the finger is impressed with a soft intestinal sensation, and can distinguish nothing but loose folds that are dissipated and lost in the surrounding softness by the slight pressure, until it comes to the neck of the uterus, when it may be felt, having consistence enough to retain its shape under considerable pressure. If we push it upward, backward, or downward, it retains the same characteristics. The finger can be carried up the side, up before, or behind it as a projection, and around it in every direction except above. This being unlike anything else in the vagina, will be easily recognized by an uneducated finger. The shape of the virgin cervix uteri is almost cylindrical, slightly compressed from before backward, and not far from three-quarters of an inch in diameter in every direction; it projects half an inch into the vagina, and the projecting or free end of it is apparently cut nearly square off, so as to present at its inferior face almost a flat surface, with a mere dimple in the centre corresponding with the os uteri. The multipara cervix uteri is generally a little lower in the pelvis, and often slightly turned to one side, does not project so much into the vagina, is about an inch wide, or often a little more, from half an inch to three-quarters in its anterior posterior diameter, and instead of being truncated, seems formed of two distinct projections at its

inferior extremity (the anterior and posterior labia of the os uteri).
Between the labia or projections is a deep fissure, with its extremities directed to the sides, large enough to partially admit the extremity of the index finger. Os tincæ is applicable to this form
of the os uteri, but in no wise is expressive as connected with the
shape of the virgin os uteri, neither is it descriptive of the senile
uterine mouth.

Os Uteri in the Aged.—The os uteri in the old is higher in the
pelvis than in the virgin or multipara, does not project into the
vagina, and feels more like a pit at the termination of the vagina.
As women advance in age this description is more applicable than
very soon after the cessation of the menstrual discharge. There
is often a cord or frænum-like projection in the vaginal walls,
which is planted into the external surface of the anterior and
posterior lips of the mouth of the uterus, and thus extends backward and forward to be lost in the anterior and posterior median
line of the walls of the vagina. This frænum is more apparent, if
not more developed, as women advance in age; but I have known
it so prominent as to be mistaken for the results of disease, even
in the middle-aged. In one case an intelligent practitioner thought
it an evidence of the injurious effect of strong caustics. The consistence of the virgin and multipara cervix uteri is the same. To
the sense of touch it gives the idea (which is a correct one) of deep
fibrous tissue, almost as hard as cartilage covered over thickly
with areolar tissue. Dr. Bennett compares it to the feel of the
cartilage of the lower extremity of the nose. It seems to me not
quite so dense, although nearly so. It is wholly insensible to
pressure with the pulp of the finger, and it requires considerable
force to produce pain with a plain round instrument. This fact
should be borne in mind in our examinations, viz., *a healthy cervix uteri is not tender to the touch.*

Corpus Uteri.—We may examine the shape, size, and sensitiveness of the body of the uterus by pressing it down well into the
pelvis with the left hand, while the fore and middle finger of the
right presses upon it as high up as possible. When the uterus is
healthy, the fundus cannot be felt above the symphysis, even by
lifting it with our fingers, so that if it can be felt by both hands
it may be considered enlarged.

A Tender Uterus is an Inflamed Uterus.—I cannot refrain from emphasizing the fact that the uterus is insensible to the handling of an ordinary examination, and that *a tender uterus is a diseased uterus;* in fact, generally inflamed. What condition converts comparatively insensible organs elsewhere—the periosteum and cartilages, for instance—into highly sensitive ones?

Examination of Urethral Canal.—If we have gained all the information we can from the use of the fingers, we may next use the probe, for the purpose of penetrating the cervical and uterine cavities. When, from the sense of touch, there is suspicion of inflammation of the urethra, the probe may be used with great propriety in examining this canal. There is almost always pain when the probe is introduced into the healthy urethra, but it is a peculiar smarting pain; if the urethra is inflamed it is a sore pain; it feels as though the probe had touched a sore place; it is soreness. Dr. Simpson first recommended and practised the use of the probe for the purpose of probing the uterus, and he has given to it a certain appropriate shape, size, and adjustment, which add very considerably to its usefulness and adaptability to this particular use. It may be found in almost any of the shops of our instrument makers, under the name of Simpson's uterine sound.

Object in Using Probe.—The main objects in examinations with the probe in such cases as I have now under consideration are, to measure the size and length of the cervical and uterine cavities, the mobility and position of the uterus, and, if need be, the connection of that organ with pelvic growths. The instrument must be adapted to these purposes; in order to this it must be long enough, of the right size, and made of flexible metal.

Size and Length of Probe.—It should be ten or twelve inches long, with one end fixed to a flat handle; the probe end should be terminated with the ordinary probe-pointed enlargement, about one-eighth of an inch in diameter. The wire behind the bulbous termination should be one line in diameter, round and smooth, and should gradually increase in size to the handle, where it ought to be about a quarter of an inch in diameter. The best material, I think, is copper, galvanized. I have not spoken of notches and other scale-marks upon it, because I like it better plain. Yet I see no objection to them as recommended by Dr. Simpson. It is always well to have two or three sizes of probes for special purposes, but the one I have here described is the one I should recommend, to arrive at any deviation from the natural uterine measurement.

Fig. 16.

Mode of Using.—After oiling the instrument, and introducing the index finger of the right hand, and placing it upon the os uteri, the probe may be carried along the palmar surface of the finger until the point arrives at the mouth of the uterus, when by elevating the point, it may be carried forward into the cavity of the cervix. In order to insure its passage through the cavity of the cervix into the cavity of the body, the probe must be bent to the same degree of curvature as a male catheter. Great gentleness must be observed in the use of this instrument, because it is an easy matter to do violence to the mucous membrane by a very little rudeness of management. After the probe has passed to the os internum, a sense of constriction is felt through the instrument, which feeling soon gives way, and the probe then goes to the fundus without further resistance.

Length of the Cervical and Uterine Cavities.—The cervical cavity in the virgin is about an inch and a quarter in depth, and the cavity of the body from a half to three-quarters of an inch; the former in the multipara is one and a half inches, and the latter an inch deep. In old age they both are nearly or wholly obliterated. I do not often use the probe in this way for the examination of the uterus in cases of inflammation and ulceration, but have adopted the suggestion of Prof. Miller, of Louisville, and use it through the spec-

Uterine Probe, or Sound.

ulum, and shall consequently have more to say about it in connection with the use of that instrument. To expose the neck of the uterus so as to examine it by the sense of sight, it is necessary to have a speculum, and we ought to have a pair of long light dressing forceps also; they will be very useful on several accounts.

Speculum.—Since the speculum has come into so general use, it has assumed a variety of shapes, and been composed of quite a number of different sorts of materials. For different purposes it is convenient, if not necessary, to be provided with different shapes, sizes, &c., of this instrument, but for ordinary use, the common cylindrical, or quadrivalve, are the best forms. My preference is for the compound called "German silver." If we use the cylinder, we ought to have three different sizes: one small, one large, and the other a medium size. With regard to the adaptation of the cylindrical instrument, the larger size we can use in the case the better, as it will the more completely expose the cervix. In selecting the cylindrical-shaped instrument, we should procure one with as great bevel at the internal end as we can find. There should also be always adapted to it a wooden director. The instrument will pass the external parts with less pain, and will not require the care to prevent it from injuring the vagina, that it will without the director.

The common glass instruments, whether plain or covered, as recommended by Fergusson, I never use, excepting for the purpose of leeching. The instrument I prefer above all others for ordinary use, is the quadrivalve made according to Charriere's plan, or what is equally as good, perhaps, Tieman's new quadrivalve. The former has a plug, or director; the latter is so arranged at the ends of the blades that they close in together, and make the internal end of the instrument smaller, and render the director unnecessary. One quadrivalve will answer almost any case that occurs, if it be of medium size. It is constructed of four blades, that are caused to open or close by the use of a nut upon a screw. After introducing the instrument, the internal end may be increased greatly in size, without the external end being enlarged at all. It is only

Fig. 17.

Quadrivalve Speculum.

14

necessary to see it to perfectly understand its construction and the mode of managing it. A good instrument of this sort may be procured at Messrs. Tolle & Dagenhardt, or Messrs. Bliss & Sharp, of Chicago.

Position of Patient for Speculum.—To be prepared to use this instrument to the best advantage, our patient should be placed in the position I have heretofore described, viz., across the bed, before a large window, through which as much daylight should be freely admitted as possible. The better light the better view, and unless we have plenty, we cannot be certain of correct results in our examinations. The bed and patient should be so placed that the light may fall straight through the instrument, and full upon the parts at its internal extremity. We should also have some cotton-wool, sweet oil, and a couple of napkins, together with the dressing forceps I have before spoken of.

Mode of Using the Speculum.—In commencing the examination, we should oil our speculum, and our middle and index fingers. Kneeling before the patient, we should introduce the index finger, and, if need be, the middle one also, to ascertain the position of the cervix uteri. This precaution will enable us to know in what direction, and how far, to introduce the speculum. After this preliminary examination, the forefinger and thumb of the left hand should be placed upon the edge of the labia, one upon each side, with which they should be gently separated; holding the speculum in the right hand, somewhat like a pen, we may introduce it by the guidance of the thumb and finger placed as above. In introducing it, we should push it forward sufficiently to reach the cervix, and direct it upward, downward, or to one side, as we may have ascertained, by digital examination, to be the position of the os and cervix. The director may now be removed, when the os uteri will be seen at the open end of the tube.

How to Find the Os Uteri.—If this is not the case, we may use our probe, and gently push the parts from one side to the other, turning the speculum in different directions, until it is found. If the neck is too large to enter the speculum, we may spread the blades still more, until it is brought into full view. Most frequently the parts are covered with some sort of secretion, and we should always, with cotton-wool or lint, remove all of it, so that the naked mucous membrane alone presents itself to our view.

Without this precaution, we may overlook an obvious and extensive ulceration; for as the parts are covered over with this thick, opaque secretion, it either completely hides them from view, or much modifies their appearance. I have often met with cases where I have observed them attentively, for the purpose, if possible, of detecting ulcerations without this step, but failed, until the cotton was used, when extensive ulceration appeared. Indeed, I never think of coming to a conclusion of any kind by the use of the speculum, without this precautionary measure. By means of the sight, we can see the color, size, shape, and some other conditions of the parts, and the color, consistence, and derivation of the secretions. When the mucus, pus, or blood comes from the mouth of the uterus, we can see it issuing from it. The shape and size of the neck and os of the uterus differ in different individuals, according as they have been impregnated or not.

Appearance of the Os and Cervix in the Virgin.—The virgin uterus is small; the cervical end is nearly round, and terminates in a truncated extremity. Through the speculum, it does not present the appearance of labial projections, and the os is either a small slit about a quarter of an inch long, or a round opening into the middle of the truncated extremity. It is about large enough to admit with facility the end of a female catheter, and the neck projects, in relief, from the bottom of the parts exposed by the speculum, something like half an inch.

Appearance in the Multiparous Uterus.—The appearance of the multiparous uterus is quite different from this; the cervix terminates in labial projections, which divide its extremities into an anterior and posterior half, and it does not project with so much prominence into the speculum. The os is represented by the cleft between these labial projections, and is large enough, in many instances, to admit the tip of the index finger.

Appearance in the Aged.—In the aged, the labial projections seem to have atrophied to obliteration, and the speculum shows a round opening, surrounded by the walls of the vagina.

Exceptions to these Appearances.—Although the above is an accurate description of these appearances under the different circumstances, there are many natural deviations from it.

Color.—The color of the mucous membrane covering the cervix, and entering the os uterus, may be compared to that of the inside of the lips of the mouth, a pale rose red.

Appearance of Secretion.—The parts are merely lubricated, not smeared or inundated, with mucus. There is just enough of this secretion to keep the membrane moist, but not enough to hide the surface from view. I speak now of the cervix uteri.

Indication of Mucus in Abundance.—An abundance of mucus must be regarded as an evidence of excitement; its constant and persistent abundance as an evidence of disease. "Remember that in spite of their name, it is not the business of mucous membranes to secrete mucus; the more perfect their condition, the more favorable the surrounding circumstances, the less they do so. . . . . The greater the diminution of their life, the greater the secretion." The more disease, the greater the secretion, until their integrity is destroyed, when the secretion becomes modified. The source whence the mucus is derived will show the point of disease; if it comes from the os uteri, the disease is in the cavity of the cervix, or body of the uterus.

Indication from Pus.—These remarks apply with greater force to the production of pus by the vagina or cervix. Pus, or purulent mucus, indicates disease; and when we find muco-pus, or clear pus, in the end of the speculum, it would be preposterous to conclude that there was no disease there, merely weakness of the parts. It is extremely doubtful whether pus can be produced by mucous membrane, without destruction of the epithelium, at least. Temporary congestion often increases the amount of mucus to be found in the vagina, but no pus. The color of the mucous membrane, in cases of congestion, is a livid or a dark purple red, instead of the scarlet or abrasive inflammation.

Probe and Speculum Conjointly.—When the neck of the uterus is exposed in the speculum, it will often be profitable to use the probe. If proper attention is paid to appearances under the use of the probe, much information may be gained. When the mucous membrane of the cavity of the cervix or body is inflamed, it is generally much more fragile than natural, so that it bleeds upon slight contact with the end of the probe. In cases where the inflammation extends to the cavity of the uterus, the probe passes the os internum without obstruction, and passes farther up than natural from the increased size of the cavity.

Characteristic Signs of Inflammation.—The signs of inflammation of the submucous tissue or substance of the neck of the uterus are, increase of size, tenderness, and generally hardness; of the mucous membrane, increased color and secretion; of ulceration, still more intense redness, purulent discharge, tenderness, and generally enlargement. The former conditions may be ascertained by the touch, the latter by the sight, and when they are mingled, by both combined. It may be superfluous to pursue the diagnostic description of these cases further; but as I believe that a great many members of the profession do not sufficiently appreciate the importance of some of the appearances and conditions I have described, and as I am thoroughly convinced of their significance, I am determined, at the risk of reiteration, to place these diseased appearances and conditions in a more prominent light. Open external ulceration of the uterine cervix, after the parts are well exposed, *and cleared of mucus and pus,* by wiping, cannot be well mistaken, or overlooked; and the only thing I shall insist upon here is that the practitioner must not be led to believe the case one of no importance because the ulceration is not very extensive. This raw scarlet surface is always indicative of mischief; and we should expect any amount of suffering from even small patches of it. There are cases where the appearances are not so obvious; where in fact all the parts exposed by the speculum and within reach of our vision have a natural appearance. No redness, rawness, or other discoloration can be detected on the neck, in the mouth of the uterus, nor on the vaginal surfaces; they are quite healthy in appearance and reality.

Diagnosis of Endocervicitis.—But there is an obvious and in many instances a copious secretion of tenacious mucus flowing from and lying in the os uteri; wipe this away and all looks right. This is a case of endocervicitis. In some instances this mucus is colored with streaks of yellow by the presence of pus, or it is wholly yellow; here there is loss of integrity in the epithelium of the cervical cavity. The mucous membrane in the cervical cavity is ulcerated. If we remember that the mucous membrane secretes only enough mucus for lubricating purposes, in the natural condition, we can arrive at no other conclusion than that the membrane is in a state of hyper-excitement when its secretion is rather abundant, or altered, or both. When we see mucus in even small, yet perceptible quantities, issuing from the anus, what is the inference? If this is abundant, persistent, and colored yellow, however healthy the anus might appear externally, we could not believe that the rectum was in a healthy condition. Why not then positively determine that the mucous membrane is inflamed, which floods the os uteri with mucus or pus, or with both? If we introduce the probe into the cavity of the cervix thus abundantly secreting, the patient will nearly always complain that we touch a "sore place; a tender spot;" that it hurts her in her back, &c. &c. And very often blood will immediately follow the withdrawal of the instrument. This, however, is not invariably the case. Another diagnostic evidence of endocervicitis, is the increase of the pain ordinarily experienced by the patient when the probe or nitrate of silver is introduced.

Diagnostic Effect of Caustic Applications.—There is not a new pain produced, but the old pain is aggravated, and the quality of the symptoms is the same while the number is increased. If the pain in the sacrum has been the one mostly complained of, the introduction of the caustic makes the back ache worse; if the pain in the iliac region has caused most suffering, it will be aggravated. The hyper-secretion, or perverted secretion of the mucous membrane, must be regarded as an indication of disease of that membrane. If we have these facts fixed in our mind, and if we act upon them, we may discover and cure disease that would otherwise escape our attention and thwart our skill. But there is another obvious and common sense sign of inflammation, which has not been applied in our investigations of diseases of the uterus, viz., tenderness. Tenderness or sensitiveness to the touch anywhere else, leads us to suspect inflammation, but in the uterus it is unaccountably set down as indicating an irritable uterus, and not an inflamed uterus.

Diagnosis of Submucous Inflammation.—I think when I touch the uterus with the finger or an instrument, and the patient shrinks from the contact, and says "she is sore," or "it is sore," that there is inflammation there. Tenderness is not an evidence of mucous inflammation, but of submucous or fibrous inflammation of the uterus. And it is a matter of importance to determine the presence or absence of submucous inflammation either as an independent affection, or complication of inflammation of the mucous membrane. It is a great error to confine our attention to the abrasions or ulcerations of the mucous membrane, and to believe we must see those abrasions or ulcerations before we can admit the presence of inflammation.

Complication of Mucous with Submucous Inflammation.—We should not shut our understanding to the fact that the uterus should be examined by the same diagnostic rules that govern our investigations of disease in other organs. Some authors tell us that ulceration results from inflammation of the submucous tissue, and others that the inflammation begins in the mucous membrane. I am sure that inflammation sometimes exists in both these tissues at the same time. In this case we shall have tenderness and hyper-secretion. At other times there is submucous without mucous inflammation; then we shall have tenderness without hyper-secretion. Again, we may have mucous without submucous inflammation, when hyper-secretion without tenderness will indicate it. These remarks will fix the importance of these two symptoms as indicating the seat of the disease.

Size of the Uterus ordinarily Increased—Exceptions.—The size of the organ would seem to be a good indication of the presence or absence of inflammation; but this may vary very much under what would appear to be the same form of disease. In endocervicitis, it is usual to find the cervical canal increased in calibre; but this is certainly not always the case, as I have met with unmistakable instances in which this cavity was increased in size and the os uteri almost closed; it was so small as to admit only a very small probe. Where there is mucous inflammation of the cervix extending toward the cavity of the body, and more particularly where the disease extends into the cavity of the body, the whole organ is likely to be enlarged. So much enlargement sometimes takes place, that the fundus may be felt considerably above the pubis: neither is this always the case, however; often there is no enlargement. This hypertrophy, or general enlargement of the organ, is more frequently indicative of mucous than submucous or fibrous inflammation.

Atrophy as the Result of Inflammation.—In fact, I think that long-continued inflammation of the substance of the body and cervix often brings about atrophy, or shrinking of the uterus. Permanent increase of size, or hardness of the cervix, must be the results of submucous inflammation, and generally coexist with it.

Examine for Complications.—Some, if not all, of the pelvic or local symptoms, or such as much resemble them, may be produced and perpetuated without inflammation of the uterus; hence it is necessary to examine the case with reference to this fact. We shall also occasionally find that, notwithstanding the complete cure of actually existing uterine inflammation, the local symptoms, in a modified form, still continue. These circumstances will be found to depend upon the independent or coexistent presence of some of the complications I have described.

Cystitis as a Complication.—Chronic cystitis, rectitis, prolapse of the rectum, piles, urethritis, cellulitis, &c., &c., are among the most common causes of these symptoms. It is only necessary to mention these facts to enable the intelligent practitioner to explain anomalous cases that occasionally occur. There can be no doubt, I think, that holding the uterus to a rigid accountability for all the pelvic symptoms enumerated as the ordinary result of its diseases, has caused a good deal of confusion, and has enabled certain writers triumphantly to assure their adherents that in a number of cases the symptoms were present, but the ulceration was absent. A number of organs commanding extensive sympathies, sensitive under inflammation, crowded together in so small a space as the pelvis, supplied to a great extent with branches from the same nerves and arteries, must all be more or less congested, inflamed, and pained together; and nothing but an intelligent and deliberate physical examination can make out the difference in their relative suffering, or certainly ascertain which of them is affected when one alone is diseased.

Almost the only disease with which chronic inflammation and ulceration of the cervix uteri are likely to be confounded, is cancer in some of its stages. The many well-marked symptoms and physical conditions which accompany this last disease are now, however, so well understood and so thoroughly described, that the novice need not be embarrassed in his diagnosis of it.

I find in Becquerel's "Traité Clinique des Maladies de Uterus," pp. 320–323, vol. i, so complete and faithful a diagnostic summary between cancer and the different conditions of chronic inflammation of the cervix, that I have given its substance for the concluding portion of this chapter. It is subjoined.

Cancer in the Scirrhous Condition.

Cervix hard, unequal; nodulated, os not always open, sometimes wrinkled or furrowed.

Scirrhus of the neck often implicates the vagina.

Hereditary influence is often traceable.

Touch is painless.

Discharge sometimes absent, in certain cases very abundant, and consisting, for the most part, of albuminous serum.

Menstruation increased, being neither more nor less painful, and passing often into the state of real hemorrhage.

Absence of special anæmia when the vagina and body of the uterus are involved. Cancerous cachexia.

Progress continuous and without cessation.

The pain in cancer is very sharp, intense, and lancinating, and not influenced by locomotion or movements of any kind.

Inflammation with Ulceration.

Neck less hard, developed regularly in one of the lips, os always open.

The induration of the neck never extends to the vagina. Mobility of uterus complete.

No hereditary influence.

Touch painful.

Discharge constant and characterized by the presence of transparent mucus, muco-pus, or purulent mucus.

Menstruation more painful, often retarded, almost always scanty.

Special anæmia as above described.

Often stationary for a long time.

Pains less severe, more dull, and perceptibly influenced by walking and other sorts of motion.

Ulcerated State.

Developed at the critical period of life generally.

Preceded and accompanied by hemorrhages.

Severe, sharp, lancinating pain.

Development essentially in sharp irregularities and nodosities.

Adhesions to other organs soon as ulceration is formed; immobility of the uterus.

The surface only slightly soft, subjacent tissue scirrhous.

Ulceration deep, unequal, essentially irregular, with thick, elevated, and hard edges.

Always granulations.

Chronic Inflammation and Softening.

Occurs earlier in life almost always.

Not preceded by hemorrhage.

Pain dull and profound.

Enlargement regular and rounded, or regularly lobulated.

Complete absence of adhesions to other organs. Entire mobility of the neck and body of the uterus.

Tissue of the cervix not hard, and easily destroyed.

When ulcerations exist, less deep, with tumefied edges.

Granulation often accompanies the other lesions.

Ulcerated State.

Discharges extremely abundant, consisting of purulent and often sanguineous serum; nauseous and often fetid odor.

Great hemorrhage from time to time, not necessarily at menstrual period.

Chronic Inflammation and Softening.

Discharges less abundant, consisting of muco-pus alone, or accompanied with a little blood, without odor.

Always hemorrhage, but often a mere prolongation of the menstrual discharge.

Cancerous Ulceration.

Developed upon an hypertrophied and scirrhous surface.

Ulceration deep, vast, unequal, grayish surface with thick edges, and easily bleeding.

Ulcerated surface hard, presenting numerous lobes and tubercles, with nodosities and great hardness.

Often great loss of substance.

Cervix and corpus uteri immovable, on account of adhesions.

Discharges sanious, fetid, sanguinolent, and of an insupportable and characteristic odor.

Cancerous cachexia always present.

Simple Ulceration.

Ulceration often on a healthy tissue, or presenting the soft or hard varieties of inflammatory injection.

Ulceration more superficial, the edges less developed, and more regular at the bottom, not always easily made to bleed.

Nothing of the sort in chronic inflammation and ulceration.

Ulceration is not always accompanied with loss of substance.

Neck and body always movable.

Discharge of muco-pus or purulent mucus; always less abundant.

Special anæmia.

CHAPTER XVIII.

GENERAL TREATMENT.

General Treatment.—I am sensible of the great difficulty of properly estimating the value of any given remedy or plan of treatment for the cure of disease. Nature does very much sometimes to aid imperfect means, and even to effect a cure under improper treatment, while at other times the circumstances inseparable from a case thoroughly thwart the best directed efforts; and very often we record cures and attribute great efficacy to our plan of management, when the favorable termination is due alone, and perhaps in spite of us, to the natural conservative energy of the system or the parts concerned. It is a mistake, therefore, to be too sanguine in our expectations even with the use of our favorite course of treatment, or to depreciate everything which has not fulfilled our hopes. We should patiently, honestly, thoroughly, and judiciously try every means within our knowledge for the benefit of our patient, let him labor under whatever disease he may. The reader is doubtless perfectly aware of the very great differences of opinion in the profession as to the treatment most beneficial in inflammation of the cervix uteri and its accompanying ailments. In alluding to these many and diverse opinions, I must record my conviction of the honesty with which they are maintained by the principal disputants of the present day, and most exhort the junior members of the profession to cautious and thorough research on the subject. There must be a right and a wrong side to every disputed question; and, as a general thing, extremists are wrong; remembering this general truth,—we cannot always be kept in doubt by the facts in the case, if, without prejudice or party bias of any kind, we earnestly set to work to learn.

Spontaneous Cures.—Are there any spontaneous cures in these cases? I think there are, and I propose inquiring into the method adopted by nature, and take it as a guide to some extent, at least, for the plan of artificial treatment. Change of circumstances frequently makes robust persons of invalids. This change is generally from irregular habits of living to regular and appropriate; from the highest state of luxury and ease to one of need, or at least economy and industry, in which the patient must exercise her mind and muscles to a proper degree. The healthy tone of the stomach, muscles, and brain thus brought about decreases the susceptibility to slight suffering, enables the patient apparently entirely to recover from disease, and bear small ills without complaint. I need not specify the various circumstances and conditions of life which improve the tone and elevate the functional activity of the whole organism; they are numerous, and will suggest themselves to the reader. How many journeys are taken, how much time spent at watering-places and places of amusement for this purpose? And often they answer the purpose, and the patient is restored to health.

Change of General Circumstances only Temporary in their Effect. —This improvement, in cases of disease of the uterus, is brought about rather by diminishing the nervous susceptibility to the wearing influence and pain of the local disease, and by fortifying the system against its advance by establishing excellent general health, than by actual cure of the ulceration and inflammation. As a consequence we find a return to the former mode of living, habits, and circumstances reproduces more or less rapidly the same train of general symptoms, and makes it necessary to resort to a repetition of the journey, or whatever other means were previously successful for their removal. This is only an apparent, not a real cure, and I hope I will be excused for saying that such is the kind of cures which always result from an exclusive general treatment. Tonics, laxatives, and alteratives put the general condition of the patient on a better footing, and the patient suffers less from her local disease, and even considers herself well; but suspend the general roborant appliances and the patient again sinks into her former state of valetudinarianism. I have often witnessed these changes as the effect of accidental mutation in the condition of the patient, intentional changes of place and circumstances, or well-advised general treatment.

Supervention of Acute Inflammation.—There is, however, another method resorted to by nature, and which sometimes results in permanent and complete cure. Chronic inflammation has very little tendency to spontaneous subsidence; its duration is at least indefinite. Situated in the neck of the uterus this is particularly the case. Acute inflammation, however, on the contrary, has a strong tendency to terminate in resolution, to subside and leave the parts in a healthy condition. And, in cases of chronic inflammation in any of the organs, the supervention of the acute form proves sometimes salutary. It absorbs the whole chronic action and takes its place in the tissues; and as it subsides, the diseased organ is left in a healthy condition. We have an opportunity of seeing this process of usurpation, displacement, or whatever else it may be termed, in diseases of the eye, and witnessing the salutary sequence.

Acute Inflammation after Parturition or Abortion sometimes works a Cure.—Some of the functions of the uterus when naturally performed are followed by acute inflammation in the neck of the uterus. I allude particularly to parturition; and while these inflammations sometimes linger and become themselves chronic, they generally under favorable circumstances subside kindly, and where the cervix had previously been affected by chronic inflammation, sometimes favorably modify, if not entirely cure it. I think that very few cases of parturition occur that do not cause sufficient violence to the cervix and os uteri, to be followed by a greater or less degree of acute inflammation. A great many are certainly thus followed by inflammation. The acute inflammation resulting from abortions occasionally has the same effect. I hope there is no danger of being misunderstood. Instances have occurred in the hands of most experienced practitioners where the uterine health of a primipara has been benefited by pregnancy and the processes of parturition.

Principles of Local Treatment.—The local treatment of these inflammations is founded on the same principle of these natural cures. In the case of obstinate inflammation of the eye, we often resort to strong stimulants to modify a chronic inflammation, *i. e.*, turn it into a moderately acute one, which, usurping the place of the chronic, causes it to subside, and leave the organ sound. And we know how successful it often is. So with the local treatment of inflammation of the cervix uteri. We awaken an acute inflammation in the tissues occupied by the chronic; and, as the former subsides, the disease is favorably modified, if not entirely cured. This is a radical cure, where a sufficiently strong impression is produced either by the natural or artificial process.

Physicians array themselves in two divisions in the treatment of uterine diseases. One division comprises those who consider the local disease as unimportant effects of the bad condition of the general health, who pay particular attention to the general condition of the patient, and who give but little, if any, local treatment. While the other division relies upon local treatment for the cure, and the general merely as accessory. Those who look upon the local as the essential treatment, are also somewhat divided as to the kind of treatment. One of these subdivisions thinks that if the uterus can be placed and sustained in its proper relative position to the other organs the inflammation will spontaneously subside; while the other party believes in the use of strong stimulants and caustics applied directly to the diseased parts. I shall not at present pay much attention to the plan of mechanical support, leaving it for a future chapter, but will proceed to give the general and local treatment which can be relied upon with most confidence for the relief of patients affected with inflammation of the cervix uteri, and I shall first give the general treatment.

Posture, Exercise, and Repose.—The young practitioner will soon learn that posture and exercise are important considerations in the general treatment, and he will be taught by most writers that the reclining posture and strict quietude must almost universally be observed. Walking generally causes an increase of pain, and, it is natural to suppose, an increase of inflammation; so that exercise on foot or in the erect position is regarded as injurious. On the other hand, confinement to the recumbent posture and the observance of strict quietude is very hard upon the general health; the patient becomes more nervous, and all her functions are performed in an irregular and imperfect manner. As a consequence, in very many instances, the symptoms are much aggravated. In the great majority of these cases, therefore, I think the patients are injured by confinement and recumbency. It would neither be scientific, sensible, nor successful, however, to lay down any absolute rule in respect to exercise and quietude. I think we may arrive at pretty accurate conclusions as to the sort of cases and the conditions under which each should be observed. More than ordinary acuteness of the symptoms, indicating a high degree of inflammation, occurring in the beginning and continuing through-out, or arising during the progress of a case as the effect of tem-porary causes, will make rest indispensable to the removal of them. Hemorrhage at the time of menstruation, or between the menstrual periods, is also a reason for strict quiet. Where neither of these conditions is presented, I think the patient will in most cases be much benefited by judiciously directed exercise. I feel like in-sisting upon the enforcement of outdoor exercise as the rule in these cases; for I have often had an opportunity of contrasting, in the same cases, the influence of quiet and exercise upon the re-covery of patients of delicate nervous constitutions. One patient who had been unable to sit up for even a short part of the day for several months, on account of the pain in the hips, dragging in the loins, and great nervous prostration, was sent to a water-cure, and in three months she returned home capable of walking several miles a day, and enjoyed comparatively robust health. In a few weeks after returning to a home in which she enjoyed the luxuries and ease so desired by all who prize good living, she became "miserable," and was obliged to abandon her exercise entirely. It is encouraging to state, that in less than six months of proper local treatment, she was permanently cured. This is but a type of many similar cases that have been benefited by the enforcement of exercise and other items of proper living, but, I must also add, not cured. It has been my constant aim for many years to induce patients of this kind to take as much exercise as they can bear. Under the mistaken notion that any local pain indicates an aggra-vation of their disease, and that to exercise when it gives them pain, even to a moderate amount, is a great evil, they confine themselves to their room, and even their bed, to the forfeiture of the healthy tone and energy of the nervous system which shield them from the intolerable and inexpressble *ennui*, melancholy, and irritability, which are so characteristic of bedridden women. Pain and weariness, that subside after a few hours' rest, should not be regarded; it is only in those cases in which the pain and weariness increase at every effort at exertion that exercise should be abandoned, and then we should insist upon its being resumed again as soon as sufficient advance in the cure has been made to justify another attempt. We should not tire, during the whole treatment, in our endeavors to institute a system of regular and gradually increasing exercise, on account of the consideration that it is indispensable to the enjoyment of useful and comfortable health. Selection of the kind of exercise will depend, of course, upon the condition of the patient in respect to pecuniary matters as well as the state of her disease. Fortunately, the best kind is such as is within the reach of every kind of patients, not excepting those who are under the necessity of earning a living. The capacities and demands of our nature are formed to answer the curse pronounced against Adam. We not only earn our bread by the sweat of our brow, but the labor necessary to procure the bread brings almost all the conditions that insure health and happiness. It is, in fact, a great evil of the present state of society, that our ladies cannot find in useful employment that healthy tonic exercise for the body and mind which they need, and that such exercise and employment are allowable and acceptable only in amusement. There is almost no variety in mental and corporeal exercise required by the highest social amusements, and it is only when we descend to the primitive sports that our demands in this respect are met. It is undignified in ladies to fish, hunt in the woods, or engage in muscular feats. They must for muscular exercise engage in the measured sameness of the quadrille, or the giddy whirl and violence of the waltz, or cramp their limbs to the steady routine of a system of calisthenics. What are all these, for variety and adaptedness to their wants, compared to the washing, ironing, sweeping, milking, churning, spinning, weaving, cooking, walking, running, of household engagements; the stimulus of need; thinking of all these things; timing them; proportioning them; calculating, economizing, nursing, doctoring, advising, correcting, teaching, and conducting little minds and bodies through the physical, moral, and intellectual discipline which capacitates, unfolds, and imbues them with what is good and useful? Woman's duties, taking them altogether, when well and appropriately performed, will do more than all the amusements that can be invented to keep woman well and healthy in every particular. In fact, it is only woman thus employed that can enjoy amusements. To the woman that seeks constantly after amusements, these very amusements become an irksome and toilsome business; they have a disagreeable sameness, and do not divert her; they simply vitiate her tastes. We all want variety, and constant employment, with a *sense of usefulness* attached to it. With this view of the usefulness of mental and bodily labor, I encourage my patients to engage in their domestic duties and labor, gauging the amount of labor by their capacity of endurance. Attention to the homes of wealthy women, as society is now constituted, requires a great deal of anxiety and mental exercise. Without a proper variety of muscular exercise, the woman, in attending to the duties connected with it, becomes more nervous; but the home of the poor industrious citizen or farmer gives enough and a healthy variety of both muscular and mental exercise to promote health and happiness. Should there be such objection in any shape as to make this course impracticable or improper, it is an interesting question to decide what sort of physical exercise is most desirable and beneficial. I am decidedly in favor of exercise on foot, outdoor, as one of the very best kind, far preferable to carriage riding or horseback. The carriage riding is not sufficient exercise for the most of such patients, and yet those who are most debilitated, and utterly unable to walk, may be much benefited by riding in an open carriage until they become vigorous enough to walk, when it should be abandoned. Convalescent patients may ride on horseback if they can have an easy-going animal; but this sort of motion is too violent; there is too much jolting for such cases until nearly or quite cured of the local trouble. We ought to induce our patient to walk more each day than the previous one, if possible, until she has plenty of exercise.

Diet.—Unless during acute suffering, or on account of dyspepsia in some shape, a good substantial or nutritious diet should be allowed; and sometimes we may, with propriety, allow stimulating drinks; but as an ordinary thing, these last should be dispensed with entirely.

Sexual Intercourse.—Young physicians have often asked me whether sexual intercourse is injurious during the time of treatment, and whether it should be permitted. I have no hesitation in insisting upon entire abstinence from this act. The recovery of our patient will be more rapid, certain, and complete, when this is observed; and I believe that failures are the result of carelessness in this respect. It is very common for our patients to enjoy more comfort when absent from their husbands; and come home from a journey, as they think entirely cured, to be assured of the contrary, by the first effort at coition, and become miserable with pain, nervousness, &c., in a short time, on account of indulging in this conjugal act.   I desire therefore to be explicit in warning my young friends in the profession not to omit the interdiction of sexual intercourse, however delicate the task.   A private interview should be sought with the husband for that special purpose.

Main Objects of General Treatment.—The main object to be gained by general treatment, is to palliate the general condition of the system, to aid the local in effecting the cure, and to remove, when practicable, the effects left after a cure of the local disease. A cure of local chronic disease, by general treatment alone, is hardly to be expected; although, in some instances, it may be indispensable to such result.   When general treatment is used as a palliative or adjunct in local diseases, it is directed to the relief of general symptoms attendant upon them.   It will be impossible for me to notice the treatment necessary in all the symptoms which attend and add to the distress of our patients in uterine diseases; but there are certain prominent and troublesome ones, on which I cannot with propriety omit to dwell.   I do so the more readily, from the embarrassment which I know, from experience, fills the mind of the inexperienced, as to the proper value to place upon general treatment, and the course to be pursued.

General Symptoms requiring Special Attention.—The symptoms, the treatment of which I propose to speak of in detail, are, 1st, general nervous prostration; 2dly, nervous excitability, exaltation of nervous excitement; 3dly, anæmia; 4thly, general plethora; 5thly, local plethora; 6thly, constipation; 7thly, indigestion.   These are generally more or less complicated with each other, and sometimes several of them coexist; but, ordinarily, some one assumes the most prominence, and occasions most distress, and consequently requires more of our attention than the others.

Nervous Prostration.—There is often great nervous prostration, and a sense of weakness, when, so far as we can judge, hæmatosis and nutrition are usually well performed.   What is the cause of this depression must be sought out in each case, as there is no uniformity in the functional deviations.   Very frequently there is a deficiency of menstrual discharge, the scantiness being very ob-

vious; at other times it is too copious. We should inquire into
the functions of all the important organs, and correct them when
disordered, as nearly as possible, by changing the habits and circumstances of the patient, and afterward, or in connection, address
remedies to the organs themselves. The stomach, liver, bowels,
skin, kidneys, and uterus, should furnish their discharges in the
most natural manner; and if they are not doing so, should be corrected by the most gentle means. If several of these organs are
in a state of functional deviation from health, we should not expect to correct them all at one time, but alternate our attention
between them; first, with our remedies influencing one, and then
another. I insist here, with reference to the plan to be pursued,
not to address all these organs, or even a large part of them, with
medicinal agents at one time. There is no question, I think, but
that complicated formulæ often nullify themselves by containing
ingredients intended for the liver, kidneys, and skin, which ought
all to act about the same time. We should act upon each of these
alternately, in quick succession, if we think best; but let each
organ feel the full impression of its remedy, before the blood and
nervous energies are directed to another. In addition to this indirect way of increasing the tone of the nervous system, it is natural for us to look about for something that will act more directly.
Our patient becomes so depressed, and suffers so much from terrible feelings of prostration, that her condition appeals to our sympathies for more direct and immediate relief. If left to themselves, or the advice of injudicious friends, they almost always
resort to stimulants, as whisky, ether, chloroform, ammonia, &c.
In some cases only are these temporary remedies advisable, and
when used, they nearly always leave the patient in a worse condition than before they were taken. They are allowable only as
necessary evils, and should be avoided when possible. These patients are usually depressed mentally also, and much good may be
done by operating upon their minds. A physician who enters the
room with a cheerful countenance, and a pleasant and gentle bearing towards the patient, and who engages her in conversation, first
about her case, and afterward about some favorite theme, will do
more toward temporarily relieving the great nervous and mental
depression, than all the ether and ammonia the stomach can be
made to bear. Earnest and kind assurances that her symptoms,

though causing her a great deal of suffering, are not of a serious nature, and will soon subside, act generally as a good cordial to the spirits and nerves. In paroxysms of excessive nervous prostration, despondency, &c., I have seen the tonic influence of very cold air do a great deal toward relieving them. These paroxysms generally occur in close and often heated rooms,—two conditions which should be removed. If it is cold weather, we should cover the patient to protect her, and let the frosty air—the colder the better—into the room, by opening *all* the windows and doors, and keep the room cleared of visitors. It will astonish anybody who has not observed the effect of a temperature near to zero, on those swooning hypochondriacs. A change almost immediately occurs for the better. If the air is not cold, it will still do much good to give it perfectly fresh to the patients in abundance. When able, they may be taken outdoors. This treatment introduces the natural stimulants, oxygen and cold, into the lungs, and brings them in contact with the nerves, and is more enlivening than medicine. How long the room should be kept open and cold, will depend upon the effect, but we should always, if possible, make these patients sleep in open, cold rooms. This is a very important item, which it will often require ingenuity as well as authority to enforce. These patients should live outdoors as nearly as possible, and be as much as they can on their feet.

.*Food.*—Their food should have reference to the condition of the abdominal functions entirely, and be regulated by them. There is generally great intestinal torpor, which should be removed if possible.* Good cheerful company, travel—if the patient will not employ her body and mind in domestic pursuits—temperate and reasonable diversions, and; above all, time and patience are requisite remedies. It is obstinate and chronic, and with the most judicious management will require time, if it does not vanish as the local treatment advances.

Nervous Excitability.—Connected with it in some manner is great nervousness, excitability, irritability, or exaltation of all the nervous phenomena. This nervous irritability shows itself in great mental excitability, want of sleep, unreasonable agitation, restlessness, dissatisfaction; in short, in almost every phase of

* See remarks on treatment of Constipation.

mental, muscular, or nervous excitement. There is also excitability of the different organs, with or without general nervousness, palpitation of the heart, nervous headache, local muscular contraction, &c. Successful management of these nervous and excitable patients requires a careful scrutiny into their general condition; the chylopoietic functions should be regulated in the most careful manner; the skin and kidneys should be attended to with great watchfulness. All that I have said as to general management in cases of nervous depression will apply to this kind of cases. As complete a revolution of the circumstances of the patient should be made as is practicable. From a life of ease, luxury, and absence of care, she should be, if possible, placed in circumstances requiring care, with muscular out-of-door exercise to the greatest extent she is capable of. If we cannot place our patients in situations which their cases require, we can send them on journeys that will require exertion, calculation, care, and the deprivation of their usual domestic luxuries. The remark is frequently made that we must temper our remedies to the delicacy of the patients; and I am afraid that this injunction is misconstrued into the necessity of too great tenderness of treatment. The better rule is to make use of such means as will raise the patient from her state of delicacy to robustness. It is the delicacy of her constitution that causes her to suffer so much. This can be strengthened only by proper physical, moral, and mental training. The moral and mental condition of our patients when so very excitable should be attended to. Improper reading and society should be avoided, and social and literary habits should be reduced to great plainness and simplicity. Above all things, books and society should not interfere with regular rest, exercise, and outdoor exposure. As I have said before, this last should be as great in amount as can be borne, accompanied with active muscular exercise, as walking, and should be practised in all weathers, sufficient protection being secured by enough clothing of the right sort. With regard to the use of medicine, it is a fact, that it is an exceedingly difficult thing to find any remedy that does not produce exaggerated and in most cases disagreeable and even injurious effects. So much excitability of the nervous system nearly always modifies the effects of remedies, and we can seldom predict the operation of any of them, nor can we determine the value of any until it has been tried. When tonics can be borne, they often very much relieve and sometimes entirely cure this great nervous excitability. Of the mineral tonics, probably bismuth, arsenic, and zinc agree best. Iron is not frequently tolerated in any shape by these very nervous patients. Quinine, nux vomica, cherry and chamomile are the best tonics in these cases, but we must not be surprised if none of them are borne. Alcoholic stimulants in general agree with them, and are the best cordials for temporary nervous excitement, but should be conscientiously avoided when possible, as not a few, I am sorry to say, of most estimable and intelligent women have used them too much, and engendered an appetite that could not be denied. Opium, as well as the narcotics generally, fails to have any good effect, but on the contrary disagrees with the patient totally. This, however, is not always the case, as it acts like a charm with some. In all it should be studiously avoided as deleterious in the long run, and there is danger of creating an appetite for it. We may the more readily be persuaded to omit the use of all these medicines, as their effects are temporary, while hygienic and regimenal remedies are permanent in their effects. The management of those cases of localized nervousness or unnatural excitability in particular organs, as palpitations of the heart, nervous headache, &c., is about the same as above, except that more attention to the stomach, from which they usually arise, may be necessary.

Anæmia.—Anæmia, with its disagreeable concomitants, sometimes also calls for separate treatment. It would be an unnecessary waste of time and space to enter minutely into the general treatment necessary, where anæmia is the prominent and troublesome symptom. This condition calls for the same treatment found useful under other circumstances, and while it may not be entirely amenable to, it will be very much benefited by the remedies indicated by the state of the blood. Iron, cod-liver oil, quinine, bitter infusions, and nutritious diet, with outdoor exercise to the extent the patient can bear, are the efficient remedies.

Plethora.—But we sometimes find general plethora instead of anæmia, a state in which there is actually an unusual amount and a too rich composition of the blood. I need not dwell upon this general state of the system, as the treatment is simple and familiar. The great fear is that, on account of the painfulness about the hips and legs, the patient may be too much inclined to an inactive life. On no account should this class of patients be allowed their ease; they must be urged to use up their surplus blood in active exercise, and the kind of exercise, next to the cares and labor of a household, best adapted to them, is walking. Every muscle in her body must be brought into action; every secretion must be kept free, and her mind ought to be taxed to continuous effort during the day by some useful occupation, while the strictest temperance, with reference to ingesta, should be her rule of living. Obesity and the troublesome and dangerous effects of plethora will be thus avoided, connected or unconnected with general plethora.

Local Congestions.—We sometimes meet with instances of violent, dangerous, and even fatal determinations of blood to particular organs, as the consequence of the general ill health which accompanies uterine disease, such as stupor, stertorous breathing, &c., indicating an oppressed condition of the brain, great dyspnœa, and sense of suffocation, showing congestion of the lungs. The treatment of these congestions does not differ from what would be appropriate under other circumstances of their occurrence, and consists in revellents, alteratives, &c. The most frequent, and perhaps obstinate, of the local congestions, are such as occur in the chylopoietic viscera, manifested by excessive secretion and discharges from the stomach and bowels. It is not uncommon for these patients to have suddenly-recurring attacks of vomiting, cramps in the stomach and bowels, diarrhœa, and consequent great distress. Aside from the local treatment, we will be called upon to exert our skill against the exhausting and depressing influences of these attacks. It will almost always be found that such attacks are preceded by constipation, with scanty secretions, furred tongue, and other evidence of unhealthy secretions. By carefully correcting this condition, we may avert these painful and exhausting occurrences. The plan recommended and so much prescribed by Abernethy, will often palliate very much, viz., six or eight grains of blue mass, at night, worked off by some saline cathartic in the morning of every fourth or fifth day. If there is more permanent diarrhœa, great care should be exercised in the choice of diet; the use of warm baths should be recommended, very warm clothing, and not much medicine, as the cure will depend upon the appropriate treatment of the local disease, instead of the treatment of the general symptoms. All these symptoms, except the diarrhœa, are apt to be moderate, and can be borne until the diseased uterus is cured; but there are two symptoms so very annoying, and which require so much patience in the treatment, and exercise so much unfavorable influence upon the uterine disease, that I hope I will be pardoned by the reader for dwelling upon them more at length.

Constipation.—I allude to constipation and indigestion, particularly the former. I have already spoken of the deleterious influence of constipation, and I think I am justified in saying that, if disregarded, it retards the cure of chronic diseases of the unimpregnated uterus more than any other sympathetic affection. And I wish to warn the practitioner to be very particular in attending to this symptom. There is probably more tendency to costiveness in females than in males, chiefly owing to difference in habits. Sedentary life, confinement to close, badly ventilated rooms, are among the circumstances that bring on this condition. Irregularity of meals, late hours, deficient sleep, concentrated diet, imperfect mastication of food, all should be corrected, as any one of them alone will do harm, and all or any of these combined—and this is frequently the case—are very deleterious to the functions of the alimentary canal. But an inexcusable and very common custom of most females is making the act of defecation a disagreeable and procrastinated necessity, instead of a pleasant and punctual duty. The most trivial excuse—the presence of friends, a little cold, hot, or wet weather, being among strangers, or a slightly inconvenient distance from a proper place—will frequently be sufficient to limit defecation to once a week; then the act is performed in a hurried manner. It is amazing to know to what lengths this negligence is often carried. I have known two weeks to have transpired, frequently, according to the history of patients, without any attempt to relieve the bowels. Now this should be corrected by persistent method. The habit of eating from hunger at certain hours depends upon lifelong practice, and, when once established, cannot be changed without violence to many functions, causing urgent and repeated demands upon the system for a resumption of it. Regular bowels come from an equally long-continued habit of going to the close-stool at particular hours of the day. Years of negligence destroy the habitual regularity with which the bowels move; hence we should not be discouraged if the habit be not re-established without long perseverance. A new habit cannot be formed, nor an old one altered, without long and persevering effort in the right direction. We should, therefore, encourage a patient that is in earnest in her search after health, to persevere for months, years, and indeed her whole life, if necessary, in going to her water-closet without fail, once every day, at a certain hour, as regularly as the clock points to it. This is indispensable to a correction of the bad habit of constipation. A very effective part of this regular endeavor is to cause the mind to dwell upon the necessity for an evacuation, and the process itself, for at least half an hour before retiring to the proper place. It is not a difficult matter, with many persons, to create a desire in this way. Let no consideration of convenience enter into this punctual effort at stool. Arrived at the proper place, the position should be an easy one; no inconvenient strain upon any muscle should be allowed, and the patient should be possessed with an entire sense of leisure to perform the act completely. The value of all these considerations, where faithfully followed, is incalculable, and very few cases can long resist them. Without them, medicine will only temporarily relieve, instead of permanently curing, obstinate cases. I should caution against severe effort, or straining, as it is called; let time, patience, and gentle effort be the plan. Another matter of great importance, when an effort is made to have an evacuation, is to have the abdomen distended by ingesta. The patient should be instructed to eat plentifully of vegetable diet, such as by its bulk is calculated to produce fulness. If the patient go to the water-closet with a sense of fulness of the abdomen, success will be much more likely. Should the regular time for making an effort be soon after breakfast, which is undoubtedly the best time, and the meal has not been sufficient to produce a sense of moderate distension, a full glass of water will complete that condition. For the purpose of giving fulness and a sense of distension, various kinds of ripe fruit may be resorted to with advantage. In prescribing fruit for constipation, we should bear in mind that there are three indications fulfilled by it, some kinds fulfilling all, while others fulfil only a part of them. They are, first and best, distension; secondly, increase of secretion, on account of their acids; and thirdly, increasing peristaltic action of the bowels by indigestible fibres, seeds, or rind. Ripe and mellow apples, without being divested of the rind, may be eaten in sufficient quantities to produce a sense of fulness, and this should always be at the conclusion of a meal, breakfast, for instance; the acids will increase the intestinal secretion, and the rind quicken the peristaltic motion of the bowels by acting directly upon the mucous membrane, and through it on the muscular structure. Very acid fruit, as lemon and orange, only produce their effect on account of the acids they contain. They are excellent as a part of the ingesta of patients whose stools are dry and hard and lumpy. Fruits containing an abundance of seeds, as figs, or of rind, as tamarind, &c., increase the peristaltic action without causing much secretion. By inquiring into the character of the stools, we will have a good guide as to the kind or mixture of fruits to be selected. There are kinds of diet, breads particularly, that act like these last fruits, and may be used in conjunction with or independent of them. Breads in which the bran, or hull of the grain, is contained in considerable quantities, are of this character. The Graham bread, as it is usually called, ordinary coarse brown corn or wheat bread, are those mostly resorted to. When this kind of bread is used for constipation, it should be eaten at breakfast, dinner, and supper, in such quantities as the experience of the patient finds necessary. I have advised patients who could not use the coarse breads, to make what may be called bran crackers. A tablespoonful of flour, one pint of wheat bran, two tablespoonfuls of white sugar, and water enough to make them all into a pasty mixture, are the ingredients. This mixture is made into cakes, small or large, as may be wished, and baked in an oven until hard. When soaked in tea, coffee, or milk, they are not unpleasant. I have known patients benefited by swallowing certain seeds, with the rind whole. A tablespoonful of wheat grains, oats, barley, white mustard seed, &c., can all be used for this purpose, and are not more disagreeable than medicines. Another kind of diet which may be used to produce the kind of effect here aimed at, consists of the various small vegetables, as celery, radishes, pepper-grass, lettuce, asparagus, cabbage, &c. These may all be taken in quantities to cause distension.

In speaking of fruits, I ought to mention the berries as an excellent means, cheap and easily procured, to accomplish all the objects attained by other fruits.

Everything should be done by habitual effort, exercise, diet, drink, &c., before resorting to the use of medicines; because, as is well known to the patients generally as well as to the practitioner, the more medicines taken the more will be necessary. They lose their influence, and the dose must be increased in order to produce a full effect. This is almost always the case. Notwithstanding this evil, we are often reduced to the necessity of using laxatives to overcome constipation. To a just and intelligent application of medicines in the treatment of constipation, it is indispensably necessary to make ourselves acquainted with the condition of the alimentary canal with reference to its secretions and muscular powers. It will be found that there are sometimes great deficiency of secretion, and torpor or want of vitality of the muscular structure, or weakness of this tissue. The want of secretion may be in the upper portion, in which case the bilious color is wanting in the stools, or the small intestines may give out less watery material, and then the stools are less fluid, or even dry. The secretions may also be deficient in the lower portion, or colon; in which case the fæces will be scybalous, dry, and lumpy. The muscular torpor, from want of irritability, is more frequent in the colon or rectum than in the small intestines. When in the colon, there is increase in size of the lower abdomen, sense of fulness and hardness, and the fæces are expelled with great difficulty. If there is sufficient activity of the colon, but the rectum is torpid, large accumulations occur there, the pelvic distress is increased, and nervousness, general and local, is exceedingly annoying. Sometimes all these conditions are combined to render the case one of the most troublesome and difficult to manage. Mechanical obstruction by stricture of the rectum, formed by pressure of the uterus, may give rise to chronic constipation, which may become permanent and almost incurable; or the uterus, by lying on the bowel and pressing it against the sacrum, often gives rise to costiveness, that can be removed only by correcting the position of that organ. It is not sufficient to know that the patient does not have regular operations from the bowels, but we must know why she is thus constipated. Whether on account of want of secretion, and if so, of what secretion; whether it is attributable to general debility, combined with muscular weakness of the intestines; whether to lack of irritability of the intestinal tube and consequent torpor, and if so, where is this lack of irritability? Does it exist in the whole length of the canal, in the colon, or the rectum? Or whether there is obstruction from stricture in the rectum, piles, thickening of the mucous membrane, rigidity of the sphincter, or pressure from the uterus bearing so heavily upon it. To give a laxative merely because it ordinarily produces a fecal discharge, is always unphilosophical, and sometimes exceedingly injurious in its effects. I think it is inattention to the exact state of the alimentary canal that makes constipation so often incurable. For constipation, attended with very dry, hard stools, showing a deficiency in all the secretions from the bowels, in addition to the course of diet, including acid fruits, &c., our object should be to administer such drugs as will most effectually stimulate to secretion. The various saline medicines are indicated. Sulphate of magnesia is a most excellent one, and a good way of administering it is in combination with sulphuric acid. From one to two drachms, or even half an ounce, given in combination with acid enough to taste somewhat sharply, will promote secretion along the whole of the small intestines, cause a large effusion of water, which will dissolve the fæces and render their evacuation easy and sure. In the morning, some time before eating, is the best time to take it. When there is reason to believe that the portal circulation is slow, and the liver furnishing less than its usual amount of secretion, some form of mercurial should be used with the salts. If the case is chronic and the constipation obstinate, we may give six to ten grains of blue mass, in pills, at bedtime every fourth or fifth night, and follow it with Epsom salts in the morning. A continuance of this alterative cathartic for four to six weeks seldom fails to cause a change in the alimentary secretions. Sometimes it is better to give these cathartics nearer, and sometimes farther apart. We must judge of this more by the susceptibility to the constitutional influence of mercury than anything else. It is almost always the case that this very scanty state of the secretions is accompanied with an impoverished state of the blood; hence iron in some shape will be beneficial in most cases. If there is much debility, a long course of tonics will be indispensable. It may often happen that this scanty condition of the secretions is attended with debility of the muscular fibre of the intestinal canal. When this is the case, we must add to the above treatment that which is applicable to this kind of intestinal torpor, which I shall not consider. Before doing so, however, I will remark that several other salts will answer as well, and sometimes even better, than sul. magnesia. The kind of tonics which are most effectual in debility of the muscular structure of the intestinal canal are such as give general strength, and it is mostly desirable to combine them with special tonics. The later are rhubarb and nux vomica. These have always seemed to me to have a special tonic influence upon the intestinal tube, and, when properly given, to increase the susceptibility to their own action. The rhubarb, although an alimentary tonic, induces less susceptibility to its own influence than the nux vomica. The best way to give the rhubarb is either in the root, without pulverization, or in the extract. When given alone in the root, the patient can take a little twice a day, by chewing it, and after mixing with the saliva swallowing it. A little experience will enable the patient to judge of the right quantity, which she can repeat as often as it is required. When the rhubarb is taken in this way, she may also take a solution of sul. ferri and strychnia, in water, one grain of the former to one-sixteenth of a grain of the latter.

A formula that is very simple and effective is as follows:

R. Strychnia, ... gr. j.
Ferri Sul., ... gr. viij.
Acid Sul., q. s.
Aqua, ... ℥ij.

Mix. Make solution. One teaspoonful three times a day after eating. Sixteen grains sul. quin. may be added to the above formula with advantage. Or,

R. Strychnia, ... gr. j.
Ext. Rhei, ... ℈iss.
Sul. Ferri, ... gr. x.

Mix. Make sixteen pills. One to be taken once, twice, or three times a day, as may be found necessary.

I have often succeeded in overcoming this constipation or debility by giving one grain of sul. quin. with five grains of powdered nux vomica after each meal. Or the same amount of nux vomica, with iron by hydrogen, two grains, after eating, each time. It is usual to use aloes in the constipation of uterine diseases; but I have found very few cases with which this drug did not disagree. But there is a torpor of the intestines where general tonics cannot be borne; where, in fact, there does not seem to be any general debility,—there is only a want of susceptibility to the stimuli which ordinarily arouse them to action. The secretions color the fæces properly and give them sufficient moisture: there seems to be no fault in their appearance, consistence, odor, or other character whatever. They are deficient only. The patient may be plethoric and florid, her general muscular strength sufficient, and her blood, so far as we can judge, good in composition. Special tonics and stimuli are indicated in such instances, and they alone should be used. Such measures should be adopted as will arouse the muscular action of the intestines. Nux vomica, in five-grain doses, with the rhubarb extract or without it, or the strychnia in solution, in sixteenth to twentieth of a grain doses, constitute our most valuable medicinal appliances. This is the kind of constipation that is most benefited and is most amenable to a persevering regimenal and dietetic course of management, such as I have above endeavored to give. In addition to the rhubarb and nux vomica treatment, we may get some good from external appliances and manipulations to the walls of the abdomen. The most valuable, when gently, perseveringly, and methodically applied is what is understood by the term kneading. The colon is the torpid portion in most cases of this sort of constipation. The process of kneading consists in handling it so as to stimulate its fibres directly. One plan is to grasp it with the hand and squeeze it from one end to the other. We should begin at the right groin, and with a knowledge of the position and direction of it, grasp it with both hands at this point, then a little higher up on the same side, and then a little higher, until we reach the right hypochondriac region. We should then follow it across the abdomen to the left hypochondriac region, and thence down to the left iliac. Or, we may double our hands as bakers do when kneading their dough, and standing over the patient, press with the knuckles of both hands, first in the right iliac region, and imitating the process of kneading, pass slowly from this to the right hypochondriac, thence across the abdomen and down as before directed. If we trust this process to a non-professional attendant, we should be sure to show how to do it, as it is important that it should be done right. When this process of kneading or squeezing the colon is first instituted, it should be practised with the utmost gentleness, but the force and rapidity of motion may be increased until great freedom may be made use of. It should be resorted to a short time before retiring to the water-closet, say half an hour. Some patients find an efficient laxative in what they sometimes call a water compress, applied to the abdomen over night. It is made by doubling a napkin several times, so as to make a thick compress, large enough to cover the entire abdomen anteriorly. This is saturated with water, and, after being placed upon the abdomen, covered with a roller or bandage so as to keep it in place. It is thus allowed to remain from the time of going to bed until the time to rise in the morning. I think this water compress is best adapted to cases in which there is a deficiency of secretion in the intestinal tube. A bandage, or, what is better, a roller applied tightly enough to press the walls strongly upon the contents of the abdomen, frequently stimulates them to proper action, both as it respects secretion and peristaltic motion. When it is determined to use the roller or bandage for its stimulating influence, it ought to be applied upon rising in the morning, or, what is perhaps better, immediately after breakfast. This bandage should not be worn constantly, nor even many hours in the day. From the time of rising until two hours after breakfast, or from breakfast for three hours thereafter, will be long enough. The constant use of the bandage would but increase the evil—lax abdominal muscles—for which it is advised. Before leaving this part of the subject, I desire to say, with reference to the freedom with which I have advised nux vomica to overcome intestinal torpor, that in all cases we should remember its effects are cumulative, and quite a difference of susceptibility to its influence is manifested by different persons, in consequence of which the patient should be watched, and the dose graduated to the least quantity necessary in the case. Although I have given nux vomica and strychnia for a considerable length of time to a great variety of persons, and for several weeks together, I have never seen anything more than slight inconvenience from it in the shape of nervous startings. Very rarely we meet persons who cannot take it at all; it disagrees with them as soon as they commence its use.—There is another species of intestinal torpor, of a very obstinate character and very distressing to the patient; I mean a lax, torpid rectum; so torpid as to allow the fæces to accumulate in large quantities and cause great inconvenience from pressure. To such an extent does this collection sometimes go as to press the posterior walls of the vagina forward and protrude it between the labia. The first indication in such cases is to dissolve the fecal mass and discharge it. Various kinds of injections are useful for this purpose, warm oil, warm water, &c., but one which I have seen do much good is composed of one ounce of fresh oxgall and four ounces warm water. This composition dissolves the fæces very readily, and the fresh bile stimulates the intestine to their expulsion. The evacuation of course will give only temporary relief, and there remains the most important indication, that of giving tone to the bowels, with a view of preventing the accumulation in future. This is difficult, and in some instances of long standing quite impossible. Much good can be done in nearly, all cases, however, and we do not discharge our duty if we do not try to relieve when we cannot cure every case. Cold water thrown into the rectum once or twice a day, in small quantities,—eight ounces,—is always good, without some special reason to the contrary. There are generally two indications to be fulfilled in these cases: relaxation of the sphincters and restoring the tonicity of the proper rectal fibres. It is a singular fact, which I think I have observed, that the sphincter muscles increase in strength with the advance of age; this is one of the causes why the fæces are voided with more difficulty in old persons. To give tone to the rectal muscles, astringent injections have been recommended and extensively used; but in my practice they have been almost uniformly useless, many times injurious, and always disagreeable. They dry up the secretions, an evil not to be compensated for by any other effect; they do not, so far as I can judge, cause contraction of the muscular fibres, but they are very apt, if persisted in for a length of time, to cause inflammation. I have derived more benefit from tonic suppositories and injections than from any other kind of medicinal treatment. A suppository of twenty grains extract gentian, or five grains sul. quin., ten grains extract cornus florida, or a mucilaginous suspension of any of these, introduced into the rectum every night at bedtime, and retained if possible until morning, are good tonics and eligible modes of using them. It will be necessary to secure the retention and efficient contact of these tonics, to first, empty the bowels with ox gall and warm water, and afterwards introduce them with as little irritation as possible. The quantity of mucilaginous material should not exceed two ounces. The tonic treatment of this kind must be varied, taking first one tonic and then another, in first one form and then a different one, and must be kept up for a long time to do much good. We cannot be too careful in all our treatment, to avoid anything to which the rectum shows any sensitiveness. When it becomes tender and sensitive, we should at once desist until all of this has subsided before we are justified in beginning again. It too frequently happens that both the physician and patient become discouraged, and desist before the remedies have had a fair trial.

Is there anything that will relax the sphincter ani? I am not aware that any means operate with efficiency in this direction, but I have used, in a few instances, with apparent benefit, the ointment of belladonna, made by mixing the extract with lard. I apply it to the anus externally upon going to bed at night, and think that it promises decided encouragement to continue, until the question against or in favor of its usefulness is fully determined. This is a good formula.

R. Ext. Belladonna, ... ℥ij.
Ung. Simplex, ... ℥j.

Mix. Make an ointment. The parts to be well smeared with it at bedtime.

This application certainly removes the irritability of the sphincter, which causes it sometimes to resist the extrusion of the fæces. As I have before remarked, there are cases in which this relaxation cannot be cured: we are then compelled to resort to palliatives, and we must be careful to palliate intelligently. We are to give the weak rectum artificial support, to enable it to retain as near as may be its ordinary size. This can be done only through the vagina. An air or sponge pessary introduced into the vagina, so as to press the rectum against the sacrum, and thus diminish its capacity, will prevent the great accumulations from taking place, and in that way prevent one source of great inconvenience. Dr. Hodge recommends the globe pessary for this condition of the rectum, which answers very well in many cases, perhaps in the majority; but each case must be studied with reference to its own peculiarities, and the shape, size, and consistency of the pessary adapted to it. When our object is palliation alone, there is no objection to wearing the pessary all the time, but if it is used to palliate what we believe to be a curable case, we ought to use it intermittingly, and the patient should not wear it at night especially. It would probably be better in a majority of the cases to introduce it before rising in the morning, and allow it to remain until noon. One thing I think essential in the size and position of the pessary, and that is, that it does not compress the rectum below its natural capacity; there should be room enough for an ordinary amount of fæces in it, lest it become a source of obstruction, which it will do when larger or improperly placed. As will be noticed, I have omitted to say anything of enemata in constipation, from inactivity of the colon or upper portion of the alimentary canal. As an occasional means, injections operate well; but, like other laxatives, when used for a length of time they lose their influence entirely. If we determine to use injections as an habitual laxative, by proper changes in kind and quantity, we may prolong their efficacy very much. To a person unused to them, half a pint of cold water will act very well. When the bowels fail to respond to this quantity, there ought to be an increase of two or three ounces, and then that amount used until its effects are not satisfactory, when a few ounces, more should be added, and so on we may increase the amount until the quantity becomes intolerable. When this is the case, we may order half a pint of water with a drachm or two of common salt, chlorate potassa, or nitrate of soda or potassa. We should increase the quantity of water or strength of solution, or both, as the susceptibility of the rectum decreased, until we cannot carry either farther. After we have thus obtained as much good from injections as we can, it is sometimes expedient to use suppositories as laxatives. Suppositories are made of laxative medicines or of any other material. Compound extract colocynth or other purgative extract may be used; or we may inclose in some of the extracts a dose of the podophyllum, or any of the purgative resinoids or alkaloids. These should be retained until absorption takes place. The common suppository of soap, tallow, and wax, sperm, stearine, &c., are of the second kind. It not unfrequently happens that the above modes of using injections and suppositories may be alternated very profitably, the full effects of each being experienced upon their resumption after having used the other for a time. But some persons cannot use injections; the rectum is too sensitive, and attempts to do so induce so much irritation that they must abandon them. In such cases suppositories are out of the question.

I have elsewhere shown that the uterus, by its wrong position, sometimes presses upon the rectum and obstructs the passage of the·fæces. This may be effected by retroversion or prolapse. The indication of course is to restore the uterus to its proper place, and as I shall have occasion to speak elsewhere of these difficulties (malpositions), I do not think it necessary to more than mention them here.

Indigestion is another very troublesome condition among the many which attend uterine disease, and it will demand much of our attention. It would not be profitable to dwell at any great length upon this symptom, as it will become the duty of the physician to study each case separately. Attention to the bowels, keeping them perfectly regular, will very much alleviate if it does not cure most cases; but sometimes we find the stomach very seriously disordered when the bowels are perfectly regular. In such cases we should inquire into the alkalinity and acidity of the urine as a good index for the administration of medicine. If the urine is highly alkaline, acids and bitters are indicated and will be well borne; if the urine is highly acid in its manifestation, alkalies must be used,—liquor potassa, lime-water, soda, &c. In the former case, animal diet may be tried; in the latter, vegetable diet as likely to be good palliatives, and under proper circumstances curative. The indigestion, like most other symptoms, however, will be obstinate and generally incurable until after the local disease is cured. It may be inferred from what I have already said, that I consider the general treatment, as I have endeavored to sketch it, of secondary importance, and the local as the essential treatment; but wishing to be perfectly clear on this point, I will reiterate what I have already said in regard to the objects of general treatment. They are—1st, To palliate the general condition of the patient before and during local treatment; 2dly, To aid local treatment in effecting the cure; and 3dly, To cure the effects which may remain after the local disease has been removed. I do not believe that a cure is ever effected by general treatment alone.

CHAPTER XIX.

LOCAL TREATMENT.

Baths.—The local treatment of inflammation of the cervix uteri is made up of several therapeutic items, varying according to the intensity, quality, and seat of disease. Of these there are, however, a few that are applicable to almost all cases; hence their description, modes of use, &c., may be considered before going farther. Baths, injections, and some minor remedies are of this kind. Water, when applied to the surface, is purely sedative in its effects if it is of the temperature of the part on which it is used. If the bath is partial, the sedative influence is for the most part confined or limited to the part to which the application is made. So with injections per rectum or vaginam. They sooth the parts contained in the pelvis. If the water is warmer than the part of the surface bathed, the effect is stimulant; if it is colder, by virtue of the physiological action brought into play, it is first sedative and then stimulant. The circulation and nervous influence of the vagina, for instance, when the cold water is first thrown into it, are depressed, but very soon after its evacuation, or withdrawal, the vessels become excited to increased circulation of blood, and increased heat takes place and the nerves become more sensitive. In all these respects baths and injections act alike. The injections are internal baths; the uterus is bathed through the vagina by injections. But the effects of baths and injections may be modified by containing medicinal substances. They may be rendered more stimulant or more sedative, or be even made to possess other qualities by impregnation with medicines: one of which in very common use is astringent in character. Another mode of using water and applying it, either simple or impregnated with medicine, is to wet a cloth or a sponge with it and bind it to the surface, or introduce it into the vagina. Several thicknesses of cotton cloth applied to the abdomen and saturated with water is what is called the water compress; and often when allowed to remain in contact with the skin for several hours it produces considerable excitement, and if persisted in for days, will cause first a vesicular, next a pustular, and finally a phlegmonous eruption. The way to render it effective is, after applying the wet cloth compress to cover it over with oil-silk, and then confine the whole with a bandage or roller, with a view to prevent evaporation. Sponge introduced into the vagina saturated with water holding some medicine in solution is a common way of affecting the uterus. I do not design giving an extended view of baths or their application and modus operandi, but so much aid is occasionally obtained by the use of baths, that I cannot refrain from speaking of the application of some forms of them to diseases of the uterus. The bath most applicable in inflammation of the cervix uteri and most commonly used is the sitz or hip bath. It is intended to allay the inflammatory irritation and pain. It is often the case that there is a great deal of suffering from pain without much inflammatory action in the parts; in these cases a sitz bath will often give great relief. In many instances the efficacy of the bath may be enhanced by having the patient introduce a speculum while in the water, so that it may pass up the vagina to the neck of the uterus, and thus directly affect the part diseased. In cases of medicated sitz baths, the organ may thus receive the full benefit of the saline, anodyne, or other medicinal impregnation. The common glass tube will do very well for this use, where we wish only to bathe the neck of the uterus; but if we wish the fluid to come in contact with the vaginal walls and remain there for a considerable time, the wire speculum is the best. While speaking of the use of the speculum in this way, I may mention that a very efficacious mode of applying medicated washes without the bath to the cervix uteri or vaginal walls, is to have the patient lie upon her back, introduce the speculum, and then pour the fluid into it. By remaining in that position she can retain the contact of the medicated solution as long as desirable.. Ice-water, ice, astringent powders, or almost any form of substance may be applied and retained in contact with the os and cervix uteri with great advantage in this way. This mode of using remedies is particularly useful in bleeding fungus or vascular tumor of any sort.

Hip Bath.—The sitz bath, when a patient is suffering with the pain and heat of uterine disease, may be used as often as necessary, twice a day at least; but three, four, or even a greater number of times will not be too often, when they are found to be soothing and useful. We may extemporize a hip or sitz bath, by putting water in a common washing tub; but the cheap vessels made for the purpose are within the command of almost all persons. There should be so much water that when the patient sits down in it, the whole pelvis will be covered.

Temperature of the Bath.—What should be the temperature of the bath? The patient's sense of comfort, or discomfort, from its use, should be our guide in this respect. We should seek a temperature that is comfortable and soothing to the patient, while in the water, and that leaves no sense of discomfort. The baths are intended for, and should add to the comfort of the patient; when they do not do this, they should at once be discontinued. As a general rule, I advise my patient to take tepid water for her first bath, and then gradually use them cooler until they are cold, unless they become disagreeable in some respect; if they do so, to continue them tepid. The colder a bath is, the more good it does, provided it be comfortable. The time for taking it may be regulated by the convenience of the patient, and the necessity for it, with the view of allaying pain, heat, &c.; probably in the majority of instances, the most advisable times for taking it are upon rising and retiring. The length of time the patient remains in the bath should also be regulated somewhat by their effects. If the patient remain too long in the water, it will debilitate her, particularly if there is considerable water and the bath is frequently repeated; on the other hand, if she does not remain long enough, she will not derive any benefit from it. She may try remaining in it fifteen minutes, if she does not find herself very much relieved before that time, and she ought to be governed in her use of subsequent baths in this particular by the effects of the first few trials. While in the bath, the intended temperature of the water may be kept up by adding hot water from time to time. The hip bath is used almost wholly with reference to the local disease, but when general baths are required, it is usually for the relief of some attendant general symptom.

Shower Bath.—The shower bath may be used as a roborant excitor of the circulation, if upon trial it can be borne, and produce a good effect. Some patients think they are very much benefited by the shower bath, and say they cannot do without it.

Sponge Bath.—The sponge bath is useful in causing a tonic and soothing reaction upon the surface. Neither of these can be tolerated by very feeble patients. The cold or tepid sponge bath, administered at bedtime, not unfrequently soothes nervous irritability, and enables restless persons to sleep soundly. I have not used baths in any other form than these, but when used as I have here indicated, I have seen such pleasant results from them, that I cannot refrain from recommending them.

Injections.—These are applicable to almost all cases of inflammation of the cervix uteri, do a great deal of good, and are believed to be sufficient to cure some cases. As I have before said, they may be used as internal baths, to get the influence of water and temperature on the vagina and uterus, as the medium for the application of medicinal substances to the mucous surface of this cavity and viscus, and also as detergents, to wash the vagina of all substances that should be removed from it, for purposes of cleanliness. In some one of these forms injections may be used in nearly every sort of cervical inflammation. The simple injection of water may, and ought to be used, by all females who have inflammation of the uterine neck. The medicated injections can be useful only in cases where the inflammation is within reach of them, as when inflammation affects the mucous membrane of the vagina, or the membrane covering the external surface of the vaginal portion of the cervix. For obvious reasons, injections containing medicines can hardly do any good, by virtue of the solution, when the inflammation is situated inside the cervical cavity. Vaginal injections cannot reach the seat of disease. I have not used intra-uterine injections, as I think there are less hazardous modes of conveying medicines into the cavity of that organ. The efficacy of injections depends very much upon the manner in which they are administered, and the kind of instrument used.

Manner of Using Injections—Kind of Syringe.—The essential quality of a syringe is its capability of receiving at one end and discharging at the other perpetually, so that any quantity of water may be used without withdrawing and reintroducing the pipe. A large number of forms of syringe have been invented, but for convenience, that form is, I think, preferable which has a vulcanized rubber hollow ball mounted in the middle of a long flexible tube; by pressing on this ball, and relaxing it, the water is drawn in at one end and forced out at the other. A pewter, britannia, or ivory tube delivers the water into the vagina, and by its length may be made to convey it to the uppermost part of that cavity, and thus completely wash the whole of its walls. A siphon may be made to answer the same purpose, by having the fountain high enough to give some force to the current. Should the patient use a syringe of the above description, she may sit over one vessel, and have the water in another in front of her. By inserting one end in the vagina, and the other in the vessel of water, the whole of it may be made to pass through the vagina and fall into the vessel beneath her, and thus do away with the inconvenience of undressing.

Quantity of Injection.—The quantity necessary to be used in an injection will vary very much in different sorts of cases; if water alone is to be used, and we wish to get the sedative influence, the quantity must generally be large, that is, from one to eight quarts; if we wish to stimulate the uterus with very warm water, a large quantity will also be necessary; if we wish the injections cold, it is better not to use so much. ·

Medicated Injections.—The medicated injections also should be large or small, according to the effect we wish to produce, and the strength of the solution. A pint, or at most a quart, will be sufficient for astringent injections. We often use anodyne injections on account of their soothing influence upon the sensitive parts. As a general rule, anodyne injections need not to be very large, say a pint, or less, but the patient can continue passing it through the vagina until its effect is attained. This may be done by using only one vessel, pumping from, and allowing it to fall into the same. Frequency must be determined also by the object of the injection. Simple water injections can be used more frequently than medicated ones, and anodyne more frequently than astringent. The simple injections, if they afford relief, may be used from three to six times a day, or oftener; narcotics three or four times, or oftener, owing to the urgency of the symptoms requiring them, and the good they are found to do.

Astringent Injections.—Astringent injections ought not to be made use of, as a general thing, oftener than twice a day, and in some cases to which they are applicable, this is entirely too often. Of all the vaginal injections used, the astringents are most commonly resorted to, and are productive of most good.

Modus Operandi.—When an astringent is thrown into the vagina, the first effect is to coagulate the mucus, pus, or blood contained in it ; after this, its contact with the mucous membrane becomes more intimate, and its influence is exerted upon the capillary bloodvessels, and the glandulæ or crypts. The vessels are constricted in size, and circulate less blood, and the calibre and functional activity of the crypts are diminished, and little congestions and inflammation are for the most part cured, or at any rate benefited. When the vessels are circulating too much blood, and the muciparous apparatus furnishing too much secretion, this astringency is desirable. We ought not, with certain exceptions, to use astringent injections when there is no hyper-secretion from the mucous membrane of the vagina or cervix uteri, nor an ulcerated or inflamed surface with which the solution can come in contact. The frequency with which they may be used must be indicated by observing these two effects, and the dryness more particularly.

Frequency of Using.—I think we may lay down a rule for repeating them, like this : never repeat an astringent injection while the vagina is dry from the effects of a preceding one. We should, after obtaining the full astringency of an injection, in the stoppage of a leucorrhœal discharge, wait until the mucus again renders the mucous membrane moist. It will be found, very often, that this requires twenty-four and even thirty-six hours to take place. A disregard of this direction will sometimes induce an increase of inflammation, and give our patient great inconvenience. In fact, too long a continuance of astringent injections is apt to cause vaginal inflammation.

Alternate different Astringent Remedies.—I think, however much an astringent may be indicated, that the same article ought not to be used more than twelve or fourteen consecutive days, and should then be alternated with another one of the same class, or simpler ones. This last I generally prefer. A permanent dryness of the vagina after any one astringent, should preclude the use of that article at least, and cause us to try another, and so on until we get one that will agree with the case ; or else we must abandon all astringents, and fall back upon simple water.

Température of Injections.—I know of no better rule to govern the temperature of injections than the comfort of the patient. After a trial of tepid, warm, cool, and cold, let her suit herself by the effect they have upon her. Any temperature that is disagreeable should be avoided. The extract of opium makes a good anodyne injection. Five grains to a pint of tepid water, used for ten minutes, a quarter or half an hour, will often allay pain, arising from inflammation within the vagina, very readily : or one grain of extract of belladonna may be used in the same way. In fact, we may choose among the narcotic extracts, remembering that the solution must be impregnated with at least three doses of the medicine. Among the astringents, alum is the most common, the most useful and efficient astringent. It possesses the advantage of having no poisonous ingredient in it. As Dr. Bennett has taught us, it sometimes produces severe inflammation ; but this is doubtless owing to the inconsiderate use of it, and arises from its very efficacy in suppressing the vaginal secretion. One drachm to the quart of water, tepid, cold or warm, as the patient may desire, is perhaps the strength of solution that will most commonly agree well ; but in this the patient should be governed by the sensation it leaves behind. There should at first be a sense of dryness, quite obvious to the patient, which should pass entirely off in less than six hours ; much better if it is entirely gone in two hours after the injection is administered. If this sense of dryness is perceptible, we should not allow the patient to use an injection for several hours after it is gone ; and the longer it continues, the longer should be the intervals between them. If it last six hours, the interval should be twenty-four ; if two hours, the interval should be twelve ; if it last twelve hours, it should be discontinued, as it will most likely do harm. Another good astringent is sugar of lead ; this is, perhaps, next in efficiency to the alum. Double the quantity may be dissolved in the same amount of water. I do not like sulphate of zinc, although highly recommended. Thirty grains of it may be dissolved in a pint of water, as an astringent injection. The sugar of lead, or zinc, ought not to be continued as long as the alum. Some of the vegetable astringents are often used to good advantage,—strong decoctions of oak bark, rhatany, kino, or solutions of pure tannic acid. This last is an admirable astringent, not less efficient than the metallic, but also less injurious. It can be used of the same strength as alum, or even in double that strength, if desired.—Injections and baths ought to be suspended during the time for menstruating; if tepid and simple, they probably do no harm at this time; but if cold or astringent, they are pretty sure to interrupt, more or less completely, this flow. During all the interval, they may be employed uninterruptedly.—Almost every practitioner that has had much experience in the treatment of uterine diseases has a favorite injection. I am disposed to adhere to the simpler forms, seeking rather for correct principles by which to be governed in administering them, than for great variety of substances.

Accident in Injection.—There is one annoying, and sometimes to the patient alarming, little accident, that occasionally occurs during the reception of an injection in the vagina. Suddenly, while injecting the fluid, she is seized with severe cramping pain in the hypogastric region, which radiates to the back and hips, down the thighs, and sometimes over the whole abdomen. She becomes sick at her stomach, is attacked with rigors, and her feet and hands often become cold. This pain continues, with exacerbations and remissions, for several minutes or hours, and when it subsides, leaves a sense of soreness, more or less considerable, corresponding with the severity of the attack. As the chilliness and rigors of the first few moments subside, there is reaction; the patient becomes warm, and sometimes decidedly feverish. In all cases in which I have witnessed these symptoms, the patients were using a syringe, in the end of which, within the vagina, were several perforations, some on the side of the bulb at the end, and one at the very extremity. I think that one of the perforations had been accidentally placed in apposition with the external os uteri, and as the water was forced through this perforation, it entered the cavity of the cervix, and passed through it into the cavity of the body of the uterus, inducing the first shock, and the pains following it were caused by the spasmodic attempts on the part of the uterus to expel it. Although I have, in a large number of instances, been called upon to witness and prescribe for these symptoms, I have not seen them proceed to dangerous extremities. I think these are cases of injection into the womb; and in this respect, they constitute my whole observation. An opiate injection per rectum, fomentations over the pubis, and quiet, are all the remedies I have found necessary. And often the symptoms subside so soon that I have not been under the necessity of prescribing at all.—We occasionally meet with patients who cannot use baths or injections. In these cases it will be found, almost invariably, that this inability arises from their producing an exaggerated effect. If it is simple tepid water used for the bath or injection, its results are too sedative. The bath debilitates the patient, instead of simply soothing her. I have seen a single tepid bath prostrate a patient so that she would have to lie in bed for several hours before its effects wore off. A cold bath induces chilliness and permanent coldness, and reaction is not established: the system recovers from its effects only after a number of hours, and that slowly. Hip, sitz, or general baths may produce these effects, and when they do so, should be abandoned as injurious. Other nervous symptoms, as difficulty of breathing, nausea, dysuria, &c., also occasionally seem to be the effects of baths. It is singular that some patients are so susceptible to the depressing effects of water that injections debilitate them very rapidly, and they are obliged to abandon them on this account. Cold water, as an injection, not unfrequently causes general coldness. But it is the medicated injections that most frequently produce an exaggerated effect. Alum injections, even when the solution is weak, with some patients produce such disagreeable and constant dryness, and sense of heat, as to make them quite intolerable. And the sensitiveness of the vagina becomes so great that some patients are forced to cease the injections of alum wholly. The same objections apply to other astringents, to a less degree, and the consequence is, that however baths and injections may seem to be indicated, in the cases where idiosyncrasy renders them so objectionable, we must forego their use entirely.

Should they be Used in Pregnancy?—Is pregnancy an objection to the use of local baths and injections? I think not, with proper care; a hot bath about the hips would be objectionable; a very cold bath that might cause much of a shock, or internal congestions, would not be advisable; but plenty of tepid water, and even cool water temperately used as baths, give the pregnant woman great comfort, and cannot generally be followed by any bad effect. Injections may be used with less caution than baths. The caution which we would administer to all is, that they should not be copious. In pregnancy the patient ought not to use more than a quart at one time. The injections should always be tepid or 'cool; not very cold nor very warm, lest they stimulate the muscular, vascular, or nervous system of the uterus too much, and induce hemorrhage, or provoke contractions. Both of these effects, I think, I have known produced by such injections: the cold causing contraction and expulsion; and the very warm, hemorrhage and death of the ovum. Strong astringents should also be avoided. Much comfort may be derived from anodyne injections, when there is neuralgic suffering about the uterus or vagina, during pregnancy. Cases of superficial inflammation, and even early ulceration of the vaginal portion of the cervix, may always be benefited by injections, baths, and the general treatment which I have heretofore detailed. In fact, most cases, if not all where there is no idiosyncratic objection to the baths and injections, will be very much benefited by them. When, however, the disease has been of long standing, or extends between the labia of the os uteri, or into the cavity of the cervix, these will only slightly benefit it. We must then seek for something that will more profoundly influence the nutritional changes, and the vascular and nervous tissues of the parts.

Principles that should Govern us in Choosing the Kind of Local Treatment.—The substances I have been in the habit of using more frequently, and in fact, almost exclusively, are, after the various depletory measures, nitrate of silver, tannin, acid nitrate of mercury, nitric acid, and caustic potassa.. Of these, the nitrate of silver is most frequently used. In fact, it has so generally answered the purpose in my hands, that I look upon the others as substitutes, and to be used only when it disagrees or fails. This, of course, refers to simple mucous inflammation, or ulceration. I shall, therefore, proceed to describe the use of nitrate of silver, as the standard treatment (if I may be allowed such a term) of inflammation and ulceration of the mucous membrane of the os and cervix uteri. Before doing so, however, I wish to draw a broad and well-defined line between cases to which these stimulants and caustics are applicable, and those to which local depletion and counter irritation are adapted, as the local means best suited to them.

And in order to be understood, I will again draw the attention of the reader to the fact, that when a mucous membrane is inflamed, touching it gives to the sense of the patient the idea of rawness; when a part is touched in which the inflammation is beneath the mucous membrane, the idea of tenderness is experienced. When the mucous membrane of the cervix, for instance, is the exclusive seat of disease, if there is any disagreeable feeling experienced upon touching it, it is that of rawness; but if the substance of the cervix or body of the uterus is inflamed, when it is touched by the finger, or an instrument, the patient complains of tenderness. We should bear in mind, too, in estimating the value of the sense of tenderness in distinguishing between mucous and submucous inflammation, that we may sometimes be deceived by the complaints of patients, when the mucous membrane of the vagina is inflamed, into the opinion that inflammation is in the uterus. We ought, therefore, successively to press upon the different parts with our finger in a digital examination, and, after the speculum is introduced, with the probe, and question the patient, when each point is touched, as to the sensitiveness at that place. When pressing upon the uterus with the finger or probe, if the patient complains of tenderness or soreness, we ought to suspect submucous disease. Now, when the uterus is very slightly if at all tender to the touch, it is not likely that there is much submucous disease. To the mucous inflammation, these stimulants, astringents, and caustics are adapted, and to a more limited extent to the submucous. We very frequently find the increased secretion, the pus-colored mucus and rawness, combined with the deep tenderness and tense pain of submucous inflammation. In these cases we should be careful to subdue this last by depletory measures, alteratives, counter-irritants, &c., before we resort very freely to caustic and stimulant applications to the mucous membrane. When, however, there is evidence of inflammation of the mucous membrane of the cervix, outside or inside of the cavity of the body of the uterus only, a judicious employment of astringents and caustics will do more good for it than any other treatment with which I am acquainted. As this is the most numerous class of cases, and as separate submucous inflammation will come up for consideration after awhile, I will describe the treatment of it first, and the others afterward, premising that in mixed cases we should, to some extent, subdue submucous before we begin to use the treatment for the mucous inflammation; and in such cases, when we do begin to treat the mucous membrane with the caustics, we should do so with caution, lest we increase the deeper, or submucous inflammation. I think this caution is not sufficiently understood, or acted upon. Too often the neck of the uterus is leeched, because it is inflamed, or it is touched with the nitrate of silver, because it is inflamed; and yet if the practitioner were to stop and think a moment, he would readily decide that leeching will not cure mucous inflammation, or that nitrate of silver is not applicable to submucous inflammation.

CHAPTER XX.

CHRONIC inflammation is an habitual and established affection, having almost no tendency to spontaneous termination; it must be subverted to be cured. This can unquestionably be best done by local means, when the part affected is accessible. Inflammation of the cervix uteri is still less prone to spontaneous termination, from circumstances already mentioned, viz., the menstrual congestions, determination of blood from its dependent position, and the excitement inseparable from the functions of the genital organs. On these accounts, the strong impression of nitrate of silver and its substitutes is required. There can be no doubt but that the stronger the impression we can produce, the more completely the chronic inflammation is swallowed up by the acute, and hence the more radical the change; but if the impression is too strong, it may lead to greater damage than the disease for which it is used would produce. Doubtless, the white-hot iron which is recommended and used by some practitioners, causes more powerful effects upon the disease, more radically influences it than any application of nitrate of silver. But I think that we might not always be able to limit the extent of its influence within proper boundaries. The strong caustics are likewise more radical than the milder, and cure inflammation of the cervix more rapidly, and with as much or even greater certainty; but their effects are sometimes fearfully active, owing to an extension of the inflammation to other tissues than those to which they are applied. In order to avoid all likelihood of bad results from such extension of inflammation, the milder caustics are used, and their lack of power is compensated by the repetition of their use. As already intimated, the nitrate of silver is by far the most effective of these, in cases of inflammation and ulceration of the mucous tissue of the cervix. When the inflammation extends to the deeper tissues it is not generally sufficient without the aid of other means.

The nitrate of silver is objected to by some as too strong and harsh a remedy to apply to so delicate an organ as the uterus, and speak of "burning the uterus" with lunar caustic as a "horrible operation." Honest observation, however, will convince every practitioner of intelligence that, with the precautions ordinarily enjoined, no more risk need be incurred by the use of nitrate of silver than by the use of any other valuable remedy. That there are cases to which it is not applicable, and in which it is too harsh, is certainly true; and it will be my endeavor to point these out, and enable the practitioner, by attention to the matter, to avoid damage from the nitrate in almost all cases. It is best that we should be aware of the fact that the nitrate is not infallible, nor always innocent; but we should also lay aside the unreasonable prejudice which arises from the term caustic, and which is hardly applicable to it, and determine, by our own observation, its title to the claim of a remedy in these cases.

Preparation for the Use of Nitrate of Silver.—All the preparation necessary, so far as the patient is concerned, will be effected in the examination for the purpose of clearly diagnosticating the disease, viz., the perfect exposition of the cervix uteri by the speculum, and the removal of all the mucus, blood, &c., by which it is often covered. This cleansing of the cervix from mucus, pus, or blood is important, from the consideration that these substances neutralize the effect of the nitrate by decomposing it.

Should be Pure.—In selecting our remedy, we should endeavor to procure a perfectly pure article, free from adulterations and impurities, as they act as diluents of it, and render the application less effective.

Forms of Application.—It may be applied in the solid or fluid form. The former I think, in the great majority of cases, preferable; while the latter, where the more concentrated solid form is too stimulant, may be made very useful.

Solid Form best.—I am desirous of expressing a decided preference for the solid form, because its application may be made more easily, certainly, and definitely, and because the peculiar impression of this substance is thus more surely produced. The solid should be in the form of cylindrical pieces of half an inch in length; the size and form usually found in the shops. In some cases, the larger will be found most convenient, while in others we will use more easily the very small pieces.

Instruments for Using Nitrate.—I think a great deal depends upon the kind of instrument employed as a porte-caustique. In fact, we cannot expect to treat these patients successfully without having instruments that will expose the parts perfectly, and make the contact of our applications thorough and complete throughout. I am sure that many failures to cure arise from imperfection of instruments and want of thoroughness in application. It has been my lot, frequently, to be called to see patients of this kind in consultation with medical men who had been treating them for months, with a glass cylinder for a speculum, and a goose-quill for a caustic-holder. In the very simplest of cases, where the inflammation or ulceration is all external to the os—an uncommon thing—it is only possible, it is certainly not probable, that success can be secured by such imperfect means of operating. To say the least of it, such treatment is clownish. Let the practitioner have the best instruments, to completely include in its exposure the whole of the cervix and the vaginal cul-de-sac: and to enable him to apply his remedy to all the inflamed surface outside the os, and inside the cervix, and, if need be, up to the fundus inside the corpus uteri. If, upon trial, his instruments do not enable him thus intelligently and thoroughly to proceed, he will do his patient and his own reputation injustice, as well as will misrepresent his profession, and will be utterly inexcusable, if he does not invent, if need be, such means as will effect these objects. A porte-caustique for the solid nitrate, which I have used for several years, and with which I am very well satisfied, is made by Messrs. Tolle & Dagenhardt, instrument makers, in this city, and by Tieman, of New York. A large number of my medical friends have furnished themselves with this kind, believing it to be preferable to any of the common ones in use. The main feature is the flexible wire of which a portion of it is made.

Flexible Caustic-holder.—This should consist of two pieces, one piece a sheath, about five or six inches long, and the other piece copper wire, about five inches long, surmounted at one end with platinum holders, into which the caustic may be fitted. These two pieces should be so made that when intended for use they can be screwed together, making an instrument ten or eleven inches long; when not used, the wire portion, holding the caustic, can be inserted into the sheath, thus making a caustic preserver as well as porte-caustique. We should be supplied with two or three sizes of these instruments, as a matter of convenience in cases where the os and cavity of the cervix differ in size. The object of having the stem made of copper or other flexible wire, is to enable us to bend it to suit the curvature of the uterus, or angle caused by a difference in the direction of the axis of the vagina and uterus. In many cases, we cannot bring the cavity of the cervix and body of the uterus to correspond with the direction of the cavity of the vagina; in such instances, a straight, inflexible porte-caustique will but very imperfectly enter and penetrate beyond the os; but if we have an instrument that will bend, and retain the flexure we produce in it, we may, as with the uterine sound, flex it so as to enter the cervix, and penetrate even to the fundus of the uterus. This flexi-

Fig. 18.   Fig. 19.

The flexible caustic-holder, in two pieces, to be screwed together when used, and sheathed by placing the wire part in the other when not used. (Full size.)

ble caustic-holder, or some other instrument that will answer the purpose of entering the uterus, I consider indispensable to success in a large number of the cases we are called upon to treat. The part to which the application is to be made, should in all cases be divested of all mucus, pus, or secretion of any kind, before the medicine is placed in contact, and then it will act with more efficiency.

Mode of Applying it.—The nitrate should be applied thoroughly. Where there is inflammation external to the os, the nitrate should

be deliberately and gently passed over the whole inflamed or ulcerated part; and should the disease extend inside of the os and cavity
of the cervix, and even to the corpus uteri, it should be fearlessly
but carefully carried up to the full extent of the disease. The contact should be perfect in every part, and sufficiently prolonged to
produce all the effect it can produce by a single touch. If we use
no more force than is necessary to keep the substance in contact
with the part, there is no danger of keeping it there too long.
This is the true "antiphlogistic touch," and it depends for this
quality upon its completeness and thoroughness. Every time the
application is made, we should try to be thus thorough in our use
of it.

Fig. 20.

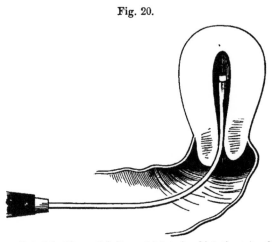

Showing the small-sized flexible caustic-holder as it is introduced into the cavity of the uterus for
endocervicitis and endometritis.

Frequency of Application.—This kind of application can be
profitably made, as an ordinary practice, about once in six days;
but we should be sure that all the perceptible influence of one
application has subsided for at least twenty-four hours—and better
if it is forty-eight hours—before another is made. This may require, in some instances, eight, or even ten days; or it may, in
other cases, take place in five days. It is desirable, in making
these applications, to avoid the period of menstrual excitement,
by not making it two days before the time for it, and waiting as
much as two days after its complete subsidence. In most pa-

tients we will be able to make four applications a month; but often only three can be tolerated. In common ulceration or mucous inflammation, external or internal to the os uteri, we may expect to be under the necessity of making twelve or fourteen applications of this sort. In many cases more applications will be necessary, and in a few cases a less number of applications will be required. Practitioners speak of curing their patients with three or four, some even with one application; but I am sure that they are nearly always deceived. Out of the large number of patients I have treated for inflammation and ulceration of the cervix, I have never known one to be cured with less than nine or ten applications. To the inexperienced I wish to say emphatically, be thorough in your applications, and be careful not to stop making them until every vestige of inflammation is gone.

Thoroughness and Perseverance in its Use.—Failures occur very frequently on account of too little being done by the caustic. Improvement is not a cure; nor are we warranted in believing that a patient, because she is better, will continue to improve until she gets well. The treatment must be persevered in until the cure is complete. I have observed, also, that regularity is important in the treatment of these cases. It will not do to visit the patient at our convenience; but we should see her and make the application at the regular time, and attend to it promptly. It is not unusual, I think, for physicians to see their patients with so much irregularity as to fail in procuring the benefit of each successive application, and, to some extent at least, lose the advantage of one application before another is made. As I have already pointed out, a large number of cases are attended with inflammation in the cavity of the cervix; and in many instances, when there is no inflammation external to the os, the cavity of the cervix is the seat of much disease. We should remember this, and watch for it. I do not think it will be time wholly lost if I call attention more particularly to the mode of using the nitrate in these cases. As I have before stated, the continued discharge of pus, mucopus, or even mucus to a considerable extent, is evidence of endocervicitis, and we should not cease treating these cases until this discharge has completely ceased. An entire cessation of the discharge from the cervix should be verified by the use of the speculum.

Application in the Cervical Cavity.—When the inflammation is in or extends to the cavity of the cervix, a flexible caustic-holder is indispensable to its successful treatment. We can be sure of making a thorough application inside the cervix—after exposing the neck in the speculum as fully and carefully as possible—by introducing the uterine sound into and through the cervix, in order to exactly measure the direction and amount of curvature, and then bending the wire of the porte-caustique so as to correspond with the curve of the sound which has passed into the uterus. After this preparation, if the caustic-holder is not too large, it will readily pass into the parts surveyed by the sound, and thus bring the nitrate in contact with the diseased surface very completely, which, if allowed to remain in contact for a few seconds, will produce its full effect upon it.

Solution, Strength, and Mode of Using.—The nitrate of silver is applied also in solution, and very often a cure is effected by it in this way. To be efficacious, the aqueous solution of the nitrate should be very strong; say one part of the silver to four parts of water. This solution is less powerful in its influence on the parts to which it is applied than the solid.

Frequency of Using Solution.—It should be made consequently more frequently, every fourth day, for instance. Every part of the diseased membrane should be touched by it. We should not expect it to run upon the parts, but we should place it in contact by the instrument used.

Acid Solution of the Nitrate of Silver.—Another solution of the nitrate may be made by dissolving it to saturation in pure nitric acid. This solution is, of course, very different to the other, and possesses the qualities of a powerful caustic. It must, therefore, be used with great caution, and subject to the rules for the government of the use of the acid nitrate of mercury, or other strong fluid caustics.

Instruments for Using Solution.—The instruments I have used for applying medicines in fluid form to those parts, are the camel's-hair pencil, and a small swab, made by wrapping and fastening with thread a little cotton to the end of a fine piece of flexible whalebone. Either of these instruments will pass into the os uteri, if necessary, and carry along with it the solution. They may, of course, be made to apply the fluid to the outside of the

uterine mouth with equal efficiency. The watery solution may be
used profusely, as there is but little danger from contact with the
sound parts. The acid solution must, on the contrary, be used
very sparingly. The treatment of ulceration with the aqueous
solution of the nitrate will require a long time, comparatively, for
a cure; certainly twenty applications will be almost always necessary. But we should not stop the use of it until the ulceration,
congestion, and hypersecretion all disappear.

Effects of the Solid.—When the nitrate of silver is applied in
the solid form to an ulcerated or inflamed cervix uteri, the first
effect is to coagulate the albuminous compounds on its surface
into a thin, very white film of the thickness of white writing paper.
This film adheres to and protects the surface from further influences that are not sufficient to destroy it; hence, after this effect,
the nitrate can produce no more impression upon it. If, however,
sufficient rudeness or force is used to separate this pellicle or film,
from its adhesions, the nitrate will produce a similar effect upon
the denuded surface, so that by forcibly passing the nitrate of
silver over the mucous membrane for a number of times, a considerable amount of surface and substance, at some depth, may
be destroyed. Taking these examples of its action, we see that
it may be made to have a gentle or powerful caustic effect; hence
the dispute as to whether it is or is not a caustic. If the nitrate
is applied to the surface of a healthy mucous membrane, it produces the same effects, but more slowly and to a less marked degree. The difference in rapidity with which this coagulum or
film is produced on the surface of an ulcer has been seized upon
by Dr. Bennett as diagnostic of ulceration. On the ulcerated
surface it is almost immediate, while on the membrane retaining
its integrity, the film is formed more slowly. The surface of an
ulcer becomes immediately white, that of the membrane slowly so.

Modus Operandi of Cure.—It has seemed to me that the application of the solid nitrate operates favorably, by two effects it
has upon the diseased surface: the first is the profound stimulant
action upon the capillaries, brought under its influence; and secondly, the formation of the film, which protects it from all foreign
influence while it lasts. When this film falls off, it leaves the surface of the ulcer raw and bleeding, or if it has been applied to the
mucous membrane, it is deprived of its epithelium. This occurs

about the end of the third day, sometimes sooner, and sometimes later.

Discharge.—In the case of the ulcer, there is, after the loss of the film, quite a discharge of bloody serum, which lasts for forty-eight hours, or more. When this discharge ceases, it is on account of the generation of a temporary or permanent epithelium. Now, if the ulcer is examined, its edges will be found more defined, while its general surface shows an improved state of the granulations. After each time the discharge becomes less, the diseased surface smaller, until a completely healthy appearance is assumed. The application usually produces some pain, which lasts ordinarily from three to twelve hours. For an application to be beneficial, the pain should not continue longer than twenty-four hours.

Kind of Pain produced by Application.—The kind of pain produced by the application is not always the same; in simple mucous inflammation, it is apt to be of a burning or smarting character, or it sometimes merely increases the pains felt before the nitrate was used; the backache, pain in the side, or any other pain which had before existed, is increased, or, as the patient often expresses it, she feels the pain and other sensations which indicate the appearance of the menstrual discharge. In very many instances the patient complains of no additional suffering. If the inflammation extends to the submucous tissue, the pain is apt to be more severe; it is soreness, a tense pain, or throbbing pain, and does not subside as readily as when the inflammation is confined to the mucous membrane.

Pain worse in Endocervicitis.—If the inflammation extends inside the cervix, and the nitrate is introduced into its cavity, the pain is apt to be somewhat more severe. Although all the local troubles are generally increased for a few hours, we meet with a few patients who are immediately and very much relieved at the time of every application. This difference of suffering after each application is like what we observe with reference to the increase or decrease of symptoms after the beginning and continuance of local treatment. Some patients will suffer more after the commencement of local treatment for several weeks, and then gradually improve, and get well, while others will go on to convalescence from the beginning. Others, again, will not improve until the local treatment is finished. In addition to the increase of local symptoms under the application of nitrate of silver, patients are often rendered very much worse in their general symptoms. They are more nervous, their headache is increased, nausea is caused or increased; in fact, all the general symptoms enumerated as being caused by uterine inflammation will be found sometimes to arise from the effect of an application.

On the other hand, very frequently the general symptoms may be, and are, permanently relieved by the local application from the beginning. If we observe through the speculum somewhat closely the effects of the application of the nitrate, we will find in the first place, and almost immediately, the ulcerated surface turn very white, from the formation of a film of coagulated albumen. A short time afterward, the mucous membrane of the vagina and neck of the uterus become livid from congestion. In two days, or less time, the albuminous pellicle begins to be detached, and the surface beneath is left of a scarlet red, and often blood may be seen exuding from this raw, uncovered surface. This exfoliation, or detachment, goes on for two days, until all the surface covered with the coagulum is left raw and bleeding; on the fourth or fifth day, this surface is again covered with a very thin epithelium, and the membrane ceases to bleed. The injected condition of the uterus and vagina, with the finishing of these processes, subsides. In four, five, or six days, the effects are all gone, and the capillaries begin to return to their old inactive state. Astringent injections do good by expediting all these processes, I think, particularly the subsidence of congestion of the vagina.

Chronic or Ultimate Effect of the Nitrate upon the Tissues.—The chronic effects of nitrate of silver—by which I mean the permanent influence it produces upon the tissues of the uterus—are worth closer study, and I should be glad to give them to some extent, but I propose at present, for want of time and space, to confine myself to a very limited view of them.

Atrophy.—Sometimes the continued application of nitrate to the mucous membrane of the uterus induces condensation of the tissues beneath it, as well as in the mucous membrane itself; hence, results, not unfrequently, true atrophy of that organ.

Contraction of the Os.—In some cases, where the application is made to the os uteri for several months, that orifice becomes so small as to be of the size of a mere pin-hole. This may some-

times take place while there is still inflammation in the cavity of
the cervix.   When this is the case, the secretions issuing from it
will sufficiently indicate it.   We need not be embarrassed in our
treatment by this occurrence, as we can easily dilate the os uteri
to almost any extent by tents of compressed sponge, or, if this is
not at hand, slippery elm bark bougies.   By using one of these
tents, or bougies, for twenty-four hours before we desire to make
the application, the opening will be large enough to answer all purposes.   This contracted condition of the os uteri, where there has
or has not been treatment, should not deceive us in reference to the
presence of inflammation in the cervix.   I have not unfrequently
been called to see cases in which the mouth of the womb was
scarcely perceptible to the eye on account of its contraction from
inflammation or the use of the nitrate; several of which had been
pronounced to be in an entirely healthy condition.   Yet, from this
minute opening, quite a large amount of muco-pus or tenacious
mucus found its way in the twenty-four hours, and could be seen
filling up the upper part of the vagina.   By dilating the os uteri
with sponge or slippery elm, and applying nitrate of silver inside
the cervix for a number of times, all the distressing symptoms and
the copious secretion subsided together.

Effect upon Menstruation.—The menses are ordinarily rendered
more easy and natural by the cure of the inflammation from the
use of the nitrate applications; but this is not always the case.
At first, the sanguineous flow is increased; this may be for the
first, and even second month, but in some instances, after this; it
then diminishes to a great extent, so as to amount almost to amenorrhœa.   I think this diminution of the menstrual flow keeps
pace with and is dependent upon the condensing or atrophizing of
the tissues of the organ.   I have noticed this to occur so often
that I regard it as a sufficient indication for the withholding of
this remedy altogether when this condition is observed.   This atrophizing and amenorrhizing influence of the nitrate is much more
apparent after its introduction into the cervix and uterine cavity.
I do not remember to have seen atrophy result from treatment
with any of the substitutes for the nitrate.   Sometimes the menstrual diminution results apparently from the effects of the frequent application of the nitrate to the mucous membrane of the
cavity of the cervix and corpus uteri; while, so far as we can

judge from examination, there is no diminution in the size of the uterus, nor where it seems to be hardened in consistence. When this is the case, it is doubtless on account of the transforming influence exerted upon the mucous membrane, perhaps a condensation of its structure to such an extent as to prevent the capillary fractures which in health allow the transudation of the menstrual blood.

Effect in Dysmenorrhœa.—Painful menstruation is modified to a greater or less extent by the application of the nitrate. For the first, and even second month, there may not be much difference, but after this the painfulness ordinarily diminishes until it ceases, or nearly so. Sometimes, however, at the first recurrence after the beginning of treatment, the pain is almost entirely relieved. This is remarkably the case in cases where the pain has been of a cramping instead of an aching or burning nature.

How are we to know when to stop its use ?—How can we know when the nitrate has been sufficiently used? We are to continue the treatment, as I have before said, until every vestige of inflammation is removed. We must continue the applications until all the ulceration is removed that is within our sight, and then continue them in the cavity of the cervix, and, if need be, the cavity of the body, until no free mucus is seen issuing from the cervix or in the vaginal cavity. It is a mistake to suppose that the inflammation is cured until the pus or mucus, or both, which are evidences of its existence, cease to appear when we make our examination. I cannot emphasize this direction sufficiently to do justice to its importance. While there is yellow or puriform mucus in any quantity issuing from the os uteri, there is ulcerated mucous membrane within the cavities above, which require the use of the applications; while there is hypersecretion or free mucus issuing from the os uteri persistently, there is inflammation or persistent congestion of the mucous membrane of the cavity of the cervix, which requires the lunar caustic for its cure. We should continue it, therefore, until these cease to flow, as well as until the obvious ulceration is healed.

The Nitrate sometimes fails.—The nitrate sometimes fails to cure these inflammations and ulcerations, and although it may not be considered necessary by the reader to inquire into the causes of its failure, yet I think we will treat these cases more successfully by rightly understanding why we do not succeed by the use of the ordinary remedies.

Not strong enough.—There are cases in which it falls short of producing the impression necessary to arouse the capillaries to a more healthy action; it is not sufficiently powerful. In these cases no apparent or real good is done, but the inflammation continues about the same all the time. The cases in which it fails in this way are generally indolent; the granulations are large and flabby, the cervix large and doughy to the touch, with very little sensibility, and the surface is inclined to bleed easily.

Substitute in such Cases.—These require some of the stronger substitutes, applied occasionally, and alternated with the nitrate, or with some of the milder substitutes. The caustic potassa is much the best substitute in such cases. My plan of applying it, in such cases, is to moisten a very small camel's-hair brush with the mucus of the vagina, and rub it over the stick of caustic potash until the brush becomes well saturated with it. I then apply the brush to the diseased part. I continue to apply the mucous solution of the potash to the surface in this way until the desired effect is produced. In this manner we may procure a strong stimulant influence, or slight caustic effect, without the destructive substances running upon the sound parts. A swab, made by tying a small piece of cotton to a small stick of whalebone, will answer the purpose equally well. We first moisten the cotton swab in the thick mucus, and pass over it the stick of caustic until it dissolves off and retains a part of it, and then apply it to the diseased part. Or we may dip the brush or swab in strong nitric acid, and apply to the parts. The swab I think the better of the two, as it does not take up the caustic fluid so freely, and hence is not likely to allow it to flow over the sound parts.

Sometimes the Nitrate fails without apparent Reasons.—But we often meet with instances in which, without any apparent reason, the nitrate fails to do any good. These cases we should study, with a view to ascertain whether the impression is not sufficiently powerful, or whether the impression is not of the right sort, and select our substitute according to our conclusions in this respect.

May cease to do Good after being Beneficial.—Again, the nitrate may do good and seem to be curing a case, but after several applications there seems to be no advance. The ulceration remains the same from week to week, without any change. It will be necessary, in these cases, almost invariably to resort to some stronger stimulant, as the acid nitrate of mercury, the acid nitrate of silver, or the caustic potash, with the brush or swab.

Acid Nitrate of Mercury, and of Silver.—The acid nitrate of mercury can be procured at the shops; the acid nitrate of silver is made by dissolving the nitrate of silver in the strongest nitric acid, to saturation. Any of these may be tried once a month, to be succeeded by milder substitutes, as tannin, sul. cupri, creasote, &c., at intervals of a week between them.

Sometimes the Nitrate of Silver does Harm.—But sometimes the nitrate of silver not only fails, but it entirely disagrees with the cases, and it has to be abandoned. I have known a number of cases in which the nitrate aggravated the inflammation every time it was applied.

In Aged Persons.—This is particularly apt to be the case in old persons, after the childbearing age has passed. In them the inflammation assumes, nearly always, a peculiar appearance; the cervix is small, the granulations minute, the surface very red, and the discharge a thin and acrid muco-pus. These are apt to be obstinate, and almost invariably made much worse by the application of the nitrate, and, what seems singular, are benefited by the stronger stimulus of potassa fusa. One application of caustic potash, with the brush or swab, every four weeks, followed every six days with tannin, usually answers very well for this kind of ulceration. Creasote generally agrees well with it.

Aphthous Inflammation.—Another sort of inflammation, attended with patches of exudation not unlike aphthæ, is almost invariably very much aggravated by the application of the nitrate of silver. This requires milder treatment. Tannin and creasote, alternated every six days, with one application, if necessary, of caustic potassa, will answer very well. On several occasions, I have found the nitrate, after having done well for several weeks, suddenly and unaccountably to disagree with cases, and the ulceration spread rapidly. These have been rendered tractable by the caustic potash, pretty freely applied.

Causes too much Discharge.—But without reference to the kind of ulceration, the nitrate of silver, so far as I am able to judge, sometimes disagrees and does harm, on account of the excessive discharge or hemorrhage it causes. Ordinarily, when the nitrate is thoroughly applied, as I have elsewhere said, there is some discharge of bloody serum, amounting to half an ounce, or double that quantity. This takes place from the second to the fourth days inclusive. Sometimes it is much less, sometimes it is more abundant. I have met with instances, however, where there was great hemorrhage, and was so exhausting as to preclude the use of the nitrate entirely. So far as I can see, there is no peculiar appearance by which we might be led to suspect the occurrence of hemorrhage, before trying the remedy. One case that I am now treating, is peculiarly susceptible in this respect. A single application of nitrate of silver, in the middle of her menstrual month, caused her to flow so copiously as to make it necessary to keep her bed, use cold applications, and acid drinks. In spite of these, she lost fifteen or twenty ounces of blood in eight or ten days. This was repeated the next month, and it became necessary to abandon the remedy altogether.

This, of course, is a remarkable case, but in many instances so much loss of blood has taken place as to cause me entirely to forego its use in those cases. In the cases in which hemorrhage forbids the use of the nitrate, the substitutes I have found most suitable are the caustic potassa and tannin. The caustic potassa may be used once in the middle of each menstrual month, with the little cotton swab I have described, so as thoroughly to stimulate the inflamed part and produce very little cauterization; and every fourth or fifth day in the intervals, completely saturate the inflamed surface with pulverized tannin, applied with the camel's-hair pencil or the swab. Before using the tannin, we should entirely remove the viscid mucus in the neck and about it. We need not be apprehensive of any severe effect from the tannin, either in the cavity of the cervix, or on its external surface; we should apply it fully and freely to the whole inflamed surface. Creasote, alternated with the tannin every fourth or fifth day, often suits such cases. When the ulceration is external and extensive, in these bleeding cases, it is best generally to apply the caustic potash in the solid form, so as to produce a more profound effect.

Nitrate sometimes causes too great Pain.—Too great pain is sometimes the result of application of nitrate of silver. The pain, after application of the nitrate, may be merely slight, the patient scarcely feeling any inconvenience whatever; or, what is usual, it may produce some pain and suffering in from six to twenty-four hours, and then subside; or, in rare, exceptional cases, cause intense pain.

The pain, when severe, may subside in a few hours, and is not worth making any change in the remedies, or the pain may be severe and protracted. When this last is the case, injurious instead of beneficial effects are the result, and we should seek for a substitute. Caustic potash, tannin, creasote, acid nitrate of mercury, or some other acknowledged substitute, should be employed. The acid nitrate of mercury is an excellent substitute in such cases, alternated with the tannin, &c.

Worse in Cases of Submucous Inflammation.—This local pain, after using the nitrate, is more common where there is some submucous inflammation; a few leeches will frequently remove the disposition entirely.

Without these local pains, or other suffering with them, there is, as the result of the application of the nitrate to the os and cervix uteri, sometimes excessive nervous symptoms. The nervous excitement sometimes becomes so great that it is very alarming. A patient upon whom I attended but a few months since, was rendered entirely sleepless, and almost insane, by the exciting influence of these applications, and it was necessary to send her off to the country for tranquillity and recuperation. In quite a number of instances which have come under my observation, the nervous symptoms were so increased, that I had to change the treatment, or use substitutes that would not produce these peculiar effects. It is singular, that these very nervous patients complain very little, if at all, of the local effects of the application, and are only rendered nervous by it.

I should hardly finish what I ought to say of the nitrate of silver, and its substitutes, were I not to state that the latter do not cause any of these symptoms of distress which I have mentioned as the occasional result of the application of the former. There is something, then, peculiar and distinctive in the influence of the nitrate of silver, as evidenced by this fact. It is not merely stimulant, astringent, or caustic, in its effects upon this inflammation, but it has its own peculiar influence. It may be asked, if the nitrate causes these bad effects sometimes, and none of its substitutes ever do, why use the nitrate at all?

In the first place, when it does agree with a case, there is no remedy that acts so kindly, so efficiently and certainly, as this. In the second place, the weaker substitutes are slower and less certain than the nitrate, and consequently, when successful, take more time to make a cure. And in the third place, the stronger caustics, as the acid nitrate of mercury, the acid nitrate of silver, and the caustic potash, require greater care, and any accident occurring from them may be much more serious, and, if carelessly or awkwardly used, are likely to do damage to parts not intended to be influenced by them. The nitrate requires almost no preparation or precautionary measures for its use; for the stronger substitutes we must prepare carefully, and use much precautionary vigilance. The nitrate in solution does not produce such decided effects as the solid, and hence, of course, will not cause hemorrhage, pain, or nervousness to the same extent that the latter does. Can we continue to use the nitrate when it causes the above inconveniences, and counteract or neutralize its effects by some other remedy? The pain and hemorrhage are apt to become less at each successive application, and hence, if the patient can bear them for a few times, we may continue to employ them, and then the cases are generally cured by them; but occasionally they disagree after having acted kindly for a time.

Remedy for the Hemorrhage.—When the hemorrhage is considerable, Dr. Bennett recommends a plan which I have followed with good results sometimes, and that is, to make the application to only a part of the ulcerated or inflamed surface. When the application is extended inside the cavity of the cervix, this direction cannot be observed. And it is in these cases that the hemorrhage is the worst. Astringent injections and cold applications, baths, &c., when the hemorrhage is not very great, will afford some relief and enable us to go on in their use. Generally, however, we will have to do with the substitutes when the hemorrhage is considerable.

Remedies for the Pain.—When the pain is great, emollient injections of linseed infusion, infusion of slippery elm bark, with laudanum, in large quantities, thrown into the vagina, or half a teaspoonful of laudanum in a little starch-water or linseed tea, per rectum, will also aid very much in quieting. It is better in all cases for the patient to remain still in the recumbent position, for some hours after an application; when there is much pain, it is indispensable. The patient should be quiet until the pain is over.

Remedies for Nervousness.—When the nervous symptoms are excessive, we should be cautious about repeating the applications. If opium does not disagree with patients, its anodyne influence may enable us to continue the treatment. Tinct. hyoscyamus and camphor may also be tried, or valerian, brandy, &c. But some of these, particularly the last, must be used sparingly. If the nervousness subsides in a few hours, either with or without the aid of the anodynes, we can still resort to the nitrate applications. But if it continue at all obstinate, we must use some of the substitutes. I can but remark again, that it is singular that the caustic potash, and all the stronger caustics, produce less pain, less hemorrhage, and less nervous excitement, than the nitrate of silver.

Is its Application allowable in Pregnancy?—Is it ever allowable to apply the nitrate to the cervix uteri, inside or out, after the commencement of pregnancy? I confess that I am afraid to do so, and if a patient becomes pregnant during treatment, I advise a discontinuance until after confinement, and complete involution has taken place; say three months after accouchement. I know that Drs. Bennett and Whitehead both advise the use of the nitrate during the first three months, for the purpose of avoiding abortion, but the great irritation it sometimes causes intimidates me from using it, or recommending others to do so. I think I have seen abortion caused by it, in cases where pregnancy was not suspected. On the other hand, I have seen cases where pregnancy was not thought to exist, treated for some time without any bad effects. Upon the whole, I think it is much the best practice to desist after conception, or not to begin if we know it has taken place.

Loss of a Piece in the Cervix.—Some object to the introduction of the nitrate of silver, in the solid form, into the cervix uteri, lest a piece of it accidentally be left in that cavity, and very bad results follow. I have had this accident to occur to me repeatedly, and as yet I have not seen any bad results from it. It is true, the pain is sometimes a little more severe and protracted in duration,

but it dissolves and runs out, or is expelled into the vagina, which
is the more probable course, and there is dissolved and neutralized
by the mucus of that cavity.   I have been so strongly impressed
with the harmlessness of the presence of a small piece of the nitrate there, that I have, in certain cases, intentionally passed some
up the cervix, and allowed it to dissolve in the fluid and distribute
itself over the surface of that cavity.

Pressure by Bougies in Endocervicitis.—Before leaving this
part of the subject, I will mention another substitute for the nitrate, which, in certain cases, I have seen do a great deal of good;
that is, pressure upon the mucous membrane of the cavity of the
cervix, by means of bougies, prepared sponge, &c.   In some cases,
with all the facilities afforded by flexible caustic-holders, our applications are imperfect, and the cure is unreasonably protracted.
This may be the case when the cavity of the cervix is small, or
tortuous from flexion or inflammatory adhesions. A bougie of slippery elm, large enough to fill the cervical cavity, introduced as
high as the inflammation extends, and allowed to remain for
twenty-four or thirty-six hours, not only prepares the way for
other applications, but favorably modifies the disease by its pressure upon the capillaries.   In order to use this bougie handily, we
may cut it about two inches long and about the right size, and
then tie a piece of thread around one end, so that it can be removed at will.   After exposing the cervix with the speculum, we
may, with the dressing forceps, introduce it as high as possible,
leaving the end with the ligature extending out of the os uteri.
If not supported it may slip out, and not remain long enough
to do any good; hence it is a good plan to place a sponge or
piece of cotton against it, to prevent it from being discharged.
This should be repeated every four or five days.   The use of the
stem pessary proves beneficial, too, I think, in some instances, on
account of the stem pressing upon the inflamed part inside the
cavity of the cervix, and thus changing the character of capillary
action.   If used intermittingly, it will act better in this respect
than if allowed to remain constantly in place.   We may use flexible gum bougies, wax or metallic.   The object to be gained, it
should be borne in mind, is pressure, intermitted and sufficiently
strong to produce a decided impression.   I think I have, on several
occasions, verified the excellent effect of pressure applied in this

way, when it was difficult to make perfect applications of the nitrate; the main point of the disease being so high up, and the
canal at a pretty sharp angle with the axis of the vagina.

Medicated Bougies.—These slippery elm bougies may be made
to carry medicated applications, and retain them in contact with
the inflamed spot, when situated high up in the cervix. Calomel
may be placed, as I have done, upon the end of the bougie, or
ointment of creasote, calamine ointment, ointment of lead, or any
other that is likely to produce a proper stimulus. The tincture of
iodine, the iodine ointment, and also the iodide of potassium, in
pieces, pushed before the bougie, to be dissolved and diffused over
the mucous membrane, are good substitutes for the nitrate, that
may be used with the bougie. The use of bougies in this way is
like the treatment sometimes instituted for inflammation in other
mucous canals, as the urethra, for instance, with salutary effect.
The danger from the bougie is less, perhaps, than any irritating
application to the part, producing its effect, not by causing acute
inflammation, as does the nitrate and other strong stimulants, but
by pressing upon the part, and thus diminishing the capillary circulation in it, reducing the inflammation.

I subjoin a summary of the treatment of Robert Ellis, Esq.,
Obstetric Surgeon to the Chelsea and Belgrave Dispensary, London Lancet for July, 1862, reprint. It is a choice tabular view of
the kinds of ulceration we meet with, and the very best mode of
treating them, and I think will be useful to the inexperienced:

VARIETY.	TREATMENT.
1. *Indolent Ulcer.*—Cervix hypertrophied, of a pale pink color, and hard. Os patulous to a small extent. Ulcer of a rose red. Granulations large, flat, insensitive, and edge of the ulcer sharply defined. Discharge: mucus, with a little pus, and occasionally a drop of blood.	For a few times the caustic pencil,— solid nitrate silver. Afterwards, the solution of nitrate of silver in strong nitric acid.
2. *Inflamed Ulcer.*—Cervix tender, hard, a little hypertrophied, hot and red. Vagina hot and tender. Ulcer of a vivid red. Granulations small and bleeding. A livid red border around the ulcer. Discharge: a muco-pus, yellow and viscid, with frequently a drop of bright red blood entangled in it.	Occasional leeching, hip bath (warm), emollient injections. Then acid nitrate of mercury several times, succeeded by the solid lunar caustic, potassa fusa, or cum calce.

VARIETY.	TREATMENT.
3. *Fungous Ulcer.*—Cervix soft, large, spongy to the touch. Os wide open, so as to admit the finger. Ulcer large, pale, studded with large and friable granulations. Discharge: glairy, brownish mucus, frequently deeply tinged with blood.	At first the caustic pencil. Subsequently, nitric acid, solution of nitrate of silver, or acid nitrate of mercury; electric, or actual cautery.
4. *Senile Ulcer.*—Cervix small, red, a little hard. Ulcer small, extremely sensitive, of a bright red color. Granulations very small, red, and irritable. Discharge: a thin muco-pus.	Potassa fusa, or strong nitric acid, with nitrate of silver once or twice at long intervals. Then solid sulphate of copper, in pencil.
5. *Diphtheritic Ulcer.*—Cervix of ordinary size, a little hot, dry, and tender. Ulcer covered in patches with a white membrane, adhering closely, irritable, and readily bleeding beneath. Discharge: a thin acrid mucus, without pus, but occasionally tinged with blood.	At first, electric cautery, potassa cum calce, or acid nitrate of mercury, two or three times at long intervals. *No nitrate of silver.* Subsequently stimulant applications, tincture of iodine or sulphate of copper.

CHAPTER XXI.

SUBMUCOUS inflammation, as has been seen, is observed under a variety of accompanying circumstances; with mucous inflammation, without mucous inflammation, and without change of size or consistency, and with fibrinous deposit, enlargement, and induration of the cervix. Of course, all these circumstances will more or less modify the treatment of the different cases in which they are observed to occur.

Submucous Inflammation, with Ulceration and Mucous Inflammation.—There is often evidence of submucous inflammation when ulceration affects the mucous surface of the cervix. When the tenderness is not considerable, nor the part enlarged and tumefied, the cure of the mucous disease by the means heretofore indicated will suffice to cure the submucous also, and hence the case will need no further treatment whatever. But if the cervix is quite tender to the touch, somewhat swollen and hot, and the ulcerated surface red and excavated, and giving out pus copiously, other remedies than those adapted for the cure of the mucous inflammation ought to be used. Leeches, in numbers to suit the intensity of the inflammation and the general condition of the patient, should be applied, and repeated every week, until the tenderness and heat have subsided to a great extent. But as local depletion will not always produce the effect, it will almost always be better to resort to internal alteratives and sedatives. Very many of these cases yield promptly to the alterative influence of mercury, gradually induced, with an occasional active cathartic of salines. Probably the best general plan is to leech the cervix, give a cathartic of calomel, to be rendered a little more active by sulph. magnesia, citrate of magnesia, Seidlitz powders, or Congress water. If, after two or three days, the local tenderness, pain, and heat continue, it will be well to give a grain of protiodide of mercury, or calomel, in similar doses, combined with opium, every four or six hours until slight ptyalism is produced. The cathartic, depletory, and alterative treatment should be continued until the submucous portion of the disease is removed; when the inflammation of the mucous membrane may be treated as I have directed when particularly speaking of that subject. As leeches are not always attainable, it becomes a matter of interest to find a substitute for them. We have this, fortunately, in scarifications; a remedy to which we may resort without apprehension when local depletion seems to be indicated.

Scarification—Mode of doing.—The mode of doing this is practised differently by different individuals. The plan which I have found most convenient and effective, is to make the incisions in the os uteri, and direct them somewhat outward. A knife with a probe point is best adapted for this operation. The blade should be about two inches long, and one-eighth of an inch wide, very thin, and mounted upon a light straight handle, seven or eight inches long. After the neck and mouth of the uterus is brought full into view by the speculum, the probe point may be introduced half an inch through the os into the cavity of the cervix; when thus placed, the handle should be carried as far to the side of the speculum to which the edge is directed as allowable, and then withdrawn with enough pressure to make the edge cut through the mucous membrane. This will allow of a considerable flow of blood. If we wish to obtain a large amount of blood, several of these little incisions may be made around the circle of the os. The copiousness of the flow may be regulated by the depth as well as number of these incisions. In a few days ordinarily all trace of these little wounds is lost, and the os resumes its usual appearance. This is the class of cases to which the depletion so often directed is very well adapted. · The indication for the depletion is in the tenderness, heat, and swelling, all of which are dependent upon submucous inflammation, and not upon the ulceration, or other signs of mucous inflammation. Depletion has, indeed, but little if any good effect upon inflammation of the mucous membrane.

Leeching.—Although it may seem hardly necessary to give any direction with regard to the mode of applying leeches to the cervix, it may be useful to the young practitioner in the treatment of these cases to do so. A common glass speculum, introduced so as to include and isolate the cervix, is what I have been in the habit of using. The leeches are thrown down to the bottom of the speculum after the parts are cleaned of mucus, and the leeches are watched until they seize the part. Three are about as many as can conveniently be used in the speculum at one time. If more are considered necessary, they must be applied after the bleeding from the bites of the first has pretty well subsided. The bleeding usually continues for several hours, and as much or more blood is lost after they are removed as they draw. Submucous inflammation sometimes outlasts the mucous, and when we have this last cured, the troublesome symptoms still continue, or very seldom the submucous begins and continues independent of the mucous part of the disease. These cases, unattended by hypertrophy or induration, as they sometimes are, cannot be diagnosticated by the speculum alone. The physician is rather surprised, perhaps, that the symptoms of uterine inflammation should continue after an examination with the speculum shows a perfectly healthy color, size, and secretion of the organ; yet this is sometimes the case. If the sound or finger be pressed against the parts, they will be found to be tender. This condition is not unfrequently left after the cure of chronic mucous inflammation, and keeps up the symptoms. When it is a sequel to the chronic mixed form of mucous and submucous inflammation, it is apt to subside spontaneously after a time. When it exists independent of mucous inflammation, and is not the sequel to the mixed form, it is often, though I think not necessarily, connected with scanty menstruation. Not unfrequently in this variety of the disease, the uterus is smaller than natural. It is quite common, when attended with scanty menstruation, to attribute it to this last circumstance; but I am inclined to think the deficient menstruation and atrophy are both attributable to the inflammation as a cause. It would be irrational to stimulate the uterus to greater congestion, to increase the flow when it was thus the subject of inflammation. The treatment should be directed to the inflammation. The remedies used will depend upon the acuteness of the symptoms: if the pain is considerable, or the tenderness great, leeching moderately or cupping upon the sacrum is quite desirable, but it should be used only to remove the tenderness and pain; the subsequent treatment should consist in the use of alteratives, counter-irritants, and tonics. Small doses of mercury, followed by the saline purgatives, as alteratives, answer admirably; six grains of blue mass every fourth or fifth night, followed in the morning with Epsom salts, Seidlitz powders, or citrate magnesia, is a good alterative. One grain of calomel may be substituted for the blue mass, if the patient is plethoric. If the patient is anæmic or weak, the bleeding should not be resorted to, but the alterative may be accompanied with tonic treatment. The preparations of iron, syrup of the iodide, syrup pyro-phosphate, iron by hydrogen, or the tincture, are good and eligible tonics. I am very partial to the chl. tinct., given in twenty-drop doses three times a day. I have seen a great deal of good done by it in removing that congestive sort of inflammation that so often keeps up the sensitiveness of the organs, after the more active symptoms had been removed, or in anæmic and weak patients. When there is not much acuteness of inflammation, or necessity for depletion, much good will result from proper counter-irritation.

Seton as a Counter-irritant.—The seton is one of the best means for this purpose. It should be introduced and allowed to remain for several months, and caused to discharge pretty freely for the most of that time by occasional turning, and if necessary, impregnation with some irritating powder, as cantharides, or savin root. It should be made of one whole, large skein of silk, or even larger, so that the impression may be powerful. The best place for it is over the symphysis pubis, or in cases where one of the iliac regions is the seat of pain, this is a desirable locality. I have sometimes directed my patients to dress the seton daily with mercurial ointment, until gentle mercurialization occurred, with much resulting benefit, I have thought. Soothing injections often diminish the sensitiveness of the uterus ; and if they do no other good, should be used for this purpose. Two teaspoonfuls of laudanum to a pint of water, to which thirty grains of acet. plumbi are added, make a very good injection. This should be passed into and through the vagina for several minutes. Belladonna, hyoscyamus, aconite, gelseminum, and cicuta, may be used also with the same view, in proper quantities.

Hardness and Enlargement—Treatment of.—After the inflammation in the substance of the cervix has continued for a great length of time, fibrinous deposit hardens the tissue, and makes an enlargement which becomes permanent and difficult of cure. This enlargement and hardness are attended with various degrees of tenderness ; sometimes the parts are not very sensitive to the touch, while in other cases the least touch causes exquisite pain and suffering.   These conditions of course will very much modify the treatment.   When there is tenderness, heat, and other signs of an acute condition of inflammation, leeching or cupping, cathartics, alteratives, anodynes, should be resorted to, until these symptoms are removed or very much relieved.   I have seen a slight ptyalism do away with these symptoms very quickly, and induce such a state of comparative comfort, that the patients believed themselves cured.   Where the increase in size is considerable, however, they will soon return, and it is necessary to apply remedies that will cause a deeper and more lasting influence.

Caustic Potash—Object in Using.—The only one to which I have ever resorted for this purpose is the caustic potash ; and when judiciously used, I think it will fulfil the indications quite completely.   The object, as stated by Dr. Bennett, is not to destroy the part, so much as to induce a change in the action of its vessels that will cause an absorption of the fibrinous deposit upon which the enlargement depends.   I would have the reader to observe that this is the kind of case to which the powerful action of the caustic is applicable, viz., when there are tenderness and other symptoms indicating a continuance of inflammation.

Mode of Applying.—Too great caution cannot be taken in applying this caustic, for fear of an unnecessary, if not mischievous extension of its effects to other parts.   We should be prepared with the ordinary quadrivalve speculum, a pair of dressing forceps, some cotton-wool, acidulated water, and sweet oil.   The parts ought to be fully exposed, the whole cervix included in the end of the speculum, and illuminated with a good light from a large window, and the patient so placed that we may operate without restraint of any kind.   When the cervix is thus included in the speculum, we should take a piece of cotton-wool, as large as may seem to be necessary, thoroughly saturated with acidulated water, and place it beneath the cervix, so as to underlie the part contained within the speculum, and come in contact with the end of the cervix below the point to which we wish to apply the caustic.   We should also pour into the speculum a small quantity of the acidulated water, enough to fill up the end of the instrument as far as it can be without being in contact with the part to be

operated upon.   It is almost if not quite impossible to apply the
caustic potassa without having it to run more or less.   This should
be remembered and be provided for.   Some liquid, as here recommended, that will immediately decompose this chemical, should
be kept in contact with all the parts where there is a possibility
of its touching by flowing upon them.   I have sometimes, where
the direction of the speculum allowed the retention of the fluid,
simply poured in the acidulated water until all the parts were
inundated that were in danger of injury from the contact of the
remedy.   After having taken these precautionary measures to secure the surrounding parts from harm, we may secure the caustic
in any way most convenient, and apply it as may seem necessary
in the case.   Ordinarily I seize it with my dressing forceps, and
use them for a caustic-holder.   The extent and duration of the
application must be determined by the appearances at the time
of using it.   The enlargement and induration sometimes include
the whole extremity of the cervix, while at others it is wholly or
nearly confined to one of the lips of it.   I do not think it desirable to apply the caustic extensively.   One slough is usually sufficient, and most beneficial, for our application.   This should be
made in the centre of the indurated part, if the induration be confined to one of the cervical labia ; but if the whole extremity is
the subject of the induration, the slough may be made in the central portion of the cervical lip, at the upper part of the included
portion.   The depth of the slough should be sufficient to destroy
the mucous membrane and penetrate the submucous tissue, say to
the depth of an eighth of an inch or more.   This may be done by
holding the caustic steadily in contact with the part sufficiently
long.   The slough should be not larger than a dime in circumference, and, in cases of moderate enlargement, the size of a half
dime will answer all purposes.   With reference to the depth of
the impression, I would say that I have oftener regretted having
made too light an application, than too prolonged.   Thoroughness, combined with carefulness, is just as necessary in the application of the caustic potassa, as of the nitrate of silver.   One application is not ordinarily sufficient ; but it should not be repeated
too near together.   The best rule by which to be guided as to the
time for repetitions, is to wait until the effect of the first has entirely subsided.   This will require from three to six weeks, owing

to the extent of slough and the curative capacity of the individual. As soon as the effects of the first application have entirely subsided, we may make a second application, as remote from the locality of the first as the size of the induration will admit.   When this has gone through the different phases of inflammation, sloughing, and healing, we may make the third, &c. We may thus repeat the applications until the induration is removed.   The time selected for making these severe caustic applications should also have reference to the general condition of the patient, and the special condition of the uterus.   It would of course be improper to make the application at a time when the patient was in any way predisposed to febrile or inflammatory action, or pregnancy; menstrual congestion should also forbid it.   Midway between the menstrual periods the uterus will most likely be in the best condition.   The patient should remain in bed for the first twenty-four hours after this sort of application, and she should abstain from fatiguing use of the lower extremities for at least a week.   If there should be much pain, heat, or febrile excitement, which, in my observation, are very rare accompaniments, soothing injections, of flaxseed infusion with laudanum, may be repeated three or four times in twenty-four hours, injections of tinct. opii in the rectum, fomentations to the hypogastrium and perineum, and warm hip baths.   In all cases, where no particular objection exists, the patient should use the sitz baths, and injections of tepid water, twice a day, between the applications.   As soon as the slough is completely detached, and suppuration indicates a good condition of the ulcerated surface, the danger sometimes attending these applications is no longer to be apprehended. I have not found it necessary, nor do I think it best, as directed by Dr. Bennett, to dress the place with nitrate of silver applications.   The dangers above alluded to, so far as I am aware, are inflammation in the cellular tissue by the side of the uterus, and an increase of the submucous inflammation in the substance of the uterus.   These may almost always be avoided by the precautionary after-treatment I have directed.   Such acute inflammations of this kind as I have seen, have seemed to me to be always produced by careless exposure to cold, or incautious exercise of the legs within two or three days after the application.   They are apt to supervene after the lapse of five or six days, or after the surface of the inflamed part begins to produce healthy pus. If severe cellulitis should occur, we should have no hesitation in promptly treating it with energetic means. The evils will thereby be diminished, if not prevented. One leeching over the sacrum, or the groin of the side most painful, would, if early resorted to, do much toward resolving the inflammation. It should be followed by cathartics, calomel and opium, fomentations, counter-irritants, &c. We must of course be governed by the acuteness of the attack in the selection of our remedies. After the complete subsidence of the inflammation, there may still be enlargement of the cervix. Should this be treated with caustic, with a view to its reduction? I think not. It is attended by very little, if any inconvenience. We will meet, however, with very few instances in which the inflammation wholly subsides while there is enlargement. The latter seems to keep up the former; at any rate, they are usually together.

In all these cases of chronic submucous inflammation, we will expedite the cure by maintaining the functions in the most healthy, or nearest to a normal condition, within our power. The bowels should be regulated particularly; they should be free and unloaded. I have never found it necessary to resort to the actual cautery for the cure of these indurated and enlarged cervices. The caustic potassa has been sufficiently powerful for all purposes. Perseverance should be a guiding principle. Twelve months is not a long time to effect a cure when this kind of organic lesion results from inflammation.

CHAPTER XXII.

HYPERTROPHY OF THE CERVIX.

Hypertrophy of the Cervix is different from enlargement caused by fibrinous accumulation, and consists of an increase in the proper tissues of the organ. It is a real hypertrophy. Although not nearly so frequent as the enlargement from chronic inflammation, it is not of very rare occurrence. The symptoms do not differ from prolapse of the uterus sufficiently to characterize it. The patient generally experiences a sense of bearing down or weight on the perineum, pain in the sacral region, leucorrhœa, sometimes menorrhagia, and the various sympathetic symptoms already sufficiently dwelt upon of uterine irritation.

Diagnosis.—Upon examination the cervix is found enlarged. There are three general forms observed so well marked as to entitle them to special mention. The first is such as we usually find in the nulliparous, an elongation of the whole cervix, and some, but not generally, very great circumferential increase of size and without much deviation from shape. This form is seen in Fig. 21. The next variety is an elongation and enlargement of the anterior or posterior labium as represented in Fig. 22. I am not certain from my own observation whether this is always a pure hypertrophy or a mixture of this process with fibrinous infiltration, probably the latter.

The last variety is an elongation of the two labia, and a devia-tion from their natural direction, as seen in Fig. 23. The anterior projects forward and sometimes shows itself through the vaginal orifice immediately beneath the urethra. The projection was so great in one instance I met with as to be constantly exposed to the air and friction of the clothing, and thus become hard and ulcerated.

The only appropriate *treatment* is amputation, and it is generally sufficient to remove all the disagreeable symptoms resulting from it.

The plan I have usually pursued in removing this growth is by *ecrassement.* The chain of the écraseur is passed around, at the

Fig. 21.   Fig. 22.

Fig. 23.

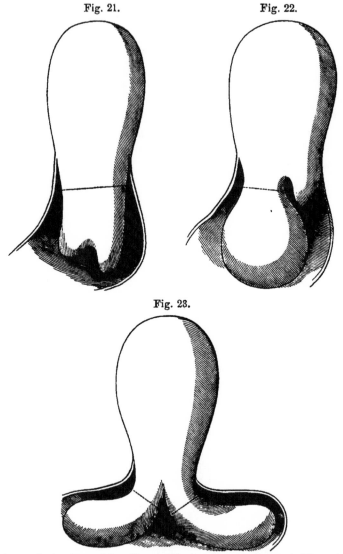

Figures showing three varieties of hypertrophic elongation and enlargement of the cervix uteri. The dotted lines show the proper place for amputation.

place where the point marked out by the dotted line is seen in the figures, and the rachet slowly worked until the division is complete. This operation is easily performed, and is perfectly safe when carefully done, and the parts cicatrize in a few days. An inconvenience mentioned by Dr. J. Marion Sims, is encountered in some instances in amputating the first variety, viz., the contraction of opening of the cervical cavity. It is an inconvenience, however, that is of no great importance generally, and may be remedied by making a small incision with a blunt-pointed bistoury immediately after the operation of amputation. Dr. Sims amputates the cervix with scissors. He exposes the organ with his speculum, cuts the parts squarely through at the dotted lines, and then draws the mucous membrane together over the cut surfaces with silver sutures. This lessens the size of the cut surfaces and the parts heal more readily.

CHAPTER XXIII.

PERIMETRITIS.

THERE is an abundance of areolar tissue in the pelvis; it is between the bladder and pubis, the bladder and vagina, the vagina and rectum, but in greater amount between the sides of the vagina, uterus, and bladder, and the pelvic bones. In a loose manner it fills up the space indicated, and is covered by and included in the folds of the lateral or broad ligaments of the uterus. Within these folds of the peritoneum, the ovaria, the Fallopian tubes, and the round ligament, are included with the cellular tissue. Inflammation attacks this areolar tissue, not unfrequently on one side, and involves the tube, the ovary, ligament, and peritoneal covering; less frequently both sides are simultaneously inflamed, and still less often that part between some of the hollow organs of the pelvis is affected, when we have a comparatively small point of disease, as, for instance, between the bladder and vagina, or this last and the rectum. This is perimetritis. There is a strong tendency when inflammation is lighted up in any part, to spread to the space at the side of the uterus and vagina covered by the broad ligament, on one or both sides. The inflammation is apt to run its course rapidly, as is usual in areolar tissue, either to resolution or suppuration, and as this tissue is abundant, and the organs in the pelvis easily moved, the effusive products are likely to be copious. In the first stage of inflammation, serum is rapidly poured out between the folds of the peritoneum by the side of the uterus and vagina; it pushes these organs to one side of the pelvis, and forms a prominent inflammatory tumefaction at the side of the pelvic cavity, within easy reach of the finger. This tumidity becomes harder in a short time, and forms a solid medium of connection between the uterus and wall of the pelvis, indicating the change from serous to fibrinous effusion. Within a week or ten days, in very acute cases, in others from two to four, or even six weeks, the areolar tissue is broken down into copious suppuration.

In some instances, the inflammation does not advance beyond the stage of serous effusion. When, after lasting for an uncertain time, the symptoms begin to subside, the tumefaction disappears, and the patient soon recovers her health; while in others it is arrested after fibrinous infiltration has cemented the parts solidly together. Although the symptoms are moderated from their first acuteness when this is the case, some of them, as undue sensitiveness, soreness, and sense of weight, and other kinds of pelvic distress, remain for a considerable time, and the patient recovers from the attack very slowly, if ever completely. When suppuration takes place, if it is completely and readily evacuated, the patient very soon regains her health and strength. In some patients of broken-down or damaged constitutions, sloughing and extensive ulceration increase the damage to the organs. I once saw a syphilitic patient in whom extensive and rapidly spreading ulceration opened the rectum, vagina, bladder, and, finally, the peritoneal cavity. Suppuration in this case was unhealthy and ichorous, smelling strongly, and produced excoriation of the parts over which it flowed. If the evacuation of the pus is imperfect on account of opening into the rectum or bladder, and even in the vagina, the symptoms may be prolonged for months and even years. And in some cases where the evacuation of the pus and subsidence of the inflammation seemed complete, the disease recurs usually with diminished acuteness a number of times. I once had a patient in whom an attack of perimetritis was contemporaneous with incipient pregnancy for four different times while under my care. In each one of these four times, the inflammation commenced at about the time the menstrual flow ought to have appeared after conception. Every time there was copious suppuration, a free discharge of the pus, and, to all appearance, a complete recovery from the inflammation. The intervals were about two years in duration. I have seen three instances in which the recurrence of the inflammation had occurred at irregular intervals from three months to a year for over six years, another ten, and one as much as eighteen years. In this last case, the abscess was situated at the left side of the uterus, and usually after a week or ten days of acute suffering, it discharged about a half ounce of fetid pus, and then disappeared, so that nothing but a slight induration at the point mentioned indicated any tendency to its recurrence. This chronic form, I think, is not very uncommon. I believe, also, that chronic induration in the spaces occupied by the pelvic areolar tissue, caused by fibrinous infiltration, not unfrequently presents itself as the effect of acute perimetritis, producing many distressing symptoms, and rendering the patient liable to a recurrence of acute attacks. The extent of the inflammation and tumefaction is governed somewhat by the condition of the patient. If she be in the puerperal state, the inflammatory excitement is likely to be greater, the swelling more extensive, and the suffering more severe, than if this condition is not present. Pregnancy increases the intensity of the disease beyond what it is in the unimpregnated condition; the fever will run higher, and the extent of the inflammation be greater. The same will be the case after abortions. The mildest form of perimetritis is that which occurs in the unimpregnated female.

Judging from my own cases, I should think that three-fourths of these attacks terminate in suppuration; and we may reasonably apprehend such a result as the rule. When pus is formed, it finds its way out through several different channels. First, and most frequently, through the vagina; the wall of the abscess nearest the vagina ulcerates through into this canal, and the pus escapes first in small quantities, and finally freely, until the whole is evacuated; a number of days and even weeks may elapse before the discharge ceases, and the cavity is filled up. The escape through the vagina is not only the most common, but this is the most favorable outlet, as the opening is generally pretty free and permanent. Second in frequency as the medium of discharge is the rectum; the pus makes its way into this intestine generally at the upper end of the septum between it and the vagina. The discharge is comparatively slow and unsatisfactory, appearing with the stools in small quantities, and continuing for a length of time. The opening into the bowels is almost, if not invariably, valvular and tortuous, permitting the escape with difficulty. If there does not occur a second opening into the vagina, the abscess will generate pus almost as fast as discharged, and we may expect times of partial relief and exacerbation for months and even years. I am acquainted with an instance in which the patient has not been entirely free from suffering from this cause for the last six years, and a number of times has been prostrated for weeks. But few days pass without the patient observing matter in fecal evacuations. The pus makes its way at other times through the inguinal regions; sometimes it points in one of the labia, or burrows through the gluteal region. It also perforates the uterus or bladder, and follows the channels leading from them outwardly. I have not seen an instance in which the uterus was penetrated. When the pus finds its way into any of these holllow organs, it causes severe irritation in them, and efforts at expulsion. Dysuria, dysentery, and vaginitis, are generally caused by it to a moderate degree, but sometimes the suffering from this cause in these organs is very great. But another mode of escape from the cavity of the abscess is into the peritoneal sac. This is comparatively infrequent, fortunately, but invariably fatal. I believe no instance is on record to contradict this statement. I have been unfortunate enough to be connected with two cases in which this untoward circumstance occurred.

One of the patients was attacked in the puerperal state, and after suffering for eight weeks with the inflammation of the tissues around the uterus, acute general peritonitis terminated her life in about thirty-six hours from the time it commenced. Upon examining the abdominal cavity, an opening was found near the left sacro-iliac junction, which communicated with the interior of the abscess, and several ounces of pus was in the cavity of the peritoneum, that had made its way through this opening. The usual lesions of extensive and acute peritonitis gave evidence of the cause of death. The other case was in a sterile married woman, about twenty-five years of age, who had been treated three weeks for typhoid fever. Dissection revealed a large pelvic abscess, with recent rupture into the peritoneal cavity, and extensive peritoneal lesions. This overwhelming peritoneal inflammation lasted only about eighteen hours before the death of the patient. When the peritoneal symptom supervened, it was regarded as the result of the intestinal ulceration which sometimes so suddenly terminates typhoid fever.

Causes.—As before intimated perimetritis occurs as a sequel to abortions, and labor at full term, and there is but little doubt but that these two conditions sometimes predispose to the disease. The menstrual congestion seems to do the same thing. Any circumstance that fills the pelvis with blood in active congestion may so predispose to it. Cold suddenly applied to the surface or to the feet and legs may excite the already congested parts into a state of inflammation. Much exercise of the limbs in walking or standing on them for a long time, when the pelvic vessels are already distended and excited, has, on some occasions, seemed to me to awaken inflammation. The incautious use of strong caustics to the cervix uteri may give rise to it. I think I saw a case in which perimetritis was brought about by severe exercise in walking immediately after the use of caustic potassa. Excessive venereal indulgence, doubtless, predisposes to this inflammation, if it does not produce it alone.

Symptoms.—The patient is attacked suddenly, usually with pain in the pelvis, hypogastrium, or iliac regions, which radiates to the sacrum, loins, and abdomen. Sometimes it passes down one extremity, or there is pain in both legs. The pain, generally at first aching and moderate, may become very severe, and darting or cramping in character. In the beginning, or after the inflammation has lasted a little while, there is pain or difficulty in urinating, and the passage of fæces through the rectum is painful, also, by pressing upon the inflamed parts. The patient usually experiences a sense of weight about the perineum, and dragging in the loins and hips. All the pains are much aggravated by motion, or assuming and continuing in the erect posture. Pressure over the epigastric and inguinal portions of the abdomen increases the pain and suffering.

At the commencement of the pain the patient is attacked with rigors of greater or less severity. The chilliness may be slight, but often it amounts to severe shaking and trembling; reaction proportionate to the intensity of the chill succeeds; the head aches, the limbs are pained, the skin is hot and dry, and the tongue coated, dry, and parched, as is also the mouth. These symptoms may come on very suddenly, and the case be well marked in a few hours from the time they commence, or so moderately and gradually as to be several days in assuming prominence. In puerperal patients they occur generally several days after confinement, and seem to be induced by undue exertion or exposure. In such cases the symptoms are likely to be more intense than in the non-puerperal cases. The pulse is rapid, the nervous system much disturbed, the heat great, and often there is delirium. The high febrile excitement is attended with severe pain, extending in various directions.

There are some instances of non-puerperal cases where the symptoms are quite intense, but as a general rule they are less so than in the former. Tumefaction and tenderness over the lower parts of the abdomen indicate a local peritoneal inflammation in many of the more severe instances, although this is not always the case. Some of these puerperal cases so closely resemble cases of metro-peritonitis—if they are not so indeed—that the cases are regarded as attacks of puerperal fever. So intense are the symptoms in some instances as apparently to jeopardize the life of the patient immediately by the gravity of the general pelvic and abdominal inflammation. And when the tumefaction and tenderness of the abdomen subside, the febrile reaction is moderated or becomes more paroxysmal, we find a hard tumor generally on one side dipping down into the pelvis and extending sometimes to the ribs and across to the umbilicus; or it may be developed in the mesial portion of the abdomen and pelvis, extending upward to a greater or less degree. Tumors of this kind are tender, and may be detected in the pelvis by a vaginal examination. They do not always suppurate, but occasionally disappear by absorption. At other times they produce copious quantities of pus. This inflammation sometimes dissects up the peritoneum, over the osseous, iliac, and lumbar muscles, to a great extent dissolving out the areolar tissue in a large space. The distension and tenderness are quite frequently confined to one side, showing the point of greatest intensity of the disease, or the side to which it may be confined, but we often find them extending entirely across, and sometimes considerably up the abdomen. These symptoms appertain to the first stage, and last for from four or five days to two weeks, and in rare cases longer, when they are gradually succeeded by those that indicate the suppurative stage. The pain becomes less acute, and changes ordinarily to a burning character, quite as distressing, if not more so than at first. It is worse at night, and prevents the patient from resting. The fever assumes something of a remitting type. It is more intense in the evening and night; toward morning a moisture is observed upon the skin, the heat becomes less, and there is some amelioration in the suffering. After a little longer, and the paroxysms are very marked, chilliness in the after part of the day is succeeded by a very rapid pulse and intense heat of the surface. This fever lasts for six or eight hours, and is resolved by a copious perspiration. These perspirations are accompanied with great languor and depression. The patient is debilitated and much worn by the continuance of these symptoms. At length, after days of this exhausting suppurative fever, the pus makes its way through the walls of the abscess, and is discharged through some of the outlets mentioned above. If the evacuation is free, and the discharge considerable, the relief is very great indeed, the fever subsides, the perspiration ceases, the spirits are good, the appetite becomes excellent, in fact the change in the patient is very great and gratifying. Convalescence is now established, and in a few days all the serious and distressing symptoms have vanished. If this discharge is not free, and but a small quantity of the matter escapes, although there is relief, it is not so complete. The patient is temporarily better, but not convalescent. The opening is not sufficient, the pus continues to increase and imperfectly discharge, and fluctuations in the intensity of suffering continue to inspire hope and cause depression, until a freer opening occurs in the same place, or another one allows the pus to escape more freely.

This description is intended to apply to cases of considerable intensity in the puerperal or non-puerperal patient. But the degrees of intensity are very different in different instances. Sometimes the symptoms are so slight as to scarcely attract attention, until the discharge begins to make its appearance. At other times there is distressing fever, but the local symptoms are so poorly marked that the case is misapprehended. I have known the fever to last for three or four weeks, ending in hectic, with its exhausting accompaniments, before the true nature of the case was discovered. Two cases of this kind were treated for typhoid fever, which they certainly very much resembled, until, upon examination, the collection of pus was discovered. These inflammations are more frequent than is generally supposed, and often overlooked. An inflammatory fever, followed by hectic symptoms, should cause us earnestly to search for the location of the inflammation.

An example of the occasional insidiousness of this affection is exhibited in the following case:

Mrs. A——, aged twenty-four, married two months, has suffered for the last four years with moderate dysmenorrhœa, and occasional leucorrhœa. Sexual intercourse has given her much pain from the first since her marriage; after three weeks the pain in the coïtus became intolerable. At this time she had severe pain in the back and pelvic region constantly, but not so severe as to prevent her being about in the attention to domestic duties and taking a short trip by rail with her husband. She had some very slight pelvic reaction, with sense of chilling, for about twenty days, when the paroxysms assumed something of a hectic character, lasting from three o'clock until seven or eight, P.M., terminating with copious diaphoresis. A little later a very severe pain in the hypogastric region was developed, attended with frequent efforts at urination. In about four days from the prevention of this pain she began to pass pus in large quantities in the urine, together with marked quantities of blood. Upon making examination at this time the pelvis on the right side and front portion was filled by a hardness very tender to the touch, which had crowded the uterus back upon the rectum and down so that the os was in contact with the perineum. These symptoms and the examination fully declared it a case of perimetritis.

When the disease becomes chronic, the symptoms become obscure, and the cases so completely resemble uterine disease, that nothing but a careful physical examination will enable us to arrive at positive conclusions. In the chronic form there are exacerbations and remissions, and during the remissions the sufferings are not very considerable. When the paroxysms occur at the menstrual periods they sometimes pass for dysmenorrhœa, which they very much resemble. Sometimes, in the chronic form, the suffering during the paroxysm is severe and prostrating, while at others it is only sufficiently intense to be inconvenient.

Diagnosis.—Although the symptoms, in most cases, are severe and sufficiently prominent, they are not often distinctive. Several other affections resemble it in many symptoms. Hence, the only way to arrive at correct diagnosis is by physical examinations. The finger will be the only instrument necessary. It is cruel to use the speculum, while it affords us no aid in the vast majority of cases. I should not think it necessary to caution the reader against the use of this instrument if I had not seen it resorted to more than once to the great torture of the patient. In making examinations for this kind of case, the patient should be so placed that we may use both hands if necessary. When one or two fingers are introduced into the vagina, they will detect unusual tumidity in the pelvis. Sometimes this tumidity extends to the bottom of the pelvis on one side, and occasionally apparently fills up the whole lower part of the pelvic cavity; at other times the tumidity is circumscribed and confined to one side high up, behind the uterus, or before it. The tumefied parts are generally hard, quite so, and very tender to the touch, so that a small amount of pressure causes great suffering; the uterine neck is almost always pushed out of its place to one side, forward or backward, upward or sometimes downward; the vagina is generally hot and dry, and all the parts sensitive. If we place one hand above the pelvis while the fingers of the other are in the vagina, we will have a consciousness of a tumor between the fingers of the two hands. It is not always the case that any tumidity may be felt above the superior strait, but generally there is tumefaction in one iliac region, or sometimes in both. The tumefaction may extend much above these regions, high up into the abdominal cavity, though I think not often. If the tumefaction is considerable, the uterus is firmly fixed in its place, but when less, this is not the case. In childbed patients we may distinguish perimetritis from peritonitis by digital examination per vaginam. There is not the hard tumefaction in the pelvis in the last as in the first. Tenderness and general distension of the abdomen are greater in peritonitis; the pulse is more rapid and is peculiar. These may and probably are often combined in puerperal fever, when the diagnosis is of less importance than when they are separate affections. The peritoneal inflammation supervenes after delivery much earlier—generally on the second day—than any of the localized inflammations do. Perimetritis is more likely to attack the patient when or after she begins to make exertion, or is exposed to cold several days, six to ten, and even more after delivery.

From acute metritis in the puerperal or non-puerperal state, it may be distinguished by examination with the finger. There is not much difference in the mode of attack and history between acute metritis and perimetritis; but by a careful tender survey of the pelvic organs, we may separate the inflamed from the sound parts. In metritis the uterus is generally and symmetrically enlarged, and extends lower down in the pelvis, and if touched at any point is tender; in perimetritis this organ is not enlarged, and if touched anywhere in such manner as not to press it against or move it on the side where the inflammation exists, is not the subject of painful impressions. The tenderness in perimetritis is generally to one side of the uterus, close to the walls of the pelvis. If the inflammation lie in the bladder, we may easily ascertain this fact, by pressing this organ between the fingers in the vagina and those above the symphysis pubis. From metatithmenia it is distinguishable by the tenderness and firmness of the tumor, the febrile symptoms, and the history of the two conditions: perimetritis being previously inflammatory, while metatithmenia, when inflammatory at all, is so some time after the commencement of the symptoms. The bloody tumor may be handled without much pain, is soft and yielding, and commences at the time of menstruating with sharp pain likened often to severe colic, without chill and fever at the beginning; sometimes with collapse more or less intense. Carcinoma filling up the lateral parts of the pelvis, is sometimes mistaken for perimetritis, but more often the latter is mistaken for the former. Carcinoma is insidious in its incipiency. It has made great advance almost always before symptoms indicate its existence, while perimetritis is heralded by inflammatory symptoms from the start. The hardness of carcinoma is greater, the tumidity more irregular and devoid of tenderness; it is not hot as in inflammation. The discharge from carcinoma is cadaverous in odor,—in the advanced stage,—thin and ichorous in character. In perimetritis the discharge is pus, and if it smells at all, the odor is faintly fecal. I have noticed this last feature in several instances of perimetritis, when the evacuation of the pus was free and copious through the vagina.

The diagnosis from chronic metritis is not always easy. When perimetritis is chronic, it causes many of the symptoms which we observe to be present in chronic metritis. It will require a careful consideration of the symptoms and history of the case, and physical examination.

Chronic perimetritis ordinarily results from an acute attack, that was accompanied with a discharge of pus more or less copious, and paroxysms of less intensity have succeeded, growing more mild, until the symptoms become obscure. Paroxysmal discharge of pus is one of the constant symptoms of chronic perimetritis. Upon a thorough and careful examination of the pelvic cavity, we may find some small spot, not in contact with the uterus, but by the side of it ordinarily, that is hard and tender to the touch. In chronic metritis there is not always tenderness.

Prognosis.—This is generally favorable. There is probably more danger in attacks during the puerperal condition, or after miscarriage, than in unimpregnated patients, although the very large majority of these cases terminate favorably. Of course, I leave out of this consideration such instances as are attended by peritonitis of simultaneous origin, and constitute only a part of the whole puerperal fever. I do not think there is much difference in the fatality of uncomplicated cases occurring under these diverse circumstances. When perimetritis proves fatal, it is generally in one of three ways. 1st. By exhaustion caused by excessive and long-continued febrile excitement, symptomatic of extensive inflammation. 2d. The exhausting effects of hectic fever, diarrhœa, diaphoresis, and want of nourishment. 3d. Severe complications, arising during the progress, as peritonitis, by extension of inflammation; or the more rapidly fatal form of peritonitis, caused by effusion of pus in its cavity. I have seen three fatal cases. Two of them resulted from rupture of the abscess, and discharge of the pus in the peritoneal cavity. One of these was puerperal, and death occurred ten weeks after confinement; the other non-puerperal, and ended in eight weeks from the attack. The one which proved fatal from exhausting hectic without evacuation of the pus, terminated in sixty days from the commencement.

A great many cases terminate in the chronic form. The cause of this sort of termination is almost invariably incomplete evacuation of the pus, and, as a consequence, imperfect obliteration of the cavity of the abscess. The pus accumulates from time to time, and fresh eruptions, attended with a greater or less exacerbation of the symptoms, every few weeks or months, occur as this result. Or the external opening, wherever it may be, does not close, and there is a constant discharge of greater or less quantity, keeping up a kind of fistulous canal, leading generally some distance to the main seat of the difficulty. Or in still another sort of cases, the pus seems to be entirely evacuated, and the cavity obliterated, and there is nothing left but a small point of indurated tissue, which is the nucleus of inflammatory action under certain circumstances, as pregnancy, unusual excitement of the sexual organs, from other reasons, &c.

Treatment.—Notwithstanding the strong and rapid tendency to suppuration, an early, energetic, and appropriate course of treatment, will enable us in many instances to avoid that sort of termination. I think we should expect suppuration, unless we can see and treat the case early, in the first forty-eight hours, for instance, in severe cases, and not longer than four or five days in any sort of case. The antiphlogistic course must be tempered by the vital powers and sanguineous condition of the patient, as well as the intensity of the disease. When the disease is intense, the treatment must be as prompt and energetic as the patient can bear. And I would say to the student, the plan I shall now lay down supposes the patient to be plethoric and full of vital energy. Venesection in the upright position, to the approach of syncope, to be followed in ten or twelve hours, unless there is a decided mitigation of the symptoms, with twelve or more leeches to the sacrum. The leeches may be again repeated in twenty-four hours, if necessary. The venesection should be followed immediately by a dose of eight or ten grains of calomel, or what would be better in strong patients, hyd. mit. chl., grs. viij, pul. Jalapa, grs. xij. This last would act copiously in a few hours. Supposing the case not to be broken down by this commencement, we may administer tinct. verat. virid. gtta. iv, every three hours, until the pulse is brought down below its natural frequency. After purgation has been free, if the pain is severe, there will be great propriety in giving pul. opii, grs. ij, hyd. mit. chl., grs. ss., every four or six hours, until a very gentle ptyalism is brought about. When the objects to be accomplished by giving the above treatment are attained, a blister over the iliac region most affected will be very appropriate. Not many instances, and these mostly puerperal, will demand or justify such energetic measures; in severe and appropriate cases, however, much suffering will be avoided by them. In a more moderate grade of cases, or where the patients are less capable of bearing such energetic treatment, leeches or cupping to the sacrum, once or more as may be required, with the active cathartic, alterative, anodyne, and tinct. verat. virid., will suffice, provided they are used with promptitude. Fomentation and poultices over the affected region are always appropriate and beneficial. They are useful in every grade and stage of the disease. Very frequently, however, it is evident from the time we first visit the patient that suppuration is unavoidable. Or else we are in doubt about the possibility of avoiding suppuration. The treatment in such instances must be very different. Here it will be very proper to give the tinct. of verat. virid. in four-drop doses, every three hours, until the excitement of circulation is allayed, and then continuing it at intervals at four or six hours, to keep it quiet. The anodynes and alteratives, as before directed, will also be in place, as well as the fomentations. As the remission of the fever with perspirations denotes actual suppuration, our treatment should be tonic and anodyne until the case terminates. In the inflammatory stages the diet must be sparing and cooling, while in the suppurative stage more nutritious and stimulating substances should be administered. We sometimes meet with patients broken down by indigestion and uterine suffering, or some other cause, that require general supporting treatment from the beginning, with the alterative, anodyne, and local, above-mentioned. They will profit by and require a sustaining diet also.

A question arises at this stage of the affection which must be decided after a careful survey of the whole case, viz., should we evacuate the pus, or should this process be wholly left to nature? As one of the disastrous terminations is a rupture in the peritoneal cavity, as nature often selects very circuitous and unsatisfactory viaducts, as the rectum, bladder, &c., and as a consequence of this last circumstance the recovery is very much protracted, I think we should, when practicable, furnish the pus an outlet of our own choosing, and as early as can conveniently be done. Soon as evidences of suppuration begin to be manifested through the general symptoms, we should make as thorough an examination as we can to ascertain where the collection has occurred. If one can discover the pus, we evacuate without apprehension of damage to any of the organs. If our first examination fails to satisfy us, it should be repeated as often as every twenty-four hours, until the discovery is made. When this is done, we institute one or two precautionary measures, which will almost preclude the possibility of doing harm by an intelligent penetration. The first is to completely evacuate the contents of the bladder and rectum by the catheter and an injection. We ought to be sure that the rectum is empty of fluid and gas. I knew fluid in the rectum to so far deceive a practitioner as to cause him to make preparation for its puncture. We ought to pass the catheter into the bladder and rectum after we sit down to operate. The next precautionary measure is to introduce the exploring trocar into the tumor, and after the pus has made its appearance, open the cavity by the side of the retained canula. In this way I think there is great safety in the operation. The patient may be prepared for the puncture by being placed on the left side before a good light, as if for operation for vesico-vaginal fistula, and anæsthetized. The part may be exposed by Sims's dilator. The instrument most convenient for making the incision is a tenotomy knife. The opening should be free and direct, so as to permit of a ready discharge. The opening should not be allowed to close. This may be prevented by keeping a tent in the wound until the pus ceases to be discharged. The objects of thus opening the cavity are to secure an external and safe outlet, and its ready evacuation, and thereby attain a speedy cure and safety against peritoneal inflammation. When the chronic form consists in frequent repetitions of the inflammation, on account, perhaps, of its imperfect subsidence, much may be done by persistent counter-irritation, and among the best kind is a seton in the groin kept running for months. An issue will have equal good effect. This permanent form of counter-irritation is better, I think, than blistering or pustulation. When the opening into the intestine or bladder becomes fistulous, as it sometimes does, and the discharge continues for months and even years if there is no vaginal opening, and the discharge is into the bowel or bladder, we should seek for a point in the tumor where it may be punctured, and the opening made free and direct through the vagina. If no such point can be found, we cannot, with propriety, interfere surgically. The openings are, however, often located so that we may easily reach them, as through the lower part of the abdominal walls, the labia, the gluteal region, the perineum, or vagina. If the orifice is accessible, we may generally succeed in obliterating the suppurating cavity and fistulous canal. Preparatory to making an effort to do so, we should try to ascertain the tortuosities of the fistulous duct, and the depth of the pus-cavity. In some instances the canal is so crooked that the straight probe will pass but a very short distance, and it becomes necessary to send it in various ways; and sometimes an elastic bougie will suit better for a probe than the ordinary metallic one. Prof. Simpson recommends leaving a wire in the track of the fistula until adhesive inflammation is excited. I have not tried this means, for I have been so well pleased with injections of iodine, that I have used them almost exclusively. I inject through a small-sized catheter. The smallest sized elastic catheter, pushed to the bottom of the cavity, will convey the fluid in its concentrated strength to the bottom, and thus produce the effect at that point. We ought, after introducing the catheter, to inject the cavity with tepid soap-suds, so as completely to cleanse the internal parts of pus, and then immediately throw up the solution.

The undiluted tincture of iodine will not usually be too strong, and may be used once in three days, unless great inflammation follows its injection. Sometimes the first injection does away with the production of pus, and produces adhesive inflammation. In order effectually to inaugurate the treatment, it sometimes, indeed generally, becomes necessary to slit up the orifice of the fistula somewhat, as it is usually smaller than any other part of the duct.

CHAPTER XXIV.

DISPLACEMENTS OF THE UTERUS.

FROM what I have already said, it will be inferred that, for the most part, I consider displacements· of the unimpregnated uterus as complications and effects of inflammation; and that any other treatment but such as will remove the causing condition, is not curative. The idea that they are primary affections, and require independent treatment, is so firmly rooted in the minds of patients and physicians, and the fact that we cannot always cure the uterine inflammation and enlargement upon which they depend, and the fact that they do, in very rare instances, occur as the effect of apparently inscrutable causes, and consequently call for special treatment, render a separate consideration of these proper and necessary. It is plain that the notions in reference to the main item of treatment, viz., mechanical support, are too vague to be profitable to the novice. It will not be possible for me to enter into a very minute detail of such treatment, nor do I wish to be understood as trying to do anything more than suggest an outline of the principles that should govern us in the use of means for the relief of them. To do this satisfactorily, I must consider succinctly the different sorts of displacements, and, if possible, arrive at their mode of producing inconvenience and suffering, and their mechanical effects upon other organs, especially those in proximity to the uterus. I need not say that such considerations must also be imperfect, and assure the reader that they are merely intended to be suggestive.

Many, if not most of the failures of mechanical support for the relief of displacements, depend upon a want of correct knowledge as to their nature in any given case, and consequently, of the right kind of instrument to be used, and the mode of applying it. It will be found, upon attentive examination of them, that displacements cause distress by pressing or dragging upon other organs, by taking away the support of other organs, and thus allowing them to be misplaced: as when it sinks low into the pelvis, the abdominal organs fall into this cavity to fill the vacuum; or when the os cervix or portion of the uterus is tender, causing distress by bearing on these places as it settles against the perineum, sacrum, or rectum. In retroversion, for instance, the round ligaments, broad ligaments, and bladder, are drawn more or less out of their place, and cause suffering; so with all other displacements, varying, however, in their respective effects. In addition to this the uterus may, in retroversion, press upon the rectum, and cause inflammation and pain in it; or, by stopping up this organ, cause costiveness and difficult defecation. Retroversion of the uterus, when the fundus or posterior wall is inflamed, causes suffering by the contact of the tender part against the rectum, sacrum, or perineum.

Nature of Displacements.—The nature or doctrine of displacements of the uterus involves the conditions of its annexed organs. The uterus cannot be retroverted while the round ligaments retain all their natural conditions as to length, position, &c. Some parts of the vagina must also be changed in their conditions. So of prolapse; the round, and particularly the broad ligaments, and the vagina, &c., must be made to deviate by the acting causes of these displacements, before these latter can occur. And it is not a question which is at fault, as though the onus of failure in the functions appertained to either the ligaments or vagina, but which is most at fault. This requires us to decide which one of them does most to hold the uterus in place, which can be done only by distinguishing the kinds of displacement, and considering them with reference to this matter. The vagina can do but little to resist causes operating to produce retroversion or anteversion, but if narrow and rigid may strongly resist the tendency to prolapse. The ligaments seem so arranged as to resist displacements in every direction, and, excepting anteversion, they are pretty firmly opposed to acting causes. Another point of importance in the doctrine of displacements is, the determination of the question, whether the displacements occur as the effect of relaxation of the sustaining organs, or whether the sustaining organs are relaxed on account of the long-continued action of the causes operating on the uterus. In most cases, I think, the last of these conditions obtains; but, undoubtedly, they are quite frequently contemporaneous and consentaneous circumstances. After labor, the ligaments and vagina must have returned to their natural and healthy dimensions and firmness before they can resist displacing influences. This they cannot do for some weeks, and during this time the weight of the uterus, on account of imperfect involution, is much greater than they are in the habit of sustaining. Although usually at fault, because overcome when in this natural condition, and forced into deficiency of function, they may, as the above view shows, be deficient because of their condition at the time the operating causes are applied. But we cannot always say that the uterus has fallen, or become displaced, because the ligaments and vagina are too weak or too lax to sustain it in place, but in many instances because the acting cause is sufficient to force them to overcome them, and carry the uterus in the course of its action in spite of their healthy resistance. I think this is the true explanation which may be given in most cases.

Depression or Lapse.—The principal and only important varieties of displacements are, first, a simple depression or falling of the uterus in the axis of the superior strait. The inconveniences resulting from this deviation are painful tenesmus, constipation, or hemorrhoids, on account of pressure upon the rectum; sciatic pains, on account of pressure upon the sacral nerves; pain in the uterus itself, on account of pressing upon its own tender cervix; and feeling of weight on the perineum, and dragging about the loins and hips. The broad ligaments are stretched; perhaps the round ligaments somewhat increased in length, less so, however, than the others. The change in the direction of the vagina from almost directly backward to backward and downward, is also quite obvious. The rectification of this deviation is usually accomplished by lifting the uterus up. It requires no change in axial direction with the pelvis, but simply an elevation to restore the uterus to its proper place.

Prolapse.—Secondly, prolapse in various degrees, from slight depression to complete extrusion from the labia. This displacement in the slightest degree, and in fact in all its degrees, pulls upon and stretches all the ligaments, the broad ligaments by far the most. The vagina suffers displacement proportionally with the prolapse. It is inverted, its walls being doubled upon themselves, and its cavity progressively shortened until entirely effaced. This displacement is always in a direction corresponding with the

20

axis of the vagina and different portions of the pelvis, and follows the curve formed by the hollow of the sacrum and continued by the perineum. The inconveniences arising from this displacement are not unlike the last, until it becomes great, when the bearing down becomes more distressing, as well as the dragging on the loins. To these are added those arising from a prolapse of the abdominal viscera into the pelvic cavity; sinking sensation in the epigastric region, and dragging upon the hypochondria, &c. The means calculated successfully to restore the uterus in this displacement must lift it up and correct its axial deviation.

Retroversion.—Thirdly, retroversion. This displacement is present when the fundus is depressed by being thrown back into the hollow of the sacrum, while the cervix is drawn forwards and upward, so as to be upon a level, or above a level with the arch of the symphysis pubis. The difference between this and prolapse is, that the fundus is thrown lower down into the hollow of the sacrum, and the axis of the uterus is almost natural, but the lower end becomes the upper. The inconveniences arising from this displacement are caused by pressure on the rectum, perineum, and sacral nerves in the posterior inferior part of the pelvis, and sometimes pressure upon the neck of the bladder or urethra in front, and dragging upon the ligaments. The ligaments most severely stretched are the round, the broad being much less so. The condition of the vagina is changed very considerably; the anterior wall is very much shortened, while in married women the posterior is elongated somewhat. The means employed for the correction will act by elevating the fundus and pressing the cervix backward towards the middle of the pelvis.

Anteversion.—Fourthly, anteversion is, in most respects, nearly the opposite in position. The cervix is turned back upon the sacrum, and elevated somewhat above its natural position, while the fundus is thrown forward upon the bladder and anterior walls of the vagina, so as to come down to a level, or nearly so, with the arch of the symphysis pubis. The inconveniences arising from this displacement are caused by pressure upon the bladder, urethra, and rectum, tenderness in sexual intercourse, and dragging about the pubis. The broad ligaments are stretched most, and the vagina is elongated and depressed somewhat at its posterior extremity. Not unfrequently the rectum is pressed upon by the cervix uteri, and distress arises as a consequence. The means for rectifying this position must lift the fundus upward, and push it backward, or draw the cervix forward and lift it slightly upward.

Causes.—Anything that will increase the weight of the uterus predisposes it to deviations and displacements. When thus predisposed by increased weight, if the patient is much in the erect posture, the uterus will settle down into displacements. It will be observed that the deviations I have mentioned are lapses in some manner or form. When the uterus is slightly enlarged and increased in weight, the erect posture is not always enough to cause displacement, but if the patient strain from the tenesmus of dysentery or dysuria, or in lifting, or is jolted so as to bring the weight of the abdominal viscera down upon the pelvic organs, she feels a distressing sense of pressure upon the perineum, rectum, or bladder, or all of them, and thenceforth she suffers from some of the disagreeable symptoms attendant upon the displacement. Should the uterus be much larger than natural, the erect posture maintained for any length of time insures a displacement. Inflammation of the cervix is almost always attended with increased size and weight of the whole uterus, and thus it predisposes the organ to displacement. This accounts for the fact that we very often find these two conditions present. I have no doubt, from ample observation, that inflammation is very frequently the cause of depression and other displacements in this way. Dr. Bennett thinks that when inflammation attacks the posterior walls of the uterus, retroversion is the result, and that anteversion is caused by inflammation of the anterior wall, and leaves us to infer that these displacements are almost always connected with inflammations as an effect. While I believe with him that these displacements may result from inflammations thus localized, they are, I am satisfied, often caused by inflammation of the cervix alone, and that without any peculiarity discoverable by an examination. I think I have seen every variety of displacement connected with cervical inflammation.

Imperfect Involution.—Imperfect involution is another cause or displacements. Involution may be imperfect and yet be progressing naturally when a co-operating cause determines displacement. If, for instance, a woman who has given birth to a child, arise from the bed in four or five days and maintains the erect posture for a length of time, or remain permanently out of bed, engaged perhaps in some arduous duties on her feet, the uterus, being still so much above its natural size and weight, falls in some manner in spite of the relaxed ligaments and vagina. This is the manner in which some persons contract uterine displacements after labor.

Arrest of Involution.—But it occasionally occurs that long after the usual time for involution to be complete, the uterus remains increased in volume and weight, because of the arrest or the tardiness of this process. In this condition, the uterus is liable to settle below its natural level, in part or as a whole. Subacute inflammation, attacking the post-parturient uterus, not unfrequently delays the return of this organ to its natural dimensions for many weeks after confinement. In all these last cases, the displacement occurs at a period not very remote from parturition or abortion. And very many cases can easily be traced to this time, or near it, seeming to be a sequence to it. I cannot forbear remarking here, with reference even to these cases, that when they have become chronic they are found to be complicated with inflammation; and that where this can be removed entirely, the restoration of the position of the uterus generally takes place spontaneously, and is easily effected by treatment, or ceases to present any indication for treatment, on account of the absence of symptoms.

Tumors.—Tumors, developed in some part of the fibrous structure, cause an increase in weight, and thus encourage, at first, displacement, and after awhile determine it. The position of the tumor will govern the nature of the deviation. If the tumor is in the anterior wall, it is apt to cause anteversion; if in the posterior walls, retroversion; if in the cavity or cervix, prolapse, or merely lapse.

Loaded Intestinal Canal.—With the causes above spoken of predisposing to it, I think the pressure of heavily loaded intestines may determine displacement. Fecal accumulations in habitual constipation would be sufficient.

Distended Bladder.—Equally, perhaps more certainly mischievous, is a bladder constantly filled with urine overriding the uterus, and pressing backward and downward. This cause would seem to favor retroversion, on account of the position of the bladder. These causes are undoubtedly not sufficient, however, to produce permanent displacement, without the co-operation of increased weight of the uterus.

Symptoms.—However the displacements may occur, the symptoms attendant upon them will not enable us to distinguish one from the other. As I have already shown, certain deviations very often give rise to particular symptoms; but this is not so constantly the case that our diagnosis can be materially influenced by them. It is true, also, that the symptoms cannot be distinguished from symptoms arising from other pelvic diseases; hence there is no alternative left us: a physical examination will be our only sufficient means for forming an accurate diagnosis. The symptoms are an expression of the sufferings of other organs, for the most part from pressure on them by the uterus. Pain, numbness, debility, formication, or change of temperature, general or partial, of the lower extremities, or one of them, on account of the pressure upon some of the large nerves, particularly the sciatic running down them, or tenesmus, constipation, hemorrhoids, and sense of heat or weight in the rectum, indicates the pressure upon that bowel. The dysuria, cutting, burning, or rending pain in the bladder, incontinence of urine, and other distressing vesical disorders, are expressive of the suffering caused by pressure on the neck of the bladder or urethra by the displaced uterus. But a general pelvic tenesmus, or feeling of bearing down, with weight and dragging upon the perineum, not unfrequently are produced or aggravated by the uterus lying heavily upon the bottom of the pelvis. This is perineal distress. There is also a general feeling of pelvic distress, such as dragging pain in the hips and loins, weight and pressure about the pubis, a feeling as of a cord drawing in one or both of the inguinal regions, or general sense of weakness and indescribable *malaise*. Another sort of difficulty seems to be produced by the uterus in its descent, dragging other organs out of their natural position. The bladder may be thus drawn down, and cause a sense of dragging from the umbilicus, or produce various sorts of trouble in the functions of holding and evacuating the urine.

Prolapse of Ovaria.—The ovaria are displaced and drawn down into the pelvis, with the feeling of tension in the broad ligaments, or iliac regions. In extreme cases of prolapse, and even sometimes in the slighter degrees of displacement, disagreeable symptoms arise from abdominal organs falling into the pelvis to fill the partial vacuum caused by the descent of the uterus. To this cause, doubtless, may sometimes be justly attributed the sense of weakness and emptiness in the epigastrium, and the pain in the side, in the region of the liver and spleen.

What is more common, however, is the suffering caused by the inflamed uterus, slightly depressed, pressing upon its own sensitive diseased parts, as the cervix, posterior wall, or even fundus. Again, there can be no doubt but that the rectum, bladder, cellular tissue within the pelvis, and parts surrounding the nerves, and even the nerves themselves, are often subjects of inflammation, and, being pressed by the inflamed, enlarged, and sensitive uterus, are much more susceptible to the above described influences than if they were all in their ordinary healthy condition.

Examinations to determine displacements should be both digital and instrumental, and thorough enough to ascertain the particulars as to the position and condition of the uterus in all other respects that can have any bearing upon the case.

When very great suffering of the character above mentioned exists, there is almost always a combination of inflammation and displacement. Should there be any obscurity in the position of the uterus, on any account, our diagnosis may be cleared up to a demonstration by introducing the probe into the cavity of the organ. The direction the instrument takes will clearly show the direction of the uterine cavity, and thus indicate the position of the organ.

Treatment of Displacements.—The efficacious mode of treating displacements, as I have before intimated, is the perfect removal of the causes of them. When this can be done completely, the inconvenience and suffering attributed to them are removed, or very materially ameliorated. But there are cases where this cannot be done, from various reasons, among which are the prejudices of patients and medical men against the treatment necessary for the cure of inflammation, the impossibility of curing the inflammation when every opportunity is enjoyed, and even sometimes, when the inflammation is removed, so far as we can judge, there may be a continuance of displacement with its symptoms. In all or most of these, the skilful treatment of the cases, as displacements, will often result in great palliation, if not in cure. All I need say here in reference to the treatment of inflammation is, that full directions may be found in the foregoing chapters on that subject.

Treatment of Deficient Involution.—Involution as one of the causes, should be treated as subacute inflammation of the whole of that organ. All the symptoms, as we usually find them, will justify this procedure. It is, in fact, subacute inflammation occurring in the uterus after accouchement, and operates in establishing displacement upon first getting up from childbed. To avoid this condition by sufficient quiet and care, is much easier than the cure after the disease is established. When it does exist, rest in the horizontal position, alteratives, laxatives, external fomentation, counter-irritation, &c., will ordinarily relieve it, and prevent or cure the mischief done or apprehended.

Removal of Tumors.—When the uterus is enlarged or depressed by tumors, their removal is the only radical way of overcoming the difficulty. If a loaded condition of the bowels is the operating cause, we must endeavor, by alteratives, laxatives, &c., to remove this condition.

Mechanical Support.—The main object I had in view in introducing this chapter on displacement, was to discuss the subject of mechanical support. And as an introduction, I may state my conviction, that very few general practitioners study these affections sufficiently to acquire the skill requisite for the best management of them. Too often we are satisfied with merely recommending some form of supporter or pessary, and leave the execution of our designs to the patient or her friends. There is too much carelessness in this respect to form a proper estimate of the nature of mechanical support for the uterus. Believing these statements to be true, I have ventured to attempt to describe, in a very concise manner, as far as they have been suggested to me by a large observation and a careful study of the subject, the conditions which give direction to the particular kinds and modes of using these contrivances.

Two Kinds of Mechanical Support.—There are two general kinds of mechanical contrivances for the purpose of support.

Abdominal Supporter.—The first which I shall mention is the abdominal supporter. It is made of various shapes, sizes, and materials. The object is to lift the abdominal organs off the uterus and its appendages, and support them so that they cannot press the latter into the pelvis.

Rationale of their Action.—It will be remembered that the inner face of the pubis looks obliquely upward and backward, forming the base of support for the abdominal organs. These latter press, on account of the inclined position of the pelvis, obliquely upon the uterus, instead of perpendicularly. The farther back the pubis is pressed toward the sacral promontory, the more completely beneath the abdominal organs, the more certainly they rest upon the upper portion of the inner face, and the less they press into the pelvis.

Abdominal supporters are intended to, and fulfil their purposes best when they aid the pubis in this function. One indispensable portion of them consists in a plate or pad, so arranged as to press the abdominal muscles immediately above the symphysis as far back toward the promontory of the sacrum as possible. This plate, pushed in above the symphysis and below the intestines, prevents them from weighing heavily upon the pelvic viscera. And when it is recollected that it requires, in most persons, a very little intrusion of this plate or pad into this part of the abdomen to intercept a line falling through the central part of the trunk above, and thus to assume a position for the support of the abdominal organs, we can see how it might be efficacious in certain instances, and nearly harmless in all. The plate or pad is pressed to its place, and held in position by springs or bandages variously arranged to suit the fancy of the contriver. Connected with the springs which pass around the hips, not unlike the springs of a truss, is a pad that bears upon the back, so as to press the loins forward, and thus incline the face of the pubis more to the horizontal. The cases to which properly formed abdominal supporters are applicable are not numerous; but there can be no question about their utility as a means of ameliorating the distress sometimes experienced on account of pressure on the top of the pelvis, and on the pelvic viscera by the abdominal organs. The observation of the profession has not settled into rules as to the manner of using these supporters, nor are they used at all by many of the most intelligent members of the profession. Yet I am persuaded that in certain instances where we are under the necessity of confining our efforts to palliation, the supporter, judiciously selected, adapted, and applied, will afford relief that cannot be obtained by any other means. There are many circumstances and conditions to be studied as indices to the kind of instrument applicable, its mode of use, and as affording data by which to judge of the practicability of any kind. These considerations should be studied in each case. Thin persons with large pelves, who stoop habitually forward, are the sort of patients, as a general thing, whose shape and condition render them applicable; while fat, erect women do not profit by them. These conditions are mentioned irrespective of the special affections for which they are recommended. Their adaptability depends, further, upon the complete absence of tenderness of the hypogastric region, or of great sensitiveness from any cause in any portion of the body upon which any part of the instrument may press. The firmness of the abdominal muscles may also prevent the supporter from having its proper effects. But notwithstanding the fact that the case and the patient may be suitable, so far as we can judge, a trial will be the only means of deciding the adaptation. I hope the profession will try more patiently and philosophically this means of palliating the sufferings of women, when we cannot cure the diseases which give origin to them.

Its Value.—It is needless, after what I have said above, to state that I consider the abdominal supporter but a make-shift, and useful as a palliative only when radical means cannot, for any reason, be used or relied upon. Unfortunately, these are too numerous to be allowed to go unprovided for.

Supporter and Pessary combined.—They are sometimes available as external attachments to pessaries, thus keeping these last in place, and making them more completely fulfil their special purposes. The combined abdominal supporter and pessary, when they can be borne, make an efficient means of rectifying displacements. The perineal pad so often attached to the supporter, although more easily worn and less likely to do damage, is also less efficient than a suitable pessary in the vagina, kept in place by a spring coming down in front from the front pad of the supporter.

Pessary.—The pessary, however, is much more commonly used alone than in combination with any other means of support. This instrument is made of various materials, and in many different shapes; the grand object of all of them is to maintain the uterus in its proper place and position. They are direct supporters of the organ by touching the uterus in some part, and by the contact holding it in place, or support it by converting the vagina into ligaments of support.

Ring Pessary.—The ring pessary, for instance, by distending the vagina all around the cervix, supports the uterus upon a level with its own circle, thus lifting it off the perineum, rectum, or nerves at the bottom of the pelvis. The ring pessary, by pressing upon the vagina, and drawing it upward and backward, is calculated to correct some inclinations of the uterus as well as to lift it up in place, as in cases of retroversion particularly.

Stem Pessary.—The stem pessary, as it is introduced in part into the cavity of the uterus, if properly adapted also corrects all sorts of deviations. It should have a perineal flat support, or be attached to some external means of maintenance. It is the most perfect mechanical support we can make use of, for if adjusted in accordance with a just knowledge of the natural place and position of the uterus, it is certain to prevent it from departing from it.

Globe Pessary.—The globe lies immediately behind the pubic bone, presses upon the anterior wall of the vagina and uterus, lifting the fundus upward and backward, and, when sufficiently large, raises the whole uterus by drawing upon the anterior wall of the vagina.

Oval.—The oval occupies the same position in the vagina, and operates the same way when it lies crosswise, immediately behind the rami of the pubis and ischium.

Disk.—The disk with convex depression in the centre merely lifts the organ up from the perineum by its thickness. It lies under the uterus, in the centre of the pelvis, and is perpendicular in its action. It is incapable of correcting any other malposition than depression or prolapse. The gum bag, distended with air, of all these shapes, globular and circular, diskal and oval, of course operates as I have described these. Notwithstanding the above is a general, and, for the most part, correct idea of the *modus operandi* of the different shaped pessaries, in no two cases will the same formed pessary have precisely the same bearing above and below, and consequently these items cannot be attended to in each different case. There is, in other words, an individuality in every case of every sort of deviation, that will require for it separate study. This instrument is not only made in different shapes, but there is a diversity of material used for its construction. Several of the metals, as gold, silver, copper, steel, &c., enter into the composition of pessaries; horn, wood, ivory, India-rubber and gutta-percha, are also made use of. These two last are worked with so much skill, of late years, that they are taking the place of other material in the manufacture of pessaries. The hard and soft rubber pessaries are assuming almost every variety of shape. Although it is quite impossible to give directions that will be applicable to all sorts of cases, and much must always be left to the judgment of the attendant, I venture to hope that the few general considerations which I shall submit will awaken intelligent reflection in the mind of the student as to the difficulties that will often present themselves, and sometimes entirely baffle him in the use of the pessary.

Preparation of the Vagina.—It should be remembered that the vagina will frequently need preparation before it will tolerate the pessary. How unreasonable it would be to introduce the globe, oval, disk or ring pessary into the vagina, and expect it to tolerate the presence of any of them, when it was in a state of inflammation, great contraction, or rigidity. These conditions, if present, should be removed before attempting the use of any pessary. When this is not practicable, we should not think of using the instrument. A condition of the vagina most tolerant of the pessary, but which often thwarts our best considered plans, is a very lax state of its walls or sphincter. This relaxation sometimes obstinately persists, in spite of every effort to remove it. By attaching external supports to the pessary, in these cases, we may keep it in position, and thus compensate for the absence of the co-operating support of the vagina.

Condition of the Uterus that modifies the Use of the Pessary.— These considerations as to the state of the vagina have but little reference to the kind of instrument, either in shape, size, or material; but there is another class of circumstances that will govern us in the selection of a pessary. These have reference to the suffering organ, and the mode in which the suffering is produced. Where is the pressure? What organ is pressed upon, and by what part of the uterus is the pressure made? Does the rectum suffer by the pressure of the cervix, as the uterus stands in the direction of the axis of the superior strait? If so, the uterus must be lifted clear of it by an instrument that will not press upon the rectum in the same place, or the symptoms will be continued, and even aggravated by the pessary. If the rectum suffers by the fundus turning backward upon it, the fundus must be raised forward, without making the rectum a point of support for the instrument which does it. In like manner, if the pressure is upon the sacral nerves, perineum, bladder, &c. In any or all of these cases the uterus may be tender or inflamed. If this is the case, the pessary must be so constructed as not only to avoid pressure upon tender points of the subjacent organs, but likewise to impinge upon the uterus at a sound portion, or support without touching it at all, else the symptoms will be only partially relieved, or changed somewhat in their character. When it is remembered that all of these conditions are to be fulfilled in order to get a perfect adaptation of the pessary, it will not be surprising that we so often fail in getting good from its use, that there is so great a variety in shape, consistence, size, &c., that definite rules are impossible, and that so few practitioners agree in regard to their usefulness and adaptation. The pessary must be studied as a mechanical instrument, while its use subserves physiological purposes. It must be governed by mechanical laws, with the infinite and inappreciable exceptions which physiology always imposes upon them.

Kind of Displacements to which it is Adapted.—Another set of considerations must have reference to the mere displacement: as to whether it is retroversion, anteversion, prolapse, lapse, &c. On these last considerations will, more than any others, depend the shape of the pessary. It will sometimes be found difficult, if not impossible, to employ an instrument that will correct displacement without its making pressure on the suffering organ at the side of, or beneath the uterus, or upon a tender point in this organ itself. When such is the case, the consistence or hardness is a matter of much importance. A pessary filled with air or stuffed with hair is a better point of support for a tender uterus than a hard rubber, glass, or metal instrument. The former kind are cushions of such softness as sometimes to be tolerated by a very tender organ. A deliberate attention to the above considerations will enable us to approximate more nearly to an adaptation than a more loose and less methodical study of each individual case;

and while it may not lead us at the first experiment to perfection in this respect, it will form a basis for intelligent experimentation. And it should be expected that not only will such study, but observation also, in each instance, be necessary to arrive at a perfect adaptation.

The instrument which, mechanically speaking, is best calculated to correct all sorts of deviations, as well as to keep clear of surrounding organs, and consequently not to cause distress in them, by pressure upon them, is the stem pessary with external support. It in fact completely fulfils every mechanical indication in any species of deviation. This is the case, because the stem fitting into the cervical cavity, and even passing up into the cavity of the body, forms a lever when fixed in a certain position by a branch passing out between the labia to be connected with a fixed support externally, and it must keep the uterus precisely in place. Unfortunately for our success in these cases, this mechanical contrivance is quite intolerable under most circumstances. The reasons why it is intolerable are that the pressure of the stem upon the mucous lining of the cavities causes inflammation, and a positively fixed state of the organ is unnatural, and in some postures and movements of the body, must be annoying to other organs by interfering with their mobility.

The elastic ring made of a watch-spring covered by gutta percha, or some other impervious material, when properly adapted in size and strength to the size of the vagina, and well applied, is also a very efficient instrument, and applicable to almost all varieties of deviation. It spreads the vagina out on all sides, and causes the walls of this tube to assume almost the same relation to the lower part of the uterus that the broad ligaments do to the upper part of the organ. This imitation of the pelvic circle by the ring and its ligaments, stretching the vagina around the uterus, keeps the cervix in position, provided the vaginal walls are made tense by the size of the ring. When the broad and round ligaments are lax, however, the fundus and body are left to topple over backward to the sides or in front, and if these are heavy, it may distress the rectum or bladder. The ring pessary may be made to replace the uterus in retroversion or prolapse, better, perhaps, than any others, but is not calculated to be of any advantage in anteversion. It is not so likely to produce intolerable irritation as the stem pessary, and, in fact, may be made to agree with cases as well as almost any other kind of instrument. The size of the ring to be used will depend upon the size of the pelvis, and the tension of the vagina. If the vagina is cylindrical, firm and elastic, not rigid, free from inflammation, and not particularly sensitive, we may hope to procure toleration for the ring. It must be placed so as to suit the case. If the deviation is prolapse, the ring should be placed so that one part of its circumference be at the arch of the symphysis, while the other side is directed to the sacrum, so as to correspond with the axis of the pelvis upon a level with the lower part of the symphysis. For retroversion, the posterior part of the circle should be directed up the sacrum toward the promontory, while the anterior part is placed below the arch of the symphysis. It will be seen that the indications are fulfilled by thus accommodating the position of the uterus by the position of the instrument. In the latter position, the fundus of the uterus is raised up by the ring pressing the posterior cul-de-sac of the vagina up behind it, while the os is drawn backward by traction on the posterior wall of the vagina. In retroversion, the stretching of the vagina from before backward is usually sufficient, and it is not necessary to distend it laterally to the same extent, if at all.

Dr. Hodge's modifications of the ring pessary, called by him the open lever, may often be substituted for the ring, and in some cases, perhaps, acts better than it. Dr. Hodge's pessary is substantially a flattened ring. It is made of firm material, and curved so that it may be made to distend the posterior vaginal cul-de-sac by curving up behind the uterus, thus lifting the fundus and drawing back the os uteri. This pessary, according to the inventor, is capable of doing more good and is of greater extent of adaptation than the elastic ring. As it is in the hands of an intelligent and discriminating profession, these assurances will be tested and decided upon no doubt correctly. I confess that I am decidedly in favor of the elastic ring, which, if not too rigid, moulds and adapts itself to the inequalities of the parts, allows a limited movement to the uterus, and yields to the passage of fæces down the rectum, none of which things are accomplished by Dr. Hodge's pessary, which is unyielding and fixed in its position, and inelastic in composition. Dewees's modification of the ring—the disk—is more clumsy, and I think less useful than the ordinary ring of some elastic material. I have seen this form of instrument made of hollow elastic material, and supplied with a tube for inflation. The instrument thus formed is introduced, placed in position, and then inflated. Or it is sometimes inflated when manufactured, and kept so permanently, and used as the hard rubber or glass made by Prof. Dewees. The common air-bag, of different forms and dimensions, is made with a supply-tube for filling. This air-bag forms a soft cushion, upon which a tender uterus may rest without much offence to its susceptibilities. It also diffuses the pressure over a large space, in such a way as to relieve the distress of other inflamed and injured organs. There are other forms of pessaries recommended for, and perhaps have, their special virtues; but it would be both profitless and tedious to enumerate more of them. The globe pessary is best adapted to the correction of anteversion, and perhaps none other will answer so well. It is not appropriate in any other form of displacement. It is necessary to understand the principles which govern their application: study well each case, and then, if there is no instrument at hand that suits, make such a one as is adapted to the case for which it is needed. There are but few persons, who will take time to think upon the instances in hand, but will be able to judge of and adapt the proper instrument. It is a question, after we have adapted an instrument in a favorable manner to a case suited, what should be the management of it. The kind of instrument, and the nature of the case, must determine this question, instead of an arbitrary rule. A pessary made of porous material, that entangles the secretions, will soon become foul with them, and hence should be often removed and cleansed. This should be done, for instance, every twenty-four hours. In case of the ring made of hard and polished material, there is no need of frequent removals, as they do not absorb or entangle the mucus, pus, or blood. A profuse discharge of blood, mucus, or pus, will render an instrument that is capable of becoming so, fit to remain in for only a short time. Much tenderness and inflammation is a good reason for keeping both the pessary and vagina clean; but the same states forbid the frequent removal and introduction of the instrument, on account of the violence thus done. The presence of the pessary need not prohibit the use of medicated injections in the vagina; or, if we think best, the pessary may be medicated, or made the medium of applying medicines to the vaginal membrane. Of course, the composition of the pessary should be taken into consideration, lest chemical reaction between it and the injected substance occurs. It should not be incompatible with the nitrate of silver, acet. lead, sul. zinc, or with whatever may be used. Carelessness or ignorance, or both, in the use of the pessary, may lead to disastrous damage from them. They should be properly attended to. When allowed to remain too long in the vagina, charged with blood, mucus, or pus, which becomes entangled in them or detained about them, its decomposition becomes a source of poisonous filth, that, by absorption, may endanger the life of the patient, and most certainly will be an intolerable annoyance. Or, by continued and prolonged pressure on some particular place, the pessary may cause ulceration to destruction of much tissue, creating fistula, urinary or fecal. It is but proper and just, however, to observe in this connection, that these results are the effects of the abuse of the pessary, and should not be brought forward as objections to its judicious use, any more than cutting the intestine should be regarded as prohibitory of operations for strangulated hernia. Carefulness will, no doubt, prevent all the evil effects which have been done by the pessary; and when the damage cannot be avoided, no judicious practitioner will persist in their use. They are not, in such cases, the means adapted to the end.

Procidentia or Protrusion of the Uterus.—What has been said about the moderate displacements is not sufficient to enable the student to understand those extreme cases in which the uterus, vagina, and the bladder, are more or less completely protruded from the pelvis through the vaginal orifice, and it will be profitable I think, to consider them more at length.

Nature and Causes.—Whatever may be our theory as to the initial condition under which procidentia arises, there can be no question that there always are three very important, indeed, indispensable items in the complete case. There must be, 1st. Great relaxation of the lateral, sacral, and round ligaments. 2d. Relaxed or deficient perineum. 3d. Hypertrophy and relaxation of the vagina. I do not contend for the order in which they are here mentioned, but it is manifest that the uterus will not appear externally where the ligaments are not longer than natural, nor can it crowd down and drag along with it the bladder and the inverted walls of the vagina, when this last canal is only of its ordinary dimensions, neither will it protrude with the perineal margin intact, and in its proper place. So that it is too restricted a view of the subject to say that protrusion depends upon a relaxed vagina alone, and rely upon a diminution of its calibre as a sufficient cure in all cases, because if both the other conditions continue, the vagina may be relaxed and the evil renewed. A similar remark may be made with reference to deficient perineum.

One of the three conditions may *preponderate* in its influence, and its correction may possibly cure the case, but it is quite likely to fail. An aggravating but not always present condition is enlarged and unduly heavy condition of the uterus: I say not always present, for sometimes the uterus is even less than the ordinary weight and size. Although the pelvis is often larger, and the arch of the pubis more expanded than usual, they are not always so. If the three essential conditions above mentioned are present in a sufficient degree there will be procidentia. It is an easy matter to see that *any one of those conditions* in an extreme degree may induce the others, and will be likely under favorably acting external causes to do so. It is very probable, also, if not certain, that a permanently enlarged uterus, weighing considerably more than natural, will bring about all the three essential conditions. Pregnancy succeeded by parturition brings about hypertrophy and elongation of all the ligaments, hypertrophy and relaxation of the vagina and perineum, and in a large per cent. of labor, the perineum is not merely relaxed and hypertrophied but torn so as to leave a deficiency of it. Such states of all the parts may become permanent and allow descent. These conditions may be brought about by many different causes. Vesical or rectal tenesmus, long continued, are powerful causes. Dr. McClintock in his recent work on Diseases of Women, details a case which seemed to have been caused by the tenesmus of stricture of the rectum. From the action of the different causes it will sometimes take place suddenly, sometimes slowly. If the uterus is heavy, as after parturition or abortion, it is more likely to be sudden.

The extent of protrusion is sometimes very great. Usually the uterus is protruded as far as the length of the vagina will permit, and within this inverted cavity are contained the uterus and part of the bladder above. At other times the tumor is immensely large, reaching half way to the knees, containing the uterus, bladder, ovaria,

Fallopian tubes, and a large amount of the intestines. The vaginal mucous membrane is almost always more or less changed. It is generally dry, rough, and dark colored. In some instances it is inflamed, sensitive, and even ulcerated. Large deep excavating ulceration of the cervix and vagina are not at all uncommon. The intensity and extent of the inflammation and the gravity of ulceration exceed what is to be found in any other chronic non-malignant condition of the organ; from which it is rational to conclude that in this instance, at least, the misplacement causes the inflammation, or at least very much aggravates it. Still stronger support to this inference is the fact that, to return and maintain the uterus in its natural position, these conditions are immediately ameliorated, and as a consequence sometimes entirely cured. The protrusion does not seem to incapacitate the organ for the discharge of its natural functions. Menstruation is often quite natural, impregnation practicable, and labor proceeds favorably. Whether the protrusion ever occurs for the first time during pregnancy, from my own observation I am unable to say, but I have seen instances when it continued during the whole period of gestation. About four years since, I attended a patient whom I saw frequently during pregnancy; and up to the time when labor began, the cervix protruded three inches beyond the vulva, and only receded after the head began to descend. It then passed entirely within the external organs, and labor was accomplished as usual. Another instance I observed in the person of a German woman, who had been the subject of this displacement for many years. She was taken in labor with her tenth child, at the end of the seventh month; the uterus did not ascend during labor, but seemed to be driven further down by the action of the abdominal muscles, until it must have protruded six inches through the vulva, and I saw the head of the child preceded by the bag of water, pass through the mouth of the uterus entirely out of the pelvis. And the whole child passed through the pelvis before it passed the mouth of the womb. I have attended probably a dozen labors in women laboring under excessive protrusion, and they have never been embarrassed in the least by this unfortunate circumstance. It is a matter of some interest to know how well these patients bear up under their suffering, sometimes being capable of much useful exertion, indeed not apparently more distressed than by the chronic inflammation, much more moderate of a uterus *in situ*. It will not be necessary for me to enter into a detailed account of the symptoms that seem to depend upon procidentia. They so much resemble those attendant upon chronic inflammation and the more moderate displacements, that the reader has but to remember, or refer to them, in another part of this work.

The Diagnosis is not generally difficult. Inspection of the protrusion is the only way we can be positive in our conclusions. We find the urethral orifice external to the arch of the symphysis, and if a probe or catheter is introduced, it passes down in the front side of the tumor instead of backwards and upwards. If we pass the finger into the rectum the uterus will not be found in its proper place in the pelvis, and the digit can be bent down forward over the perineum into the centre of the tumor and there feel the fundus. At the lower end of the tumor there is an opening into which the probe will enter and pass upwards, in the central line of the mass. These marks are distinctive, and are not all present in any known condition beside.

The Prognosis may be set forth by answering two questions. Is it disastrous in its effects? Is it curable? Procidentia is not fatal in itself, but may so impair the general health of the patient as that she may fall an easy prey to intercurrent diseases. The amount of suffering resulting from it cannot be told in language, but is generally protracted and torturing; equal to anything else to which woman is subject. The second question does not admit of a definite answer. It is certainly obstinate under most judicious treatment, if not in some rare cases entirely incurable.

The Treatment is founded upon the three indications derived from the nature of the case. Restore the perineum, remove a portion or cause contraction of the hypertrophied vagina, and strengthen the relaxed ligaments. My own experience is decidedly favorable to the use of artificial support, and in a great many instances it will be practicable and effective. If the uterus can be kept in its proper place, the ligaments will contract and become more resistent, the vagina also diminishes in size, and if the perineum is not partially lost it will assume its tone, and relative form and position. Such pessaries as may be made to sustain the organ without distension or pain are best adapted to the work. In fact the vagina ought to be distended as little as possible. They should not rest on the perineum for support. Those supported externally are most successful when they can be tolerated. It is true that we sometimes succeed with globe pessaries, or discs, or lever or ring pessaries. When the perineum preserves much of its tone these instruments will fill the indications, but not otherwise. A stem pessary, with an air-bag globe of small dimensions at the upper end of the stem, will be tolerated often and prove very useful. The stem may be planted upon a shield that sets upon the external organs outside, and there retained by straps or bandages. This is a better way than to have a stiff rod reaching out and up to the top of the pelvis, or even up the abdomen. So long a rod under all movements of the body, bears with rigid fixedness upon the uterus. To fix the instrument at the vulva with a cord or band, places it where it is not subject to every form or great latitude of motion, while it does somewhat yield to internal pressure. The great trouble in the use of these instruments is that sometimes, after our best efforts to secure the results, they are not tolerated, too much sensitiveness of the parts preventing them from being worn. The pessary ought to be worn only when the patient is in the erect posture. It should be taken out after lying down and reintroduced before rising in the morning. I am sure that patience in selecting and modifying the shape of the pessary, with a clear view of the indications to be fulfilled, will enable us to succeed perfectly after having made a discouraging number of trials. We should study the case, and learn why the instrument is not tolerated, and correct the difficulty by changing or correcting the qualities of the instrument. It is remarkable how the vagina and perineum will contract and become strong, when the uterus is kept in its place for some years. I can but think that an ingenious use of artificial support will cure as many if not more cases than any other one sort of treatment. Astringent injections should be perseveringly used in connection with the artificial support. Saturated or very strong solutions of sul. acid, tannin, acetate of lead, &c., and decoctions of astringent bark, as oak, are the most eligible and effective forms for them.

The mode of management and objections to pessaries have been considered. But for reasons, all of which it is not necessary or possible to mention as applicable to every case, surgical operations become preferable. When they succeed, the cure is more quickly

accomplished, at least they are more under the control of the
surgeon than other processes, and do not entail so long an amount
of perseverance under suffering.

I beg the reader to remark that surgeons have generally in
their operations addressed themselves to but one item in the case.
One party operates upon the perineum, restoring or lengthening
it, more or less completely to close up the vaginal orifice, while
another party lessens the diameter of the vagina itself, and condensing its walls into cicatricial or undistensible tissue ; and it is
feared that the success of one procedure too frequently leads the

Fig. 24.

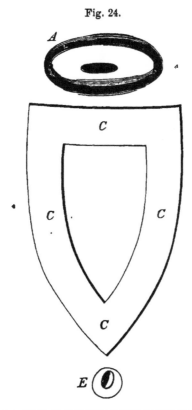

A, cervix uteri; E, urethra; C C C C. denuded surface.

operator to almost indiscriminate repetition of one kind of operation, instead of acknowledging the importance of another, and
the necessity of meeting it with a different sort of surgery. It

is not necessary to trace the history of these operations.   Two,
quite different in their nature, have been perfected and practised
by the two greatest living representatives of female surgery, viz.,
Dr. J. Marion Sims and Mr. I. Baker Brown.   Dr. Sims operates
on the walls of the vagina.   His operation consists in removing

Fig. 25.

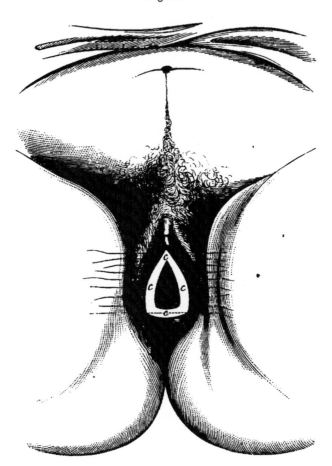

Figure 25, showing the uterus entirely protruded from the external organs. *A*, urethra ; *B*, os uteri ;
c c c c, the denuded parts with the wire sutures ready to approximate the denuded edges.

the epithelium of the mucous membrane, so as to denude the latter
thoroughly, around a triangular space on the anterior wall of the
vagina.   The base of the triangle is at the cervix, and the apex

near the urethra. It is represented by Fig. 24. Dr. Sims recommends this to be done with the uterus returned into the vagina,
but I cannot understand why the operation may not be more
easily done with the uterus in its procident state. I have never
done the operation, but I certainly would denude the membrane
and insert the silver wires as they are seen in Fig. 25, then return the uterus, and afterwards bring the parts in apposition,
and keep them so by twisting the wires. Dr. Emmet prefers the
scissors to remove the epithelium to the knife; he thinks there is
less bleeding. The patient is prepared for the operation by thoroughly evacuating the bowels the day before, and administering, an hour before its commencement, half a grain of morphia.
Chloroform ought to be given so as to keep the patient unconscious. Then placing the patient in position on her back, with
the thighs well separated, the uterus is drawn down so as completely to invert the vagina, and held by a tenaculum in the hands
of an assistant. The surgeon, by means of the scissors and tenaculum, removes the membrane, as represented in Fig. 24. This
being done and the bleeding having ceased, he may proceed to the
introduction of the sutures, being careful to cause the needle to enter
at equal distances from the margin of the cut surface outside of the
triangle, pass well into the substance of the membrane, and come
out close to the margin of the cut surface inside of the triangle,
and in the same manner to dip under the other limb of the triangle. At the base they should be brought out every quarter
of an inch in the cut, crossing from the longer limb of the figure.
Drs. Sims and Emmet pass silk sutures through with the needle,
and thus bring the wires through by attaching them to the thread.
After this much of the operation is completed, the patient may be
turned on the left side, and the vagina distended as for the operation for vesico-vaginal fistula, the parts carefully coapted, the upper
two wires requiring great care to bring the whole of the elongated
denuded surface together. The rest of the stitches from above downward may be drawn and twisted so that the denuded surfaces lie
in even contact. The patient must be kept quiet by opium for ten
days, the bladder emptied with the catheter every four or six
hours, to prevent the urine from running on the wound, and the
vagina should be syringed twice a day, after the third day. Dr.
Emmet advises us to remove the sutures on the tenth day, but

says they may be allowed to remain longer. The sutures should be sufficiently numerous—every quarter of an inch—to keep the parts thoroughly in contact, and they must be drawn tight enough to bring them well together, without strangulating them. For direction as to twisting the wires, the reader is referred to the remarks, on this subject, in the article on vesico-vaginal fistula. They should be cut and arranged after being twisted, as in the operation for that accident. This operation is applicable to cases where the hypertrophy of the vagina is very great, and the perineum entire but much distended.

Mr. I. Baker Brown's operation is applicable to those cases where there is deficiency of perineum from laceration. It consists in denuding the posterior wall of the vagina an inch above the raphe of the perineum, and up the sides of the orifice two-thirds of the inner surface. The mucous membrane should be pretty thoroughly removed in order to give a solid substance for adhesion, deep stitches as for restoration of the ruptured perineum passed, and the parts evenly adjusted. Fig. 26 shows the surfaces prepared and the sutures inserted.

Fig. 26.

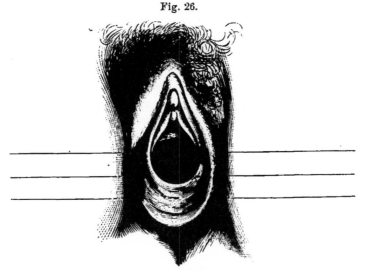

Figure 26, showing the parts, c c, denuded and the sutures passed.

There can be no doubt but that cases might be cured by a combination of these two operations, where either one alone would fail. In such cases, Sims's operation should be done first, and after the patient is entirely recovered from it, the deficient perineum can be restored.

Retroversion and Retroflexion of the Uterus during Pregnancy. —The uterus is sometimes found retroverted or retroflected during pregnancy. When small during the first few weeks of pregnancy, its existence is not observed because it produces no inconvenience, and it is not until it grows large enough to partly or completely fill up the pelvis, that anything is known of it unless discovered by accident. If it is examined at such time, the os uteri will be found against the symphysis pubis, sometimes but little above the arch, but occasionally as high as the top of that junction. If the uterus is retroverted fully, the mouth looks upward and forward; if retroflexion exists, the os is still at the symphysis, but its opening is directed *downward* and forward. In this last case the cervix is bent upon itself at a sharp angle, the lower extremity as before remarked looking downward and forward, and the uterine extremity turned backward and downward. So that the difference in these two conditions consists in the bent state of the cervix, and not in the position of the uterus. The body of this organ has its axis reversed almost completely, the fundus extremity running through the lower bone of the sacrum, while the upper extremity of the axial line passes out of the abdomen above the symphysis. The body lies in the hollow of the sacrum included in the peritoneal cul de sac between the vagina and the rectum. Both these canals are compressed, the rectum hard against the sacrum and the vagina up against the pelvic bone. The direction of the vagina is upward and forward instead of backward, its usual course. The finger cannot be made to sink deep into the vagina except behind the pubis; in introducing, it turns upward and forward. The urethra runs up in close contact with the symphysis pubis, and is narrowed very materially by extension and pressure, so that it very imperfectly performs the function of a viaduct from the bladder.

Causes.—Although pregnancy usually corrects misplacements of the uterus, such is not always the case, for this condition is sometimes a mere continuation of its unimpregnated position. It is well understood by accoucheurs also, that in the early months of pregnancy the normal position of the organ is depression, and that prolapse and retroversion are not unusual effects of recent

impregnation.   Under certain circumstances this last deviation is
not corrected by the advance of growth in the organ.   Where
other causes co-operate, a distended bladder may aid in causing
the uterus to assume and retain this position, as may also loaded
intestines pressing upon the fundus and anterior face.   These
causes and perhaps others operate to bring about a gradual displacement, but there are some that cause the condition suddenly.
It should be remembered that it is only at a certain time that these
sudden causes can produce the effect, and that is after the end of
the third month and before the beginning of the fifth month.   It
is about this time that the uterus attains a bulk sufficient to partly
or entirely fill up the pelvic cavity.   If when it has attained this size
a sudden impulse is imparted to the fundus and anterior face of
the organ, the fundus may be crowded so low into the hollow of
the sacrum as to reverse the axis.   In this state the forces acting
in favor of correction are feeble and may fail to bring it about.
Strong abdominal pressure upon the intestines and bladder under
tenesmus, falls upon the feet or breech, lifting heavy weights,
and even severe sneezing and coughing, are occasionally causative.   In the cases where the efficient causes are suddenly applied,
the symptoms are acute and established at once.   In the other
cases the train of symptoms gradually make their appearance.

Symptoms.—When induced suddenly the patient is seized with
great pain in the back, with a sense of weight upon the perineum,
constipation, retention of urine, tenesmus, dragging sensation in
the loins, and often though not always, sickness of stomach and
vomiting.   If gradually established, the pains, constipation, and
retention of urine are slowly established, it requiring from seven
to twenty-one days or more to render them intolerable.   I knew
a case caused by a woman riding all day in railroad cars without
urinating.

There are two important symptoms, viz., retention of the urine
and fæces; from these result most of the distress complained of.
Great distension of the bladder and the terrible suffering thereby
produced is the worst.   The student should bear in mind that
quite frequently this symptom is deceptive.   The urine is constantly dribbling from the meatus, and the patient thinks, and
will say, she passes plenty of urine.   The fact of this constant
slight discharge should cause us to suspect that the bladder is dis-

tended; it does not occur when the bladder is empty; it is not sufficient to prevent it from being distended. Indeed I do not now recollect any circumstances but over-distension that causes it. Retention of fæces is not productive of so great trouble as the other, but is attended with more or less inconvenience.

Great pelvic distress with *stillicidium urinæ* are almost characteristic of retroflexion or retroversion, when recent pregnancy exists.

Diagnosis.—This is usually not difficult. The first, a very important consideration, is the existence of pregnancy. Upon making vaginal examination, immediately upon introducing the finger it comes in contact with a tumor. The pelvis is filled up by it, in the posterior and lower part so that the finger is directed upward and forward. Very high up, the vaginal cavity is quite small from pressure at its extremity; in contact with the pubis is the *os tincæ*, very firmly held in its place. The tumor is *round*, elastic and smooth; not so hard as fibrous tumor, more central than ovarian, and more uniformly round than extra-uterine pregnancy. It may be ascertained in most instances, also, that the tumor is larger toward the sacrum than the symphysis.

Termination.—When left to itself retroversion will terminate in abortion, when the contents of the uterus will be expelled and the symptoms thus relieved; or the bladder may be ruptured, the urine being discharged in the peritoneal cavity, causing painful death; or the uterus may be ruptured, and its contents discharged in the cavity of the peritoneum, giving rise to fatal peritonitis; or the fœtus and its membranes may be surrounded by fibrinous material, the patient recover, and these substances remain there enveloped, or, inducing local suppurative inflammation, be discharged by exulceration. Sometimes the tenesmus becomes so great as, by the violence of the efforts, to break through the posterior walls of the vagina and uterus, and discharge the contents through the vulva from this artificial opening. Inflammation sometimes arises without being initiated by any of these disastrous accidents, and less suddenly causes the death of the patient. I think there can be no doubt but that there are very rarely cases of spontaneous reposition, recovery, and completion of the term of gestation.

The Prognosis is unqualifiedly bad if left to nature, but equally favorable if intelligently treated at the proper time.

Treatment.—The main thing to be done is to replace the uterus. This can very generally be accomplished. The attempt should not be delayed, as the uterus is constantly increasing in size, and the impaction becoming more certainly greater, increasing the difficulties as well as dangers. To facilitate the replacement, the bladder should be emptied by the catheter when practicable, and the fæces removed from the rectum. This takes away some of the obstacles. Sometimes the urethra is so tortuous in its course, and the walls compressed so completely together, that a catheter will not enter the bladder. An elastic catheter will sometimes pass the obstruction when the metallic will not; whichever we may use should be urged forward with the utmost gentleness, bearing in mind the great danger of perforating the attenuated urethra. The patient should be placed upon her knees and chest, or on the left side with the left arm behind her, the thighs strongly flexed and the right drawn up close to the abdomen and thrown forward. She should be placed on a table or the edge of a bed, so that the genital organs are easily controlled by the operator. In this position we may often succeed in replacement by the hand alone. The right hand should be well lubricated, and all the fingers be introduced into the vagina, so that the palmar surface is turned to the sacrum. The tumor is thus pushed up very gently and slowly with the pulps of the fingers pressed closely upon the face of the sacrum, as high as the hand may be made to reach. There are not many cases in which the fingers will fail to carry the fundus above the promontory of the sacrum. When thus elevated, it suddenly starts up and assumes the normal position. If, however, the fingers do not reach high enough for this purpose, a collapsed gum-elastic bag or bladder, may be carried up between the fingers and the uterus, and when elevated as much as we can reach, the bag may be inflated sufficiently to raise the uterus high enough. I have succeeded in all the cases I have tried with this method, and I think when the impaction is not so great as to preclude dislodgement that it will almost invariably succeed. Some surgeons recommend the introduction of the empty bag into the rectum, and inflating it there, and pushing it up; others introduce a drumstick, with the end cushioned and lubricated, into the rectum, and pressing it against the uterus, elevating it in that way. Again, an instrument is used, not unlike two drumsticks, somewhat curved, attached together. The attachment confines the ends very near each other. The end of one of the branches goes into the rectum and the other into the vagina. Thus arranged, they pass up and carry before them the uterus. These expedients are very sure, but rough and not very safe means of arriving at the results. I think as much force in a proper direction can be applied by the fingers and elastic bag, as it is judicious to employ in such cases. There are other methods of proceeding, but I do not think it necessary to mention any other, as these will suffice when reduction is practicable.

In all these efforts to elevate the fundus we may fail, and then we may evacuate the uterus. This can generally be done by passing a bent probe through the mouth of the uterus, far enough to rupture the membranes, and permit the escape of the liquor amnii. This being done, abortion will soon ensue, I can conscientiously only mention, for I can hardly think the operation of puncturing the uterus with a trocar through the vaginal wall ever commendable or necessary. The cervix is probably hardly ever so inaccessible but that some form of bent instrument can be made to enter it.

Inversion of the Uterus.—Inversion is the turning of the uterus inside out, with the fundus down and the cervix up, a reversion of its surfaces and ends. It is partial or complete. When partial, the fundus is depressed in all degrees, from a mere indentation to a considerable protrusion through the cervix and os uteri. The depression of the fundus, or partial inversion, passes into complete when the whole organ, fundus, body, and neck, have passed through the mouth, and hang down below it. It presents a recent and a chronic form. The recent may be regarded as extending 'through the first two weeks; after which, the circumstances and condition of the uterus and patient become what they remain in the future, however long it lasts. The uterus, in that time, has been condensed by contraction and involution to such an extent as to make the case permanent and difficult of change, except to diminution and further condensation. Inversion almost invariably occurs anterior to or at the time of the removal of the placenta, but several hours, and, in very rare cases, several days may elapse before it is complete and discovered; for it is quite probable that in these instances partial inversion or greater or less depression of the fundus had existed from the time of delivery. It is believed by different parties that there are two modes observed in the process of inversion. Sometimes the fundus is indented or depressed in the cavity of the body like the bottom of a "junk bottle," the depression rapidly or slowly increasing until it is completely down. At others, the whole of the fundus, and, more or less, the whole of the body, are firmly contracted, while the cervix remains flabby and relaxed. In this condition a slight amount of abdominal tenesmus will drive the contracted part down through the relaxed cervix; and thus initiated, it requires but a continued action of the fibres of the organ and abdominal muscles to finish the process. The causes of inversion are not always obvious, as cases have occurred under circumstances when least expected from any discoverable reasons, and inversion fails to be brought about by circumstances that are usually enumerated as sufficient. We occasionally meet with instances that have no history, and neither patient nor physician can give us a clear idea of the time or manner of the occurrence. Such a case was the subject of litigation in this city a few years since. And other cases are recorded in virgins, and consequently referred to congenital origin. In a large majority, however, we may trace the history back to accouchement. The predisposing causes are enlargements and partial or complete passiveness of a part or the whole of the muscular fibres of the uterus. These are the conditions in confinement at full term, or abortion or premature labor, also enlargement from hydatids, hydrometra, tumors, &c., &c. When the uterus is thus enlarged and lax after a greater or less loss of its contents, traction on the cord or placenta, or contained tumor, or injudicious or accidental pressure on the fundus by the hand of some person, or the action of the abdominal muscles thrusting the contents of the abdomen downward upon that part of the organ, it may be inverted. It is possible, I think, also, that powerful, irregular action of the fibres of the uterus, may cause the initiation and completion of the process of inversion. It is then said to be spontaneous. The weight of the placenta, or the contraction to expel a polypus, may commence inversion, and even complete it. The irregular contractions that result in inversion may commence before the expulsion of the child. After the liquor amnii has been discharged for a long time, the uterus contracts to suit the inequalities of the fœtal surface; the globular shape of the organ being replaced by inequalities in a number of places. Much is yet to be learned on this subject. It would seem clear from statistics brought forward by Drs. West and McClintock that it is exceedingly rare, if it ever occurs, under good management of labor cases. It has not been encountered in patients confined in the London Maternity Charity, nor the Lying-in Hospital of Dublin in 140,000 cases. The student is not to consider from this that it is impossible for it to occur in the hands of the ablest of accoucheurs.

Symptoms.—Usually these are appalling in the extreme. Without warning the patient is seized with faintness, coldness of the extremities, sense of great prostration, rapid and very feeble pulse, oppression about the heart, copious perspiration, hurried breathing, often vomiting, ringing in the ears, and blindness. Soon these symptoms increase, until the patient lies in a profound state of collapse, indifferent to everything transpiring around her, or throwing herself in every direction in paroxysms of agony inexpressible. This condition of collapse is not always the result of copious hemorrhage, but seems to be of nervous origin. A shock not unlike that caused by severe accidents, as falls, strokes, &c. But, generally mingled with this sort of impression, there is profound exhaustion from loss of blood. From this state of collapse the patient may very slowly rally, until she enters a tedious and imperfect convalescence. Or, in the cases where the exhaustion from hemorrhage is added to the great depression of the shock, the patient may be overwhelmed, and in a hour, or very few hours, her sufferings end in death. Imperfect recovery from the great effects of the first shock may enable the patient to live for several days, and at last, in five to ten days, die. In case the patient recovers from the first symptoms, after some weeks she may regain a fair degree of health, and retain it or even improve until lactation gives place to ovulation, or until this last function supervenes upon the first. The first menstrual discharge is preceded by copious mucous evacuation, and when the menses begin, they are more than ordinarily profuse, and generally before they cease amount to prostrating hemorrhage. This hemorrhage is repeated monthly, more frequently, or is continuous, while the leucorrhœal discharges become very profuse. Functional derangement of other and important organs enters the list of morbid impressions: the bowels are constipated, the heart palpitates, the stomach cannot digest with its former vigor and completeness, the head aches, the eyes become weak; the disposition of the patient changes; the memory fails her; she is pale, cold, and anæmic; in short, she enters a decadence that is continuous, until, after several months, or a few years, she is exhausted and dies. Although this is the course usually pursued by cases of inversion, it must be remembered that there is a class of them in which the patients do not suffer even much inconvenience, and their condition is discovered only by accident during their life, or on the dissecting-table.

Diagnosis.—When the symptoms present themselves so as to awaken suspicion, the diagnosis of recent cases may be made out quite clearly, by the descent of a tumor into or entirely through the vagina, and the absence of the uterine globe above the symphysis pubis. The diagnosis, after a few days or weeks have elapsed, and the case becomes chronic, is not quite so simple and ready. The tumor is felt in the vagina, and is more sensitive than polypus. It is easily surrounded by the fingers, and by introducing two fingers in the vagina to the upper end of the tumor, the depression formed by the junction of the vagina and uterus may generally be easily surveyed. If this is not entirely satisfactory, the sound should be introduced into the vagina before the fingers are withdrawn, and, guided by them, be made to sink as deeply into this depression as it will go without too much force. If the uterus is inverted, the probe will not pass beyond the fingers any distance, but if the vaginal tumor be a polypus, the sound will pass up at some point some inches above the fingers into the uterine cavity. The operator may test the position of the uterus in another way, by introducing the finger high up into the rectum, so that the end may reach above the tumor, and retaining it there, he may pass a catheter or sound into the bladder, and approximate the two: if the womb is in place, its thickness will be perceived interposed between the two, but if inverted, the extremity of the catheter can be brought down upon the finger, with nothing but the membranous walls of the bladder and rectum intervening.

Prognosis.—No more serious complication of labor can occur than inversion of the uterus. The danger is great and imminent; in a considerable majority of cases proving fatal, the patient dies within a few hours. Mr. Crosse says, " In seventy-two out of one hundred and nine fatal cases, the patients died within a few hours, eight of the remainder within a week, and six more within four weeks; another at five months, the result of an operation which had an unsuccessful issue, one died at eight months, three at nine months, and the others at various periods of from one to twenty years." (*West.*) Death in the first place soon after delivery seems to be the result of rapid exhaustion of the vital forces by the terrible shock to the nervous system and the profuse hemorrhage that often complicates it. Death in subsequent times, however remote in the chronic form, is brought about by impairment of the vital functions by the same means, operating more slowly but as surely. The patient dies from exhaustion in both forms. Accordingly, we find that while inflammation has something to do in affecting the issue in rare instances, those cases in which there is no uncommon hemorrhage or leucorrhœal discharge last longest, and sometimes do not prove fatal at all, the patient enjoying fair health for many years. I know one patient, fifty-six years of age, whose uterus was inverted sixteen years ago, and yet remains in that condition, as I have verified by examination, who is in the enjoyment of as good health as the majority of women of her time of life.

Treatment.—The management of recent cases will be the easier the sooner after the accident it is commenced. Its reduction is generally successfully accomplished within the first hour or two if intelligently attempted. It is more difficult as time elapses, but it should *never* be considered impracticable until proper and persevering efforts have been made. The first item for consideration and action, is to dispose of an attached placenta, when the uterus has not detached it before, during, or after its descent. If the placenta is wholly adherent, its attachment should in nowise be interfered with until the uterus is returned to its former position, but if it is partially detached, it should be immediately separated by gently "peeling" it off with the fingers. This instruction has reference solely to the prevention or lessening the amount of hemorrhage. If the placenta is attached throughout, the hemorrhage will be trifling; if partially separated, the condition most likely to be accompanied with fatal hemorrhage exists,—relaxation of the uterus and partial separation of the placenta. It is well known that sufficient contraction of the uterus will separate the

22

placenta, and when not contracted enough to do so, it is in too lax a state for us to desire its detachment. If the placenta is partially separated, the completion of it by the fingers, as in the case when included in the uterus, will enable and stimulate this organ to contraction, and thus to the suppression of the hemorrhage. I do not think the question of convenience of return, or the possibility of being foiled in the reduction by the continued attachment, should be entertained. The want of contraction enough to throw off the placenta is an evidence of such profound inertia as to insure easy reduction of the uterus.

It being decided what course to pursue with the placenta, immediate efforts should be made to revert. And before beginning these efforts, we should remind ourselves of some facts in the case, that are apt to be lost sight of in the hurry and confusion of such an appalling occasion. One fact is, that immediately after the occurrence of the accident, the uterus is in the same flaccid condition in which it was incapable of resisting the action of the cause; another is, that it soon begins to contract, becomes firm, and, consequently, more difficult to affect by counter influences; and a third, that the more the uterus is stimulated, by handling or otherwise, the sooner and more firm the contraction becomes, and, consequently, the greater difficulty in reduction.

No operator has complained to us of the bulk being too great to return, but all of the resistance caused by contraction. The experience of Dr. Meigs is conclusive on this point. He found that upon attempting to reduce the size of the uterus, by squeezing it to expel the blood, he caused it to contract, and it became so hard as to resist his efforts to push it up within the os; but as soon as he pressed upon the fundus, he would depress it, or rather elevate it, until by continuing pressure, he made it ascend first into the body, and through it into the neck, and finally up to its proper place. Dr. White, of Buffalo, although he did not mention with the same distinctness the effects of the two sorts of pressure, was enabled, by indenting first and then following up the vantage, finally to push the fundus up the same way through the os and body of the uterus, after he had in vain tried to reduce it by squeezing, &c. Dr. White's case was reduced in this way eight days after delivery. And I must be allowed to express the opinion, that it increases the difficulties in recent cases of inversion to try to lessen the bulk of the uterus. A great bulk indicates a flabby reducible state, and is favorable to success, instead of otherwise. Do not squeeze the uterus to lessen its size in these cases.

The two cases I have referred to, of Drs. White and Meigs, so intelligently and deliberately observed, and so clearly explained, furnish us with more intelligible means of arriving at correct ideas of the steps by which inversion of the uterus is reversed, than any I am able to find on record. They both concur in showing the usefulness of one hand in the vagina to steady the uterus, and direct the force applied to the fundus by the other hand, and the injurious effects of compressing the body of the organ. The most appropriate mode of operating in recent inversion, therefore, is to introduce the left hand into the vagina behind the uterus, while with the fingers of the right the fundus is indented, and gently, but steadily and perseveringly, reverted entirely above the os and cervix, until it assumes the globular shape, and proper position above the symphysis. If the fingers of the right hand cannot be used to advantage, or are too weak to accomplish the desired elevation, we may use an instrument resorted to by Dr. White, a large elastic rectum bougie, or by Dr. Beers, shaped like the end of a walking-cane, with a round smooth head upon a staff. The indentation and elevation may be more efficiently effected by this latter instrument, perhaps.

The fact cannot be too forcibly impressed upon our minds, in undertaking this operation, that gentle firmness is the proper expression for the force to be employed. Perseverance, instead of violence, is both more certain, successful, and secure, in overcoming the resistance of muscular fibre anywhere. This is especially true with the uterus, the strongest muscle in the body. As nearly as may be, we should act in the absence of uterine contractions. During and after the time we are attempting the return of the organ, the strength of the patient must be supported by stimulants, tonics, and nutrients. Brandy will, perhaps, serve best to restore the circulation and heat; it may be aided by the use of the aromatic spirits of ammonia, and laudanum. In addition to the stimulant and supporting influence which laudanum exerts, it allays the irritable condition, so frequently present, of the stomach, the uterus, &c. After the urgency of the symptoms has passed by, the tinct. of iron, quinia, beef essence, and nutritious diet generally, will be necessary to restore the impaired condition of the vital energies. The energy with which the stimulants are to be urged during the shock, must be regulated by the urgency of the danger. Large doses of brandy, laudanum, and spirits of ammonia, will not only be borne, but often called for to meet the symptoms.

The treatment of the chronic form is palliative and curative. The palliative is for the purpose, as far as possible, to check the drain which is so constantly exhausting the patient, to support the system as well as we can, and use any other means suggested by the circumstances for the relief of distressing symptoms.

The hemorrhage is from the mucous membrane of the uterus, its outer surface as it lies in the vagina, as also the profuse mucous discharge. I think much may be done to moderate, if not stop, these evacuations, by astringents introduced into the vagina, so as to surround and lie in contact with the uterus. Pledgets of lint, saturated with the persul. of iron, passed up into the vagina, and allowed to remain on the bleeding surface of the uterus until the bleeding ceases, will be of great service. The tinct. ferri chlo. on lint is an excellent application for the same purpose. Other astringents may be tried in the same manner. If these should fail, the vagina may be tamponed fully with cotton dipped in astringents, or not, as the physician may think best. Severe paroxysms of hemorrhage should be carefully treated in this way until they terminate, it being desirable to save as much blood as possible. It is not necessary to suggest to the intelligent reader the necessity of rest in the horizontal position. Between these paroxysms the patient should use astringent injections of concentrated strength, saturated solutions of alum, acetate of lead, tannin, &c., with a view to condense the mucous membrane, and render it less vascular, and in this way abate the urgency of the losses. The tinct. ferri chl., one part to four of water, twice or thrice a day, will have an efficient astringent effect upon the uterus. When the organ extends through the vulva, it is irritated by contact with the limbs and clothing and it is very desirable to return it into the vagina, and keep it within that cavity. The gum elastic air pessary, supported by a T bandage, will keep it in the vagina, and may render it more easy of a radical cure, by reduction or reversion. I would urge the attendant to personal attention to this treatment, to such an extent, at least, as is necessary to have it efficiently tried. Very few patients have the intelligence to appreciate the importance of it, or to know when proper trial of it has been made.

The radical treatment has for its objects either a restoration of the organ or its amputation and removal. So far as we can judge, although both operations are attended with danger, that of amputation the more. And I think it clearly the duty of the practitioner, when driven to a choice between the two, to give preference to attempts at restoration. We have not only greater safety as an argument in favor of it, but successful restoration reinstates the patient in all her sexual capacities, while amputation, if not disastrous in other respects, renders her forever sexually neuter. It is to be hoped that before long the operation of amputation will be regarded as unjustifiable, because of the certainty of restoration. Great improvement in our means and the mode of effecting this must be made, however, before this conclusion can be reached. There is no longer room for doubting that restoration of the inverted uterus occurs spontaneously. I think it is proven by the case of Dr. Hatch, published in Dr. Meigs's Obstetrics. The case of Madame Beauchardat, published by Baudelocque, is also, I think, conclusive on the point of restoration. Other cases, less clearly and circumstantially reported, may be found scattered through medical literature for the last century. There are two methods, if they may be so denominated, that have been successful in reducing chronic inversion of the uterus. Two representative cases are published in the American Journal of Medical Sciences for July, 1858; one by Prof. White, of Buffalo (it was his second case), and one by Dr. Tyler Smith, of London. It will be observed, by examining the reports of these cases, that the restoration began by the cervix passing through the os uteri first, then the body, and finally the fundus. This is different from what I think is the common mode of restoration in recent cases. The operation for reversion in Dr. White's second case was completed, we are led to suppose, in something more than an hour, and at one sitting. The uterus had been inverted five months. Dr. White operated by introducing the hand into the vagina while the patient was in a state of anæsthesia from chloroform, squeezing the uterus so as to lessen the size as much as possible, and at the same time pressing the organ upwards by means of the large rectum bougie. Success followed a somewhat protracted manipulation. The uterus was restored by the lips of the os uteri beginning to fold outward, and the neck to pass up through this opening, next the body, and afterwards the fundus. There is nothing in this case said about the fundus being indented from beginning to end. This is no more than might be expected by considering the anatomical circumstances. The fundus and corpus uteri are firmer and more solid than the cervix, and hence less likely to yield to the same amount of force. The force applied to the fundus, when the organ is strongly pressed upward, acts more efficiently upon the cervix than any other part, from the fact that the vagina, attached all around the mouth, has not merely the effect of resisting the upward pressure of the uterus, but being upon the outer surface, it initiates and keeps up the funnel-shape expansion of the os necessary to permit the other parts to pass through it, as well as to draw it down over the part entering it from below. I believe that, in some respects, this is the best manner of operating for immediate restoration, yet one thing done seems to me to be superfluous if not mischievous, viz., the squeezing the uterus. Dr. Sims recommends that the uterus be supported by one hand above the pubis to prevent too great extension upon the vagina. While the uterus is being pushed up from below, the cup-shaped cavity formed by the inverted cervix may be felt if we forcibly press the fingers down into the pelvis from above over the pubis. This manipulation affords us valuable aid in forming our diagnosis, while it gives the opportunity of assisting in the reversion. The great thing to be gained is the commencement. After the neck is one-half reverted, the restoration proceeds with more rapidity and ease than before, until complete. A better instrument than the bougie used by Dr. White would be a cup on a strong handle, large enough to safely lodge the fundus of the uterus. I hope I may not be understood as criticizing the operation of Dr. White; on the contrary, I award to him the highest praise for his success, and the very lucid manner of tracing the steps of the restoration, and delineating the means and manner of doing it. His experience has taught the profession one of the most valuable lessons on this subject. Then, as seen now in the light of the experience of others, the steps in the operation for immediate restoration are, first to introduce the hand into the vagina, and, embracing the uterus with it, hold the organ steady, with the fundus and cervix nearly parallel with the axis of the superior strait; second, place the fundus of the uterus in the cup of the instrument held by the other hand, and then press gently upward, increasing the firmness of it until it is as great as the parts will bear without violence, and continuing it with such force until the parts yield and pass up. The time required may be considerable, and it is an object to continue it for a long time, increasing the pressure so slowly as not to be perceived, except by comparing it at considerable intervals. The patient should be under the influence of chloroform to insensibility, and placed on her back, with the limbs widely separated across the bed, and with the hips very near it; or, what would be better, an operating table of convenient height about two feet wide and five long. Greater facility would be afforded for attendants on such a table. The surgeon should kneel or seat himself in front of the patient, so as to have free use of both hands and perfect command of the parts. The second mode of restoring the inverted uterus, as practised by Dr. Tyler Smith, is to apply the force so gradually as to require several days for the completion of it. The means used were, first, the frequent introduction, I think twice a day, of the hand into the vagina to squeeze the uterus; and, second, to keep a gum-elastic air-bag distended in the vagina, which constantly pressed the fundus upward, certainly, however, with no great force. He succeeded in restoring a uterus that had been inverted for fifteen years. With proper apparatus I should very much prefer this gradual method, as requiring less violence, being less hazardous, and perhaps less painful. A small boxwood cup, with a thin air-bag in it, to act as a cushion, or cushioned with something else, mounted on a strong gum elastic stem, fastened to outside support, I think may be made to exert more force upon the fundus, and keep it up more continuously than an air-bag in the vagina. The amount of pressure should be regulated by the tolerance of the patient. The more she can bear the better. By means of the cup we may give a better direction to the force used, as well as the other advantages. It is but right to inform the reader that I have had no experience in either of these modes of restoring the uterus, and that these directions are deducible from facts derived from others. In attempting this last mode we ought to give the patient enough personal attention to assure ourselves that the instrument is fulfilling its object. I should not consider that I was gaining anything by the introduction of the hand as recommended by Dr. Smith, but that I was causing needless irritation and pain.

As before remarked, it is to be hoped that restoration will become a practicable thing, almost if not quite always, and hence amputation no longer necessary. In the present state of our knowledge, however, I think we may regard amputation as preferable to trusting the recovery of the patient to palliative means only. And if restoration is impracticable—but not otherwise— and the patient clearly going downward through the different grades of chronic exhaustion to final dissolution, we are in duty bound to remove the organ. This may be done by the knife or ecraseur. The operation is sufficiently simple. By means of vulsel forceps the uterus may be drawn down to the vulva, transfixed in the middle of the cervix with a large needle, armed with a double hemp-twine ligature, tied on each side very tightly, and then cut through by the knife just below the ligature. Or, without the ligature, the chain of the ecraseur may be crushed through at the same point. I think amputation by the ligature alone entirely unjustifiable, and would prefer the ecraseur. If the ecraseur is used, the inexperienced surgeon should bear in mind the all-important necessity of not working the instrument too rapidly. The chain should not be carried through it under fifteen or twenty minutes, resting after each motion of the instrument. Hemorrhage is the immediate danger in this operation, and this mode of using the ecraseur I think safe. The ligature, when the knife is employed, is indispensable to safety. The subsequent dangers are peritonitis and toxæmia, which must be treated as elsewhere directed. When the uterus is reverted, it is a question whether there is danger of, and consequent necessity for precautions against, the reproduction of the inversion. Although the danger is very remote, the patient should be closely watched for one or two days to avoid it, and if doubtful symptoms arise, the probe or a female catheter should be introduced, and, if need be, kept there for two or three days.

CHAPTER XXV.

DISEASED DEVIATIONS OF INVOLUTION OF THE UTERUS.

THE uterus is very much hypertrophied by the processes of gestation, so that after its contents are expelled by labor, the organ weighs from one and a half to two pounds. An atrophizing process, called involution, serves to reduce the organ to its original conditions in size and weight.

Involution is a physiological change, so much so as evolution; but not unfrequently disease invades the tissues, and renders it abortive; 1st, causing it to be temporarily "delayed;" 2dly, to fall short of completion after it has been commenced; or, 3dly, to proceed entirely beyond the limits compatible with the healthy functions of the uterus, reducing it below its usual weight and size.

I mean by the term "delayed involution," to designate a condition of the uterus in which this process does not begin for a number of days—from ten to fourteen—after parturition. I think this is not very unfrequently the case in a state of disease.

The generality of physiologists think the contractions which immediately succeed and continue after labor, by interrupting the circulation in the substance of the uterus, initiate that process, and that by the end of a fortnight it is half finished. Should these contractions be rendered inefficient, involution is at a stand, the uterus remains large, the circulation too great for safety to the patient, and sufficient to keep up the nutrition in the muscular fibres, which are still capable of a good degree of energetic action. For a number of days, the uterus is felt to be as large as a child's head above the pubis, and not very firm.

Causes.—The most common cause of this delay is inflammation attacking the substance of the uterine walls. The inflammation may be acute, and the patient's suffering such as to demand attention, or so slight as to pass without much notice. Cases of puerperal metritis, for a week or ten days immediately succeeding delivery, not unfrequently prevent this enlarged condition of the organ.

Another cause which probably operates to prevent involution, is atony of the uterine muscular fibres. The contractions are feeble, and so inefficient, as to delay for a long time, and render very slow, the early stages of involution. Too early assumption of the erect posture, and undue exercise on foot, keeping the bloodvessels of the uterus distended unduly, and thus overcoming the muscular contraction, is not unfrequently the cause of delayed involution.

Symptoms.—The symptoms of delayed involution, separate from the inflammation, are not always very well marked. Weight, heat, and aching in the back, are the most frequent symptoms we have in many cases, especially if inflammation is the cause. There is always great danger, however, of a very alarming symptom, while this state of the uterus exists, and that is great flooding. Where the delayed involution is dependent on atony of the muscular fibres, hemorrhage is sure to take place if the patient exerts herself considerably. Sometimes, as the first indication of any seriously wrong condition of the uterus, the patient is suddenly seized with copious hemorrhage, which subsides under the influence of rest, cold, and astringents, but suddenly and unexpectedly recurs without adequate cause. These recurrent hemorrhages, occurring in the first fourteen days, are always due, according to my observation, to delayed involution. When suspected, the diagnosis is not difficult, by an examination with one finger of the right hand, per vaginam while with the left hand pressure is made above the pubis. The uterus, thus examined, is found to be as large as immediately after labor is ended.

The soft uncertain condition of the uterine globe will not always enable us to discover it by placing a hand upon the lower part of the abdomen alone, but by including the organ between the two, there will be no danger of mistake. If the organ retains sufficient firmness to be easily distinguished above the pubis by the single hand, there will be but little danger of hemorrhage. The local distress will then be the only indication of the necessity of a diagnostic examination, when the greatly enlarged condition will be easily detected by the examination above directed. The fingers may be easily made to enter the mouth of the organ, and move the whole mass, while the hand above will easily recognize the movement; or the hand above may be made to press it down upon the fingers below. In this way we will easily satisfy ourselves of the

increased size of the organ. As I have before intimated, in some
cases the uterine globe is quite firm, in others very flabby. The
firm condition is a surety against danger from hemorrhage, while
it is an evidence of inflammation; while the soft state of it indicates
a strong hemorrhagic tendency.

Prognosis.—There is imminent danger of serious, if not fatal
hemorrhage. I have known as many as two cases of sudden fatality from flooding after the seventh day from the time of labor.
It is always a serious condition, and should be watched diligently,
and treated efficiently. Even in cases where the delay is caused
by acute inflammation, great hemorrhage may take place, although
not so likely as when caused by muscular atony alone. If the
delay is for a very considerable length of time, the involution is
pretty sure not to be completed, but the uterus remains in a state
of subinvolution for an indefinite time. Very often the causes
which effect delays continue to act, and finally produce subinvolution.

Treatment.—The treatment will depend upon the accompanying
or causing conditions. If there is inflammation of the uterus, the
antiphlogistic measures necessary to combat it will be demanded,
with counter-irritation, fomentations, &c. Should atony, unattended with inflammation, exist, ergot in large doses will be demanded imperatively until ergotism is brought about.

I usually give ℨss. pul. secale corn. in infusion every half hour,
until contractions are brought about. When this is done, the
effect of the drug may so subside, that it will be necessary to administer it again in twelve or twenty-four hours, until all disposition to relax has passed away. When the atony and inflammatory
condition coexist, which may be known by the tenderness, fever,
and hemorrhage occurring together, the ergot and other treatment
should be combined. Hemorrhage is not likely to come on until
after the inflammation is pretty well subsided, in this class of cases,
and it aids usually in removing the last of it.

I subjoin two cases as representatives of the two conditions of
the uterus, and the mode of treating them :

Case 1st was furnished me by Dr. S. Wickersham of this city.
He was called to Mrs. E——, an Irish woman, aged 28, in her
fourth labor, May 7th, 1863, 4 o'clock P. M. She had been in
labor, attended by a midwife, for the most of the day. At 1

o'clock A. M. of the 8th pains had entirely ceased, from atony or exhaustion of the uterus. Constitutional symptoms began to show the necessity for relief. The forceps were used, and the child was delivered. The placenta was delivered in due time without difficulty, and the uterus contracted well. No hemorrhage more than usual. The pulse was unusually frequent at and after the time of delivery. The labor was followed in two days with puerperal fever, in which the uterus and peritoneum were both involved. Up to the 20th, she had improved very much, so as to be considered by the doctor as convalescent. In the early part of the day, sudden and violent hemorrhage prostrated the patient to what was at the time considered a moribund condition; but by active stimulation and external warmth to her cold extremities, she rallied very much, and appeared to be slowly recovering. At 6 o'clock P. M., on the 24th, the hemorrhage returned with "terrible violence," and she was thought again to be dying. Notwithstanding the most energetic use of stimulants, she could hardly rally from this last attack. On the 26th, in consultation with Dr. Wickersham, I found the patient so prostrated as to leave but little hope of her recovery. Suspecting that the uterus was in a state similar to what is found immediately after delivery, I insisted upon making an examination, which was resisted by the patient and friends. Through the kind perseverance of Dr. Wickersham I was permitted to do so. The uterus was so flaccid that I could not discover it above the pubis until after introducing the finger into the vagina and moving it about, when the fundus could be felt as high as the umbilicus with the regular globular form. The mouth and cervix were large and flabby, and easily admitted two fingers. After this examination the indication seemed plain to cause the contraction of the uterus. Large doses of ergot were given in addition to the stimulating and supporting treatment. Hemorrhage was very slight on the morning of the 27th. She continued to improve slowly until the 9th of June. At 5 o'clock A. M. the hemorrhage returned, and lasted until 10 o'clock A. M., but in so moderate a degree as to produce but little effect upon the patient. I was not in attendance after the first consultation, and could not trace the steps of condensation, but after the 9th of June the hemorrhage did not recur.

It will be seen that on the twelfth day after confinement dangcrous hemorrhage took place; that it again returned on the sixteenth day after delivery to a very alarming extent; and that after the liberal use of ergot the hemorrhage returned but slightly. It should be noted, also, that the cessation of the hemorrhage was sudden, and probably resulted from faintness, and that it returned so soon as the arterial reaction amounted to any considerable power. The faintness, doubtless, was the cause of stoppage in both the terrible attacks before ergot was given, but the hemorrhage was effec-. tually checked by contractions, the effect of the ergot.

Case 2d. Mrs. E—— is the mother of nine children. She is 33 years of age, and a German Jewess. Of robust, almost athletic make and habits, she always enjoys excellent health. In the last three confinements she has almost lost her life from loss of blood both before and after the delivery of the placenta. I attended her in the eighth labor, the last before this one. There was nothing peculiar in it until after the child was delivered, the labor having lasted but about four hours. The pains were ordinarily vigorous and propulsive. The liquor amnii was not evacuated until ten minutes before the head was distending the labia. After the child was expelled the uterus did not contract thoroughly. It seemed large and rather soft. This state lasted for half an hour, when a feeble contraction detached but did not expel the placenta. From this time hemorrhage became excessive. I waited for half an hour,—using friction, kneading, and pressure over the uterus, with application of ice to the vulva,—for contraction of the uterus and expulsion of the placenta, but although there were occasional pains, they were so feeble as to produce no effect upon the hemorrhage. About this time the ergot I had sent for arrived, and I gave immediately ʒss. in a little wine and water. Fearing the prostration which was rapidly coming over the patient, I introduced my hand into the uterus, grasped the placenta, and irritated the organ by moving the whole around in it. This brought on contractions enough to expel my hand and placenta, and deluge the bed with coagula and fluid blood. Very soon the ergot began to act, and the hemorrhage ceased. I give this description of her eighth labor to show her predisposition to *inertia uterina*. As the ninth labor approached, I determined I would administer the ergot as soon as the parts were well dilated, and the head began to pass the os uteri. I was sent for at 8 o'clock P. M., June 30th, 1864, to attend her. I found the pains active and the os uteri fully dilated, and the membranes distending the labia. I at once gave her ergot 3ss. in infusion, making her swallow the ergot as well as the water. This was repeated in half an hour. By this time ergotism was fairly established. In three-quarters of an hour from the time I arrived the child was born, and in a few minutes the placenta was expelled from the uterus into the vagina whence it was removed. No hemorrhage followed. The uterus was well contracted. I considered her condition very favorable, and at the end of another hour took my leave. Her condition for the first forty-eight hours was in no respect unusual, except that the lochial discharge was rather free. From this time I saw but little of her until the 10th of July. I returned from the country at 5 o'clock P. M., and found she had been flooding since early in the morning, not very greatly, but sufficient to begin to produce faintness. The uterus could be felt above the symphysis pubis as large as a child's head, and not very hard. I ordered cold to the pubis, and twenty drops aromatic sul. acid in some water every four hours, expecting soon to have the hemorrhage checked; but to my surprise, at 8 o'clock on the 11th, the hemorrhage still continued, being but slightly moderated by the means used. I now ordered two teaspoonfuls of vin. ergoti every half hour until the hemorrhage ceased. But the nurse said that the "second dose put her in so much pain and caused such large clots of blood to come from her that she dare not give it again." The hemorrhage ceased entirely from this time until the afternoon of the 13th, when it returned with considerable violence. The ergot was again given, and from this time forward the patient had a favorable convalescence, and is now in the enjoyment of good health.

Subinvolution is the condition in which involution commences and proceeds to a certain extent, and then is arrested, and thus leaves the uterus larger than it ought to be, the organ being more vascular and heavier. There is quite a difference in the degree of subinvolution: sometimes the organ remains three, four, or even five inches long, and proportionally broad, while in other instances it is scarcely over the natural size.

Causes.—The most common cause of subinvolution is inflammation supervening during the time the process of involution is in progress. Often the inflammation commences immediately after delivery, and involution proceeds very slowly and with great difficulty, until, finally, it ceases before being completed; but other times the inflammation commences during the process, probably on account of exposure, or prematurely rising from the bed and resuming usual exercise. On account of the increased vascularity of the organ, its greater size and weight, there are almost always more than ordinary sanguineous discharges at the menstrual periods, and sometimes even between them. Superadded to these we may, and often do have an extensive array of the local and sympathetic symptoms of chronic inflammation of the uterus, for which I will refer the reader to the subject of inflammation. The diagnosis of subinvolution is not difficult. By embracing the uterus between the fingers of the left hand above the symphysis, and those of the right within the vagina, the size will be found increased; we may also ascertain by introducing the sound that it is larger than usual. But inflammation may sometimes increase the size of the uterus in virgin females, so that it will present the same symptoms and answer to the same measurement with the sound and fingers. And while the diagnosis between subinvolution with inflammation, and inflammation with tumefaction of the uterus, is a matter of no practical importance, I think the distinction may generally be made. In cases of subinvolution the uterus partakes more of the shape of the recently impregnated organ. It is thicker antero-posteriorly and from side to side, than when tumefied by inflammation. In inflammation it is apt to be proportionably longer than broad and thick. . The neck is narrow and long in the last, and broad and comparatively thick and short in the first state.

But I have found that these cases are curable by the same means that remove inflammation. Remove the inflammation, and the involution is spontaneously completed. If attended with symptoms of acute inflammation, antiphlogistic treatment at first is demanded; but should the symptoms be chronic, nitrate of silver, bathing, injections, alteratives, tonics, and laxatives, as recommended elsewhere for chronic inflammation in this work, are indicated and will pretty surely succeed.

Hyperinvolution is the state of the organ where the involution has proceeded to such a degree as to condense the tissues beyond their ordinary density. The condensation thus accomplished renders it less vascular and erectile, and the fibrous structure is paler and harder than natural.   As the result of this condensation and diminution in the quantity of the circulation, the uterus as a whole is smaller and lighter than common.   The degree to which hyperinvolution may be carried varies greatly; sometimes it is so slight as to require great care to distinguish it, at another the uterus is reduced to half its ordinary weight and dimensions.

Causes.—Inflammation seems here to be more concerned in the production of hyperinvolution than any other morbid process. From examinations during the progressive steps of morbid states of involution, I am inclined to think that in cases where inflammation of the mucous structures exists exclusively, or where inflammation of the mucous membrane preponderates, the involution is arrested, and hence we have subinvolution; but when the inflammation is mostly confined to the submucous tissue it proceeds to hyperinvolution.

Symptoms.—The condensation of the tissue and reduction of the vascularity of the organ always diminish the menstrual flow; and hence we have decreased menstruation in a moderate degree, and obstinate amenorrhœa in the more extreme condition.   The symptoms attendant upon hyperinvolution are very similar to those enumerated in the description of chronic inflammation.   They are sometimes very distressing, rendering the patient thoroughly miserable for many years.   The worst cases of this form of diseased involution I have met with have been traced to inflammation resulting from abortions.   But it likewise takes place as the effect of inflammation after ordinary or full term parturition.

The *diagnosis* is easy with the aid of the uterine sound.   This instrument will not enter the uterus as far as it does into a healthy organ.   The uterus is lighter and more easily moved, also, by the finger introduced into the vagina.

One of the almost invariable effects of hyperinvolution is sterility.   I have met with a number of cases of sterility occurring soon after marriage, on account of abortion, in the first three or four months, being followed by inflammation and hyperinvolution: the patient ever afterwards remaining sterile.

The successful *treatment* of these cases requires a great deal of patience and well-adapted measures.   If the change in the condition of the uterus is slight, we may sometimes succeed by introducing a bougie of slippery elm bark, large enough to distend the cavity of the cervix as much as practicable, three or four days before the expected menstrual discharge. This seldom fails to increase the discharge, and if used perseveringly for several months, will sometimes cure the case. The bougie should be cut out of the bark so as to be about an inch and three-quarters long, for cases of moderate contraction, and secured by a thread, and then introduced. It should be allowed to remain until the discharge begins, and then removed. If, however, it is of long standing, and the diminution in size very considerable, we will be under the necessity of using the intra-uterine pessary recommended by Prof. Simpson. It may be made of zinc and copper, in order to galvanize the parts. A description of an apparatus quite appropriate to the purpose, will be found on page 82, on the subject of amenorrhœa. The only thing I need say in reference to the instrument here is, that the size will have to be adapted to the degree of aberration in each individual case. In extreme cases, the instrument must be small, and after wearing it for one or two months, a slightly larger one should be introduced ; and again, after another lapse of time, it should be replaced by a still larger one. These instruments, sometimes, cannot be borne for a great length of time; when this is the case, they may be introduced just before the menstrual period, say from four or five days, and then withdrawn when the discharge is freely established. We meet with instances which it is necessary to patiently treat in this way for years, and yet succeed in making a cure. Perseverance is an invaluable quality in treating all the chronic diseases of the uterus, and in none will our patience be taxed to a greater degree than in this.

CHAPTER XXVI.

CANCER OF THE UTERUS.

"THOSE growths may be termed cancerous which destroy the natural structure of all the tissues, which are constitutional from their very commencement, or become so in the natural process of their development, and which, when once they have infected the constitution, if extirpated, invariably return, and conduct the person who is affected by them to inevitable destruction." (Miller, as quoted by West.)

This general definition of cancer will include all its varieties, which are usually divided into four: 1st. Medullary; 2dly. Epithelial; 3dly. Colloid; 4thly. Scirrhus. I have mentioned these varieties in the order of frequency in which they usually occur in the uterine tissues. It will not be thought strange for me to say, that I have not seen either a case of colloid or scirrhus in the uterus. There can be little doubt, however, that both are rarely met with. The medullary variety is by far the most common form with which this organ is affected, the epithelial being also quite common. Cancer of the uterus is of very frequent occurrence, and the deaths from it, compared to death from the same disease occurring elsewhere in women, predominate over all other localities. It invades the cervical portion of the uterus more frequently than all other parts of the organ, yet it begins in every other portion; for instance, in the fundus, body, or cavities of the body, or in the cervix. In some rare instances it runs its course to fatal results without involving all these parts. When it begins in the cervix, it usually, either gradually or suddenly, passes upward to the fundus; or if beginning in the fundus, or body, it creeps downward to the os tinca. I have seen two instances where the lower portion of the cervix was but slightly, if at all, changed, while all the other parts of the organ were infiltrated by cancerous deposit. The material of cancer, particularly the medullary, is deposited in the tissues, supplanting them more or less perfectly.

This is so completely the case, that even the vascular tissue is supplanted. I think the obliteration of the vessels by the intruding deposit, thus destroying the nutrition of the tissues, is what produces the numerous sloughs, which commingle with the discharges, and give them their peculiar odor.

The epithelial appears to engraft itself upon certain points of tissues, and generally springs forward into fungi. It seems to cause, indeed, an exaggeration of the vascular tissue at first; and in the beginning, its tendency is self-isolation instead of diffusion. This is only the case in the beginning, for after it has lasted for a time, the adjacent structures become changed, and more or less true cancerous deposit may be found. But at no time is it so rapidly destructive as the medullary variety.

The tissue most commonly attacked by all the varieties except the epithelial is the fibrous or fibro-cellular substance. The mucous is not frequently attacked but by the epithelial variety. When the submucous tissues are the parts attacked, they are thickened and indurated, the thickening and induration being very irregular in shape and size. The enlargement and induration of cancer differ from these conditions caused by inflammatory effusion, by being always uneven, to such an extent as to distort the organ. If one of the lips of the os uteri is hardened from cancerous deposit, the elevated points are sharp and angular, and the hardened parts terminate abruptly, and in a manner unlike the induration from any other cause. The hardening from inflammatory fibrinous deposit is more globular than angular, and less abrupt in its termination in the sound parts. If one lip is the subject, it is swollen so as to develop it as a whole generally instead of one of its edges. The induration terminates by fading away, as it were, into the surrounding parts. If the cancerous deposit is in the body or side on any part of the wall, it is enlarged into an irregular shape and there are pits and points in several places. If enlarged by fibrinous deposit, there is not more than one point, and it is ordinarily round, and apt to occupy as one even lump the whole of the wall of the side or posterior surface. The epithelial variety enlarges the whole organ somewhat equally, apparently by an increased development of the vascular system and general tissues of the organ. The epithelial most frequently sprouts out into a tumor that enlarges, persists, sloughs, and is reproduced with great rapidity.

In the fibrous or fibro-cellular tissue, the infiltration and induration increase for an uncertain length of time, until, perhaps, the cancerous deposit so far displaces and replaces the ordinary tissues, that the nutrition of the parts is disturbed by the destruction of the bloodvessels, and sloughing takes place over a small or large space, owing to circumstances, but always over an irregular space, thus leaving a greater or less chasm. This is ulceration,—cancerous ulceration. The absorbents do not remove the parts, and thus cause ulceration, but there is sloughing and denudation by death of many minute parts, the absorbents having but little to do in the process. The sloughing causes the smell and putrilaginous character of the discharges. This process widens and deepens the chasm, sometimes quite rapidly, at others very slowly. As I have before observed, in the epithelial or fungous variety the sloughing and reproduction are constantly going on, the one compensating for the other, so that there is not much, if any, permanent diminution in the parts upon which it is seated. The medullary variety, on the other hand, removes more or less of the organ involved. In the case of the medullary variety, after induration and enlargement have fairly begun to advance in the uterus, the nutrition of the surrounding organs and tissues is disturbed, and the deposit is infiltrated into all the surrounding parts,—the bladder, the rectum, the areolar tissue by the side of the uterus, the peritoneum, in fact, into everything in the neighborhood. This general deposit is not limited by the coverings or divisions of the parts, but all become united, so that the whole pelvic tissues become one agglomerated mass of cancer; or, if it take one direction, the bladder and uterus may be glued together, or the rectum may be bound thus to the uterus. This disposition of the deposit very soon becomes sufficient to fix the uterus immovably in its place. In the epithelial outgrowths the most of the changes for a long time, comparatively, from the beginning of the disease, is the crowding upon the surrounding parts. The serous membrane covering the organs ordinarily resists these changes longer than any other tissues, and sometimes may be dissected out from among the wreck of tissues quite intact. It does, however, after a while, undergo sloughing, if not cancerous change.

After the ulcerative process has fairly begun, it advances more or less rapidly, until much of the surrounding parts is destroyed; the bladder and uterus become one continuous cavity, and sooner or later the rectum also is laid open, and then the pelvic viscera are involved in one confused excavation, from which the putrilage of cancerous degeneration is poured out, commingled with urine, fæces, and menstrual fluid.

I have observed that there is some proportion between the rapidity of the destructive progress of cancer and the age of the patient. It is slower in the aged, and destroys the young patient most readily. Of three cases in which cancerous deposit began in the body or fundus of the uterus instead of the neck, two were in patients beyond the climacteric period, one being sixty-four years of age and the other fifty-seven when the symptoms first attracted their attention. The other patient was forty-three. In this last patient, simultaneously with the evidence of deposit in the body of the uterus, signs of it appeared in the bladder, vagina, and clitoris, in the pelvis and the duodenum, and in the pyloric orifice of the stomach. From these and other facts in my possession, I am inclined not only to look for a more rapid degeneration of the tissues invaded by cancer in comparatively young patients, but to expect that when the deposit does occur in the body or fundus of the uterus, it is more likely to do so in the aged.

The morbid and microscopic anatomy of cancer of the uterus do not differ from the disease elsewhere, and as my design is to furnish a guide to students of as practical a character as possible in an economical space, I will refer them to Walsh, Bennett, Rokitansky, and other authors who have treated on the subject, for particulars respecting it.

Symptoms.—Discharges, pain, and fœtor are the symptoms that usually attract our attention in cases of cancer of the uterus. When a patient complains of any of these, however, the case is generally an advanced one. Pain, perhaps, is the symptom first experienced, and is caused earlier than any other. Unfortunately, pain is so common to women, they suffer so often in the region of the uterus and hips, that this symptom is not heeded by them until some other symptom makes its appearance. The pain is not generally intense nor troublesome until the disease is recognized. Nor is it peculiar. It is described as lancinating, darting, twinging, and very correctly, too, but there is often nothing of this kind of pain during the whole course of uterine cancer. And I think I express the truth in saying that cancer is almost the least painful of all the diseases of the uterus. There is less local suffering in cancer than in inflammation and ordinary ulceration, until ulceration and discharge denote its fully developed condition. Yet there is generally pain in all the stages of cancer, and it is very proper to place it among the symptoms; and the pain is generally darting, twinging, recurring at irregular intervals, darting from one part of the pelvic portion of the body to another, but not, I think, very severe.

The discharges in cancer are of three sorts, and the mixture of these sorts in different proportions. They are, 1st, blood; 2d, limpid serum; 3d, sloughs, generally minute. The first two are not offensive to the smell when pure or mixed together, as they often are, and they only become so by being mingled with the last, dissolving or holding in suspension, or being merely mixed with greater or less pieces of dead tissue. In the earlier stages of cancer the blood or serum may be and generally is effused, while the latter is reserved to the open or ulcerated stage. In this open or ulcerated stage all three kinds of discharges are almost always mixed together. In women who are still menstruating, the discharge first experienced is of blood. There is, at first, an increase in the amount of menstrual discharge; a little later, and blood is lost between the times of menstruation. The blood thus lost is derived from the same source as the menstrual blood,—the vessels of the mucous membrane of the corpus uteri. Later, when hemorrhage is so constant and attended with fœtor, it is effused from eroded vessels upon the ulcerated surface.

The blood in the former case, is produced as the result of constant turgescence; in the latter, on account of the disintegration of tissue. Limpid, unoffensive serum is almost always observed in the cases of old women, after the menstrual period of life has passed, and generally coming from the os uteri, which may be for a long time unchanged, indicating that it comes from some distance up in the organ. In fact, if the same serum was effused from the surface of the vaginal portion of the cervix, it would most likely be mixed with blood, because the parts producing it would not be sufficiently protected to insure the integrity of such frail tissue. In two remarkable instances the copious discharge of this limpid serum was for many months the only sign of disease presented by the patients. One of my patients, sixty-one years old, had been under the necessity of wearing napkins for six or more months before calling my attention to her condition. The discharge was so copious, when I saw her for the first time, that I collected about two drachms from the speculum in ten minutes. When examined, it was found to resemble distilled water in appearance, it was so clear and colorless. There was no smell, nor other offensive quality to it. When examined by the microscope, no solid substances were found except a very few natural epithelial scales. In a very gradual manner this transparent liquid became colored with blood. It was sometimes clear and sometimes bloody, for several months before becoming fœtid, and only for a few weeks before the patient died was it constantly bloody and fœtid. The cervix uteri, in this case, was not attacked at all, and the mouth and lips of the neck were natural. The body of the uterus, as high as the fundus, was enlarged more than double its natural size, indurated and nodulated; and when examined after death, the walls presented the peculiar fibrous hardness of cancer, but there was no excrescence in the cavity, as I had expected to find.

Whether the discharge is blood or serum, at first, or a mixture of both, it is generally odorless; but after a time it becomes fœtid, and remains so persistently. The fœtor appears, from the testimony of most observers, to be peculiar; but I have not been able to distinguish it from the smell of putrilage of other productions. Perhaps it is for the want of a sufficiently acute olfactory. I have no doubt that the smell is caused by the minute, numerous sloughs, constantly detached, and undergoing decomposition in the fluid. When all these symptoms unite, they form a case almost unmistakable. Lancinating pain, sero-sanguineous discharge, and peculiar fœtor, continuing persistently for days and weeks, are almost distinctive of cancer.

I cannot lay much stress on either one of these symptoms; but of the three, the most importance should be attached to the fœtor. Persisting for weeks, if not traceable to the decomposition of an ovum—and this is exceedingly rare—it should at least cause us to suspect a cancer. Contemporaneous with the complete establish-

ment of these symptoms we have constitutional suffering. It is
not often, I think, that general suffering precedes the local symptoms of cancer, and it has always seemed to me to follow as the
effect of local disease. It has not been my lot to meet with the
broken-down constitution sometimes said to be generated by the
cancerous diathesis. Cancerous anæmia, causing the straw-colored
translucency of the skin, considered characteristic of the malignant cachexia, is not distinguishable from the hemorrhagic anæmia
occurring sometimes in persons of the same age, produced by the
drain upon the blood. I make these statements just as the facts
strike me, and leave them with a scrutinizing profession to adopt
or reject.

In the fully developed condition of carcinoma the constitution
suffers, and the collection of symptoms are such as arise from the
embarrassment and failure of the functions in a long struggle
with pain, loss of blood, anxiety, and inaction. Debility, with indigestion, palpitations, restlessness, neuralgia, constipation at first,
colliquative diarrhœa and aphthæ toward the end, night-sweats,
wandering of mind, unsteadiness of purpose, succeeded by delirium
and apathy; in fact, all the train of symptoms which precede dissolution, when it approaches through protracted struggles, in which
pain and exhausting discharges are the destroying agencies.

Causes.—But little can be said profitably as to the causes of
cancer of the uterus. The general opinion that it is hereditary in
most cases is doubtless true; and yet a great many instances occur
that cannot be traced to such a cause. This is no reason why they
may not be hereditary; because sometimes the circumstances which
permit the hereditary taint to show itself do not exist for a number
of generations. And, again, the taint may be so dilute as to require
very favorable circumstances or co-operating causes to bring it
out. If a mother dies of cancer at the age of forty-five, and impart the same morbid tendency to her daughters, the laws of cell-
development would bring it about at the same age in the child.
If, therefore, the daughter dies a year too soon of some other disease, the taint is inoperative though present. Two or three generations of cancer-bearing persons, cut off by other diseases, lose
the history of its inheritance. Or if a mother be the subject of
cancer at the end of a life of active, nay excessive childbearing,
while her daughter leads a life of celibacy, or has but a single

child, the physiological life of the two is so different that we would naturally expect some modification of consecutive cell-development to result. So that although the hereditary taint is the same in the two, their pathological ages may differ, and the daughter may not have cancer until a later period, and die before that time arrives. We should, I think, allow much for influences that may modify hereditary taints, and only regard them as hereditary tendencies, to be brought out in mother and daughter under similar circumstances, and which may be postponed, or produced earlier in the one or the other by certain conditions. We will see good reason for these remarks by examining the circumstances under which cancer of the uterus most frequently occurs. Married women are affected more frequently than the single, and the fruitful than the barren. When we consider how many more married than single women there are in civilized communities, and how few married women are sterile, we may not be able to attach much importance to these facts. A much more significant fact is that a very large majority occur during the menstrual years of a woman's life. It is true that there may be nothing more than a mere coincidence in these facts, and that, after all, the hereditary mutations in the system during these years may bring about cancerous deposit, independently of any connection with the menstrual function. But it certainly is a coincidence, if not an etiological coincidence. As to the connection of cancer with chronic inflammation and ulceration of the uterus, much has been and may be said. I cannot lay my hand on statistics upon this subject, but I have never observed the coincidence of inflammation and cancer, or cancer to be a consequence of inflammation. I am aware that this statement amounts to very little; but I wish to record it as the result of my observation up to this time. It is unnecessary to point out causes of fallacy in former years, and appeal to more correct observers to settle this question. However, if they may be occasionally connected, there are but few at the present day who believe cancer to be the result of long-continued inflammation.

Diagnosis.—It would seem that the diagnosis of a disease so marked as cancer would be an easy matter, and so it is when all or even most of the peculiarities of the disease have been fully developed; but in the very beginning there may be much obscurity.

A patient complaining of nothing more than a perfectly clear, in-odorous, watery discharge, seemingly in the enjoyment of good health, would hardly be regarded as a victim to one of the most surely fatal and loathsome diseases incident to the human race; and yet, it is almost invariably so, when the patient is advanced beyond the epoch allotted to menstruation. The cancerous disease, as it usually occurs, advances beyond the period of doubtful symp-toms in a very short time, and in the majority of cases our attend-ance is not requested until a scrutinizing examination will enable us to decide very positively on the nature of the case. There is such a clear, concise, and yet thorough array of diagnostic particu-lars on pp. 217–18, presented in a manner to make it easy of refer-ence, that I should trespass on the time of the reader by repetition. It would not make it clearer, hence, I shall merely refer to them. Our attention will be attracted by the unusual amount and char-acter of discharge, pain, and smell. Then by a physical exami-nation, guided by the tabular view there given, we will not be easily deceived.

Prognosis.—The prognosis of cancer is a gloomy one. Indeed, there is no disease that so uniformly terminates fatally as can-cer of the uterus. Notwithstanding this fact forces itself upon our observation, there will sometimes, in the course of a large ob-servation, occur a recovery from it spontaneously and unexpect-edly. In ordinary medullary cancer there is less hope than in the variety called epithelial. I need not enter into the discussion of the causes of this fatality. Whether the disease is essentially a blood-disease, or whether primarily local, there are but few in-stances in which it is not multilocular. It exists from the begin-ning, or very soon afterwards, in more than one place. Yet again, this is not invariably the case. We very seldom meet with an in-stance in which the area of deposit is small and confined to one locality. If this locality is accessible, the case probably is cura-ble. I say probably, because the pathology is treacherous, and we cannot be positive. This gloomy picture is in part relieved by the greatly improved palliative means we now possess. Very much may be done to allay the agonizing state of body and mind under its ravages.

The epithelial form is not quite so fatal. It is also a very fatal form of disease, but it is now generally admitted that it is local in its existence, and that it never becomes otherwise except by travelling along contiguous tissue. If in the cervix, and particularly if in the lower extremity of the cervix, it may possibly be cured. As a drawback upon the favorable consideration here suggested, it should be said that it is not often confined to the vaginal cervix, but when we are called upon to treat it, it extends entirely beyond our reach.

The *treatment* is mostly palliative and but slightly curative. The only curative means we possess are surgical. No medicine can be relied upon to eradicate the disease from the system, although very many have temporarily enjoyed the reputation of doing so. The most that medicine can do is to defend the patient against its rapidly prostrating influence, and palliating the sufferings connected with it.

Propositions for extirpating the whole uterus have been followed out, we are told, but I need not say are regarded now as utterly unjustifiable. The operation of amputating the vaginal cervix is now regarded, however, as demanded under certain circumstances of cancroid and even cancerous disease. If the disease is clearly confined to this portion of the uterus, and the patient in a fair condition of general health, we may with propriety recommend and perform the operation. The operation is not a difficult one or necessarily very dangerous. The cervix should be drawn down to the external organs, so as to be fairly exposed to view, encircled by the chain of an ecraseur, and crushed off. This is but the work of a few moments by a skilful surgeon. The cervix may be drawn down by two vulsel forceps, and held in place by an assistant, while the operator uses the ecraseur. The dangers to be apprehended in performing the operating are the wounding of the bladder in front, or the peritoneum behind, and of leaving a portion of the disease. The dangerous consequences of a proper operation are shock and hemorrhage.

If these do not occur, the patient will usually rally, and recover from the effects of the operation. Inflammation remains to be mentioned as a consequence that may prove fatal also.

The actual cautery is recommended as very favorably modifying the ulcer of uterine cancer, and was thought to have afforded a prospect of cure in some instances. I have not had any observation in the use of it, and consequently, will not express an opinion upon its merits.   Dr. West speaks very favorably of it in certain cases.   I have used caustic potassa, but have not gained any good results from it, and do not now think of applying it to this disease. The cases in which these surgical means promise hope are very few, and their success is effected in but a moiety of these.

Palliation of the pain, smell, and debility, is the object of the most of our treatment.   We use local remedies for pain as well as general, introduced into the vagina.   Of course, the anodyne and anæsthetic remedial agents constitute our resources for combating pain.   Opium, belladonna, cicuta, hyoscyamus, and Indian hemp, may all be used locally for the pain.   The best form for their application locally, is that of a bolus of five grains of pul. opii.   I have also had the patient to introduce the finely powdered opium through a small glass tube, with a small piston of whalebone and cotton. It is applied thus to the ulcerated part and walls of the vagina in the neighborhood, and very effectually acts as an anodyne.   Ten grains of the extract of hyoscyamus may be used as a bolus, or two grains of ext. belladonna; and so on with all the anodynes. A grain of morphia may be mixed with the ext. hyos. to great advantage.

Medicated injections often soothe the diseased part very much also.   The watery extract of opium may be thrown into the vagina by a small syringe, and allowed to remain, the patient lying on her back for a length of time.   Hydrocyanic acid in solution, gtt. xx to a pint of water, passed through the vagina, has a very pleasant effect sometimes.   Injections of vapors of the anæsthetics are highly recommended, particularly by Prof. Simpson.   Carbonic acid gas and chloroform are those most used.

The chloroform vapor may be passed through the vagina by the ordinary perpetual syringe, made by the Union Rubber Company. The chloroform should be placed in the bottom of a large bottle, while the receiving-tube of the syringe may be passed through the cork and made air-tight with wax.   The other end being inserted in the vagina, high enough to almost come in contact with the disease, the pumping may be commenced.   The vapor will be caused to rise in the bottle quite rapidly under the exhausting influence of the syringe.   Care should be taken not to let the tube deep enough in the bottle to come in contact with the chloroform, lest this fluid, instead of its vapor, pass through the instrument.   The vapor thus delivered into the vagina causes a sense of heat and glow, which very soon seems to replace the pain. When properly done, patients experience great relief from this gaseous injection. The same apparatus will do to convey carbonic acid gas to the parts. The gas is generated by mixing in the bottle carb. soda and tart. acid, and then pouring a little water upon it. Although I have never yet tried the effect of great cold to the part, I have no doubt it would be very effective in relieving the pain. It should be applied through the speculum directly to the parts diseased, and no other. A small amount of the freezing mixture, of two parts pounded ice and one part common salt, in a small muslin bag, is the means used by Prof. Simpson. It is thought this cold not only relieves the pain, but that it retards the advance of the disease somewhat. The contact should be continued until the parts assume a pale, bloodless appearance, when this is practicable, and may be used twice or three times in twenty-four hours. With the local remedies for pain, may be mentioned the subcutaneous injection of morphia over the sacrum, or in the iliac region.

All these local remedies for pain will, after awhile, fall short of the relief demanded by our suffering patients, and we will be under the necessity of introducing them into the system in a more effective manner. We must resort to their internal use. I need not mention the anodynes to which we would resort in such cases; they are well known to the profession. I would, however, caution the student not to use opium, when any of the others will answer the purpose. Indian hemp will be found to do this more frequently than any of the others. They will all fail, eventually, and opium will prove the great blessing in such cases. And let me add the further caution, to commence with as small doses as will answer the purpose, and while we deal liberally enough with the drug to get its good effects, increase it slowly as possible; for with all our precautions in this respect, we will be under the necessity of giving it enormously. The anæsthetics are too evanescent to be relied upon for main remedies, but they will render the influence of opium more prompt, and, perhaps, lasting.

The hemorrhage of cancer will sometimes require prompt interference. I think, however, that although the bleeding is always ultimately exhausting, that it is seldom immediately dangerous from its copiousness. I have generally, when the hemorrhage required interference, depended upon the introduction of small pieces of ice frequently repeated. It is often very grateful to the patient, as well as hæmostatic. Dr. Simpson recommends powdered tannin, introduced through the speculum, and placed on the part; but he places more dependence on a paste made of perchloride of iron and glycerine. If the bleeding should be very alarming, notwithstanding these means, the tampon would be our last resort.

In fungus and cauliflower excrescences, if the bleeding becomes troublesome, they should be broken up with the finger, and an injection of tinct. ferri chl. made to permeate the mass thoroughly. This is generally successful, and attended with very little inconvenience of any kind. The whole sloughs off, and leaves a better state of the parts beneath for a time. The fungus is generally reproduced, however, and in a short time it is necessary to resort to the same expedient.

The offensive odor emanating from the disease makes it very desirable to have some means of correcting it. I should remark, with reference to the plans often resorted to, that they are more or less injurious to the patient and attendant, viz., the burning of sugar, myrrh, &c., in the room. This should be done very sparingly. For the air, chloride of lime and good ventilation will do better than all other expedients. We do not wish to make a stronger smell less offensive, to be sure, but we desire to remove the effluvia. Burnt sugar simply fills the room with various other less offensive gases which we breathe with them, the original cause of the trouble. Chlorine, disengaged from the chloride of lime, probably destroys the material floating in the air that offends the sense of smell. But the emanation may be lessened by the use of chlorinated water used as a wash and injection. The chlorides of zinc and soda are used mostly for this purpose, in the proportion of one grain to the ounce of water, injected into the vagina copiously. Frequent changes of the linen and bedding of the patient are matters of cleanliness that, of course, will readily suggest themselves.

Such is the melancholy paucity of our resources in cancer of the uterus. Scarce as they are, however, they may afford the sufferer great comfort; and we should fall short of our duty if we did not industriously employ them to their utmost benefit, as being the best the profession can afford.

CHAPTER XXVII

ANY organized growth within the substance of the uterine walls, or depending from or connected with any of its surfaces, may be called a tumor. This definition will include polypi of all varieties and sizes, from the mere granule that renders the mucous surface irregular by its protrusion, to the growth which fills up the uterine cavity; it also includes fibrous tumors of every description, small and encysted tumors, &c. These tumors are divided into varieties of structure, as the fibrous, fibro-cellular, vesicular, cellular or mucous, vascular, and encysted. They are also distinguished by their position and relationship with the different parts of the uterus. Pendulous tumors, of whatever structure or size, are called polypi; while the tumors that remain imbedded in the substance, or attached to the peritoneal surface of the uterus at the fundus, body, or cervix, are, in ordinary professional language, denominated tumors. It is necessary to bear in mind these usages in nomenclature, in order to rightly understand each other in our conferences upon the subject of tumors.

I shall commence a general view of the nature of these tumors by dividing them into varieties according to their structural composition, and then offer some observations on their differences and on the effects resulting from them. The fibrous variety is, by far, the most common, the most difficult of management, and dangerous, of all the benign tumors of the uterus. It is always connected with and springs from the fibrous tissue of the uterus. As it grows, it either remains imbedded in the wall of the uterus, causing the growth of all the tissues in its proximity to an extent sufficient for its accommodation, or, if its commencing nucleus is nearer the peritoneal surface, as it increases in size, it pushes that membrane before it, and extends itself with its serous covering into the abdominal cavity; or, again, if its point of origin should be near the mucous membrane, that yields and allows the projection of the tumor, as it increases in size, into the uterine or cervical cavity, whence it escapes, as it becomes larger, into the vagina. The first condition is called intramural, the second extramural, and the last polypous tumors of the uterus. In all these conditions the tumor may be connected with the fundus, corpus, or cervix uteri. They are probably more frequently situated in relation to the neck, next in frequency the body is invaded by the tumor, and, least of all, the fundus. They occur more commonly in the posterior wall of the body and cervix, not nearly so frequently in the side, and still less in the anterior wall. This variety of tumor does not often remain imbedded in the substance of the cervix, or bulge its external surface, but generally projects from the mucous aspects of that part of the uterus, and hence is polypous in denomination. Fibrous tumors of the uterus are almost always round or oval in shape, the intramural most frequently round, the polypous variety oftener oval.

They are organized, at least, so far as to have a complete circulation of blood; it is doubtful whether nerves ever penetrate them. There is quite a difference in the manner of the supply of blood in the varieties of fibrous tumors. In the intramural, minute vessels enter the substance of the tumors in many places, through the loosely connecting tissue; hence, the arteries and veins are very small individually compared to the size of the growth, but on account of their numbers afford a sufficient supply. The polypous variety and the extramural are supplied somewhat differently. The bloodvessels enter the tumor through the stalk, and as the support of the whole tumor depends upon a few of them, they are comparatively large. This arrangement renders the circulation and consequent organization of the polypus more energetic and enduring. The attachment is also different. The intramural is attached to the surrounding tissues by feeble projections of connective tissue with their vascular accompaniments, and are semiencysted in their imbedded positions, while the projecting tumors are surrounded by their investing membrane, and their connections with these membranes more intimate, although of the same nature. From these considerations, it results that the vitality of the intramural is less in force than either of the other varieties. They are less capable of sustaining themselves under perturbating influences, and more frequently than the other varieties undergo degeneration. It is particularly in them that we find copious deposit of inorganic substances. Their vitality is so low as to disable them from effectively resisting the destructive influence of inflammation, in whatever manner it may be set up. Inflammation in them results almost always in gangrene or suppuration, or both. We find no fibrinous deposits or other attempts at the reparative or protective processes of inflammation; they give way, melt down, and are discharged under even what, in other substances, would be regarded as a moderate degree of inflammation; hence, the pressure to which they are subjected by the uterine contractions during labor sometimes awakens certain destruction in their substance. These facts are fortunate and instructive. From them has originated the great operation of enucleation, as it is called, by which a part of the tumor is removed, and inflammation is started up in the mass, that speedily terminates in gangrenous solution of it. The polypus requires more embarrassment of its circulation to cause it to slough off, and when artificial destruction is too low down, they are almost invariably reproduced.

These tumors are "made up of fibres resembling those of very dense cellular tissue, or of tendinous substances, or of elastic tissue, presenting various degrees of development, and intermingled with cytoblasts, and a granular substance, the abundance of which is in inverse proportion to the perfection of the fibrous element of the growth." (*West.*) This dense fibrous structure is sometimes interrupted by spaces or cavities containing fluid.

The fibrous bands traverse the tumor in various directions, in some specimens so irregularly that they can scarcely be followed; in others they are arranged in concentric rings, &c. They are in density and appearance, as they are in structure, very much like the hypertrophied fibrous tissue of the uterus.

Changes of structure are likely to take place; these changes are always degeneracy, in some shape or other, and put an end to the vital endowments of the tumor more or less completely. One change observed as the effect of want of nutritive energy, perhaps, is the conversion of the tumor partially or completely into an earthy mass. Calcareous deposits more or less completely occupy the whole mass, which remain stationary during the life of the patient. These deposits are seldom perfectly continuous with each other, but are placed in points and nodules of various sizes; but specimens are said to exist that are so devoid of animal substance and so solid as to take a polish by proper friction.

The tumor after this calcareous degeneration may remain stationary, an innocent occupant of its position, or it may be expelled wholly from the organ. This is not likely to take place in any but the polypous variety. Calcareous degeneration takes place most frequently in the extramural variety; next in frequency the intramural is thus affected. The polypous forms seldom undergo the chalky change of structure. As I have before remarked, this change merely brings the development of the tumor to a standstill; it is a degeneration from animal to inorganic substance. It occasionally happens that in the polypous variety this process of degeneration is followed by expulsion; but the probability is that the uterus will become habituated to its presence by the time it is thus calcified, and allow it ever afterwards to remain. Nature not only brings imperfect relief to the patient by stopping the growth of the tumor, but she institutes other processes for a radical cure. The extramural variety is cut off, at its point of communication, from the uterus, and is maintained as a foreign independent substance in the abdominal cavity. So far as I can understand the facts of this mode of freeing the uterus of the tumor, it consists of two processes: the first, an investment, more or less complete, with a fibrinous or false membrane thrown out from the peritoneal membrane; and the second, an absorption of the pedicular connection between the tumor and the uterus. A sufficient amount of nutrition is carried on through these fibrinous effusions to keep up vitality in the tumor. Were it not for this previous plastic union with other parts, this tumor would be a source of danger by decomposition. The polypous variety is often detached from the connection with the internal surface of the uterus and expelled wholly, or it decomposes and comes away in putrilage and shreds. The detachment is caused by various agencies, viz., the powerful expulsive efforts of the uterus, ulcerative inflammation, strangulation in the os or cervix, or rough handling, by foreign interference, with the finger, or by efforts in coition, &c. The intramural fibrous tumor of the uterus, formidable as it is, may be eradicated by the powers of nature. First of all, it may cease to grow, begin to decrease in size, and in a few months disappear, without any explainable cause, and by an unknown process, probably on account of such a change of structure as will permit its absorption. Secondly, ulceration of the wall of the uterus, either as the first part of the process or a subsequent item in the case, takes place; the tumor is dissolved down into gangrenous putrilage and evacuated through the vagina. Thirdly, inflammation—caused perhaps by the pressure of the uterine fibres themselves, in labor or at other times—invades the tumor, and results in a mingled process of suppuration and gangrene, ulceration of the uterine parietes, and elimination. Speculation will hardly avail us in an explanation of all the minutiæ of these natural processes of elimination, and whether we call it eremacausis, enucleation, or by any other term, we are not in possession of facts of the right kind to elucidate it; but I am inclined to think that in almost all instances the initial process is inflammation, and the destruction of the tumor is brought about by suppuration or gangrene, or both conjoined.

Fibrous tumors of the uterus are of slow growth compared with other growths in the pelvic region. Sometimes years will elapse before they attain a size capable of mischief, and probably before they are detected. Often numerous fibrous tumors exist, and are discovered after death, that caused no suspicion of their presence. They are seldom solitary; the intramural and extramural varieties, in particular, generally exist in numbers. A specimen in the museum of the Chicago Medical College exhibits dozens of them, intramural and extramural. The uterus and its tumors weigh about twenty pounds, and they had attained that size in fourteen years after the first discovery of them. Some of the tumors in this specimen contain a small amount of calcareous deposit, and many of them are almost cartilaginous in hardness.

The uterus scarcely ever retains all its healthy properties and proportions with these tumors connected with it. It is almost always increased in size or hypertrophied. Sometimes, however, there is no perceptible increase noted. Prof. Francis, many years ago, reported a post-mortem case of a fibrous tumor weighing one hundred pounds, that was developed from a small pedicle connected with the peritoneal surface of the fundus of a uterus healthy in every other respect. The size of the uterus is less influenced by extramural tumors than intramural, and the uterine hypertrophy is proportionate to the size of the attachment of the growth. A tumor attached by a large pedicle exercises greater influence

in this respect, than one with a very small one. In the intramural
variety, the cavity of the uterus is always increased in length, and
generally in breadth. The increase will bear a close proportion
to the size of the tumor. I have known the cavity of the uterus
to receive the probe to the depth of eight inches in intramural
fibrous tumors of not very large size. The polypous variety causes
hypertrophy of the uterus to a less extent than the intramural, but
greater than the extramural. When situated in the cavity of the
organ, of course, development sufficient to contain it will necessarily occur. Should it arise from the cervix inside, less increase
in proportion to the size of the tumor will take place; and if it
arise from the cervix outside the cavity, there will not necessarily
be any enlargement of the organ. Even where the attachment is
upon the extremity of the cervix, there is generally some enlargement.

The functions of the uterus, as might be supposed, are nearly
always disturbed. Although menstruation is sometimes arrested,
much more frequently it is rendered more than naturally profuse,
protracted, and frequent. Pregnancy is prevented or influenced
variously as the location and size of the tumor varies. Extramural tumors exert comparatively little influence upon this function unless they are large, when, by preventing the evolution of
the uterus, they may cause miscarriage. Conception undoubtedly
takes place less frequently than if they were not at all present.
Cervical polypi do not very materially embarrass the functions of
generation. It is to intramural tumors and intrauterine polypi,
that deviations from correct generative functions are more frequently attributable. Whether conception can take place when
the cavity of the body of the uterus is occupied by a polypus, I
am not informed ; but I should not be surprised if, in rare instances, this might occur. Conception and perfect gestation are
accomplished in the presence of very large intramural fibrous tumors in rare instances. Not long since I examined a patient with
an intramural tumor of large size, probably as large as a man's
head; the uterine cavity lengthened, so as to allow the end of the
probe to be felt above the umbilicus, through the uterine and abdominal walls, and cause the appearance of seven months' pregnancy ; and in less than twelve months after the examination, the
subject of the tumor was delivered of a fine large healthy child, at

full term, without accident or difficulty. The pregnancy seemed to have arrested the growth of the tumor in this instance.

. Although I have included fibrous polypus in the above general description of the fibrous tumors of the uterus, there are a few things that it seems necessary to say in reference to it separately. The size of the pedicle of the polypus becomes larger, as we leave the extremity of the cervix, and proceed up the cavity to the fundus. Generally, the attachment on the lower end of the cervix is quite slender, but the size of attachment at the fundus is generally large, in fact, almost as large as any other portion of the tumor. Of course, in different specimens, we will find considerable difference in this respect. The firmness of these polypi are governed also by their locality. They are less firm below, and more above, the cavity of the cervix.

Other tumors of the uterus are softer than the foregoing, and, generally, have the mucous membrane for their base of origin and support, and hence are not improperly designated mucous polypi. An exception to this remark may be found in a rare sort of polypus, called the fibro-cystic, arising beneath the mucous membrane of the cervix. It is formed of fibrous tissue, including small cysts, containing a transparent, tenacious fluid, in interspaces, and grows to a considerable size. It is always, as far as I know, attached in the cervical cavity. The soft or mucous variety of polypus is rendered more or less firm, owing to the amount of fibrous substance it carries with it. It is often so soft as to become detached by slight handling; sometimes it is firm enough to require considerable force to destroy or separate it from its attachment.

The shape and size of these tumors vary very considerably. Sometimes they are cylindrical and very long.

An instance occurred in my practice, where the polypus was about the size of the finger, and extended to the vulva from the inside of the cervix, where it was attached. It was of equal size in its whole length. They are sometimes pear-shaped, and as large as an egg, with a small pedicle; again they are sessile, and not larger than a wheat grain. They are often quite numerous in the same uterus, but they also occur singly. They may be found on any part of the lining membrane of the cavities, but are oftenest seated in the cervical cavity, or on the extremity of the neck. I think they are more frequently cured by the efforts of nature, than the fibrous variety. They certainly are removed by much slighter means than the fibrous. On account of the facility with which these tumors are removed, or destroyed, much less importance attaches to them.

Another form of cervical tumor is sometimes found, which is described and illustrated in Dr. Meigs's work on the uterus. It is a thin sack, situated in the submucous substance of the cervix, which contains an albuminous fluid. It may be felt as an elevation, or seen through the speculum, as a bright, translucent tumor, the size of a cherry, or even larger. In the only case I ever saw, it was on the anterior surface of the cervix, just above the extremity of the labium.

I do not profess to have given a full description of all the tumors of the uterus, but I think that the above description is sufficiently full and minute to answer all practical purposes. The philosophy of their formation, and their effects upon the uterine tissue, are to be explained. These tumors are without exception, hypertrophic enlargements of some portion of some of the tissues of the uterus. A vortex of the fibrous tissue commences growing, and as it grows, it assumes something of an individuality that enables it to appropriate a larger amount of nutrition than is necessary for the maintenance of its usual size; and hence has begun a tumor that will, as a general rule, increase continuously. Unlike the ordinary tissues of the body, it grows as an individual, independent part, and not as a member of the economy, which is bounded in its development by the general harmony of the system. A portion of the mucous membrane likewise may set up an unusually energetic and perpetual organization, which enables it to pass beyond the rest of the membrane in growth, and thus protrude and hang from it; or the glands of Naboth may become hypertrophied and developed into hygromatous cysts, and remain *in situ*, or protrude, carrying with them enough fibrous material to form hydrocystic polypi.

This assumption of individuality of energy and life in particular parts of certain tissues, is always accompanied by other general changes in the organ, whether as an effect, concomitant, or cause, is yet undetermined. The circulation in the whole uterus is increased, the quantity of tissue is greater; in fact, the uterus becomes hypertrophied. Of course, the general hypertrophy is much less than the partial, which gives rise to the tumor. The uterus is longer, broader, thicker, and heavier. The vascularity of the mucous membrane is greatly increased, particularly in intrauterine tumors. This augumentation of bulk is not always in a direct ratio with the size of the tumor, although the fibrous kind of tumor generally has more of this effect.

There is another outgrowth from the uterine neck and inner surface which is sometimes classed as polypus, but which is heterologous in its nature. I allude to the vascular polypus or fungus, cauliflower excrescence. The other forms of growth are all, as I have before said, hypertrophic developments of existing tissues, fibrous, mucous, glandular, &c. This is a true epithelioma or fungus, and does not afford in construction any analogy to the different uterine tissues. They differ in different specimens as to firmness or solidity. Sometimes firm enough to be mistaken by the inexperienced for a fibrous growth, at other times they are so fragile as to give way with very little handling. There are those of intermediate solidity of almost all grades. Although epithelioma, I do not know that these fungi are necessarily, or even generally, connected with cancerous degeneration in the surrounding organs or tissues, and that their malignancy is propagable as in other forms of cancer in this locality. Unlike fibrous or other benign polypi, they are very difficult of eradication. The destruction of the whole tumor and the removal of the whole mass are only temporary in their effect. If the nidus of origin remains, the growth will be reproduced. And the tumor attains its former size more rapidly after each amputation or removal of any kind. They grow quite rapidly as a general thing, and in their progress slough off in pieces and grow again. This constant loss and reproduction of substance is one of the remarkable features of the disease. These growths consist very largely of loops of fine arterial tubes, connected together by a very loose connective tissue, and are covered with epithelium merely.

Early in their growth they may be seen as a slightly elevated vascular eminence on some part of one of the lips of the os; a little later, papillary projections show themselves; still later, the whole elevation forms into a protrusion, which, at last, becomes a polypous tumor of various size. Sometimes the vagina is so distended by the tumor as to completely fill the lower half of the pelvic cavity, and the growth protrudes in bleeding shreds from the vaginal orifice.

Symptoms.—Two kinds of symptoms first arrest the attention of the patient and medical attendant, viz., discharges and pressure. The former kind originates, for the most part, in the unnatural vascularity of the uterus, and the other in the embarrassment to pelvic organs, caused by the pressure of the morbid growth upon them. These symptoms are not merely the indices to disease, but the effect of it upon the general health, and they require interference for their amelioration or cure. Thus the hemorrhage exhausts the patient, and the pressure upon the bladder or rectum does great mischief in various ways; and after the tumor rises out of the pelvis, pressure may be the cause of much damage to the abdominal organs. Hemorrhage is a symptom which is common to all uterine tumors. It is much more common in the polypous variety, rather less so in the intramural form, and not very frequent in the extramural fibrous tumors. Hemorrhage is so constant a symptom in polypus of the uterus that we expect it; and it is the almost invariable cause of our advice being sought for. It is remarkable, too, that the size of the polypous growth makes but little difference in this respect. A very minute tumor in the cavity of the cervix or corpus uteri may exsanguine our patient, and the larger produce quite a moderate amount of discharge. The hemorrhage indicating uterine growths is apt to show itself first by increase in the quantity of discharge at the menstrual period, then by the frequency or rather protracted duration of the discharge, so that the interval becomes shorter; or there may be constant hemorrhage, the menstrual epoch being designated only by the profusion compared to other times in the month. Sometimes the hemorrhage of menstruation is not greater than ordinary, but there is bleeding at intervals between the monthly periods, slight or profuse, owing to other circumstances.

Again, we meet with instances when, without premonition, hemorrhage bursts forth in large and exhausting quantities, to be followed by cessation of long duration, reappearance, &c. The period of the growth or advance of the tumor at which hemorrhage may begin varies greatly. The growth may develop to a pretty large size before any sign of its existence appears, so that when the symptoms lead us to suspect a tumor, we are surprised to find it so large. Finally, tumors attain to great size, both in the walls of the uterus and depending from the cavities and cervix of the organ, without causing any hemorrhage. These cases, however, are not common. We should always suspect the presence of a growth where there is protracted or frequently recurring hemorrhage; and if there is none in the vagina, we ought to remember the frequency with which intrauterine polypi or intramural growths cause it, and examine for them.

Although polypous growths sometimes cause tenesmus, they do not often cause serious pelvic distress., The large intramural growths often give rise to symptoms from pressure as the first evidence of their presence. Difficulty of urinating or defecating causes the patient to suspect something wrong. If the tumor is in the anterior wall, the bladder is first affected; if in the posterior wall, the rectum suffers, &c.; and finally, if the tumor continues to grow, both these canals may be pressed so firmly against the bones of the pelvis as to arrest excretion through them. A distressing instance of this kind recently came to my notice, in which the patient was being rapidly prostrated by the anxiety, pain, &c., of protracted efforts, aided by medicine and instruments to evacuate the bladder and rectum, both of which were almost completely occluded by pressure. Upon examination, the pelvis was found impacted by one of the offshoots from a large fibrous tumor that reached nearly to the umbilicus. By judicious efforts the tumor was lifted above the inferior strait, and kept so for some days, when there was no further tendency to return to its original situation, and the relief experienced immediately after it was lifted up became permanent. Pelvic distress from pressure, however, does not always take place; some of these tumors rise out of the pelvis without causing any inconvenience. When this is the case, and there is no pressure, a long time ordinarily elapses before the tumor gives the patient much inconvenience. Sometimes the amount of pressure is so slight upon any particular organ that the tumor is not announced by any symptom until the increase in the size of the abdomen gives rise to suspicion. Not unfrequently before the tumor rises out of the pelvis, or soon afterwards, it exerts great pressure upon the nerves passing through that cavity, causing pain, paralysis, or cramp in the parts to which the nerves are distributed; or pressing upon the veins induces œdema, and, in some instances, a state of things closely resembling phlegmasia dolens. An instance of this latter kind of trouble is now in charge of a friend of mine in this city. The limb is so large and unwieldy that the patient is scarcely able to walk. As the tumor attains to a still higher position, the abdominal organs suffer from pressure, and the patient's distress is increased by indigestion, vomiting, jaundice, pains in different parts of the abdomen, great distension, and pain from the great muscular tension.

After a greater or less time, the continued and prolonged embarrassment to those vital organs which are centred in the abdomen, produces marasmus, which is succeeded by death. Sometimes the tumor does not rise out of the pelvis, but a part of it grows up from that contained in this cavity, and to a greater or less extent fills up the abdomen. In such cases, we may have all the symptoms of great pressure in both the pelvis and the abdomen.

Diagnosis.—We are seldom called upon to render a diagnosis in a case of uterine tumor, until some one of the distressing symptoms enumerated above have harassed the patient very considerably. The symptoms of hemorrhage and pressure are very likely to be the first. Of all the symptoms above mentioned, hemorrhage is the only one that points to uterine growths; pressure may be caused by any other sort of tumor as well as tumor of the uterus. I do not think that position is at all distinctive, as movable tumors are likely to assume a central position. Fibrous tumors are harder than almost all other tumors in the pelvis or abdomen that are movable. They are of greater consistence than ovarian growths or hydatids of the abdominal organs. The only enlargements or growths likely to equal in denseness the fibrous uterine tumor are scirrhus and hypertrophy of the liver and spleen, or concretions in the intestinal canal or other hollow viscera. Their contour is generally globular, and when not so they may be discovered to consist of globular bodies, joined together and separated by well-defined fissures, penetrating not very deeply into the substance, their edges being rounded instead of sharp. The great item of diagnosis in these cases is the determination of the attachment to the uterus or not. By careful and diligent manipulation and the use of the probe, we can almost invariably determine whether the tumor is connected by continuous tissue

with the uterus. If it is connected with the uterus, it is of course
a uterine tumor; if not, its attachment elsewhere removes it from
the category of these tumors which are under consideration at
present. In making this statement, it is hardly worth considering the possibility of preternatural adhesion between tumors
springing from other organs. If there is the close attachment
we always find between this organ and its outgrowths, there will be
consentaneous motion of the uterus and tumor in every direction.
In moving the tumor in any direction the uterus will follow or be
moved by it; if the uterus is moved in any direction the tumor is
impressed similarly. If the growth is not attached to the uterus,
and consequently is not uterine, this organ may be moved in almost every direction without imparting motion to the tumor, and
vice versa. In using this test we should not be satisfied by moving either in any one direction, but should move them in every
practical manner,—up, down, back and forward, sidewise and
diagonally. The particular means and modes of doing this must
now be reviewed. We should first, if possible, by digital examination find the uterus and ascertain its relative position to the tumor;
if the tumor reaches above the symphysis, we should next lay hold
of it with the other hand, and move it from side to side, from before backward, and, when practicable, lift it up, in the meanwhile
closely watching, with the introduced fingers, the effect of these
movements upon the uterus. If the tumor is not attached, and
consequently not uterine, we may easily feel the motion of the
tumor through the uterus, while we can be perfectly satisfied
generally that the uterus itself, although moved to some extent,
is not dragged with the tumor when its position is changed. In
some instances, I have had good reason to be well satisfied with
an effort at rotation of the tumor; when not too small, an assistant,
properly instructed, will often succeed in partially twisting the
tumor upon its vertical or horizontal axis. When this motion is
imparted to the tumor it will be quite sure to give a correct sensation to the finger in contact with the uterus. If the uterus is
in a position to be controlled by the finger, we may move it upward and downward, from side to side, backward and forward, and
rotate it, while we keep close watch, by the senses of touch and
sight, of the movements of the tumor. It is not unusually the
case that we can satisfactorily ascertain the whereabouts and rel-

ative position of the os uteri, while we cannot define the position of the body of this organ. In these cases the uterine sound or probe will aid us very materially. There are several things to be remembered by the inexperienced in the use of this instrument, to insure satisfactory results. The sound should be soft enough to bend to any required shape, and yet firm enough to retain that shape under the application of moderate force. I think the sounds for sale at the instrument makers are generally too hard; they should be small too,—the bulbous extremity not more than the eighth of an inch in diameter; even slightly under this size will often do better.

When we have ascertained the position of the os uteri, the probe should be introduced into it as far as it will go without force, varying its direction every way. This trial, if it does not succeed in sounding the uterine cavity, will give us some idea of the direction of it. If we do not succeed, we should bend the instrument slightly, and give it another trial similar to the first; and still if we do not succeed, we should flex it still more and try again. By patience, perseverance, and gentleness, we will, after awhile, pass it to the fundus.

In most cases of fibrous tumor, the probe will pass farther into the uterus than in a natural state of that organ, the cavity being elongated. The probe will pass three, four, six, and sometimes eight inches. On more than one occasion, I have felt the end of the probe above the umbilicus, through the abdominal and uterine walls. When once introduced to the end of the cavity of the uterus, we may move the organ in every direction, which will affect the tumor if attachment exists. When the probe is thus introduced, we should place our finger on the tumor per rectum, if it is small, or over the symphysis pubis, if large enough, while the uterus is made to move in every direction, and rotate, or nearly so. I think, in some cases, the finger in the rectum will appreciate movement imparted to the tumor with more certainty than when placed on the abdominal walls, or even in the vagina. There is very little doubt left by an investigation conducted in the manner above directed, with the finger and probe, &c.

Pregnancy should be excluded from the questions occurring in order to clear up a diagnosis, before the probe is made use of, lest the membranes be ruptured, and abortion occur. This may generally be done by remembering the signs of that condition, and making thorough examinations with the hand, fingers, &c.

Although the difficulties of diagnosis are very much greater in some instances of large intramural fibrous tumors, yet we occasionally meet with perplexities in deciding whether a vaginal tumor is polypous. Perhaps the only condition that is likely to be mistaken for a polypus, is chronic inversion of the uterus; and, as the history is not always to be relied upon, we are under the necessity often of manipulating for that purpose. The probe and finger will readily make the distinction. If the uterus is inverted, the probe will pass up the same distance all around the tumor,—before it, behind, and at the sides; if it is polypus, it will not pass above the reach of the finger on but one side, and in that direction will sink to the depth of the uterine cavity. The finger, in the case of inversion, is obstructed alike on all sides; while in the case of the polypus, it will mount higher upon the tumor in one direction, which ordinarily is in front. Probably the most difficult of detection are small intrauterine polypi.

The symptoms which always induce us to suspect the presence of these bodies is persistent, exhausting, and unaccountable hemorrhage. There are no symptoms by which we can arrive at certainty in this respect; but when hemorrhage persists, in spite of judicious remedies diligently and rationally applied, we should not fail to examine the cavity of the uterus in every possible way.

The shape and size of the uterus should be investigated by the finger and probe as completely as possible, and if they yield no satisfactory results, we must open the os and cervix, so that we may see the contents of all the inside of the organ. This may be done by a sponge properly prepared. This plan was perfected by Prof. Simpson, of Edinburgh, to whom the profession is indebted for many ingenious and useful suggestions in physical examinations. The sponge should be fine, and saturated with a solution of gum-arabic; after which it must be wrapped with twine in such a way as to compress it to its smallest possible dimension. With a little care, we may make the piece take a conical shape, more particularly if we trim it with that view. In this way the sponge can be compressed so as to occupy a space of probably one-sixth the size when expanded by moisture. After applying the twine, we should let it lie for several days, until it is dried and solid. If not of the right shape, the sponge thus prepared should be cut into a long cone with a sharp knife. It is a good plan, and will facilitate the wrapping to insert a probe or stiletto into the large end of the sponge, and let it there remain until dried. When we wish to dilate the os and cervix uteri, we insert a small cone of this sponge, and push it up as far as we can, and then allow it to remain until the moisture of the parts expands the sponge; the cavities will gradually dilate under this expansive influence. It takes about twenty-four hours for the sponge to expand the parts as much as it is capable, if it is tightly embraced by them; at the end of this time, it will be so loose that it may be easily removed with forceps. To render the removal easier, it is usual to attach a thread or twine to it before inserting. Upon the removal of one piece, it should be replaced by another considerably larger, pushed up as far as it can be made to go with reasonable force. It will require but a few days, perhaps only two, sometimes more, to bring the cavities in a condition for thorough inspection, when, if there are any growths, they will be apparent, and subject to such treatment as may be necessary. In the use of these means for dilating the os and cervix, we will be very much aided by the speculum.

Although encysted tumors of the cervix may sometimes be discovered by the finger, the diagnosis will hardly be complete until we inspect the part thoroughly through the speculum, when they will present themselves in an unmistakable manner. Small polypi in the mouth, and even protruding slightly from it, may not always be discovered by the finger. We should, in all obscure cases of hemorrhage, make use of the speculum, if our diagnosis should not be clear without it. The tyro need not fear these modes and means, as the parts are not damaged by them if conducted with reasonable care.

Prognosis.—The prognosis of uterine tumors, considered in a general way, may be regarded as unfavorable. Once started into active growth their tendency is to grow constantly, so long as the energies of the patient last, and with their growth the production of destructive symptoms, more or less rapidly, varying with attendant circumstances and complications in different cases. Although this may be said to be the condition of things generally, there are some happy exceptions to the rule. Some tumors, after growing for a time, cease to increase in size, and remain stationary for the remainder of the patient's life, or they grow so slowly as not to incommode the sufferer to a dangerous degree until the end of life is accomplished by age, or terminated by some other cause. Again, nature accomplishes a cure by inflammation, evacuation, expulsion, or absorption. These exceptions to the rule of a constant tendency to death are, it is true, not very numerous, but sufficiently so to be taken into consideration in making up our judgment. In particular instances there are often circumstances that will enable us to determine the rapidity with which the fatal issue is approaching, or the competency of nature to avoid it entirely. On one occasion, at least, I met with a case that seemed to be rapidly undermining the powers of the patient by exhausting hemorrhage, and the end appeared to be near at hand, when, without any apparent reason, it suddenly ceased to give any further inconvenience from the hemorrhage, although it still grew with tolerable rapidity.

The modes by which these tumors produce death are through hemorrhage and great pressure. The patient dies of hemorrhage suddenly, on account of a severe and continuous drainage for days, or hemorrhage proves fatal in a chronic way. An eruption of hemorrhage, so severe as to cause death the first time, is very rare, though such cases do occasionally occur. Repeated recurrent attacks, at intervals of weeks or even months, induce organic diseases by vitiating the blood primarily, or so impoverish the blood as to render it incapable of supporting the vital functions, and fatal prostration occurs. We judge of the probable fatality of each case from the copiousness and frequency of the attacks, and by the robustness or delicacy of the constitution of the patient. After all, however, we should be very cautious in our prognostic calculations as to the effect of this symptom, since sudden changes are liable to take place at any time. Death comes sometimes as a consequence of the pressure on some organ or organs. When this is the case, its approach is very gradual, and the various steps in that direction are often obscured, so that we may not perceive them. The bladder may be pressed upon, irritated, and eventually mucous inflammation takes place, and spreads up through the ureters to the kidneys, and disorganization in these important excretory organs be the more immediate cause of death. Or again, the indigestion, vitiation of nutrition, and their consequences, caused by pressure upon the rectum, may bring about fatal disaster to the organs of the abdomen. Pressure upon the veins may cause fatal stoppage of their circulation, or induce phlebitis and its direful accompaniments. In some of these, and, in fact, many others, pressure may lead to fatal addenda and complications.

But complications independent of the tumor may act very unfavorably on the course of the tumor, or be rendered dangerous by the presence of the growth. Pregnancy is one of these, and although a natural function, its influence on the uterus is so great—the organ in which the tumor grows—that the case often may be much aggravated by its supervention. Pregnancy, by increasing the afflux of blood to the uterus, and its general hypertrophizing effects upon that organ, causes the tumor to grow sometimes very rapidly. It is true that in certain places, where great pressure is exerted upon it by the developing ovum, the pressure may counteract these effects. Where this is the case, at the end of pregnancy the patient is much worse off than at its beginning. But, perhaps the worst effects of the coexistence of pregnancy and a fibrous uterine tumor result from a violent termination to the progress of gestation in abortion, and the inflammation that ensues. The determination of blood to the uterus on account of the presence of the tumor, the increased size and number of arteries and veins, greatly add to the intensity, extent, and consequences of the inflammation so generally attendant upon miscarriages. There can be no doubt, too, I think, that the peritoneum is more susceptible of destructive inflammation when the uterine envelope derived from it is rendered more vascular, or at least hypertrophied, by the enlargement of the uterus. In view of these considerations, we should regard the supervention of pregnancy with anxiety. It should be remembered, however, that the inflammation invading the tumor may have a salutary effect upon it by arresting its growth, or by causing it to be eliminated by suppurative inflammation. Should pregnancy not be interrupted by the growth, and proceed to full term, the apprehensions usually indulged are increased to a manifold degree, as there is much more danger of inflammation post partum. The tumor must be very violently compressed by the throes of labor, and likely to be attacked by inflammation, which may spread to other parts, and cause great mischief and suffering, if not death.

Hemorrhage is more likely to occur and more dangerous than in ordinary cases of labor. Gangrene or suppuration in a large tumor may lead to death on account of the shock, exhaustion, or phlebitic inflammation which may be produced. Unfortunately, the prognosis is not rendered much better, so far as the ultimate termination of the case is concerned. Relief may be afforded from some of the distressing symptoms by proper palliative treatment in intramural tumors, but very few of the cases are susceptible of radical or surgical treatment. The polypous variety, while they may do nearly or quite as much damage as the intramural in one way, viz., by the hemorrhage they cause, are nearly always curable by surgical means.

Epitheliomatous growths from the cervix are much more dangerous than the fibrous or other benign polypi. They almost invariably prove fatal without treatment, wearing the system out by the exhausting discharges from them; and, by vitiating the blood, they bring about diseases arising from poison in that fluid. Treatment is much less certain than in the other growths, and a radical cure by surgical means is the exception to the general rule. There are unquestionable cases of cure after amputation of the cervix, but many more cases of failure could be recorded.

The hygroma and other encysted growths of the neck of the uterus are quite amenable to treatment. If left to pursue their own course, they cause a great deal of suffering; and, in persons of feeble constitution, exhaust by the draining hemorrhage they cause, until the patient falls an easy prey to some intercurrent disease, or they may cause death directly by the anæmic condition of the system they produce.

Treatment of tumors of the uterus should be palliative and radical; the palliative combats the symptoms arising from the presence of the tumor, ameliorates the condition of the patient, and often enables her to live longer than she could without it. It is sometimes the only treatment practicable. The radical treatment has for its object the cure of the patient, by removing the tumor, or the arrest of its growth. It is quite impossible to foresee and provide for the treatment of every inconvenient or dangerous symptom that may arise during the existence of these tumors, and I do not intend to make the attempt; but there are symptoms of a very grave character, which occur so frequently under such circum-

25

stances, that we should anticipate them, and be prepared for their occurrence. Hemorrhage, for instance, as a general thing, is present in cases of uterine tumors ; and that its efficient treatment is a matter of the greatest importance, the student will understand. Fortunately, its control, as it takes place from the uterine cavity, is as certain almost as the employment of the means. These are, for the most part, mechanical, and should be resorted to with boldness and energy where the urgency of the case demands them. I shall leave their description until after I have mentioned the medicinal treatment of hemorrhage.

Hemorrhage does not always proceed to a dangerous degree, although it may be sufficiently copious to make it a symptom of great distress and inconvenience. The medicines for hemorrhage are the whole class of astringents. Some of them have gained quite a distinction for efficiency, while others are not used at all. When hemorrhage comes on in a threatening manner, the patient should be confined to the recumbent posture, in a cool room, allowed nothing for ingesta but cold acidulated drinks, and kept as free as possible from all causes of excitement. Cold should be applied over the symphysis pubis in such a way as to keep up a steady low temperature. Ice in a beef's bladder, laid over the part, or small pieces of it may be placed between two thin cloths, and allowed slowly to melt, and run over the person of the patient. The hips should be elevated by an air or water cushion, while the head and shoulders should be kept low as is compatible with the comfort of the patient. In moderate cases, it will be sufficient to give, in addition to the above treatment, twenty drops of the aromatic sul. acid, in sufficient water to largely dilute it, every four hours. Sugar of lead in two-grain doses, combined with one grain of opium in pill, every four hours, is regarded as an efficient astringent by most physicians, in such cases. Gallic acid, in doses of eight to ten grains, administered as often, will sometimes answer an excellent purpose. Oil of turpentine, in grave cases, may be given in drachm doses, in an emulsion, every hour, with most excellent effect. But astringents may be administered by injection with very good effect. The tr. ferri chl., for instance, injected into the cavity of the uterus, so as to come in contact with the bleeding surface, will seldom fail to answer the purpose of greatly lessening, if not entirely stopping, the hemorrhage. A good way of operating is to introduce a small male elastic catheter into the cavity, and, if possible, to the fundus, and through this, with a hydrocele syringe of hard rubber, throw a drachm of the tincture. The injection should be made with very little force. A strong solution of nitrate silver, tannin, or other astringent, may be used in place of the tincture.

An expedient of great efficiency is the introduction of the powder of tannin, or gallic acid. This can be done through a catheter, with a stylet covered with a piece of lint. These modes of using astringents are far more efficacious than taking them into the stomach.

The mechanical treatment consists in the various sorts of tampon. In a great many instances, to stop the vagina entirely by a plug or tampon will be all that is necessary to check the hemorrhage. And this should be resorted to without delay, as it is not difficult to do, is attended with no danger, and very little inconvenience.

In the present advanced condition of the mechanics of medicine and surgery, I need not say that the handkerchief, the old clothes, cotton batting, and sponge, should be, and I think generally are, discarded. The gum elastic air-bag, is the only tampon justified by anything but extraordinary circumstances; and when not within reach at the time of our great necessity, should be procured at once, and the other plugging expedients be replaced by it as soon as it can be procured. The application of the air-bag is so simple, that it may be left with the patient to be used until the physician can be called.

The patient, or the nurse, should be directed to introduce it— upon the supervention of the hemorrhage—in a collapsed state, and then, by means of the inflator, to blow it up until the distension produces slight pain. This will prevent the blood from flowing out of the vagina, and fill the upper part of the vagina and uterine cavity with coagula, and thus stop the hemorrhage. This distension should be increased in an hour or two, if the hemorrhage occurs; if not, this will not be necessary. By managing the colpeurynter, or elastic air-bag, in this way, the blood may be saved, as a general thing, with very little loss. But in cases where the danger is imminent, the os uteri may be plugged, and the flow stopped with still less loss of blood, and, as I have thought, with more ease and certainty. There are two instruments used for this purpose; one is a small cylinder of gum-elastic tube, with a small bulb on the end of it, large enough to fill the cavity of the cervix. This may be introduced, by means of a probe, entirely within the os, and then inflated. The other is a cone of compressed sponge, just large enough to pass into the cavity of the cervix. When well introduced, the compressed sponge will certainly check the hemorrhage. It very soon fills the cavity of the neck so closely, that the blood will not pass it; coagulation inside of the uterus takes place, and the bleeding is stopped.

All these resources against hemorrhage are mentioned more particularly with reference to the intramural variety of tumors, as, indeed, it is seldom that they are needed in any other variety. Polypous tumors, in any of their forms, seldom induce the sudden, copious, and almost uncontrollable bleeding that occur in connection with the above-mentioned form. The injection, the tampon, and the compressed sponge, may, however, be used even in some cases of these to great advantage.

It is a nice point, as well as a very important one, to know when we ought to depend upon palliative remedies for hemorrhage, and when we should abandon them for the radical surgical treatment. Our object being to save the life of the patient, we cannot afford to see it sacrificed by the loss of a fatal amount of blood; nor can we resort to the dangerous expedient of removing the tumor, until satisfied that it is our only resort against otherwise inevitable disaster.

Unfortunately, no absolute rule can be found by which we may be governed in all cases, but we have some considerations that ought to influence us. If the hemorrhage is so serious a symptom as to threaten the existence of the patient with immediate danger, if the tumor is in the most favorable condition for an operation that intramural tumors present, if the constitution of the patient is not embarrassed by any other severe or depressing disease, and if our temporary treatment fails or is impracticable,— and this last objection will sometimes show itself,—we should endeavor to get rid of the growth by surgical means. But while the hemorrhage is moderate and controllable, or controllable when sudden and severe in its eruptions, the patient not much depressed by it, and cheerful to bear her burden, we may be satisfied with a skilfully applied palliative treatment.

But another state of things renders curative treatment inadmissible, and makes the palliative the only practicable means; I mean when the tumor has grown so large as to preclude the possibility of its removal. It often happens that the pressure of a tumor on the pelvic organs causes a great deal of suffering. In these cases it is generally because the tumor is large enough to almost or wholly fill up the cavity. When this is the state of things we may often lift it up to and above the superior strait, and thus lighten the pressure, affording relief. It may be necessary to make use of artificial support in the vagina sometimes, but at others, by enjoining the recumbent posture for a few days, watching and keeping the growth lifted up, it will not again settle back. I have seen more than one instance where this kind of palliation made the patient comfortable for several years. When the tumor is not so large as to fill up the whole pelvic cavity, it may be so situated as to press upon the bladder or rectum, sciatic nerves or urethra. We may often relieve these cases by changing the position of the tumor with the hand, or with some instrument, for the purpose of keeping it placed artificially, causing the patient to assume and remain in such a position as may be necessary to fulfil these indications. This idea in the treatment of these tumors will serve a good purpose after the tumor gets so large as in a great measure to fill up the abdomen. The recumbent posture, on the side, will relieve the pressure on the vessels near the spine and the kidneys. And when the tumor is irregular in shape, so as to press upon the viscera upon one or the other side, a position may be assumed that will take the pressure off of the part suffering.

In addition to these considerations there are others which have reference to the general health of the patient. The secretions, excretions and nutrition should be corrected and promoted by the remedies suggested by the particular case. The patient should be kept as comfortable or as free from suffering as possible, and in as vigorous a state of health as practicable.

The radical or curative treatment of tumors of the uterus have for their object sometimes the complete removal of them. In the nature of things this is for the most part necessarily surgical.

These organized growths are engrafted upon the system, and are supported by the same processes of nutrition that other parts of the body are sustained by; so that no special medicinal treatment exercises much influence over them. They cannot be singled out as the objects at which medicines can be directed through the stomach, and thus poisoned or etched away by the absorbents, and removed. It should be stated, however, that the profession are not all agreed upon this point, and cases occasionally occur that would give color to the idea that sorbefacient influences can be exerted upon them. A certain number of remedies are credited with curative effects upon almost all sorts of tumors; and among them uterine tumors. Iodine, I think, stands at the head of the list; either combined with or administered alternately with mercury, or alone. Less confidence is awarded to it now, however, than formerly. Mercury probably stands next in the esteem of the profession. These remedies to have any good effect must of necessity be continued a long time. I need not speak of the caution with which mercury particularly should be administered. When injudiciously used it is an agent capable of great mischief. Its full sorbefacient effects may be obtained by very small doses: say the eighth of a grain of calomel twice a day until it begins to produce a perceptible effect upon the bowels or gums, and then wholly intermitted for as long a time as it had been administered; when it may be again resumed for the same effect. Continued for several months in this way it must have all the effect of which it is capable. During the administration of the mercurial thus, care must be taken to watch its effects upon nutrition, and if the patient becomes prostrated, tonics, nutritious diet, and even stimulants, should supplant it, at least for a time, if not permanently. Iodine does not often affect the patient very injuriously, so that we may give it in much larger doses and continue it longer with less apprehension. Yet occasionally it is not tolerated, and produces very disagreeable effects. The alkalies have had a solvent reputation with reference to these tumors. Liquor potassa is a favorite article, and many other preparations of potassa are also used, as is soda in various forms. During the course, whatever be the sorbefacient article used, great care should be taken to keep the general health to its standard at least. When it is remembered that these tumors disappear, so far as we can judge, spontaneously, and that they are interrupted in their growth by trifling and sometimes inappreciable mechanical causes, there will, I think, be room to doubt whether the above-mentioned medicines have any effect upon them. For my part, in a limited experience, I have not learned anything favorable to them in this respect, and place no reliance upon them whatever. Some of these medicines aid us in subduing the inflammatory complications which sometimes spring up and render the danger greater.

We are very much more successful in the radical treatment of these tumors by surgical means. Some of them are safely and certainly cured by surgery; others are sometimes amenable to the same agency, but their cure is attended with great danger; while there remain a large number that are wholly beyond the reach of the surgeon. I am not aware that any living surgeon advises the removal by surgery of an extra-uterine fibrous tumor, even when we can decide that the pedicle or point of attachment is small; although this operation has been performed. The operation would consist in cutting down through the abdominal walls until the tumor is exposed, and then, after tying the pedicle with a double ligature, cut it off, and then close the wound. The intramural variety is becoming, under certain conditions and circumstances, the subject of the surgeon's knife. What are the conditions which demand and justify an attempt at removal?

Since writing the above, Prof. H. R. Storer, of Boston, in the American Journal, for January, 1866, publishes an account of the removal of a large fibrous tumor with a part of the uterus by abdominal section, and by a very ingenious and learned argument, tries to convince the profession that this operation should be admitted as legitimate. I confess, however, that I am not convinced, nor can either follow his example or advise others to do so. Dr. Koebeale, of Strasburg, teaches the advocates the legitimacy of the operation.

Dr. J. Wood, Surgeon to the Commercial Hospital, Cincinnati, records in the January (1867) No. of the Lancet and Observer, a case in which he describes an operation by himself for the removal of a fibrous tumor. The tumor and uterus were both removed, and the patient recovered.

The surgeon should well consider all the circumstances of his patient before he decides what to do. The general conclusions to which he must be driven before he should decide to submit the fate of the woman to so dangerous a procedure, are, first, that the removal of the tumor is certainly practicable; and second, that without its removal the patient is in imminent danger of death, and that she can assuredly live but a short time if the tumor is allowed to remain. To come to the first conclusion, the growth should be low and within reach of the finger and instruments, with which to completely examine it. If it is in the fundus or upper part of the walls, and does not occupy the pelvis at all, the certainty of being able to remove it by the operations I shall describe cannot be entertained. We must be able by a proper examination precisely to define its relationship with the cavity of the uterus and with other viscera, as the bladder and rectum. Without this can be done, the danger will be great when it is *not* very large. The latter proposition cannot be true unless it is a hemorrhagic case, or the tumor has begun to cause destructive pressure upon some of the pelvic organs.

It is well known that there are cases in which hemorrhage proves fatal before the tumor attains a very great size, and that in other cases there never occurs any serious amount of hemorrhage. If an operation is determined upon as necessary, the tumor should be studied with reference to it. This may be done with the probe and finger. If the tumor is in the anterior wall, the probe ought to be introduced into the bladder, and the finger in the vagina; or while one probe is in the cavity of the uterus, another may be introduced into the bladder, and the finger into the rectum. Sometimes the tumor, when near the cervix, is developed downward, so as to deploy into the vagina at the top and intrude into it, pushing the wall of the uterus before it. When this is the case, it may be attacked at the most dependent part. In many other cases, the cervix is not at all developed, and the vagina is attached to it without being opened out. We cannot operate in such cases without dilating the os and cervix, so as to get an exposure through them. This may be done to almost any extent with compressed sponge. After this exposure is ample, we may attack the tumor through the cervix.

It is always desirable to have the patient in as good a state of health as possible, and perhaps this is all that is needed, but it is now becoming a point with some very capable surgeons to create a special diathesis as nearly as they can to insure plastic inflammation afterwards. They give the muriated tincture of iron and liberal diet for several days beforehand with this view. The operations now advised and practised are, 1st, enucleation (or extirpation); 2d, exciting destructive inflammation in the tumor; or 3d, a combination of these two.

For the performance of enucleation, the patient may be placed on the left side, with the hips elevated, before a large window. The vagina may be dilated with Sims's duck-bill speculum. When the tumor is well exposed,—if the upper part of the vagina is developed so as to allow the tumor with the posterior or anterior wall of the uterus to enter it,—we may make a crucial incision as large as we can without wounding any of the pelvic viscera, and deep enough to lay open the cyst or external envelope. Half an inch is generally deep enough. We may then withdraw the dilator, and introduce our finger so as to separate the tumor from the wall of its envelope as far as we can reach. This being done, the tumor should be seized by a vulsel forceps, and drawn down toward the vulva as far as possible with one hand, while with one or two fingers of the other we should extend the separation or enucleation of the tumor from the cyst or sac in which it is contained. Operating in this way, we may sometimes succeed in directly enucleating and removing the whole mass. We have an account of a very interesting case in which this operation succeeded by Dr. B. Fordyce Barker. Operations of this kind are not always so simple as this description would seem to indicate, and much must be left to the discretion and ingenuity of the operator in contriving the instruments and mode of performing them. When the tumor cannot be thus wholly enucleated, as much should be dragged down and removed as practicable, without too much damage to the uterus and danger to the surrounding organs; and then the wound stuffed with lint or other tent until inflammation is established in the mass of the tumor. This will probably not require more than twenty-four or thirty-six hours. The tent or lint must then be removed, and the uterus kept under the influence of ergot for several days. Sometimes the tumor inflames and sloughs or suppurates away, and disappears. This process may be accomplished in a few days, being much accelerated by the expulsive efforts awakened in the uterus by the ergot, or it may require weeks for its completion. In this last case the patient suffers much from the hectic, toxæmic, or inflammatory symptoms which arise ; and in not a few the exhaustion overcomes the powers of nature, and she succumbs ; or she is overwhelmed in the excessive and wide-spreading inflammation of neighboring organs.

Mr. I. Baker Brown, the learned and ingenious author of the Surgery of Women, operates a little differently. After dilating the vagina when necessary with elastic bags filled with air or hair, used for several days previous to operating, and exposing the tumor as extensively as possible by means of the duck-bill speculum or elevator, he draws the cervix down with vulsel forceps so as to get it completely in view. The preliminary operation, as he terms it, is then performed, which consists in making two or three incisions, so as to split up the cervix in as many places, completely to the envelope of the tumor. This operation may be best and most handily performed by a pair of long, curved scissors, with one blade thin enough to enter the cavity of the dilated cervix. In order to prevent the adhesion of the cut surfaces, and avoid dangerous or inconvenient hemorrhage, Mr. Brown directs that pledgets of lint, well oiled, be placed deep in the incisions, one upon another, until thoroughly filled, and to tampon the vagina with larger pieces. The tampon may be removed at the end of twenty-four hours, and the vagina well cleansed by injections of soap and water. The incisions, I think, may be kept open by passing the finger through them once a day, as well as by further using the lint tampon. This is what Mr. Brown calls his preparatory operation, and the essential parts consist, as may be seen, in removing the constricted condition of the vagina by gradual dilatation, and the cervix uteri by freely splitting it up. Nothing further is attempted until the results of this procedure are fully realized. Mr. Brown assures us that in many of the cases the dangerous hemorrhage is entirely cured, and the development of the tumor arrested. He records the results of fourteen cases occurring in the London Surgical Home. In ten the hemorrhage was cured ; in one it was relieved. In six cases the tumor had entirely disappeared or materially decreased. When none of these results are realized, or but partially, and further operative procedure is necessary or desirable, and after the cut edges of the incisions have cicatrized, Mr.

Brown pierces the tumor towards its centre, and by turning the knife takes out a piece of the tumor, much like coring an apple. He then plugs the vagina with oiled lint, to avoid alarming hemorrhage. The coring part of Mr. Brown's operation is not so easily performed with the neatness that his simile would lead us to suppose, but we can almost always break down a portion of the tumor within the capsule, and thus cause its dissolution and elimination.

Reasoning from the same basis of Mr. Brown, I have operated in two instances upon these fibrous tumors with success, in a manner, I think, more simple and easy of performance, and, consequently, more likely to be adopted by the general practitioner. The operation is done in the following manner : After having exposed the parts and isolated the tumor, by introducing a catheter into the bladder and a probe into the cavity of the uterus, if the tumor is in the anterior wall, or the finger in the rectum and the probe into the uterus, if the tumor is in the posterior wall, and thus exactly learning its relationship, we plunge a large-sized trocar into the centre of the tumor, withdraw the stylet, and then push up through the canula as large a dossil of cotton or lint, with thread tied to it, as will pass it. We hold the cotton in the wound by a probe through the canula until this last is withdrawn. The cotton is thus left in the central part of the tumor, while the string or thread, which ought to be large and strong, keeps up a communication with the vagina. Any blood effused passes readily out at this opening, and when the purulent or ichorous discharge of the tumor begins to be produced, it finds its way out along the string. This is really forming a fistulous opening into the tumor, after exciting inflammation in its centre. Should great inflammation result, we may withdraw the cotton by pulling upon the string, and thus remove one source of irritation. Should there not be inflammation enough to disorganize the growth, a second operation may be performed, or at the first operation we may puncture the tumor in more than one place, and thus have two points of irritation. It will be seen that the operation may be easily varied by saturating the cotton with some irritating or caustic liquor, by using compressed sponge instead of cotton or lint, &c. I subjoin one of my cases as illustrative of the operation, &c. :

Miss E——, aged 29 years, has been subject to severe hemorrhage at menstrual periods, and for two years they have been alarmingly copious. She has become very much debilitated and nervous. An examination revealed an intramural fibrous tumor occupying the anterior portion of the uterus, apparently about the size of a hen's egg. It was quite low, so as to encroach upon the vaginal portion of the cervix. After sounding the bladder and uterus together, and learning something of the thickness of the tumor, I pushed a small-sized trocar up an inch or more into the interior of the tumor, and through the canula the cotton, with the thread tied to it. I left the cotton in place, and withdrew the canula. In about four hours the patient became alarmed, and sent for me on account of some hemorrhage. It was very slight, however, and was not sufficient cause for interference, but to allay her anxiety I introduced some more cotton by means of a probe into the track of the trocar. Nothing further occurred worth noting until the fourth day, when the patient became slightly feverish, and complained of pain and heat in the pelvis. There were thirst, dry tongue, nausea, and heat of the surface. The pulse was about one hundred to the minute, and rather sharp. A moderate degree of fever kept up for ten days or two weeks, with varying degrees of intensity. About the sixth day after the cotton was introduced, a sero-purulent discharge made its appearance. It became more copious and grew offensive, and the tenth day some of the cotton passed away. The offensive discharge and feverishness, with copious perspiration and some diarrhœa, continued for four or five weeks. After this they both subsided. The treatment consisted almost wholly of tepid water injections in the vagina, with a little soap in the water, three times a day, the solution of the sul. quin. in some acidulated water, an occasional opiate at bedtime, and laxatives when needed for the constipated state of the bowels.

Mr. Brown's preliminary operation may or may not precede my operation. The two patients upon whom I have operated have both recovered their health since the operation. This operation is left in the hands of the profession for future trial and adoption or rejection.

Dr. Simpson, of Edinburgh, has, in at least one instance, effected the destruction of one of these tumors by cauterizing its interior with potassa fusa. While he guarded the vagina and other parts from the effects of the caustic, he held it in contact with the tumor until it was penetrated to some extent by the caustic. In a few days the tumor was found to be in a state of inflammation, which quickly passed into gangrene and suppuration, and the mass was expelled in shreds and half-broken-down pieces. The cure soon followed.

Of all these modes of operating, that of enucleation, when practicable, is most to be desired. It affords more promise of success on account of the less amount and of the shorter continuance of inflammation and fever. There are several things all-important to the safety of the patient, in the cases of suppuration and gangrene resulting from partial enucleation or cauterization. I think one of these important matters is cleanliness, by frequent and copious injections of tepid soapsuds, nourishing diet, and tonic treatment, this last merging into stimulants, as the depression becomes greater. But of course, the practitioner will be the best judge of the after-treatment, and will have obvious indications from the symptoms which may be present. I apprehend that as much if not more success in such operations will depend upon the judicious management of the patient during these exhausting processes of suppuration, gangrene, &c., than upon the skill with which the operation is performed. I have thought that all severe operations result more favorably in the country than in cities, and especially those in hospitals.

Next in difficulty of management and cure to the intramural variety of fibrous tumors, is that variety of fibrous polypus which grows from the fundus of the uterus and extends down into the cavity of the body. One indispensable preparation to any operation that may be considered desirable in these cases is the complete dilatation of the mouth of the womb. The dilatation should be so great as to admit a free use of any instrument we should desire to employ. This may be accomplished in two ways; by slitting it up with scissors or a knife, and allowing its edges to cicatrize before proceeding further, or by introducing prepared sponge. If we dilate with a sponge we must introduce a piece every day large enough to fill up the os. When the parts are sufficiently dilated we may proceed to operate. There are several modes of operating for these polypi. The object should be to remove the whole mass when practicable. If the tumor is small, this may be done by torsion, or evulsion, both of which are

safe modes; but if the neck is large and strong we cannot remove it in this way without very greatly endangering the integrity of the uterus. In that case, if we can get the chain of the
ecraseur around the neck we should remove it in that way. But,
as I have sometimes experienced, this cannot always be done.
When this is impossible, there are three other modes of proceeding. The first is to crush the growth as completely as may be
done with strong forceps; a large lithotomy forceps will do very
well for some cases, or we may have a forceps made for the purpose. When crushed entirely up to the point of attachment, inflammation and sloughing are soon followed by the expulsion of
the mass, either piecemeal or wholly, and the patient recovers.
Mr. I. Baker Brown recommends his operation of coring, that is,
gouging out as much of the centre and as high up as possible, and
then leaving the remainder to the powers of nature. The old-
fashioned mode of operating by ligature is the most effectual in
some specimens; but as it is the most tedious, and probably also
the most dangerous method, unless there are special reasons in its
favor, I prefer one of the above as being easier performed, more
expeditious, and as effective. We perform evulsion and torsion
with strong vulsel forceps, twisting and drawing alternately.

Torsion sometimes may be successfully applied to very large
polypoid growths within the uterus. I recently met with a case
where the polypus was attached high up the posterior wall of
the uterus, almost as high as the fundus. It was very large, filling up the vagina and pelvis quite completely. I attempted to
draw it out with vulsel forceps, thinking that I might thus be able
to bring the attachment under my sight, but this was found difficult, and I introduced my hand by the side of the tumor, until I
could measure the size and discover the position of the attachment.
The neck appeared to be an inch and a half in diameter and quite
short. With the hand thus introduced, I seized and rotated the
polypus upon its axis several times, until I could feel that the attachment was entirely overcome and the tumor free in the vagina.
In less than ten minutes from the time the hand was introduced,
I drew the tumor out of the vulva. It weighed 32 oz. avoirdupois
weight. There was not much hemorrhage,—perhaps not six ounces
of blood was lost. The uterus was quite large, reaching up to the
umbilicus, before the polypus was removed. In ten days the

patient went to her home in the country, and two months after appeared to be well in every respect. Now the uterus is but slightly larger than natural, and the hemorrhages, which were very exhausting, have entirely ceased. I was prepared for hemorrhage in this case, but was glad to observe that it did not occur.

The operation of ligating the neck of the tumor can be performed with a double canula, described by Gooch, and is so familiar to every student that I need not describe it. When any of the tedious processes are adopted, it will be best to wash out the vagina and uterus with a good syringe several times during the twenty-four hours, to prevent as nearly as possible—by dislodging them—the absorption of any of the noxious products of decomposition. We should be aware of this danger, and guard against it by every available means.

The fibrous polypi springing from the cervix, and occupying more or less completely the vagina, *should never be ligated*, crushed, or dealt with in any other than in a summary way. They should be immediately and completely removed. The preferable plan of removal, when small, is to seize them with a vulsel forceps, and twist them off. This mode, by torsion, almost always proves entirely successful when the polypus is not large. When the neck is so large that it cannot be twisted off, we should draw the tumor down with the forceps, until the attachment appears at the vulva, when the chain of the ecraseur should be thrown around it, and the connection severed with this instrument. This operation is simple, safe, and practicable. Nobody need fail in it, or fear any subsequent bad effects. But another equally simple and practicable, and almost as safe a method, is to divide the neck of the polypus close to the attachment with the scissors or knife. To do this handily, the polypus should be drawn down as before.

In about two and a half per cent., hemorrhage is troublesome after cutting off with the knife, but it may be easily controlled by the tampon; or what is the pleasanter way, to expose the part with a speculum, and apply the persulphate of iron to it. Sometimes we find instances where a polypus has an attachment to the cervix by an ordinary neck, and on account of inflammation, contracts adhesions to the vaginal walls besides. These adhesions, so far as I am aware, are not very firm, and judging from a limited experience, may be overcome by the finger.

Whether this be the case or not, the neck of the growth may be exposed by traction, and severed as before, and these adhesions be overcome by the finger, the handle or blade of the scalpel, as may seem necessary.

The soft variety of the polypus may be destroyed by torsion and evulsion always, so far as I am aware. The operation consists in seizing it with flat-bladed forceps and twisting it off. I have, on some occasions, taken it off with curved scissors, without any bad results, and with perfect success, so far as the ultimate cure was concerned.

The removal of the fungous polypus is a more serious and difficult operation. It is also much more uncertain in its results, in two respects. It is often reproduced in spite of all the care we can summon, and again produces malignant degeneration in the uterus and surrounding organs. In removing the other forms of growth, we are pretty sure of success if we remove nearly all the tumor, a small part of the neck disappearing spontaneously when left to itself.

In the fungous or epitheliomatous growth, on the contrary, it is not only necessary to remove all the fungus, but to remove the tissue whence it springs. We should be so liberal in its removal and the seat of growth, as to include all the cervix that may be amputated without damage to other organs. There are two ways of doing this. In the first place, the cervix should be drawn down by means of strong hooked forceps, so as to show at the vulva as much as we desire to remove. After this exposure, we may surround the cervix with the ecraseur, or we may fairly cut through the sound tissue entirely above the disease.

This is a serious and not always safe operation in its general results, although simple, and may be easily performed. It is apt to be followed by inflammation of the uterus and peritoneum, and thereby endanger the life of the patient. Any bleeding that may occur, may be stopped, peradventure, with the persulphate of iron and the tampon.

CHAPTER XXVIII.

OVARIAN TUMORS.

OVARIAN tumors spring from and are for the most part formed of the hypertrophied tissues of the ovaries. There are tumors very much resembling and often mistaken for them, however, developed in the lateral ligaments. These latter are generally distended and hypertrophied sacs of the parovarium, and contain thin serum, merely; are cured sometimes by tapping, or they spontaneously burst, or are broken by accident in the peritoneal cavity, and disappear, never again to annoy the patient. The Fallopian tubes sometimes are the seat of enlargement. The tubal canal becomes obliterated at each end, and fills up with the hilus usually appropriated to the lubrication of its inner surface, hypertrophies in tissue, and thus constitute a morbid development. Although rare, these two forms of tumors are observed. Doubtless, other enlargements, of a nature not yet properly understood, are sometimes originated by unknown causes. The anatomical distinctions can be made in most instances by careful dissection, so that in cadaverous investigation they need not be confounded with the ovarian tumors, except it be where all tissual distinctions are obliterated by an intercurrent or supernatant disease. It is not my purpose now to pursue this subject further than the mere mention here made. In the proper ovarian tumors, we may trace three coats or layers of tissue forming their walls. The external is the serous or peritoneal. It is shining and smooth as this membrane is elsewhere, and seldom changed in any way, except it may be thickened and hypertrophied. It can be traced into the peritoneal covering of the viscera and abdominal parietes, and consequently needs no elaborate description. The internal coat or lining membrane, is doubtless the membrana granulosa of the ovisac, very much hypertrophied. When small, something like epithelium seems to be its entire composition. As they grow and develop, the epithelial arrangement is less perfect,

26

until, when very large, we can observe it only in patches. In many cases when thus large, this membrane has a smooth, lustrous appearance, but in others, it is more or less thickly studded with granular projections, varying from almost imperceptible minuteness, to the size of peas, or even larger. Regarding the main sac as an hypertrophied ovisac, I think these little granular sacs (for they prove to be sacs upon examination), are also of the same nature, and are the origin of the numerous endogenous or supplementary growths which constitute one of the polycystic varieties.

The middle coat is made up from the stroma of the ovary. Its strength depends upon quite a considerable amount of fibres, which enter into its composition. As the tumor develops, these fibres are enlarged, and apparently, if not really, increase in numbers, until they constitute the most of the thickness of the walls, and in some parts make quite a thickness, density, and toughness of tissue. These qualities are greater in old large sacs than in the smaller and younger ones. At the pedicle and for some distance up the sides, they are greater than in other portions, being in these parts sometimes a quarter of an inch thick, while at the fundus or distal portion, they may be thin and fragile. The whole of this coat may be very tough and thick, so as to resist great force, or it may be thin throughout, so as to be easily ruptured at almost any point. Entangled in the meshes of these fibres may be discovered, in many cases, the minute microscope points so numerously scattered through the substance of the ovaria. These points are believed to be the origin of the germinal spot in the ovum by some physiologists, and around which are developed the ovum and progressively the whole ovisacs and their contents; and I believe that their presence in the walls of the tumors, over much if not the whole of their extent, accounts for the development of the minute granular internal projections above described. In a tumor recently removed from the body, by holding it up to the light we may not unfrequently discover the peculiar buffy tinge seen in the stroma. The vessels are situated in this coat. They are numerous and some of them large, so large that great care is necessary to prevent them from bleeding when the peduncle is divided. They are developed, it is hardly necessary to say, to this great size, from the minute twigs which penetrate the substance of the ovary.

The shape of ovarian tumors may vary much. They may be regularly globular, polyglobular, angular, or irregular in almost every way. When small, the ovary may be seen as constituting a considerable portion of the tumor. When large, the ovary may be almost lost in the walls, or observed as a mere tubercle sticking to or imbedded in its side. Yet generally but one ovary is the seat of disease, but in a few instances both are affected. Not often do they become consecutively the nidus of these growths; one being first the subject of disease, and then followed by the other. And when such is the case, we are not warranted in supposing the one to be the cause of the other, either remotely or directly. Notwithstanding the above assertion, I do not wish to be understood to say, that sympathetic degeneration between these two bodies is impossible. These tumors divide themselves anatomically into monocystic and polycystic,—the one having a single cystic cavity, the other several. The polycystic variety is formed by the development of several cysts adjoining or by the side of each other, and independently attached to or springing from each other on the external surface, or within the cavity of one large one. The instances of polycysts growing by the side of each other, and being independently attached, resembles at first the single. At an early stage of development they may stand free of contact one with the other, but as they grow in size, in consequence of the small surface of the ovary to which they are attached, they crowd together, so that it is not always easy to say whether they were not developed from each other. The cysts from which smaller ones grow, are called proliferous. They are doubtless single for some time in their early development, but carrying up, as they increase in size, the proper substance of the ovary, with its rudimentary ovisacs, after a while the inner or outer surface is bulged by the maturity of these last, which, if they do not dehisce and allow the escape of the ovum, grow into a subordinate tumor. This process is separate, until there is a glomeration of cysts to quite a number, from four to fifty, of various sizes,—from the size of a man's head, down to that of a pin's head. Small ones may be so numerous as to stud a large part of the inner surface with granulated elevations. This is the most frequent variety met with in practice.

There is a great difference in the sensible qualities of the con-

tents of the cysts in different cases, and of the different cysts in
the same case. In some, it is very thin, in others, very thick
and tenacious, while the color shades from black, inky, to limpid
clearness. The monocystic, as a general thing, affords thinner,
clearer fluid than the polycystic, though this is not invariably so.
It is in the monocystic variety, however, that we generally find
the solid contents. These solid contents are, for the most part,
formed of tegumentary, adipose, hairy, osseous, and dental tissues.
All these are irregularly developed and distributed, so that no
semblance in shape or other conditions can be discovered to an
independent being,—as a fœtus. Sometimes the tegumentary
substance is small and gives attachment to a few hairs; sometimes
the quantity of hair is long and entirely isolated, and either in-
extricably knotted together, or straight and few in number. They
may be long—ten or twelve inches—or very short. Irregular
bones, or the enamel of teeth, without the bony structures, are
often attached to the side of the cavity; but no organic order,
symmetry, or completeness, has yet been seen. In fact, should a
femur, a scapula, a complete maxilla, or other complete organ, be
developed, the tumor would range itself under the head of extra-
uterine fœtation, and not ovarian disease proper. The tumor
containing solid materials of this kind is usually small. Not unfrequently large fibroid growths are observed in the ovary at the
base of a single or multiple cystic tumor. These solid fibroid or
fibrous growths may be simple or benign in their nature, or malignant. This complication of ovarian dropsy I think more frequent in persons advanced in years—over forty—than younger
ones. The contained fluid of the polycystic tumor is ordinarily
highly albuminous, of high specific gravity, tenacious, and more
or less colored. The fluid is so thick sometimes, as not to flow
through a small canula. Blood and pus are the coloring matters
of this fluid ordinarily. From one tumor of several cysts, I drew
pus from one cyst; dark coffee-grounds sanguineo-serous fluid from
another; a beautiful straw color from another; and lastly, from another, fluid of a delicate azure tint. After tapping, more or less
alteration is observed in the fluid, each operation withdrawing
fluid affected by chemical or pathological circumstances. In the
former, putridity or acridity would result; in the latter, the purulent productions of inflammation might be expected, or fibrous

concretions, or serum, changing the tenacity and thickness of the fluid.

There are some chemical and microscopic resemblances in the fluid from almost all varieties of ovarian tumor. Albumen seems to be almost always present. In some specimens of fluid, strong acids, or heat, causes it to assume a solid form, coagulating and adhering like the white of an egg when cooked in boiling water; in others a small precipitate is all that is observed. Between these extremes all shades of difference exist, but undoubtedly, nearly all ovarian tumors yield highly albuminous fluid. The reaction is alkaline. Mr. Nunn says that, " As the results of many examinations (microscopic) of different specimens of ovarian fluid, the most constant characteristic of such fluid is its containing, in greater or less abundance, cells gorged with granules; and, in addition, circumambient granules, having the same measurement, encompassed by the cell. The size of the gorged cells and included granules varies greatly, even in fluid from different cysts in the same ovary." This description of fluid could, with certainty, remain good of the first evacuation only, as pus- and blood-globules are not unfrequently found in subsequent evacuations.

I have already stated what I believe to be the nature of the sacs or cysts of ovarian tumors, and it will not be necessary to repeat it; but there is one curious question, as to the origin of the solid substances sometimes found in them, which so very much resemble the tissues of a fœtus, upon which I feel at liberty to add my conjectures to those of many others who have written upon ovarian disease. I do not desire to argue elaborately in favor of the opinion I embrace, or against those who adopt a different one, nor to give instances to any extent in proof for or against. I do not think that the discovery of hair, fat, teguments, bone, or teeth in the ovary, in the imperfect state of development and irregular relationship above described, affords any evidence of impregnation or sexual connection. In explanation of the formation of these tissues, it may be conjectured that the ovaria have such independent cell capacities as will enable them not to form a living being independent of sexual influence, but under certain circumstances to imperfectly produce tissues of a very elementary nature, resembling those found in the products of impregnation. The tissues of ovarian disease are of the lower grade in a formative sense. We do not find muscle, nerve, or vessel in these places and under these circumstances. There does not seem to be generic energy enough in the ovum, unaided by seminal impression, to give the *direction* to cell action for the formation of these more complicated tissues. We should observe, too, that although the cell action of the ovum may start up the formation of these singular tissues, there is no generic order or completeness in that action. In all this we see another of nature's distinct modes of formation under peculiar circumstances. I am not aware that in any other part of the body such a condition of development ever takes place inside of a cyst,— a development which gives origin to so irregular an assemblage of tissues without the formation of any organ out of them. When fœtal tissues are found elsewhere in the body, or in any part of the male, there is something of organic completeness about them, and perfect bone, tooth, or some or many organs may be distinctly traced. Nerve-matter, muscle, &c., and, in fact, all the more complicated tissues are met with. These facts, I think, establish a broad difference between the products of conception and the irregular cell or tissual development of the ovum, and I cannot see why the theory I have mentioned may not explain all the circumstances fully.

I have not much doubt that the state of the ovaries has much to do in starting up these tumors, and it is most likely that inflammatory thickening or condensation of the peritoneal or fibrous envelopes of this organ is the condition. If we imagine the fibrous indusium to be thickened and condensed, so as to be less fragile or less amenable to the influence of the absorbents, and failing to yield to the distension of the maturing organ so confine it that the follicular fluid cannot escape, we have incipient ovarian dropsy.

It is only necessary that the unruptured coats of the ovisac should continue to secrete its fluid, and the strong indusium refuse to crack open and allow its escape, to have all the conditions of an ovarian dropsy, that will be limited only by circumstances foreign to those of their origin. However this may be, there can be no question but that the beginning is the failure to discharge the fluid of the ovisac, and that this last is, in the early stages, the fluid of the ovarian tumor. There is, also, a kind of independent pathology of ovarian tumors. After they are developed to a certain extent, they become subject to accidents and disease, and play an impor-

tant part, in consequence of this fact, in the sanitary conditions
of patients in whom they exist. Inflammation, for instance, attacks them, and causes ulceration in their walls so as even to perforate them, making a communication between the cavities of contiguous cysts, or with the peritoneal cavity. Without perforating
the walls of the tumor, the ulceration may produce a good deal of
pus, which is mingled with the other contents of the cyst in which
it occurs. General inflammation of its walls may proceed to a
fatally exhaustive extent, or spread to the peritoneum, and thus
indirectly cause death. Gangrene may also result, which may be
confined to the cavity of some of the cysts, and induce a putrid,
offensive state of the contents, or perforate the dividing partitions,
and thus make a communication between cysts, or open them into
the peritoneal cavity. The walls may also rupture from distension
in consequence of their becoming attenuated, or by a violent stroke
or fall, or other shock, the tumor may be ruptured, and the contents escape into the peritoneal cavity. By means of ulcerative
communication with the Fallopian tubes, evacuation of the fluid
occurs. Adhesion to the walls from inflammation and ulceration
through the parts thus agglomerated sometimes results, and the
fluid so discharged. Inflammation also causes adhesion at various
parts. The fibrine effused glues it to the surrounding parts,—the
abdominal walls, the intestinal canal, bladder, and other viscera.
Slight inflammation is supposed to increase the effusion in their
cavities, and cause them to grow very rapidly. Inflammation,
also, sometimes, no doubt, causes obliteration of the cavity from
adhesion of the walls. This is more frequently the case when it
results from external causes, as blows, tapping, pressure, injection, &c. Now, it hardly ever happens that these diseased conditions and accidents of the tumors fail to produce their effects upon
the health of the patient. No doubt but that death occurs from
extensive disease in the sac, without any organ being involved in
the trouble directly. A large production of pus would exhaust the
patient; gangrene, to a large extent, would cause death, as extensive gangrene of unimportant organs generally does. But an extension of disease to the peritoneum and surrounding viscera, or
by the effusion of the acrid contents of a diseased cyst, is more
likely to be the mode of progress to constitutional disturbances inaugurated by inflammation in the tumors. When the tumor bursts,

and its contents are effused into the peritoneal cavity, the peritoneum seldom escapes without inflammation ; but the degree will depend upon the nature of its contents. If they are not vitiated, but consist of the bland albuminous fluid found there ordinarily, it is very slight indeed, and lasts for a very short time only. But should pus, or the ichor of decomposition, be mingled with it, we should be prepared to expect serious if not fatal results.

I once had an opportunity of observing the progress of a case for several months, where this rupture and effusion were frequently repeated. About every three weeks the woman would attain to a large size, and a well-defined, large cyst could be felt filling up the whole abdomen and distending it greatly ; when suddenly, without premonition or apparent cause, the cyst would give way, the swelling would become more diffuse, fluctuation very evident, and the cyst could be no longer defined by the touch ; slight fever and some tenderness of the abdomen would last for two or three days, when copious perspiration and diuresis would evacuate the fluid in a few days more. After this process was completed, the abdomen would be lank, and a small cyst could be felt rising up from the left ilium ; it would increase and burst at the end of three weeks, as the other had done before. I saw the patient frequently while this process was repeated six or seven times, when, as she would not submit to the operative procedure which I insisted upon, I was dismissed, and an irregular practitioner, who was sure he could cure her, installed in my place. Not long (perhaps three months) after I was discharged she died from the inflammation resulting from one of these effusions, probably because the contents of the cyst had become vitiated by inflammation within its cavity.

But these growths may produce a pathological condition of the system, without becoming themselves the seat of disease, by the great size they may attain, mechanically interfering with the functions of the pelvic and abdominal viscera. Before rising out of the pelvis it may displace the uterus, and cause inconvenience from this effect ; it may press upon and obstruct the rectum, bladder, and urethra, or upon the iliac veins, causing obstruction to the flow of blood, and varicose veins in the legs, phlebitis or phlegmasia dolens ; or, pressing upon the nerves, cause neuralgic pains in the limbs, hips, &c. It is plain that such pathological effects, when induced, would be serious, in proportion with the greater or less impaction in the pelvis by its continued growth. Ordinarily, these inconveniences do not prove very embarrassing to the functions of the important vital organs, but sometimes the case is far otherwise, and life is very much shortened and health rendered miserable. As it rises into the abdomen, these mechanical troubles are apt to be lessened; and as the room is comparatively so great in that cavity, quite a while elapses before any great disturbance results from mechanical pressure. After awhile, however, the abdominal muscles are distended beyond convenient size, and the tumor is strongly pressed among the viscera. The kidneys, liver, stomach, intestinal tube, in fact, all the abdominal organs may become the subject of great and even fatal pressure. In many instances, however, enormous size is attained before fatal damage results. One hundred and fifty pints of fluid have been taken at a single tapping. A much less amount, in most cases, would produce very grave results by pressure. When the growth is rapid, its mechanical effects will be more distressing; and, on the contrary, the organs accommodate themselves to a great deal more pressure if gradually brought about.

Besides the inflammatory changes that take place in the tumor, chronic degeneration is occasionally observed. Deposits of earthy substances in the walls, bony spiculæ, &c., are the most frequent. Small tumors, containing solid material, are more commonly thus affected.

The modes of termination are worthy of some consideration. Many cases last through a great many years without materially influencing the general health, and up to the death of the patient, at an advanced age, prove to be nothing more than an inconvenient burden when large, and when small not the cause of even this kind of trouble. But cases of this class are not very numerous, and by a large majority they terminate, and leave the patient in the enjoyment of real or comparatively good health, or by their effect upon the constitution shorten her existence. Spontaneously favorable terminations are so rare that we can base no calculation upon them; but were it possible, it would be interesting to follow nature through her resources in this respect; and I am sorry that my means for reference are so limited as to prevent me from thoroughly examining this branch of the subject. Perhaps rupture of the sac into the peritoneal cavity, collapse and adhesion of its walls, is the most common and favorable spontaneous termination. After the rupture, in cases where cure follows, it is probable that the opening in the sac continues, and that a permanent fistula, so to speak, from the cyst into the peritoneum, places the fluid in contact with a more active absorbing surface, until, by the elasticity of its walls, it contracts to annihilation, or, at the first shock of the rupture, inflammation is originated that causes an obliteration of the cavity of the sac. Dr. Simpson speaks of instances of evacuation through the vagina. The same thing might occur in connection with the bladder or alimentary canal. I have already spoken of adhesion to and rupture through the walls of the abdomen, and consequent recovery. Inflammation in its proper tissues, no doubt, sometimes arrests the development of and obliterates the tumor without materially affecting the patient's general health. It is not improbable that other circumstances with which we are not acquainted may likewise operate to cause the arrest and cure of them, inasmuch as they unquestionably do sometimes disappear in an unaccountable manner.

The local pressure interfering with the functions of the bladder and rectum may induce complicating diseases that lead to death, and consequently cause death before the tumor is very largely developed. Inflammation will spread upon these organs to their more vital connections and relative organs; or, by interfering with excretion from the bowels or bladder, produce disease of the blood, and thus gradually undermine the health of the patient.

After the tumor has arrived into and greatly distended the abdominal cavity, pressure upon the viscera will sometimes produce disastrous terminations. The stomach is crowded into a very small space, and food can be taken but sparingly, and is often rejected before digestion is completed. The vascular supply of this organ is cramped, and its secretions vitiated and embarrassed, and in this way digestion is interfered with, the appetite destroyed, and loathing of food takes its place.

Pressure upon the vena porta embarrasses the secretion of the liver. Pressure upon the ductus choledochus, gall-bladder, and duodenum, stops the excretion of bile; it is dammed back upon the gland, absorbed, and thrown into the blood to poison the nervous centres.

There is no doubt, also, that the general compression of the organs, by pressure upon the chyle absorbents, prevents that fluid from passing as freely as usual into the blood, and thus by degrees starves the patient. But probably no more disastrous effects of the pressure of the tumor in the abdomen is noticed, than such as is produced through the kidneys. Pressure upon the emulgent veins causes congestion of the kidney tissues, retention of urea, and other matters that should be excreted, and drains off the albumen with the urine, until the blood becomes thinned enough to infiltrate into the cellular tissue, generally in the form of œdema of the extremities, or into the peritoneal cavity, giving rise to ascites. But this is not the worst mischief, perhaps, caused by the pressure on the kidneys. The poisoning of the blood with urea, and its effect on the nerves and vital organs, is too well known to require more than mere mention to suggest the rapidly fatal tendencies which result from it.

Inflammation in any of the important abdominal organs may be caused by the pressure, which will terminate fatally in a greater or less time, owing to its acuteness or slowness of progress. It will be seen by the above very short description that ovarian disease usually terminates by inducing a long train of distressing constitutional symptoms. They are not uniform in different cases, some persons suffering from one mode of complication and some from another; but nearly all are pretty sure to experience those terrible sufferings which are connected with secondary disturbances in the vital organs.

The presence of the tumor, when not large enough to press upon the organs sufficiently to do very much damage, sometimes leads to copious dropsical effusion in the peritoneal cavity. This is, at least sometimes, the result of an influence exerted upon the peritoneum, causing it to secrete more than an ordinary amount of serum.

One case upon which I operated and evacuated a large amount of serum from the peritoneal sac, recovered completely from the operation, but died about two months after from extreme abdominal distension, in spite of alteratives and diuretics.

Causes.—It is extremely doubtful whether there is anything in the general condition of the patients that predisposes to the development of ovarian tumors. There is quite a disposition, however, with certain authors, as will be apparent to any careful

reader, to trace most chronic enlargements to scrofulous taint in
the system; and these gentlemen express the belief that scrofula
predisposes to ovarian disease. The term of life during which
menstrual influences are exerted upon the general organization, is
very much more frequently the time when it is most active; so
that we may very safely conclude that in the function of menstruation we have a predisposing cause of ovarian disease. It is true
that ovarian tumors have been found in the ovaria of infants and
fœti, and originating in very aged females; but this probably is as
rare an exception to the general rule, that they occur during menstrual life, as the occurrence of menstruation in infancy and old
age is to the general rule that this takes place at the usual time.
And probably, too, a careful examination of the anatomy of these
exceptional cases will prove them to be as exceptional in nature
to the ordinary ovarian tumor, as the time in which they occur
would seem to indicate. Some circumstances connected with menstrual life appear also to increase the predisposition. Sixty-one
per cent., according to Dr. West, of the patients were married,
while only twenty-nine had never been married. After making
allowances for the greater proportion of women at twenty-five
who are married, I think we may fairly infer that marriage adds
somewhat to the chances of the occurrence of ovarian dropsy.

That patients who are the subjects of this disease should be less
likely to have children than those in whom ovulation is more perfect and complete, will not, I think, justify us in setting down sterility as the cause of it in any way, but it is more probably connected as an effect. During menstrual life, the most obnoxious
time is between the ages of twenty-five and forty, the time when
the sexual functions are exercised with more activity than any
other; so that it would be right to say that menstruation and
marriage both predispose to ovarian disease.

Unhealthy menstruation seems to be more commonly coincident
with it than healthy. Abortions and premature labor are so,
likewise.

We should attach sufficient importance to the fact that it occurs
in unmarried persons as often as twenty-nine per cent. This induces Dr. West to remark, that " it occurs in the unmarried oftener than any other organic disease of the sexual organs."

The exciting or proximate causes are such as excite the ovaria and induce abortive efforts at ovulation. What does so, we are not able to say with certainty; but we can say what might and consequently, probably does, produce this effect.

Inflammation of a low grade and somewhat chronic duration, might cause induration or thickening of the indusium, so that it would not yield to the upheaving pressure of the ovisac, so as to burst or become absorbed and permit dehiscence, or the peripheral portion of the proper substance of the ovary might be condensed by it, and thus rendered unfit for a covering for the tumor.

The probabilities, I think, are quite in favor of this mode of initiating these morbid growths, or merging a healthy into an unhealthy accumulation. When once thus commenced, the stimulus of increased incretion of fluid would carry on a kind of hypertrophy in the involucra that would permit of a further enlargement. Now the local circumstances that are regarded as the causes of the disease, would favor the occurrence of inflammation, and are very frequently attended with some of the symptoms of it. The local condition of the ovary and uterus, during each menstrual period, is often attended with pain in the ovarian region of just such a character as we would expect to indicate inflammation. This ovarian pain is present in other excited conditions of the sexual organs also, thus showing that they are often the focus of painful vascular turgescence, if not inflammation. How easy it is to conceive the possibility, nay the probability, of nutritional vitiation to an extent sufficient to prevent dehiscence of the ovum after it is matured in the ovisac. If this state of things may take place in consequence of menstrual congestion, or congestions from other transient or mild causes, how much more likely is it to be induced by the inflammations which succeed to abortions or parturitions. While inflammation is probably the cause of the beginning of the development of ovarian tumors, it does not seem necessary to their continued development, as the accumulation of fluid in a shut cavity with a secreting internal surface is a matter of course, and the limit of its amount, for the most part, does not depend upon anything but the capacity of the involucra to grow, which they usually do until interrupted by external circumstances.

Although inflammation may, in most cases, be the cause of the toughness of the covering to the ovary, which prevents the escape of the ovum, this condition may result from some other local cir-

cumstances. Congenital formation may be such as to give rise to
this difficulty,—possibly, may allow development of the ovisac;
may allow the involucra to increase as fast as the demand for more
room becomes necessary.

Prognosis.—Our knowledge with regard to the prognosis is unfortunately too definite. There is no need of much conjecture
with reference to this matter : the termination is too frequently
demonstrated. In arriving at prognosis with reference to any
disease, we ought to consider whether its ordinary course is, after
a time, to a termination in health, as is the case with many diseases, or, there being no such favorable tendency, what are the
probabilities of a cure. Unfortunately, there is no tendency to
spontaneous recovery, worth taking into consideration, in ovarian
dropsy ; probably not two per cent. but would, after a longer or
shorter time, terminate in the death of the patient. While this is
the case, it does not properly represent the value of a life threatened by this affection. Some patients live a great many years in
comparative comfort; but, by large odds, the case is generally
very different,—only a few years being sufficient to finish the
course in a downward direction. The average duration of life is
probably about three years from the time it is first perceived. So
much difference of opinion exists among observers, and so unsatisfactory are the statistics, that the average number of recoveries
under treatment is quite doubtful. The mortality, taken altogether,
I should say, is not far from fifty per cent.,—a fearful fatality ;
and the average duration of life under treatment, and in spontaneous cures, will not exceed very much, if at all, seven years. Yet,
in *selected* cases, probably two-thirds of the patients may enjoy
perfect health. Much of the mortality, in my estimation, depends
upon the injudicious selection of cases for the different modes of
treatment which have of late years proved successful. However
discouraging the prognosis in general may appear in this kind
of examination, we are not justified in extinguishing the hope
of our patient, by applying it to her case, without particular investigation into the peculiarities of it. In fact, individual prognosis is too often left out of view, and general results are applied
to individual cases. In certain instances, either too favorable or
unfavorable an opinion is expressed, to the undue elation of our
patient, or the perfect annihilation of all her hopes. All instances

of the most unfavorable diseases are not to be unconditionally consigned to death ; nor every case of the more favorable classes of disease expected to terminate in health, without question.

We should, therefore, carefully examine every individual case, with reference to its own peculiarities, its nature, and the character and condition of the patient. Is the disease simple, or a compound of cyst and solid, polycystic or monocystic ? The monocystic is very much more favorable for treatment, and terminates in spontaneous recovery oftener than the polycystic. The duration of life is greater, also, in the monocystic. The cases in which there is no great amount of solid matter, are more favorable for treatment than those in which the solid is predominant ; but ordinarily, the more of solid material the less rapid the growth, and, consequently, the longer the duration of life in the absence of interference. How long has the tumor been in attaining the present size ? If several years have elapsed since the patient was aware of its presence, it will probably continue to increase slowly, unless, as is sometimes the case, more activity has lately been observed, and a tumor that had formerly grown very slowly, and required a number of years to acquire half its size, has grown the rest in a few months. In this last, there is every probability of a rapidly fatal course. Again, if the patient has not known any increase of size until within a few months past, and yet is quite large, the prognosis is bad. Our prognosis is influenced by age to a considerable extent; occurring in young persons, it is more likely to advance rapidly, than in old ones. A woman at forty is not apt to develop an ovarian dropsy so rapidly as one at from sixteen to twenty.

Ovarian dropsy will advance less rapidly after menstruation ceases than before, and the earlier in menstrual life, the more rapidly it will advance. The prognosis, as a general thing, therefore, is worse in the young than the old. If we should decide the question how long will she live by age, we should speak more favorably to the woman advanced in age. The fact, too, of the ordinary chances for the length of life being in favor of the young, she is more likely to fall a victim to the advance of the disease than the older. We must take into consideration the circumstances that complicate it in the last stages, when we examine particular instances.

The inflammation, the pressure upon the rectum, bladder, stomach, bowels, and, above all, the kidneys, the nervous system, the vascular system, nutrition, as shown by the signs of emaciation or otherwise, should all be carefully scrutinized.

Diagnosis.—The diagnosis of ovarian tumors, when tolerably large, and not complicated with more than ordinarily embarrassing circumstances, is not difficult; but instances do occur where the . matter is far otherwise, and a positive opinion cannot, with propriety, be given. For the most part, the diagnosis is sought to be perfected with a view to treatment, and we ought not *always* to be too solicitous about exactitude in this respect, until the urgency of the case demands it *for the sake of treatment.* When it is desirable to inquire particularly into the diagnosis, as a general thing we wish to know, 1st. Whether the case under examination *is* an ovarian tumor; 2d. What sort it is,—solid, fluid, or mixed; 3d. If fluid, whether simple or multiple,—multilocular or unilocular, polycystic or monocystic; 4th. Whether adherent or not. All these questions are very important, so far as our attempts to cure the case are concerned, and without an intelligent solution of them, our treatment will be experimental. Almost all our diagnostic means are physical; the history and rational symptoms, although valuable, are much less so than personal inspection with various physical appliances. The utmost freedom of examination, therefore, should be insisted upon, before our patient can, of right, demand our opinion. This will neither be asked nor given until the case has assumed something of gravity and importance. I have known medical men to give an opinion on imperfect examinations, that they were afterwards under the necessity of reversing; and, *en passant,* I think, as a general thing, more errors of diagnosis in female diseases are committed by the imperfect use of our now highly improved measures of exploration, than originate in any other way.

The history will afford us in many cases, however, very valuable aid in arriving at correct conclusions. It is now pretty well determined that the average duration is about three years. In this time, it will spontaneously produce fatal effects, by great size and extreme distension, and the resulting damage. This is longer than pregnancy lasts, and a shorter time than is required for solid fibrous growths to reach the same results. The age at which they are most likely to occur is an average of twenty-six years, according to Mr. Brown,—although they may occur at any time during the active condition of the sexual functions, while the ovaria are subject to menstrual congestions and their effects. Quite a large number of cases make their first appearance in early menstrual life. I knew one in which the beginning of the tumor must have been simultaneous with, if not antecedent to, the commencement of the function of menstruation. Fibrous growths of the uterus are not likely to begin so soon. Their increase after being first observed is comparatively rapid, more so in the young than those somewhat advanced in age. They are not usually attended with pain in their own proper substance; this is not always true, for the congestion and hyper-excitement may be attended with pain and soreness. The functional disturbance, in their early stages, occurs in the pelvic viscera; first, on account of pressure, such as tenesmus, dysuria, dragging, or weight in the pelvis; and secondly, imperfect menstruation. Sometimes the menses are suppressed, scanty, and painful, but often no deviation is observed. The main thing in the history of the case, in this respect, is to remember that the symptoms point in the beginning to trouble in the pelvis. It is generally, or at least sometimes, stated that the tumor rises from one iliac region and continues to occupy one side for some time. This, I think, is the exception to the rule, that they are at first central; and, by Dr. Frederick Bird, this is considered an evidence of adhesion. They probably, when large enough to overcome the support of their peritoneal envelope, fall into the cul-de-sac of Douglas, behind the uterus; and then, as they grow, come up in front of the promontory of the sacrum, until they are large enough to be felt above the pubis, having their point of support in the hollow of the sacrum, instead of one of the iliac fossæ. The patient will usually speak of it as a lump, instead of saying that she is swollen, as in pregnancy. The abdomen does not enlarge generally, according to her statements, but by a tumor rising in it. She has watched it coming up out of the pelvis, and not starting from above or from one side, and encroaching upon the abdomen from either of those directions. The knowledge derived by physical examination, as already stated, is the most valuable; and while the modes of procedure are the same, and applicable to all stages of growth and

27

enlargements of the tumor, we will be able better to describe and understand them, as made use of for one that has arisen from the pelvis and pretty thoroughly filled the abdominal cavity,—a tumor that has become obvious, and from which our patient is solicitous of being relieved. The means afforded us for physical examination are—1st, palpation; 2d, percussion; 3d, auscultation; 4th, vaginal and rectal digital examination; 5th, examination with the sound or uterine probe. These may be used separately, or combined in any given case, some being more valuable in some cases, and others in different ones. Exploring needles, chemical tests, and the microscope, may also be used to great advantage. Palpation is of very little use while the tumor is still in the pelvis, and can only be of avail in conjunction with the vaginal touch or the uterine probe; as it rises in the abdomen, however, this process of examination comes into use independently. In this condition, we can examine the consistence, size, shape and mobility of the growth, and form some opinion as to its adhesion to the walls of the abdomen, and its primary attachments. In conjunction with some other processes of examination, it is more useful at this stage and size.

In the ordinary condition of the contents of the abdomen, the intestines lie in contact with the anterior and lateral walls, except in the right and left hypochondria, where the liver, over a considerable space, and the spleen, a smaller, displace them. In consequence of this state of things, the resonance caused by the gas in the alimentary tube extends all over the anterior and lateral walls, save the above exceptions. Dulness upon percussion, therefore, indicates the presence of a tumor. The mesenteric attachments between the posterior wall of the abdomen and intestinal tube prevent them from being separated to any considerable extent, hence tumors occupying much space are apt to displace and get anterior to the latter. If the tumor springs from the pelvis, this is particularly likely to be the case, as well from the above facts as the direction given to it by the axis of the superior strait; thus it is with the gravid uterus, uterine fibrous growths, and ovarian enlargements. Growths from the pelvis perhaps more completely gain the anterior position than any other sort, unless it be such as are attached to the anterior wall originally. It may be observed, too, that it takes a larger growth to disengage itself from intestinal resonance when arising from the posterior wall than from any other situation in that cavity.

By percussion we may make out the boundaries, positions, and to some extent, attachment and contents of an abdominal tumor. We should begin at the pubis, and follow a line upward to the ensiform cartilage; by so doing we will ascertain the central perpendicular extent. A good plan is to make four or five perpendicular explorations of this kind each side of the median line, extending the whole length of the abdominal cavity. After this has been done, we may proceed, by right angles, to these lines, to examine the abdomen crosswise, from its lower to its upper boundary. We will seldom miss any important growth by this mode of proceeding. If there is any doubt or obscurity, our pressure should be sufficient to bring out something of the flatness of sound from the spine, kidneys, &c. If we discover any point of sufficiently defined dulness to impress us with the idea of a tumor, we should, by percussing explorations, proceed from the point of greatest dulness to its circumference in every direction. In this way of examining, we will be able to trace it up the side to the hypochondriac regions, down into the pelvis, or define it so perfectly as to decide what must be its place of origin. Percussion and palpation will often enable us to determine the contents of a tumor as to its solidity or fluidity. Placing the finger on one side of the tumor, while we percuss the other, if the contents are wholly fluid, a wave of liquid will be set in motion on the side struck, and traverse the space to the one of the opposite; if solid, of course nothing of this kind will take place, and the impulse will be given to the whole substance of the growth. Should the contents be fluid, separated by a number of partitions, the wave or fluctuation will be less distinct than in the one where no such division exists; but in fact the obscurity is so great, that we will be at a loss by this management to decide whether the contents are solid or fluid. A slight variation of this combination of tact and percussion will often clear it up, however. When we wish to ascertain whether the fluid is contained in several cysts, we should place the pulp of the fingers of the left hand in the centre of the tumor, and then percuss with those of the right, first very near, then gradually increase the distance between them, until we find a point at which the fluctuation becomes less distinct; this is the margin of the cyst over which our left fingers are placed. Still keeping them in position, we percuss around in every direction, until we have made out the boundary and size of the cyst under examination, when we may move the fixed fingers to its margin, and commence the same process around this point. Proceeding in this way from one point in the abdomen to another, in most instances we may trace the outline of all the cysts superficially situated, and thus enumerate them, and learn their relation and absolute size. If solid bodies of whatever structure are incorporated in the mass and superficially situated, they may be detected, and their relative position, size, &c., determined with some accuracy. After tapping, when the abdomen is lessened, its walls lax and soft, palpation and percussion, singly or combined, become more demonstrative than before this operation. It not unfrequently is necessary, on account of the sensitiveness of the patient, when the tumor is small and the abdominal muscles not much under control of the will or reflex excitation, to administer chloroform until unconsciousness is induced, and the influence should often be so profound as to abolish reflex sensibility. In cases where there is inflammatory tenderness, the chloroform will give us a freedom of examination which is indispensable to an accurate diagnosis, and that we could not without it by any possibility obtain. Palpation and percussion should both be practised ordinarily with the patient in the recumbent position on the back, with knees drawn up, shoulders elevated, and the abdomen stripped quite bare of covering; in many instances, however, variation of posture is indispensable to definite results,—the standing, prone, &c. Very little need be said in this place about auscultation, as it is only applicable to the diagnosis between it and pregnancy, and will be dwelt upon when I come to speak of that more particularly. Vaginal and rectal digital examinations in ovarian disease are proper, and should not be dispensed with. The pelvis should be carefully surveyed by this method. The attachments, consistence, and relations of the diseased mass to the various organs in this cavity should be carefully noted. The uterus, rectum, and bladder, so far as practicable, ought to be examined with reference to their healthy condition, position, and involvement. Combined with external palpation, we may examine the tumor more thoroughly than with either one alone. Two fingers introduced into the vagina and pressed firmly upward against it will perceive any impulse imparted to the tumor above. With the left hand, if we press downward toward the pelvis, we may feel the motion of the diseased accumulation downward, and if the sudden impulse of percussion is applied above, we may feel an impression from its contents; if fluid, a wave or sense of fluctuation; if solid, the deadened impulse always given in such cases. When the tumor is small, and occupies the posterior peritoneal cul-desac, by introducing one finger in the rectum and the other into the vagina, the tumor may be included between them, and thus examined with more accuracy than with either alone.

Dr. Simpson has taught us how to extend our examinations into the uterus, so that our information in this direction is very materially increased, by the use of the probe, mounted upon a handle. Members of the profession who appreciate the labors of Dr. Simpson, have, by consent, named the instrument, the improvements and uses of which he has so ably promulgated, "Simpson's sound." The most useful form, I think, of the sound is round and plain, without any notches or edges, with the button-end about two lines in diameter. Back of the head or button-end the wire should be about half the thickness of the end. From this the wire should gradually increase in size towards the handle, until it is near a quarter of an inch thick. The material of which it is made should be soft enough to be bent into any shape we choose, and yet sufficiently firm to retain the flexure thus given.

The sound may be introduced into the uterus, and varied in its direction while we gently urge it forward to the extremity of the uterine cavity. The only obstacle a sound of the dimensions I have mentioned will meet with in a uterus of ordinary size, will arise from want of correspondence with the direction of the cavity. The most simple and ready revelation of the sound or probe is the direction and length of the uterine cavity. From this knowledge much valuable deduction may be drawn. But it is employed for determining the relation of the uterus to pelvic tumors, according to the ingenious directions of Dr. Simpson, very handily and to excellent purpose. While the sound is in the cavity of the uterus, this organ may be fixed by holding the instrument firmly in one position, or be moved in any direction if not restrained by adhesion or accretional attachment to the diseased mass, or to some other organ. If the uterus be fixed, and the tumor moved by its side or from it, with the fingers introduced for the purpose, the motion will be felt affecting the uterus through the attachments. On the other hand, if we watch the motion of the tumor with the fingers while the uterus is moved, the attachment or not will be determined, or the uterus may be moved in one direction and the tumor in another. In this way their attachments may be pretty certainly diagnosticated. The sound may be employed in the uterus with one hand, while palpation on the abdominal surface is effected with the other ; and if the uterus reaches above the pubis, the distance the probe is separated from the external hand, or its relation with the median line of the abdomen, or the main bulk of the growth, will enable us to determine some interesting problems. The motion received by the sound from the pressure of the hand without, or *vice versa*, is of important significance, as will be more apparent as we advance. When, however, from all these sources of inquiry we fail to get a sufficiently definite answer, there is still another physical means of diagnosis which we are justified in employing, viz., exploration. If by means of an exploring needle or trocar we draw off a small quantity of fluid, it may be subjected to microscopic and chemical tests, that will often enable us to determine the nature of the disease. Dr. G. Hughes Bennett, in a paper on Ovarian Disease, in the "Edinburgh Medical and Surgical Journal," quoted by Mr. Brown, says, as the result of his microscopic "examinations of different specimens of ovarian fluid, that the most constant characteristic of such fluid is its containing in greater or less abundance cells gorged with granules ; and in addition, circumambient granules, having the same measurement as those encompassed by the cell-wall. At one time I considered the size of these granules (if they can properly be so called) was constant, but subsequent observations have convinced me of the incorrectness of this conclusion : the size of the gorged cells and granules varies greatly, even in the fluids from different cysts of the same ovary." There can be no question but that the nature of the fluid contained in these cysts is, in all its essential features, pretty constantly the same in the early stages of progress ; but it is equally true that as they grow large enough to be influenced by pressure or other external causes, their composition, microscopically, must vary. The chemical nature of this fluid is more constant. It is alkaline in reaction and highly albuminous, always coagulating when boiled or submitted to the action of strong acids. Another sort of exploratory examination is the evacuation of a part or the whole of the fluid, provided fluid flows through the tube. If so much fluid is drawn off as to greatly relax the abdominal parietes, the contents of this cavity may be much more thoroughly examined by palpation and percussion, and the nature of the growth ascertained quite definitely. We can then handle the abdominal contents so as to trace any tumor within it to its attachments, get its precise size and shape, consistence, &c. Or if there should be no morbid growth, it will become quite evident after a greater or less evacuation of the fluid by tapping.

After having passed in review, as above, the items of general diagnosis of ovarian tumors, I propose to enter upon a differential view of the subject, for there are conditions of disease and health of the contents of the female pelvis and abdomen which they may be mistaken for. The following list is given by Mr. I. Baker Brown of conditions that may be mistaken for ovarian tumor: "1st. Retroversion and retroflexion. 2d. Tumors of the uterus, —solid, fibrous, or fibro-cystic. 3d. Pregnancy. 4th. Pregnancy complicating ovarian dropsy. 5th. Cystic tumors of the abdomen. 6th. Distended bladder. 7th. Accumulation of gas in the intestines. 8th. Accumulation of fæces in the intestines. 9th. Enlargement of the liver, spleen, or kidneys, or tumors connected with the viscera. 10th. Recto-vaginal hernia and displacement of the ovary. 11th. Pelvic abscess. 12th. Retention of menstrual fluid from imperforate hymen or closure of the os uteri. 13th. Hydrometra."

In cases of retroversion or retroflexion, if minute examination with the finger per vaginam and rectum fail, and the symptoms are of a character to make a correct diagnosis important, the uterine probe will at once determine the distinction. In some instances we might be quite unable to distinguish a small ovarian tumor from an impregnated retroverted uterus. Our proper plan in such cases is to await the peremptory demand for the knowledge, and then take the risk of introducing the probe, remembering the position of the mouth of the womb in retroversion, that it is not only near the pubis, but directed upwards as well as forwards, and that the os, in cases of misplacement by the tumor, is not directed upward, but nearly always downward,—certainly never, so far as my experience and reading goes, above the horizontal position. The probe will be equally available in examining the retroflected organ, and I think the probe should always be used where pregnancy is not strongly suspected. Should we feel much doubt of the existence of pregnancy in connection with retroversion, it would be better to lift the tumor out of the pelvis; when, if it were retroversion, the uterus would be restored to its natural position, with the os near the centre of the pelvis. In endeavoring to distinguish between ovarian and uterine tumors, we should bear in mind that the latter almost invariably change the length and size of the cavity of the uterus. Where the sound is used, it will pass further than if the uterus was not involved. The rationale of this increase of size of the uterus, so generally found to be present, is, I presume, connected with the fact that the development of a tumor in or from the walls of that organ induces general hypertrophy to some extent, as these growths are found to be a hypertrophy of some one of the uterine tissues. The tissues generally involved are the fibrous or mucous, as in hard or soft polypi from the internal, or hard from the external walls or intramural fibrous tumors. Uterine tumors are so intimately connected with the uterus that this organ cannot be moved without imparting more or less motion to the tumor, nor can the tumor, on the other hand, be moved without, in a similar way, affecting that organ. This is not the case with ovarian tumors. They are so loosely connected with the womb that considerable motion is allowable without the other partaking of it. In the sound we have the means of moving or fixing the uterus, and with the finger may watch the effect of motion upon the one or the other, as the case may be. When a fibro-cystic tumor is developed upon the uterus containing fluid, the examination to ascertain whether there is an attachment with the uterus, and with a view to learn the length of the cavity, will give us clear notions of the matter. When we are satisfied that pregnancy cannot be the condition, we may explore or tap it as an additional means of accuracy.

Hard or fluid tumors arising from a distant organ or part of the abdomen would have a different history from the ovarian tumor. If our patient is intelligent, her observation as to the place where first noticed should be relied upon as valuable knowledge respecting the probable point of origin. Ascites, when excessive, may sometimes be mistaken for ovarian tumor, but the latter is more frequently taken for the former. When the patient lies on her back, with the knees drawn up, so as much as possible to relax the muscles, and the abdomen is entirely exposed, in ascites the tumidity will be rotund, filling out in every direction, and will particularly bulge the depending portions. The flanks will both be full; the abdominal protrusion commences at the edges of the ribs, and will be equally soft at every point; fluctuation will be greatest at the most dependent parts, and resonance entirely absent; fluctuation will scarcely be perceptible in the highest part of the abdomen. These circumstances will remain the same under any change of position. If the patient stand up, the dulness is in the hypogastric and iliac regions. If she lie on her side, the dulness and fluctuation on the lower side; resonance on the upper side. All this results from the water freely settling into the lowest points, let them be what they may. In ovarian tumor, alteration of position from erect to recumbent, or from supine to prone, makes no difference in the places where resonance and fluctuation are found. They are manifested always in the same places. When the patient lies on the back, the flanks are resonant, the umbilical region dull. Fluctuation is not observed in the flank in any position; it is apt to be greatest under any posture in the middle of the abdomen. When the abdomen is exposed for inspection there is marked irregularity in its rotundity, and I think, ordinarily, the flanks, one or both, are flat. One side is apt to bulge more than the other. Probably there is more than one rather prominent region,—it may be several. There is more hardness and tension; not the flabby swaying under slight influences, so common in ascites. An important class of circumstances is the pathological condition almost always present in ascites. It seldom occurs in persons in the enjoyment of good health in every other respect. There is organic disease of the kidneys, liver, spleen, heart, lungs, or subacute peritonitis. Or there may be some cachexia from miasma, poison, or other bad influence of particular places of residence, occupation, habits, or time of life, &c. There is some notable and grave pathological accompaniment of abdominal dropsy which precedes the swelling; whereas, the ill health in ovarian dropsy is the effect and not the cause. We generally find that women preserve a good condition of health in ovarian disease until far advanced, and disordered functions come almost always as the result of great pressure upon the suffering organ. A complication of ascites with ovarian dropsy obscures our diagnosis very much. If the ascites is great, and the ovarian disease not so considerable, the tumor will be felt floating about, as it were, in the abundant fluid, when the patient changes position. Excluding by our diagnostic examination every other disease, and leaving the question between them alone, we are justified in exploration and even tapping. By the former, we come in possession of a specimen fluid, which, when submitted to chemical and microscopical investigation, is almost conclusive. By the latter, we partially empty the abdominal cavity and relax the walls, so that we can examine its contents with great freedom. If the fluid is ovarian, it will be highly albuminous, and possess the microscopical qualities I have before mentioned. If it be ascitic, the properties are those of serum found exuded anywhere from pressure or inflammation. There will be very little, if any, albumen, no epithelial cells, and the corpuscles described by Dr. Bennett.

It will occur very seldom that the question between pregnancy and ovarian disease will become so urgent that it may not safely be left to time. I can conceive no time or circumstances under which great doubt as to which of these two conditions were present but in the early stages of either, while in the pelvic cavity; and unless great pressure on the organs contained in it make delay hazardous, we should not interfere, but content ourselves to wait until the obvious evidences, as quickening and motions of the child, declare the existence of pregnancy, or until so much time has elapsed without any such signs as to throw great doubt upon the subject. At such times the tumor is high above the pelvis, and may be subjected to any searching examination we may choose. Auscultation then becomes valuable and perfectly reliable, when properly practised, in determining the presence of normal pregnancy.

Frequent examinations with the stethoscope or ear, in various positions, should be patiently and perseveringly practised before we should be satisfied to risk means of a hazardous nature that will enable us positively to decide the question. After having repeatedly thus explored the abdomen without any sign of a live fœtus, we may use the probe to examine the whereabouts and size of the uterus. No mistake will survive the test of this instrument. If I were not to explain myself a little more upon this point, I might incur the charge of rashness for recommending the sound where any doubts exist. It would be rash to use the sound until all the differential signs of pregnancy had failed, and even then, unless the urgent demand caused by the influence upon the health forbids us to wait longer for a decision. It is only in extreme cases, where the symptoms and signs derived from the breasts, menstruation, nausea, pigmentary deposits, and auscultation, had all failed, and yet I was obliged to act at once for the safety of the patient, that I should consent to use the sound. Then I would use it as the more innocent of the demonstrative tests, and as a *dernier ressort*. It is certainly more innocent than the exploring needle or evacuating trocar, and equally demonstrative. The worst effect its careful use could have would be to produce abortion or premature birth, either of which would be more likely to remove the urgency of the symptoms than do harm. I have recently seen an instance of obscurity of diagnosis, from the existence of a pregnancy of eight and a half months' duration, decided by the probe, which caused the discharge of a mummified fœtus of less than four months' growth, and, as a matter of course, almost, cured the patient.

Pregnancy complicated with ovarian dropsy, may be very perplexing to diagnosticate. At a sufficiently advanced stage of pregnancy, auscultation will reveal the fact, and it cannot be a matter of great importance to decide the nature of the tumor until the termination of the process of gestation, as a general thing, and if it should, we may proceed in its examination according to the plan heretofore indicated. This complication may continue up to the time of labor, and embarrass that process. Such embarrassment will not arise, if the tumor is large enough to rise out of the pelvis, but if it is still within that cavity, and gets below the child's head, it may arrest its progress. There are very few collections or growths that can be, in such conditions, mistaken for this. In pelvic abscess, there will be inflammatory tenderness and heat. The most likely of all others, is a prolapsed bladder. Our diagnosis, however, will be easily effected by using the catheter, when, if it is the bladder, emptying causes its collapse and the entire disappearance of the tumor. But if, after the complete evacuation of the bladder, there is yet a tumor containing fluid, exploration should be resorted to. This will clear up the diagnosis, provided the exploring trocar is large enough to evacuate a part or the whole of its contents. There are other fluid tumors, arising from the broad ligaments near the ovary, probably dependent upon a great increase of one or more of those transparent cells of serum, so generally seen by looking through this peritoneal duplicature, towards the light. These may be mistaken for actual ovarian cysts, and these are doubtless the cases of ovarian disease that are permanently cured by a single tapping. I think no means of diagnosis now known would enable us to decide, with any certainty, between the two. Exploration and chemical and microscopic examination of the fluid would throw some light upon, but not necessarily clear up the case. Cystic tumors of the abdomen, arising from other points, and hydatids of the peritoneal cavity, can be distinguished with certainty in no way except by exploration and examination of the contents. The history will, if carefully and intelligently detailed, show something, perhaps, that we may seize upon to aid us. The case should commence, if ovarian, in a tumor arising from the pelvis gradually *ascending* into the abdomen. If abdominal, it is first noticed in that cavity, and may descend until it occupies all the abdomen, and then the pelvis also. If hydatid, the increase is mere tumidity, not a well-defined tumor, and it commences in the abdomen.

The distended bladder, accumulation of gas in the intestines, or of faeces, ought not, in the present state of our science, to embarrass us any longer than the catheter or a cathartic could be brought to bear upon the case. As soon as the bladder is emptied it will collapse. The gas in the bowels causes tympanitis of the abdomen, and thus ought to be detected. The accumulation of faeces can be removed, when the tumor will be gone. Hysterical distension of the abdomen, said to simulate pregnancy, ovarian, uterine, and other tumors, entirely disappears under the influence of chloroform, as shown by Prof. Simpson, on many occasions.

Visceral enlargement, as liver, spleen, kidneys, and tumors growing from them, are not unfrequently mistaken for these tumors. I have a patient now laboring under enlargement of the spleen, who has been told more than once, that she had ovarian disease. Unless the enlargement of the liver or spleen is excessive, I cannot see how a mistake can be possible. The history as to where the tumor was first observed should be carefully traced. If either of these, it has descended. I have not seen a liver or spleen occupying the cavity of the abdomen so completely, but that its well-defined edge could be felt for a considerable distance, and this edge is always below, while the upper boundary is less defined or traceable beneath the ribs. I have, on several occasions, seen the spleen enlarged and dislocated, occupying the left iliac region, and reaching up towards the hypochondriac, but there are always sharp edges somewhere. This is not the case in ovarian dropsy; it is round, somewhat even, and elastic to the touch.

Mr. Brown mentions recto-vaginal hernia and dislocation of the ovary into the cul-de-sac of Douglas. The diagnosis would be difficult, and unimportant, unless in exceptional cases. The great importance of a correct diagnosis is based upon the urgent symptoms and fatal tendency of the disease. If the tumor is small, and situated in the cul-de-sac, we can not only afford to wait for further developments, but it is our duty to do so.

Retention of menstrual fluid, from imperforate hymen (or other obstruction to its outlet),—hydrometra,—as soon as we have by physical examination, history, and the rational symptoms, decided that the patient is not pregnant, the finger and sound will clear up all doubts in a short time. Obstructions will be ascertained or overcome by them, and our misgiving dispelled.

Supposing our diagnosis complete, as to its being an ovarian tumor, we have yet to learn, for the more intelligent treatment, several other things; among these are,—What are the contents and construction of it? Is it monocystic or polycystic? Are its contents partly solid, or wholly fluid? Although, probably, not always possible to decide these questions without exploratory operations, we have some means of clearing them up. A diligent and careful examination by percussion and inspection will enable us to judge correctly, in most cases, whether the tumor is monocystic or polycystic, or otherwise. If monocystic, the tumor is regular in its rotundity and outline; if polycystic, there is some irregularity of elevation, made out best by sliding the hand over the surface. Fluctuation, caused by percussion, is the same in all directions and from all points of it in monocystic. In poly-cystic, it is very obscure, except over partial measurements. The fingers placed near each other over the same cyst feel the fluctua-tion very sensibly; but when one is removed so as to pass over the partition between it and the next cyst, the fluctuation becomes more obscure. By examining all parts with both hands, separat-ing and approximating each other, we may make out the dimen-sions and situation of the cyst, which lies in contact with the ab-dominal walls. The fluctuation, or its absence, will determine whether a given part of the tumor is solid or fluid. The hard parts of an ovarian tumor, are, almost invariably, at the bottom of the tumor, and may be reached by the finger per vaginam. While our fingers are in contact with the base of the tumor in the pelvis, if it is wholly fluid, we may feel fluctuation, if the top of the tumor is struck with the other hand. If a solid part inter-venes between our two hands, fluctuation would not be experienced.

A very important and difficult point in our diagnosis, is the presence or absence of adhesions to the viscera or walls of the ab-domen. The nature of our practice will depend very much upon the determination of this question. Adhesions are very much more frequent in front than behind the tumor; to the walls of the abdomen than to the viscera. There is no way, so far as I am informed, to decide whether there exist adhesions to the mova-ble viscera, and unless when the tumor is small and movable, whether it is adherent behind or not. If there is none in front, there is not likely to be any behind. If there has been no peri-toneal inflammation nor signs of other inflammation in the abdo-men, we should be encouraged to hope that there is no adhesion, for there can be little doubt that adhesions are more frequently the result of organized inflammatory effusions between the two surfaces of the peritoneum, than all other causes. This is the ex-planation of adhesions, following punctures with the trocar or exploring needle. If the patient has always enjoyed good health, and not suffered from pains in the abdomen, there has probably been no inflammation, and therefore no adhesion. It should not be inferred, however, that there is certainly no adhesion because there has not been any inflammatory symptom; for either these symptoms may have been so obscure as to have escaped observation, or the adhesion may have resulted from some other cause. Certain it is, adhesions do exist in many instances without anything in the history of the case leading us to suspect them. We should inquire whether any operation, pressure or other curative measures of a local nature, have been used, as almost all of them endanger adhesions to the front wall of the abdomen.

If there is not too great tension we may gather the walls of the abdomen up, and move them over the tumor to a sufficient extent to establish their freedom from adhesions. In fact, there is not much adhesion where we can thus raise the muscles in folds. Another way is to place one hand on each side of the abdomen, with the palms applied flatly to its surface, and, pressing with some firmness against the tumor, we may attempt to slide the walls over the growth. As adhesions are not apt to be universal over the anterior part, a difference in the mobility of certain parts would lead us to point out the probable localities and extent of these complications. The hands should be moved from place to place until all the surface has been thus examined. This, when properly and carefully done, is a valuable means of examination. When the patient is lying on her back, with the walls as much relaxed as possible, if she make several deep inspirations we may observe the muscles moving over the growth, or see that the whole mass is carried down together. When in the same position, if the patient will make an effort by the abdominal muscles alone to elevate the chest, or throw her head forward, if there is no adhesion under the rectus muscles, these organs will show their bellies distinctly by swelling up in their middle portion. If there is adhesion on one side and not on the other, the muscle that is free will bulge, while the other remains flat and undefined in any such way. If both are adherent, the whole will remain smooth under the efforts. If the adhesion is at the lower part and not at the upper, the bulging will take place above, and not at all below.

Treatment.—It is not necessary to interfere, in any manner, with some cases of ovarian dropsy. Indeed, it is right to let all cases alone that do not threaten the life of the patient. There are many instances which advance slowly, or remain stationary for a great many years, and prove but an inconvenience. We would not be justified in active interference in these cases; much less should we do anything directly for cases in which independent complications of a fatal character exist, *e. g.*, phthisis or cancer. When, however, the disease is making obvious progress, and particularly when the advance is sufficiently rapid to leave but little doubt of its proving fatal within the average time of their duration, we are bound to make every effort within our power to save or prolong, as much as possible, the life of our patient.

When treatment becomes necessary, and we cast about for remedies, we will find that there are two kinds of treatment, strongly distinguished from each other, applicable to certain classes of cases respectively. General hygienic treatment is applicable to all, and this will consist, for the most part, in an abstinence from everything that stimulates the ovaria,—heated rooms, high living, lascivious practices, marriage, or if married, sexual intercourse and pregnancy; and the use of appropriate outdoor exercise, proper diet, and, indeed, everything calculated to promote sound health, and an avoidance of all debilitating influences.

The two kinds of treatment alluded to are the *palliative* and the *curative*. The one intended to relieve, as far as possible, the sufferings of the patient under the disease, or to retard the rapidity of its progress; the other to remove or destroy the tumor, and thus do away with the cause of the evil entirely. When practicable, I think this last treatment is always desirable; when not practicable, as, unfortunately, is too frequently the case, our only resource is the former.

So far as my influence goes to enforce the rule, I should emphatically insist upon determining in the beginning, before we make a move toward operative treatment, the kind of means adapted to the case under consideration; institute that treatment, and persevere with it to the end, or until some circumstance occurs to convince us of our mistake in this respect. We should not begin by trying what is usually regarded as the least hazardous, and be ready to use the more dangerous, but yet more effective, *because the former had failed.* There is no treatment worth relying upon but must produce its effects; and in nine cases out of ten, if it does not cure, it will cause disagreeable and even dangerous complications, and so render the subsequent treatment less effective. No case in which tapping, injections, or efficient pressure has been used ineffectually, is ever left in as good a condition for extirpation as before any such interference. Hence, I would not tap a tumor that ought to be extirpated, nor allow it to be injected or compressed. It may be said, and it is unquestionably true, that we may not always be able to determine the character and complications of the tumor, so as to be certain which treatment is best adapted to it. In answer to which, I would say, we must do the very best we can in this respect; and I think by thus forming and acting upon our judgment, more lives will be saved than by confounding the remedies in the manner heretofore done, using them more with reference to their danger and efficiency than their adaptation to the particular case. It would seem that practitioners are divisible into two classes as to their mode of treating ovarian tumor, viz., those who believe that radical treatment should almost always be avoided, and those who believe in operating in nearly every instance.

Two prolific sources of embarrassment in the treatment of ovarian diseases are, first, the knowledge on our part of their almost invariably fatal tendency; and, second, the inability of our patient to appreciate the appalling nature of her condition until such constitutional effects have been produced as would render any efficient treatment powerless to save,—at least this has been the case with my patients to some extent; and hence we should make a clear and decided statement to them so soon as we become able by our investigations to do so. When doubt exists as to the propriety of instituting radical treatment, we should continue to pursue the palliative until that doubt is dispelled. There are three sorts of cases to which the palliative is indisputably adapted. They are, first, those in which it is not desirable to use radical means in consequence of the absence and probable great distance of urgent symptoms, while there is a steady advance. It may as well be remarked here that no palliative measure, such as tapping, great pressure, or other means that will render radical treatment more hazardous, should be employed, where we are likely in the future to desire to resort to operative procedure. The second class of cases is that in which the symptoms are urgent, but in which it is not desirable to use radical means, in consequence of the slight chances of success. The third set are such as, in their nature and condition, would call for curative means, but the patient will not consent to their employment from fear of the danger or pain they inflict. The first set of cases is not very frequently met with compared to either of the others; yet we do occasionally meet with these slowly marching cases, in which we have an opportunity to try the effect of medicines; and it is precisely in this kind of cases that we *appear* to derive most benefit from medicines internally administered. We are prone to believe that the tardy development is dependent upon the virtue of some favorite remedy used, and deceive ourselves as to its efficiency, when really all depends on the natural slowness of the tumor. The alteratives, as mercury, iodine, sarsaparilla, chlorine, &c., have all had their advocates. It was at one time, and even now is the practice of some men of ability to give mercury to very slight ptyalism, with the hope of bringing about absorption. Iodine, administered frequently, so as to induce its specific influence upon the organism, has been, and is still, by some, highly lauded as capable of curing ovarian dropsy. A chronic administration of either of these remedies is sure to affect unfavorably the general health; and as it is extremely doubtful whether there is any efficacy in them, we should not be too profuse in their use. Effusion into the peritoneal sac, or subacute inflammatory complications, are often very much benefited by a moderately protracted course of these remedies. For the same purpose, local depletion, counter-irritants, such as iodine ointment, strong enough to induce irritation of the skin, are often useful; so are diaphoretics, diuretics, and cathartics. But, probably, the best course to pursue in these cases is to keep up the general health by appropriate exercise, diet, and tonics when necessary, and to keep the abdomen constantly but gently compressed by a flannel roller, which should be applied all over the abdomen, but applied with more stringency over the tumor, and, to add to its efficiency, a compress of sufficient thickness may also be employed. I must be permitted again to say, with reference to all these cases, that until the necessity for active interference, in consequence of the urgent effects of the growing tumor, is fairly presented, we should carefully avoid all very strong pressure or operative procedure. In the second class of cases we need not feel so restricted in our efforts at palliation. It is best, however, to bear in mind that too great activity of medication will often do more harm than good. Our object should be to promote such functions as are obstructed or restricted; the kidneys, for instance, will need especial attention, as will the intestinal canal.

The acids have always seemed to me to be particularly applicable to these cases. The nitric, nitro-muriatic, sulphuric, phosphoric, acetic, are all good, and may be alternated often, with the hope of relieving the distressing indigestion attendant upon great distension and imperfect performance of the renal functions. They also very much moderate the distressing exudations from the skin, which are often present. The chlorinated tincture of iron is an excellent tonic in such cases. These remedies may all very properly be administered in some of the bitter infusions, quassia, chamomile, wild-cherry bark, &c. The best time to give them is immediately after eating. Stimulants ought not to be too freely used, as they encourage the establishment of complications. Brandy I think the best of the stimulants, and it should be given more for the purpose of inducing sleep than anything else; and this it will often do when taken in a sufficient dose on an empty stomach at bedtime. When great restlessness and want of sleep are wearing out the patient, we must, as in all similar circumstances in other diseases, resort to the assortment of anodynes, beginning with the less disturbing, being sure to be under the necessity of ending with opium. Chloroform, internally administered, is, I am confident, not sufficiently relied upon. Teaspoonful doses, given in milk, will seldom fail to induce a fine anodyne effect. There is greater necessity, perhaps, for a gradual increase of the dose in using it than opium, or most other efficient anodynes. Hyoscyamus, belladonna, cicuta, should all be tried before opium.

We must be on the alert for complications, and ready for their appropriate treatment. The distressing constipation, which often annoys the patient and physician, will demand a great share of our attention. Injections of water and various substances will, of course, suggest themselves. It has occurred to me to be able to induce free movements of the bowels by having a pint of warm lard thrown high up in the bowels when they are very obstinate; the longer the lard is retained the better. This, administered once a day, will act excellently well sometimes. An ounce of fresh beef's gall, with three or four ounces of water, often does as well. But the time comes, sooner or later, with the steadily increasing pressure of the tumor, when to lessen its size is indispensable to the further extension of life. Tapping suggests itself as the only palliative means in this state of things. This operation is more beneficial in unilocular tumors than in any other sort, but is applicable as a palliative measure in any tumor containing fluid, when demanded by the supervention of urgent symptoms indicating the necessity of immediate relief.   Under the desperate circumstances mentioned, there can be no question about the propriety of tapping the patient; yet this apparently trifling operation is not devoid of inconveniences and dangers that should be weighed deliberately; and if they do not deter us from resorting to it, will at least make us particular not to use it as anything but an indispensable remedy.   One serious inconvenience connected with tapping is the readiness with which the fluid accumulates in the sac; indeed, the oftener the patient is tapped, the more rapidly will accumulation take place.   If, therefore, we commence using it too early, we are likely to shorten the life of the patient, if in no other way than by the increasing emaciation dependent upon the immense drain of fluid from the system.

The dangers of tapping are both immediate and remote.   The immediate are such as are connected with and occur immediately upon the performance of the operation.   Dr. Simpson sums up five that are more frequent, and against which we should be upon our guard.   First, the chance of wounding the urinary bladder.   This may be avoided by evacuating the organ, unless it is tied to the abdominal wall by adhesions.   Second, the puncture of the uterus when it is drawn up with the tumor.   By introducing the sound into its cavity we may learn its whereabouts, and thus be enabled to avoid it.   Third, the front part of the tumor may be traversed by the Fallopian tube, and this last be wounded by the trocar,— an accident which should, of course, be avoided when we can, but which I cannot think of much importance, nor do I know how it can certainly be avoided.   Fourth, the internal venous circulation, on account of the pressure, is obstructed sometimes, and the blood is directed to the veins in the walls of the abdomen, so that these veins may be wounded; but generally they are large, and may be seen, and thus avoided.   Fifth, the epigastric artery is sometimes wounded.   We should carefully feel for the pulsation of arteries in the thin walls before the trocar is plunged into the tumor.   As may be seen, these dangers may, for the most part, be provided against; but the second class of dangers, namely, the remote,— those that follow the operation some time after its performance, and are not dependent on the manner or place of the puncture,—are unavoidable, and cannot be foreseen.

The dangers and benefits of tapping cannot, and ought not, to be estimated by comparison with other operations. Each operation, of whatever kind, has its place, and is followed by its good or bad effects, for the reason, among others, that it is appropriate, or inappropriate. Generally, no two operations are applicable to any one condition of things; and we should not allow the question of danger to decide between them, unless in very rare and exceptional cases. The statistics, as far as I have been able to collect them, may be well summed up, as Dr. West has done, and I shall rely upon his figures: "The chief, indeed, almost the only numerical data of which we are possessed, bearing on this subject, are derived from a table of 20 cases, compiled by Mr. Southam, of 45 cases collected by the late Mr. Lee, and of 64, the results of which are given by Professor Kiwisch. Of these 130 cases, 22 terminated fatally within a few hours or days after tapping, and 25 more in the following six months; or in other words, 34.7 per cent. of the cases ended in the patient's death in the course of half a year after the performance of tapping. In 114, of the 130, death is stated to have taken place: 22 within less than ten days, 25 within six months, 22 within one year, 21 within two years, 11 within three years, 13 after a period exceeding three, and in some amounting to several years.

"In 109 of these cases, we are further informed how often the patients had been tapped. It appears that 46 died after the first tapping, 10 after the second, 25 after from three to six tappings, 15 after seven to twelve, 13 after more than twelve." It would appear that the first tapping is very much more dangerous than subsequent ones. Dr. West says further: "Unfavorable, however, as are the conclusions to which we are irresistibly led by such facts as those which have just been mentioned with reference to the ultimate issue of tapping, it is yet very questionable whether they represent the whole of the truth concerning this matter." Dr. Atlee, of Philadelphia, thinks tapping not a very dangerous operation. Mr. Brown thinks its dangers greatly overrated.

There can be but little doubt that much of the mortality of tapping is due to the fact of the desperate character of the cases in which it is used; and the reason why so many die in so short a time after the first operation is, that in many instances the patient is almost moribund before it is resorted to. When not attended with the immediate dangers above enumerated, tapping is either followed by great relief from suffering or by the remote or sequential dangers. They are, for the most part, prostration or inflammation. The prostration is sometimes so great, that no management can prevent the patient from dying in a very short time. Such great prostration is, however, exceedingly rare; it is more common to have it in a more moderate degree. The patient will feel faint for an hour or two, and then gradually rally, or she may continue to be pale and languid for several days. For such slight cases, the horizontal position, rest, and good, digestible, somewhat stimulating food, is all that will be needed. When the prostration is great, and danger of fatal sinking present, the case must be treated energetically. The means calculated to bring about reaction must have reference to the causes of the prostration. The evacuation from the general vascular system is not a cause, because the fluid in the tumor is extravascular; but it is a sudden change in the distribution of the blood. The evacuation of the abdominal cavity of so large a bulk of its contents, and the inability of the abdominal muscles to contract sufficiently to keep up the pressure to which the viscera have been habituated, are the causes of the irregular distribution of the blood. The want of pressure upon the abdominal viscera, allows a large accumulation of blood in the veins, and it is there retained. In proportion to the amount thus collected in the abdomen, will the blood be withdrawn from other parts and organs. The brain will partake of this temporary anæmia, and consequently be incapable of discharging its functions with its wonted efficiency. This is the condition,—not a want, but an irregular distribution of blood. Our first object should be to, as nearly as possible, re-establish the previous condition of the abdomen. This can be, to some extent, accomplished by pressure, with compresses and rollers. The compresses should be as large as the space covered by the muscles of the abdomen, and thick enough to fill up much above the level of the ribs and iliac bones on the side. The roller should be applied from the pubis to the ensiform cartilage, with as much power as the patient can bear without great discomfort. Then the head should be persistently kept below the level of the body. This simple treatment, instituted early, will do more than all other means without it. We can very properly, however, give stimulants, in addition, when necessary. When this danger is passed, inflammation of the sac or peritoneal cavity is next to be apprehended. The sac undergoes every degree of inflammation, from the slow, subacute, unobserved degree, which vitiates the fluid effused into it, either by causing decomposition in it, or by the production of pus, or effusion of blood inside, or fibrine on the external surface—in this last case causing adhesion,—or such degeneration of the walls of the sac as to cause an obliteration of the cavity, a cessation of its secreting powers, or a perforation, and consequent peritoneal communication; or, what is perhaps more common, an acute degree, announced by severe pain, referred to the point most intensely affected, or to the whole abdominal region, thus showing the probable involvement of the peritoneum. Indeed, I think it very probable that the sharp pain ordinarily present in these cases, indicates peritoneal inflammation, and that there is but little pain in the case of inflammation of the fibrous and internal coats of the sac. Fever, of a somewhat high grade, is apt to attend upon the degree of inflammation last mentioned, accompanied by headache, weariness, aching in the back, limbs, &c. But in the inflammation of the inner coats, in which pus or fibrinous products are effused in the fluid of the tumor, there is generally but slight fever, perhaps none, at first; but the vital powers are more or less depressed, copious perspirations at night, possibly delirium, and in bad cases, all the symptoms of pyæmia, hectic, exhaustion, and death. Now all morbid conditions resulting from tapping should be met promptly by the remedies appropriate to them when they occur under other circumstances,—antiphlogistic regimen, depletion, fomentations, cathartics, anodynes, alteratives, &c. In pyæmia, tonics, stimulants, good diet, and time will be our resort.

The operation of tapping is simple, and easily performed generally. To avoid the depression which follows the evacuation of so large a quantity of fluid as is contained in the abdomen sometimes, we should have our patient on the side, very near the edge of the bed, with her head and shoulders low. Two large and long hand-towels should be passed around her body, with the edges close together upon a level with the point where we wish to introduce the trocar, and these ends given to an assistant, who stands behind the patient. The assistant having in charge these hand-towels should be directed to draw upon them so as to keep up a state of tension as the fluid is being evacuated. To avoid the dangers enumerated as immediate, we should assure ourselves that the bladder is empty, and if we mistrust that it is not in its proper place, we should introduce a sound, so as to assure ourselves of the whereabouts of the fundus. If we have not already done so, we must sound the uterus, also, and thus be sure of its harmless position. After these precautions, the best rule, perhaps, is that given by Dr. Simpson, and that is, to feel for the most fluctuating point, the place where the walls are thinnest,—look for veins and feel for the pulsation of arteries. The thinnest part, where fluctuation is most evident, is usually the right place to make the puncture; but there is not always any such point, there being but little difference in this respect over the whole of the front surface of the tumor. In such case we may be governed by the ordinary rules for the place for tapping. The linea alba, between the symphysis and umbilicus, is the most eligible in the greatest number of cases. If any objection to this arises, a point midway between the umbilicus and the anterior superior spine of the ilium is, as a general thing, safe and effectual as any. Some surgeons recommend other places as free from the objections that are sometimes urged against these points. They say that tapping through the vagina is quite safe from the immediate, and not so likely to be followed by some of the sequential disasters. The rectum is thought to be still better by some. The vagina is quite a commendable place, if we are careful to ascertain well the position of the bladder and uterus, and avoid them. Our instrument (the trocar) should be large, four or five lines in diameter; the point should be sharp, and a little longer than they are usually made. The canula if not large will not freely discharge the fibrinous concretions or thick treacle-like fluid, and if the point is not long and sharp, we inflict considerable unnecessary suffering in the introduction of the instrument. We may plunge the instrument in towards the central axis of the tumor, until sent home to the rim of the canula. If, however, our instrument is not pretty sharp, it will be very much better to make an opening with a very sharp, thin bistoury, which will cause less suffering, and answer every purpose as well.

The third sort of cases to which palliative treatment is applicable, those in which our patient will not submit to radical means, must be managed in almost every particular as I have described the treatment for the other two kinds. Remembering the rules and rationale, it will not be difficult to adapt our means to the end in view.

The curative treatment of ovarian disease is believed by many, almost all authorities, to be practicable only by instrumental means. There are some very respectable writers, however, who believe that there are cases in which we may hope for success from medicinal and mechanical treatment without the use of surgical instruments, and they think that there is enough virtue in such means to warrant a trial in very many instances.

The immediate objects to be accomplished are, first, to arrest the growth of the tumor, bring it to a stand-still, and thus avoid the disastrous results which attend the attainment of very large size, with its consequent pressure, ruptures, &c. Second, when this is not practiceable, to obliterate the sac or sacs. The sac is sometimes reduced by contraction to a mere knot of compressed tissues, which are more and more atrophied and wasted, until very slight traces of their existence are left; or, by inflammation and contraction, the tumor is converted into a fibrinous mass, enveloped in a fibrinous sac, which remains the same throughout life with very little alteration; or suppuration may accompany inflammation, the whole tumor be softened down into pus, and discharged by ulcerating through the vagina, rectum, abdominal walls, or bladder, and all traces of it disappear. Or again, the walls may collapse without shrinking much, and adhere by adhesive inflammation, and in this way its effusive surface be destroyed.

When neither of the above two immediate objects is practicable, or it may not be desirable or advisable to attempt them, we may, thirdly, remove the whole or a part of the tumor from the abdomen, and thus either get rid of the whole of the offending growth, or, after a part is removed, hope to effect, by one of the processes of obliteration above alluded to, the destruction of the balance. The means used for the arrest sometimes cause an obliteration of the sac, and do more than merely stop its growth, so that it will not be the best plan to formally treat of those means by which we attain the first object desired. I shall consequently

feel at liberty to introduce and speak of such as sometimes arrest
and sometimes cause a disappearance of the tumor.

Three general ideas seem to govern individuals who rely largely
upon the use of internal remedies for the cure of ovarian dropsy,
viz., that the disease is inflammatory in its origin and continuance,.
and that antiphlogistic and alterative remedies, by arresting this
morbid process, will stop its growth, and that the conservative
powers of the system, aided by sorbefacients and secernents, will
remove it; that the tumor is developed in consequence of the presence of some one of the cachexiæ,—scrofula, for instance, as Dr.
Bird distinctly avers.   Physicians who entertain these notions of
its origin hope to make a cure by changing the general action of
the system by all the means usually recommended for the correction of scrofulous tendencies to disease, by tonics, good diet,
properly regulated exercise, clothing, bathing, and specific medication; that its origin is entirely independent of either inflammation or scrofula,—an hypertrophy, in the strict sense of the term,
of certain normal conditions,—a nutritional development of tissues
similar to the production of nutrition elsewhere.   Those who entertain this doctrine believe, also, in the atrophicating qualities of
certain medicines and mechanical appliances, and hope, by the
well-directed employment of them, to at least arrest their growth,
if not cause their removal by absorption.   A number of cases are
on record that encourage the hope of doing something by medicines internally administered; and while I am free to state that I
have really hardly any faith in them as curative means, the recollection of the discouraging results of any kind of treatment forbids
me too strongly deciding against their use in properly selected
cases.   Nor do I think it fair to say, as has been said, that cases
treated by internal medication and recovering are but instances of
spontaneous cures, and would have done as well or better without
the treatment.   Dr. Denman, in his Midwifery, by Dr. Francis,
at page 151, says: "In the beginning of this dropsy, when the
increasing ovarium is first perceptible through the integuments of
the abdomen, and sometimes in its progress, there is often so much
pain as to require repeated local bleeding by scarification or
leeches, blisters, fomentations, laxative medicines, and opiates to
appease it.   I have also endeavored to prevent or remove the first
enlargement by a course of medicines, the principal of which was

the unguentum hydrargyri rubbed upon the part, or calomel given for a considerable time in small quantities with an infusion of burnt sponge, or the ferrum tartarisatum or ammoniacal, trying occasionally what advantage was to be obtained from blisters, from a plaster of gum ammoniacum, dissolved in the acetum scillæ, or, lastly, from electricity. From all, or some of these means, I have frequently had occasion to believe some present advantage was obtained or mischief prevented; but when the disease has made a certain progress, though a variety of medicines and of local applications have been tried, no method of treatment has been discovered sufficiently efficacious to remove it or prevent its increase."

Colombat is of the opinion that "though we ought not to place much confidence in the means derived from medicine, strictly so called, we are of opinion that they ought always to be employed before recurring to those offered by surgery. Consequently, sudorifics ought first to be prescribed; for example, guaiac, sarsaparilla, and vapor baths, resolvents, and, amongst them, mercurial frictions, successfully employed by Clark and M. Nauche; hydriodate of potash, with the internal use of iodine in small doses; sea-bathing, or salt-water baths, from which M. Laennec, of Nantes, says he has obtained most excellent effects; the thermal baths of Aix, in Savoy, or those at Barège, and, lastly, antimonial frictions, cauteries, moxas, and blisters, applied upon the abdomen. Diuretics, such as squills, nitre, &c., which according to Haller, were usefully employed by Willis, a decoction of ashes in the proportion of a handful to the quart of water employed by Petit Radel, and from which he obtained a cure after having punctured the cyst. Lastly, purgatives in divided doses, as, for instance, aloes, rhubarb, croton oil, calomel, combined with castile soap and sulphate of potash, &c., are other means which, in conjunction with abstinence and compression of the abdomen, may be prescribed at the commencement of the disease for the purpose of assisting the absorption of the fluids, at first small in quantity." After trying all these, however, surgical treatment, he thinks, will be our only resort in a vast majority of cases. The efficient application of pressure seems to promise more than internal remedies. Well-regulated, efficient, and long-continued pressure may produce obliterating inflammation of the sacs and consequent cure;

or, by affording great resistance to the expansion of the tumor, arrest its growth.

Resolution and absorption of an ovarian tumor is a very doubtful fact, however, and notwithstanding their unaccountable disappearance, should not be counted prognostically. The second object in our treatment, that of obliterating the sac *in situ*, affords more reason for hope in properly selected cases. The means used consist of tapping, with pressure, with injections of stimulants to induce inflammation of the sac, and with injections and pressure combined; or, what is sometimes successful, the establishment of a fistulous opening in the sac that either communicates externally through the abdominal walls, through the vagina or rectum, or simply with the peritoneal cavity. The above-mentioned treatment is applicable, properly, to the unilocular or single cyst cases only, as it is impracticable to tap, inject, or establish a fistula when there are many sacs; and what is still more discouraging in the multilocular variety, the sacs are not only filled again after tapping, as is generally the case with the monocyst, but there is a constant reproduction, or, perhaps, it would be more correct to say that they are continuously developed from the ovisacs that are matured every month. Tapping, followed by pressure or injection, is very apt to change the condition of the tumor in one respect, at least, and that is, to cause adhesions to the surrounding peritoneal surface. In one case of unilocular tumor, in which an external fistulous opening was made after the patient had been tapped six times, and had iodine injections three times, the sac, so far as we could determine, was universally adherent; no portion of it could be brought out of the wound.

Very fortunate instances sometimes occur in which the evacuation of the tumor by tapping is followed by a speedy and permanent obliteration of the sac. It is exceedingly doubtful, however, whether these were not cysts developed from the broad ligament, and not involving the ovarian tissues at all. Certainly, they are exceptional, and cannot be expected in any given case, so that we ought never to be satisfied with tapping when our object is the obliteration of the cyst. Pressure, in conjunction with tapping, is applicable, perhaps, to a larger number of cases than any of the other modes of treatment. It is very much more successful in cases of the monocystic than in any other variety. The application of pressure to a tapped sac has for its object a complete closure of the cavity of the cyst in such a manner as to bring its walls, as nearly as practicable, in contact throughout. This at once, if thoroughly effected, modifies the secerning capacity of its surface, and perhaps, from the time of its application, arrests more or less completely the effusion of the fluid. Now, if this cannot be done so as to operate upon all the surface of the walls, we can almost always bring some portion of the collapsed walls in contact. The continuous and prolonged contact of these surfaces brings about a low, and in some cases a pretty high grade of inflammation, causing adhesion or a change in their structure, so that they are no longer of the same ovisac nature, and hence they do not effuse the thick albumen previously produced, and the tumor remains inactive or shrinks, and nearly or entirely disappears; or suppurative inflammation may dissolve down, and discharge the mass through some adventitious or natural outlet.

The manner of applying the pressure is of the greatest importance. The apparatus should be permanent, and exert as much force as the patient can bear without too great pain, fever, derangement of the abdominal viscera, or other indications of too acute a degree of inflammation in the cyst or damage to some organ. It should be applied to the tumor as nearly as possible, and the forcible pressure should be exerted alone upon the collapsed mass, so as to crowd it back against the sacrum, lumbar muscles, spine, and other hard parts of the posterior wall of the abdomen. In order to do this properly, after the fluid is evacuated as completely as possible, we should examine the abdomen minutely, so as to ascertain as clearly as possible the position of the collapsed cyst. This will usually be a little more to one side than the other, and we may generally easily define its shape and get a good idea of its size. We should now construct a compact compress, corresponding in shape and size with the shape and size of the evacuated sac. The compress should be embraced by solid wood or tin outside. The compress can be made of hair, gum-elastic material, or napkins. If of the latter, they should be well stitched together, so that there can be no shifting in their position. After attaching the soft portion of the compress to the hard firmly, so that any pressure upon the latter may be exerted unvaryingly upon the former, it may be placed immediately over the tapped tumor, and pressure applied from a direction to press it against the hardest part, bearing on the posterior walls of the abdomen or pelvis. An attentive examination of the tumor under the pressure of the instrument will inform us pretty accurately as to the efficiency, completeness, and direction of the pressure of the compress. The compress may be managed better by a belt of soft but firm leather, to surround the body in such a place as to press over the centre of the compress. The power and direction of the pressure may be regulated thoroughly and at will by subjecting it to a tourniquet screw pressure from the belt. Of course there must be thigh and shoulder straps to the belt, in order to keep it from slipping up or down. When we have adapted these simple contrivances, we should turn the screw to such a degree as to press strongly as the patient can bear, and with it thereafter regulate the pressure as we may judge best. Having thoroughly satisfied ourselves of the appropriate adaptation of the apparatus, we should wrap the whole abdomen agreeably tight, from pubis to sternum, with a flannel roller. We should every day remove the flannel roller, and examine the compress and belt to be sure that they are not disarranged, and if in the least so, we should readapt them. We may tighten the screw or loosen it each time, or allow it to remain untouched, as the case may be. The greatest care should be taken not to produce too great pressure with this compress. It should be loosened when chilliness, febrile excitement, or other general signs of distress are added to local pain; it may be tightened as soon as the symptoms decline.

This mode of applying pressure, I think, is much more efficient and manageable than the plan recommended by Mr. I. B. Brown, the accomplished surgeon of female diseases and injuries, of London. His plan is to make a graduated compress of napkins so as to fit the top of the pelvis, and after applying it over the tumor so as to press it down into the pelvic cavity and against its back part, place over the whole a broad bandage tightly fastened from pubes to sternum. With this appliance we cannot always be accurate in the extent, position, and rate of the pressure, and, consequently, much more skill and experience are necessary in its application. Its success, hence, is much more frequent in Mr. Brown's hands than it has been with the profession generally. I am not aware that Mr. Brown teaches the necessity of pressure to all the collapsed tumor, but understand him to make most of his pressure at the origin of the tumor,—the ovarian region. The tumor, when collapsed by tapping after great distension, seldom sinks anything more than partially into the pelvis; the long-exercised traction upwards generally lifts the ovary of that side above the pelvis, and thus we may generally somewhat accurately fit our means to its slope and position. An objection, Mr. Brown thinks, sometimes applied to pressure, is the presence and great aggravation of prolapsus uteri. This objection, it will be apparent, is very much more applicable to his mode of causing it than the one I recommend. Multilocular tumors may be cured in this way perhaps more frequently than any other besides extirpation; for the pressure may be made to bear upon and greatly influence the development of the small cysts that are not evacuated by pressure. I have more than once evacuated several sacs through one opening in the abdominal walls by partially withdrawing the trocar, and directing the point toward a full sac after the one first pierced had been evacuated. This should be attempted in a multilocular tumor before we use pressure; and it is allowable, I think, to introduce the trocar in several places where there are a number of cysts that cannot be reached by the instrument from one point. I would not be understood as advising a reckless use of the trocar in these many-cysted ovarian tumors, but after we have decided from the circumstances of a careful examination of a given case that tapping and pressure is the treatment, we risk nothing, I think, in being thorough in our efforts to evacuate as nearly as possible all the sacs.—The bad effects arising from tapping and pressure are inflammation and its consequences. When there are symptoms of severe acute inflammation, the pressure should be removed, and leeches, cathartics, &c., should be employed to moderate or remove it. If the inflammation is in the sac, we should wait until all the acute symptoms subside before the pad or compress is placed again. If, however, we can satisfy ourselves that the inflammation is in some other part distressed by the pressure, by varying the direction of the pressure, provided we can include the tumor under it, we need not wait until all the acute symptoms have vanished. I have a better opinion of this kind of treatment, when carefully managed and watched, than any other, except the complete extirpation of the ovary. Another plan of obliterating the sac of ovarian tumors is to first evacuate, and then inject it with some substance calculated to induce inflammation in it, which, by its adhesive or destructive processes, may completely effect this object. A large number of cases are reported cured by this plan of treatment. For obvious reasons it is almost exclusively confined in its usefulness to the unilocular variety. Under certain circumstances only can we expect to reach more than one cyst at a time with the trocar and injections. When a cyst is simple, the patient in good health, and we succeed in properly managing the operation, there is not a great deal of danger in it, and we may reasonably hope for benefit from it. The most simple, and I think effective mode of operating, is to first draw off nearly all the fluid, except, say, one or two pounds, as well as we can judge of it, with a large trocar. After this is accomplished, we should pass an elastic catheter or other flexible tube through the canula of the trocar to the bottom of the cavity. With a hard-rubber syringe we may inject the medicine, whatever that may be, through the catheter into the interior of the cyst. By using this elastic tube there is no danger of failing to carry the material to the part we desire to reach without its coming in contact with anything else, or being decomposed before it arrives at its destination. The formulæ for this kind of injections are numerous, and several different substances used. Iodine seems now to be the substance generally employed. Dr. Simpson recommends several ounces of the tincture. Six ounces is probably enough to use at one time. I have used on several occasions six ounces of a mixture containing one scruple of iodine, two scruples of iod. potass to the ounce of water. This is certainly iodine enough, if specific in its influence, to cure any tumor. My plan is to allow it to remain in the sac instead of removing any of it.

Iodism is likely to occur to a slight extent, but to be the source of no considerable inconvenience. If it should be thought best to remove a part, or the whole of the iodine, the better way to do it is to pump it out through the tube, by means of which it was introduced, instead of squeezing it back through the canula of the trocar. This plan of extracting it, precludes the possibility of allowing any contact with the peritoneum; which in the event of disarrangement of the canula, might otherwise take place. Although, ordinarily, no great amount of acute inflammation takes place as the effect of this injection, yet we should remember that it sometimes does proceed to a dangerous extent, and be upon our guard with the means necessary to prevent a fatal degree. In fact, it would be negligence on our part not to watch with solicitude all the most trifling operations upon an ovarian cyst. It may be asked whether iodine is the best substance to use as an injection in such cases? Although I have to some extent fallen in with the fashion of using iodine, I cannot resist the conviction that there are substances that would do as well, against which some objections that apply to iodine could not be urged. Iodine operates promptly upon the organism when introduced in this way, by being absorbed and taken into the circulation; yet, I think there can be but few who desire anything more than its local effect upon the inner surface of the sac. Alcohol, wine, brandy, in fact any local stimulant whose general effect after absorption is more transient, as well as less powerful, would perhaps answer just as well. It cannot be that the internal effect of iodine upon the kidneys and other organs of excretion can enter largely into its good effects, for if such were the case, it would be better given by the stomach. Injection of iodine was regarded several years ago as the most eligible mode of treating this affection, because of its comparative safety and frequent success; but there can be no doubt that it was overrated, and now the profession is less ready to trust it. I believe it to be both more dangerous and less efficient than pressure after tapping. This is not in accordance with the opinion of Dr. Simpson, I believe. I have lately known of a case having been treated with iodine injections combined with pressure. I speak of this case to warn against similar proceeding, for it is plain, upon a little reflection, that if the pressure is properly applied, it will so lessen the cavity of the cyst as to endanger the effusion of the iodine, through the puncture in the sac, into the peritoneal cavity, and thus induce a fatal peritonitis. And if pressure is to be used, we should wait for two or three days after the injection.

The last and doubtless most effectual plan for obliterating the sac, is the establishment of a fistulous opening, communicating with the peritoneal cavity, or the external surface, directly or indirectly, through the vagina or rectum. This plan is also the most dangerous plan, resulting in a large number fatally. Quite

29

a difference in the effects, both remedial and morbid, may be remarked in the different places for the fistulous opening. When
properly and carefully managed, the opening in the peritoneal
cavity is productive of least harm, and less likely to be followed by
a cure. The opening in the vagina is more effective, and the direct
opening through the abdominal walls both more efficacious and
more hazardous, than any of the others. When a communication
is perfected and perpetuated between the cavities of the tumor
and the peritoneum, the surface of the latter being a better absorbing surface, the contents are absorbed, thrown into the circulation, and eliminated by excretion through the kidneys and alimentary canal. This process being carried on more rapidly than the
secretion by the tumor, the latter is allowed to contract more and
more, until its secreting surface is wholly lost, and indurated tissue is all that is left behind to mark its former existence. Some
very important precautions are necessary to such happy results,
as will appear by an attentive consideration of the subject. It is
found, for instance, that sometimes the contents of the tumor are
poison to the peritoneal lining of the abdomen, and therefore fatal
inflammation may result from its effusion into the cavity. We
cannot say without an inspection of the fluid, whether this is
likely to occur upon performance of an operation or not, and I
fear that we can by that means arrive at only a presumption upon
the subject. In evacuating for the first time these growths, we
find, occasionally, clear, transparent, good, innocent-looking fluid,
begin to flow, when as the flow continues, the latter part looks
darker, grumous, and ill-conditioned; now, whether we might not
be deceived upon inspection is a matter of question, and really
furnish a virus to the surface of the peritoneum, instead of the
bland albumen of the healthy ovarian tumors. However this
may be, we do know from cases placed on record, by Dr. Simpson
particularly, and observed, not unfrequently, that these tumors
do sometimes burst into the abdominal cavity, and disappear,
without any bad symptoms, so that we are justifiable in hoping
the artificial opening may result well. Dr. Simpson recommends
(and it is certainly the most sure way, although, as I have remarked, we must under all circumstances, be in doubt), prior to
opening communication with the peritoneal cavity, that we tap
the tumor, and remove some of the fluid for examination, and if it

is the ordinary bland, mucilaginous, transparent substance found generally after first tapping, he assures us we may proceed to the operation unhesitatingly, or rather, may keep the puncture in the sac open afterwards, instead of allowing it to close up, as it usually does. This is done by, in the first place, not removing nearly all the fluid from the sac by tapping, but allowing enough to remain to keep it partially distended; and in the second place, every twenty-four hours so to press upon the tumor as to well up the fluid through the opening in the sac, and thus break the slight adhesions which may have formed between the edges of the wound, and allow it to escape into the peritoneum. Dr. Simpson thinks this is the safer way, so far as the danger from the operation is concerned, but as will be seen, not so certain of accomplishing the object. He has cured cases in this way. The most effectual and the most dangerous way is, to cut down upon the tumor, and remove a piece from its wall large enough to insure patency, withdraw a part of the fluid, and then close the wound in the abdomen, and allow the rest of the fluid to flow into the peritoneal cavity thence to be absorbed. The immediate danger in this operation is, that of dividing some of the bloodvessels which ramify through the walls of the tumor, and thus allow internal hemorrhage to take place. To avoid this, it is recommended by Mr. Brown to draw out, examine, and divide, only that portion which is clear of vascular ramifications. Others have recommended to tie any branch large enough to bleed. There is but little doubt that the precaution recommended by Mr. Brown would be sufficient to avoid that difficulty. The large wound through the peritoneum makes the chance of inflammation in that membrane greater than the mere puncture of the trocar. Upon the whole, I think I should prefer Dr. Simpson's plan of keeping the opening made by the trocar in the tumor patent, by frequent well-directed manipulation. It ought to be practised I think oftener than every twenty-four hours; as often as every twelve, for the first two days. It will probably be found upon extensive trial, that it may not always be practicable. Should there be adhesion at the point where the trocar passes, it would necessarily fail.

The plan for making a fistulous opening externally, is more practicable perhaps than the one just detailed, from the consideration that it is more manageable.

The operation is simple, and not attended with much immediate danger; the danger coming in the shape of acute inflammation soon after the operation, or exhausting suppurative inflammation and its attendants. Mr. Brown, who has given it a more extensive trial than anybody else, selects a point midway between the umbilicus and the anterior superior spines of the ilium of the side in which the tumor originated. His plan is to make an angular incision at this point down to the peritoneum, dissect up the angle from that membrane so as to completely expose it, evacuate the tumor through this exposed part with a trocar, stitch the sac to the sides of the opening, enlarge the puncture in the cyst, and keep it open by a pledget of lint or other substance as he finds most convenient. Others cut down to the peritoneum, at a point midway between the umbilicus and symphysis pubis, stitch the sac to the sides of the wound, and keep open by lint or stomach-tube. Care should be taken, especially if the contents of the sac should have a suspicious appearance, to prevent it escaping into the peritoneal cavity. Often there is adhesion at this part, when the stitches will not be necessary. This opening should be kept patent until the cavity of the cyst is lost by contraction, inflammatory adhesion, or granulation, or all these combined, which is probably the common mode of their disappearance. Some difficulty will be found in doing this, there is such a strong tendency in the wound to contract and heal up by granulation. If necessary, we may from time to time somewhat enlarge it with the knife, and we should not allow it to close until the discharge has entirely ceased. From what I can see of the dangers of this operation, they are very little, if any, less than from ovariotomy; and I should not feel induced to resort to it, unless it were in a simple cyst, where tapping, injection of iodine, or the use of pressure had entirely failed, or where after exposing the cyst ovariotomy was found impracticable, from extensive adhesions. This I have done in one instance. The adhesions were so extensive, that the cyst could not be removed, in fact, they seemed to be about universal; the incision was small, only admitting two fingers; the sac had adhered at the point where the opening was made, so the incision was all that was necessary in the way of an operation. The patient died of acute peritoneal inflammation in three days afterwards. A post-mortem examination revealed extensive inflammation of the sac and peritoneum.

Professors Kiwisch and Scanzoni, of Wurtzburg, are warm advocates of a fistulous opening through the vagina into the tumor, to be kept open until the same obliteration takes place that was spoken of as occurring in the case of opening through the front walls of the abdomen. Scanzoni operated on fourteen cases: eight resulted in a perfect cure; in two, the fluid collected again in a few weeks, one died of typhus fever two months after, and three were lost sight of. In none of the fourteen did death occur as a consequence of the proceeding. He mentions one case only, in his whole experience, in which death occurred from peritonitis, and that was Prof. Kiwisch's case. Scanzoni admits its danger, but shows quite a favorable opinion of it. Dr. West gives three cases of his own, two of which were cured, but had formidable inflammation; the third died, not as an effect of the operation, but from something else, which he does not state. Scanzoni taps with a trocar through the vagina, and allows the canula to remain until the cure is effected.* This, of course, occupies a variable time; the tube is withdrawn by Scanzoni by the eighth or tenth day in some cases. He says that some of his cases recovered without any sign of inflammation, or other inconvenience. Dr. West operates by introducing the trocar, and withdrawing the fluid, passing a number twelve catheter through, and removing the canula over the catheter. The catheter is allowed to remain until the cure is complete. Simple cysts are the only kind that can be cured by the fistula method, and the recommendation to tap once, to be sure that there is but one cyst, is a good suggestion. The cyst cannot always be reached from the vagina, and only in such cases as it is crowded down into the pelvis, so as to give obvious fluctuation in that canal, should we think of this operation.

The third object in the treatment, partial or complete removal of the diseased mass, remains to be considered. Could the tumor always be removed when the operation is once commenced, one of the greatest objections to ovariotomy would be removed, and the

---

* The only case I have operated on in this way died of pyæmia from suppuration of the cyst. The canula remained for fifteen days.

question of the propriety of attempting it would be very much simplified; but, unfortunately, the operation cannot only not be finished in many instances when attempted, but it is utterly impossible in our present state of knowledge to predict with the most favorable opportunities whether difficulty will arise or not. The only obstacle worth considering, if not the only one, to completion of the operation of ovariotomy, is the adhesion of the tumor to the abdominal walls or viscera so strongly and largely that it is impossible to separate them, and if separated, to add very much to the dangers of inflammation. This obstacle and its dangers are so great as to entirely deter many courageous surgeons from performing ovariotomy, or allowing it to be considered a legitimate operation; but, notwithstanding the above facts, when confined to appropriate cases, with appropriate tentative measures, care, and the avoidance as far as possible of the objectionable features of the operation itself, in the modes of performance more particularly, and the desistance from or change in the operative procedure when much adhesion exists, I should unhesitatingly advise a resort to it.

Some of the effects which generally intervene between the surgical violence of the operation and fatal termination are, 1st, peritoneal inflammation; 2d, hemorrhage from the peduncle; 3d, pyæmia; 4th, abdominal shock; 5th, exhaustion.

The danger from the above conditions is somewhat in comparison with the order in which they are named. The most danger is to be expected from peritoneal inflammation, next from hemorrhage, &c. Now, an operation ought never to be performed when possible to do otherwise, in such a manner as to *increase* the risks of any one of the above conditions. Everything should be avoided, so far as at all practicable, that irritates the peritoneum in any shape; perfect security from hemorrhage should be an indispensable consideration. It is not supposable that with every possible improvement the operation will ever become safe, but there is every rational position in favor of the assumption, that many deaths may be avoided by thus rigidly adhering to a determination to leave nothing undone to lessen all the above-mentioned causes of death. Success in a great undertaking, notwithstanding bad management, should not encourage us to hope for a like purely fortunate result. The more good sense and science we can combine in an operation, the more likely we will succeed. Many good men do things they know hazardous, *when they could as well avoid them*, because they have been successful in the same way before. Dr. Tyler Smith read four cases of ovariotomy before the London Obstetrical Society, which are published in the Lancet, for September, 1861, in one of which he says: "The ligatures and stump were cut off as closely as possible, and the whole returned into the pelvis, with the expectation that they would be quickly enveloped in coagulable lymph, *so as to prevent injury*. The patient recovered without a single bad symptom, taking no medicine but thirty drops of laudanum." This woman recovered well, and will justify similar procedure when unavoidable, but the treatment was bad, and cannot be justified. Mr. Spencer Wells very justly reprimanded him for it in a gentlemanly way, by saying that the procedure ought not to be imitated without more discussion. Before undertaking to avoid or prevent any of these untoward circumstances which result from ovariotomy, it will be necessary to examine more at length the immediate causes of them. And first of peritoneal inflammation. One of these causes, to a great extent unavoidable, is exposure of the peritoneal cavity to the contact of air, and, in many instances, unnecessary handling or contact with the fingers, sponges, cloths, &c. The greater in amount, the longer duration either of these sources of irritation are applied, the more danger, notwithstanding immunity observed in cases where both were extreme. If contact of the atmosphere is slight, and its duration short, we must have less danger to apprehend than if the whole abdominal cavity were exposed for a long time; and it cannot be a philosophical conclusion that a small wound, two inches long, exposing but a small extent of the peritoneum, open but ten minutes, is as dangerous as an incision from sternum to pubis open for an hour. · Much less is it rational to handle all the viscera, wipe them, and return them into the abdomen, and perhaps sponge out the whole cavity. That women have recovered from such rash acts is no reason why they are proper or allowable when it is possible to avoid them. While this kind of causes of peritoneal inflammation cannot be wholly avoided, all that is possible should be. The incision should be no larger than necessary to permit the extraction of the collapsed tumor, and the hand should never touch the peritoneum when adhesions do not make an enlargement of the

incision necessary and the introduction of the fingers indispensable; then the enlargement of the opening should be carried to
the least possible degree, and the handling as slight as necessary
to accomplish the object, either to separate the adhesions, or satisfy
us that this is impracticable. Perhaps the return of the ligature
into the abdomen is fraught with more danger than any other of
the ordinary parts of the operation. Its contact with the peritoneum will irritate it, and awaken inflammation, unless in very fortunate cases it is surrounded with fibrine, and thus the danger
averted. It should be remembered, however, always, that it is a
danger averted, and that the limit of the inflammation is the aver-
sion of the danger, and that the presence of the ligature is the
danger. The inclusion of the ligature, although it may not light
up extensive and rapidly fatal peritonitis, as a focus of phlegmonous inflammation may generate pus enough to poison the blood,
and induce pyæmic fever and exhaustion. It is most likely that
pyæmia results more frequently in this than any other way. But
the inclusion of the ligature is the cause of the next most frequent
condition effecting death, after ovariotomy, that is, hemorrhage.
However secure the ligature may be applied apparently, there is
no certainty that bleeding will not occur from the peduncle.

Dr. West introduces a table which, perhaps, shows the subject
in its proper proportion. In fifty-nine cases,

29 proved fatal from peritonitis.
13   "   "   hemorrhage.
8   "   "   exhaustion.
2   "   "   shock.
3   "   "   abscess.
2   "   "   ulceration of intestines.
1   "   "   tetanus.
1   "   "   phlebitis.
—
59

It will be seen that a large proportion of the deaths occur as the
consequence of hemorrhage in the peritoneal sac; the patient
bleeds to death internally. But if the hemorrhage should not be
great enough to exhaust and cause death as a hemorrhage, as an
irritating substance it may produce peritoneal inflammation, and
in that way cause death. Ulceration of the intestines, in conse-

quence of the ligature lying in contact with them, is another evil, resulting from the presence of the ligature in the abdominal cavity. The principal objections to the ligature, and which should warn us against its use either internally or externally, are, that it causes peritoneal inflammation by lying in contact with it in the cavity; by inducing strangulation and sending forth the inflammation along this very susceptible membrane, thus causing it to become general; by inducing phlegmonous inflammation, exhausting abscess, and, perhaps, also causing pyæmia; and, lastly, the great uncertainty in securing the vessels of the stump against hemorrhage. I regard the dangers here mentioned against the ligature as adding so much hazard to the operation that I could not think of making use of it, unless it seemed entirely indispensable in the particular case. More can be done, fortunately, to prevent peritoneal inflammation and internal hemorrhage than any other conditions which are found to prove fatal by their presence. The shock which succeeds the operation, now that chloroform can be made to shield the nervous system against the effects of such extensive violence, is not so much to be feared, but should always be carefully guarded against. If inflammation and hemorrhage can be prevented by improvements in the operation, even in fifty per cent. of the cases, fatality will be very materially decreased. Whatever may be the teaching of experience as to the tolerance of and recuperation from violent and rash procedure in the performance of so dangerous an operation, it cannot justify us in making use of means or methods that common sense, physiology, and pathology, all combine in declaring more hazardous than other means and methods which we can employ. Let, therefore, no more incision and exposure, or handling of the viscera, be allowed than is absolutely necessary to extract the collapsed tumor, and discard scrupulously the ligature and clamp. After so much of an introduction by way of expressing my conviction that too much preventive measures cannot be employed, and that the operation is too often more hazardous than it need be, by following the examples of men who have succeeded in spite of rashness and carelessness, I am ready to describe what I consider the best method of operating. The best time for operating is about the middle of the menstrual month of our patient, and if consistent with the circumstances of the case between the first of October

and the first of June in this climate; but, of course, all will depend upon the urgency of the symptoms in the case as to this last
item of time.   The patient should not be too much prepared, as it
is easy to get up functional disturbance, and thus with medicines
intended to place the patient in more favorable condition for operation, get the system in a state the least calculated to resist the
attack of traumatic disease.   Above all, we ought not to disturb
the abdominal organs with a cathartic; and if the stomach, liver,
or other organs require a purgative, it should be given three or
four days before the time to operate, and we should give them
time to recover their tranquil tenor of function before we venture
upon it.   Should the bowels be constipated, we may give an enema
six or eight hours beforehand, but not oil or other cathartics.   It
is good practice to give a pretty full dose of opium an hour before
we put the patient upon the table.   This is all the preparation she
needs, provided she is in good condition of general health.   Of
this last we should be sure when practicable.   We should at least
satisfy ourselves that there is no predisposition to the suppurative,
erysipelatous, or inflammatory diathesis.   For three hours she
should be in a state of perfect physical and mental tranquillity, if
possible to secure it, and at the time placed in a perfect anæsthetic
condition before being removed from her bed, and without seeing
any of our preparations.   *We* should be prepared by carefully
passing in review the successive steps of the operation minutely
and deliberately, performing the operation mentally first, making
provision for all possible complications or embarrassments, and
carefully placing everything we shall need, as suggested by this
kind of reflection, just where it can be most handily available. We
should next state to our assistants what we expect to do in each
successive step in the operation, and the part we desire them severally to perform under all given circumstances and times of the
operation.   This complete understanding between all the parties
who partake with us in the operation, will preclude the danger of
confusion and possible imperfection in the procedure.   We want a
good scalpel, blunt-pointed bistoury, large trocar, two large ligatures in needles, an ecraseur, half a dozen silver pins, three inches
long, with movable steel-points to them, several large waxed ligatures, three or four small needles, armed with silver wire, a tenaculum, silk for artery ligatures, and it is well to have small artery

forceps. This will constitute a pretty full operative armament. We shall also want two or three fine sponges of different sizes, a flannel roller large enough to make a binder for the patient, one pan of cold and one of warm water, some adhesive plaster, and old cloths of linen for compresses. Our table should be firm, about two feet wide, six feet long, and conveniently high, covered with sufficient soft quilts not to be too hard, and placed near the lightest part of a large and airy room, brought to a temperature of about 70° Fahrenheit. When all the above preparations are made, we are ready to operate. We should have at least three intelligent physicians as assistants,—one to act as main assistant, one to use the chloroform, and one as general handy man.

Supposing our diagnosis to be as complete in all respects as possible, especially with reference to the nature of the contents of the tumor and the adhesions, &c., we may cause the bladder to be evacuated, etherize and place our patient, on the table, on her back, with the limbs hanging over one end to the knee. We should stand at the end of the table in front of the patient, when the large protuberant abdomen will rise before us, and the part below the umbilicus be completely exposed. With the scalpel, the operator should make an incision in the linea alba, midway between the umbilicus and symphysis pubis, about two inches long, through the walls of the abdomen down to the tumor, but he should avoid opening the cyst with the knife. It will require some care to avoid the cyst, particularly as the aponeurosis is thin at this place. The strokes should be light, and an examination made for the peritoneum after each touch of the knife, as soon as near the proper depth has been reached. When the sac is exposed, we should plunge the trocar into it and evacuate it, and, so far as possible, extract it through the wound. If the tumor is multilocular this procedure will soon bring another cyst to the incision, when it should be punctured and evacuated and thrown out through the wound, the next punctured, &c., until the whole is reduced sufficiently in size to be drawn entirely out. Should there be a solid portion at the base of the tumor so large as not to come out of the wound easily, we may enlarge the latter sufficiently. We occasionally meet with instances in which the contents of the different sacs are so thick and tenacious that they will not flow through a canula of any size. It would be improper, when found thus thick and gelatinous in consistence to continue our efforts to evacuate them, but we should enlarge the wound sufficiently to remove the whole mass, there being less risk in this procedure than to allow the peritoneal cavity to be irritated by the contact of such material in the vain hope of getting it entirely away. All this being done, the tumor lying outside the abdomen, the pedicle passing through the opening should be pierced on each side with one of the needles, with the large ligature as near as possible to the tumor. This is for the purpose of giving us perfect command of the stump after the tumor is separated. If the peduncle is thin, the chain of the ecraseur should now be thrown around it at the base of the tumor, leaving the stump as long as possible, and slowly crushed through it. We should next place the stump in the wound, leaving its edge a little above the level of the skin, and by means of the silver pins one inch apart surround them with the thread, and secure perfect adaptation of the lips of the wound. The pins should be made to enter the skin about an inch from the edge of the incision on one side, dip down as near to the peritoneum without touching it as we can well effect, and, piercing the opposite lip, come out about an inch beyond the other side of the wound. If the edges gape at all, some superficial stitches with the fine silver wire should close this more completely. A compress of wet lint should be placed over the wound, and all surrounded by the flannel roller, the patient removed and placed in bed, and allowed to come from under the effect of the chloroform. This operation is simple, easily performed, and, I think, inflicts as little dangerous violence, if not less, than any other operation the details of which are given. The extent of the wound is the least possible for the purposes of it. No handling or even rough contact of the viscera is required. The peritoneal cavity is exposed in the smallest extent, and for the shortest time, and all the objections and dangers urged against the ligature around the stump or in the abdominal cavity are obviated and avoided. The separation of the peduncle with the ecraseur, and the placing the edge outside the wound, so that if hemorrhage does occur it is outside, and is within the control as well as observation of the medical attendant, lessen, it seems to me, very materially the *causes* of inflammation at least, and as a logical, if not actual sequence, the dangers of the operation.

When the peduncle is thick or vascular, particularly the latter, we ought not to trust to the ecraseur merely, but the stump ought to be embraced in a clamp. This instrument may be found made by most of our instrument-makers. It consists of two light steel bars, about four inches long, serrated on the sides intended for coaptation. These two steel bars are placed one on each side of the peduncle close to the tumor, approximated and held firmly together by one screw at each end passing through them. The screws are tightened sufficiently to prevent hemorrhage, and the whole left outside the wound, when this last may be closed as before recommended. The most convenient way of arranging the clamp is to turn it across the wound, and, when the peduncle is long enough, to place it at the lower angle.

But sometimes the attachment is so very short that it is not practicable to bring the stump out of the wound. When such is the case, we may pass a needle armed with a double strong hempen ligature, and tie it very tightly on both sides of the peduncle, cut off, and drop it back into the cavity of the abdomen, leaving the ligature through the lower angle of the wound. We may then close it as before recommended.

Very many cases, however, which promise to be favorable for the above procedure, so far as we can judge from examination beforehand, as we proceed in the operation we will find cannot be terminated without much more difficulty, and, in some, our procedure must be very materially varied from the above detailed plan.

The main and almost only obstacle to the performance of this operation is adhesion of the tumor to the abdominal walls or viscera. The amount and firmness of adhesion will determine the extent and nature of the varying steps. The incision through the abdominal wall should be the same in all cases, and unless adhesion be at the point where the incision is made, I would evacuate the presenting cyst, and remove it through the opening as far as practicable, and if the adhesions were small and easily overcome, do so by traction, or if to the intestines or omentum, they might be separated, after bringing them under the eye through the wound, and we should thus proceed to evacuate separate cysts as far as practicable. Should this plan prove ineffectual after persevering trial, the opening should be enlarged to four or five inches in extent, and then a careful and thorough inspection of the condition of the adhesions be made. When adhesions are present we find them, 1st, in long bands of false membrane, reaching from the abdominal walls to the tumor; 2d, delicate flat adhesions, fastening the omentum to the tumor; or, 3d, close, firm adhesions of the flat surface of the abdominal walls to the sides of the tumor. When a tumor grows rapidly, the adhesions are apt to be drawn into long bands. These bands are sometimes so tough that great force is required to separate them. They are sometimes four or five inches in width, and thick and strong. They may be separated by the ecraseur. I should not think of tying them, as they are poorly supplied with vessels, and separation by this instrument is sufficient to prevent hemorrhage. Not unfrequently there are five or six of these bands. The second sort of adhesions are next in frequency to the first. The adhesions of this form should be overcome by the fingers. The fingers may be passed between the tumor and omentum, and with great gentleness overcome. We can often do this at the points of attachment without tearing the vessels or substance of the omentum. Should any vessels bleed, we may tie them with fine silk ligatures and cut them close. The broad, flat adhesions between the peritoneal covering of the tumor and the walls of the abdomen occur in tumors of slow growth, and are the least frequent and most difficult to manage. They may be so loose that we can tear them apart. When this is practicable, it is very much the best mode of overcoming them. But often they are so firm that the peritoneum would be separated in the attempt to do this. When such is the case, we may cut them off at the edge of the attachment with a knife or scissors, being careful to look for any source of hemorrhage, and tie it when found. Ordinarily, the inflammation which caused the adhesions also closes the vessels in such a way as to make them less liable to hemorrhage. If it should be practicable to withdraw a considerable portion, or even if not, we should lay open that part at or outside the incision, and turn the sides over each edge of the wound, and through the incision in the cysts try, and if possible evacuate the fluid from all the remaining cysts, and in this way as perfectly collapse the tumor as possible. If practicable, we ought not only to puncture the cysts that we cannot remove in this way, but remove a part of the partition between those remaining, so as to convert them all into one cavity. We should now separate all the tumor outside with the ecraseur, first securing it so that the portions remaining do not pass beyond our reach into the abdominal cavity. Bringing the edges of the amputated sac above the surface, it should then be secured by the pins, as in the first proceeding, and the wound should be closed up, except to the extent of half an inch at the lower angle, which should be kept open by lint, or the insertion of a flexible tube. If none of the sacs can be removed after puncturing and destroying as far as possible the partitions, we should keep the wound open as above directed. The after treatment should have reference to the danger from the shock, from which the system may not recover itself. Peritoneal inflammation is by far the most dangerous; and where a part or the whole of the tumor is left, inflammation of the cysts or hemorrhage from the amputated stump.

Mr. Brown recommends small pieces of ice in the mouth to suck, as one of the means preventive of inflammation; and as they are very grateful, particularly the second day, I would join him in it. The bladder must be watched, and evacuated at least every eight hours. If the patient can do this spontaneously, it is well; if not, the catheter ought to be resorted to. Every two or three hours the dressing in the region of the wound should be inspected, lest a slow but exhausting hemorrhage occur, and the chances of our patient be lessened.

There cannot be any very definite directions given as to the treatment of inflammation, or any other of the effects of the operation, other than may be found with reference to them in any well-digested treatise upon the subject.

We must be watchful to perceive the very beginning of them, and energetic in the treatment. An incipient inflammation may be subdued much easier than a well-established one. In instituting measures to combat inflammation or any of these effects, we must carefully consider the condition of our patient, there being no question that tonics and even stimulants, with anodynes, are better adapted to cure inflammation than depletion in patients of this kind under certain circumstances. It is only in the young and robust, well nourished at the time, that depletion can be extensively used; in moderation, when early used, there are but few who will not be benefited by it, even when we must follow it with quinine and stimulants. There are a few chronic cases in broken-down constitutions, emaciated, highly debilitated, and particularly dyspeptic, with depraved secretions generally, that we must sustain from the time of the completion of the operation, in order to prevent inflammation and exhaustion, one or both.

In illustration of the after treatment of and the course pursued by these cases after the operation, I subjoin three cases.

Case 1. Mrs. H——, aged 38 years, has been gradually increasing in size with ovarian dropsy, for the last four years. She first observed a small tumor in the right iliac region. Before removal, it filled up the pelvis and abdomen, until the patient was as large as at full term of pregnancy. Her suffering for several months had been very great, on account of impaction of the pelvis, and her general health was rapidly deteriorating. The tumor was multilocular, and, with its contents, weighed nineteen pounds.

The operation for its extirpation was performed in presence and with the assistance of Drs. Shumway, Cheeney, Davis, Bevan, and Jones. An incision between three and four inches in length in the linea alba, about equal distances from the umbilicus and pubis, enabled us to draw off the contents of the numerous cysts with the trocar, and extract the whole tumor with great facility and despatch. The pedicle was transfixed by a needle carrying a large double hempen ligature, and tied in two parts firmly as I could draw the twine. The wound was closed by three silver pins, the lower of which was passed through the middle of the pedicle, so as to confine the cut edge upon a level with the skin of the abdomen. I should have mentioned that, after being tied, the pedicle was divided between the ligature and the tumor with the knife. When the tumor was first exposed, there were within view, ramifying over its surface, a number of large veins; several of them were larger than a goose-quill. The pedicle was large and fleshy, showing several large venous trunks. There were no adhesions anywhere; and the only obstacle to a ready removal of the tumor, when lessened by evacuation, was caused by a cyst that completely filled up the cavity of the pelvis. It was so completely moulded into the inequalities of that cavity, that some force and address in manœuvring were requisite to lift it out. The wound was covered with a water compress, and the abdomen encircled with a broad woollen binder. The operation and dressing were finished in eighteen minutes after the patient was completely under the influence of chloroform. Dr. Shumway, whose patient she was, and to whom much credit is due, for his assiduity and skill in the after-management of the case, kept full and accurate notes; they are highly interesting on several accounts, and will be read with profit by those who are studying the subject. They are subjoined.

Operation was completed at 3 o'clock and 33 min., Oct. 29th. 4 P.M., vomited once a little water; pulse 76; complains of smarting pain at seat of wound, and severe pain in the bowels, "like the pain of colic;" ordered tinct. opii, gtt. 40. 6 P.M., much more comfortable; says that she has not a particle of pain in the right side, but complains of pain in the left hypogastric region and down the left thigh; pulse 88, full and soft. 6.30 P.M., pain continuing, gave tinct. opii, 30 gtt. 8.30 P.M., still complains of severe pain in left hypogastrium and thigh; pulse 88; gave tinct. opii, 40 gtt. 11.30 P.M., patient easier; has slept, with a few minutes' intermission now and then, since 9 P.M.; pulse 88, as before. 12 P.M., return of pain; gave tr. opii, 30 gtt. Thursday, 30th, 3 A.M., used catheter, at patient's request; comfortable; pulse 88. 9 A.M., pulse 90, full and soft; complains of return of pains in the left thigh; gave tinct. opii, 30 gtt. 12 M., complains of thirst, otherwise very comfortable; ordered ice, *ad libitum;* repeat tr. opii, 30 gtt. 3 P.M., symptoms unchanged; relieved bladder with catheter; tinct. opii, 30 gtt. 5.30 no change, patient comfortable. 9 P.M., patient cheerful; ordered tr. opii, continued every three hours if she is restless or complains of pain. Friday, 31st, 8.30 A.M., found patient comfortable; has slept at intervals during the night; pulse 95; skin cool; does not complain of pain, except a slight smarting at the seat of wound; catheter. 8.30 P.M., finds patient feverish; pulse 100; countenance anxious; skin dry and hot; very restless and irritable; much troubled by secretion of mucus in the trachea, with desire to cough; slight fulness of the abdomen; had taken 40 drops tinct. opii; at 4 P.M., ordered 40 drops more; as she was still unable to pass water, her bladder was relieved by use of catheter. 10 P. M., less bronchial irritation; continue tr. opii, 30 gtt. every three hours. Saturday, Nov. 1st, 8.30 A.M., more comfortable; pulse 96, softer and fuller; has slept at intervals during the night; tympanitis increased; ordered em. terebinth.; continue tr. opii. 12 M., considerable tympanitis; complains of pain in the bowels, with desire to go to stool; upper part of the wound looking well; slight phlegmonous inflammation about the pedicle; pulse 100: continue em. terebinth. and tr. opii. 6 P.M., tympanitis increased, but less tenderness over the abdomen than there was last night; more cheerful; pulse 96, soft and full; gave 5 grs. calomel: continue em. terebinth. and tr. opii, with b. c. soda. Sunday, Nov. 2d, 8.30 A.M., found patient cheerful; she says that "she is almost well, and meant to have had her clothes changed before the doctor came;" nurse was engaged making her toilet; ordered all operations of that kind to be suspended; pulse 100; tympanitis increased, but no tenderness over the abdomen; continue em. terebinth., tinct. opii, and *perfect rest.* 4 P.M., found patient much worse; was taken soon after we left in the morning with severe pain in the left hypogastrium; pulse 128, small and quick; skin hot and dry; tympanitis or flatulence much increased; gave tr. opii, 40 gtt., and ordered enema of

Ol. Terebinth., ...	℥ss.
Ol. Ricini, ...	℥j.
Yolks of two eggs,	
Gruel, ...	Oj. M.

6 P.M., enema retained; gave another of soapsuds, which brought away a large amount of gas, but no feculent matter; pulse 120; pain subsided; abdomen tender; wound looking tolerably well; union appears firm at the upper part; considerable phlegmonous inflammation about the pedicle; ordered the em. terebinth. and tr. opii, continued every three hours, and one grain sulph. quina to be added to each dose, and the abdomen covered with a poultice of flaxseed meal, after being rubbed with ol. terebinth., care being taken not to get any of the oil into the wound; gave fl. ext. rhei and senna, ℥ij. Monday, 3d Nov., 8.30 A.M., patient rested tolerably well through the night; pulse 116; abdomen very much distended; wound firmly united at the upper part, but looking unhealthy about the ligatures; cut the ligatures, and removed two of the needles; a serous discharge followed the withdrawal of the needles; ordered beef-tea and milk-punch freely; pinned a bandage tightly around the abdomen; she does not complain of pain upon pressure; repeat fl. ext. rhei et senna. 12 M., abdomen

enormously distended; patient restless; pulse 120; gave enema, which brought away a large amount of gas, which much relieved the distension of the abdomen; repeat the rhei and senna. 6 P.M., pulse 120; distension very great; applied adhesive straps to support the wound; complains of no pain or tenderness; bowels have not moved; ordered enema, which only brought a discharge of gas; gave 12 grs. calomel. Tuesday, 4th, 8.30 A.M., bowels have not moved; pulse 116; only complains of the distension of the abdomen; ordered bot. citrate magnesia. 2 P.M., the bowels not having moved, gave enema of ol. terebinth., which produced a free movement of the bowels, with discharge of a large quantity of gas, almost entirely relieving the abdominal distension. 5.30 P.M., patient expresses herself as "almost well;" pulse 108; skin cool; ordered injection repeated in three hours if the bowels do not move again in the meantime; continue milk-punch and beef-tea, and give tinct. opii, 30 gtt., after the bowels move again. Wednesday, 5th, 9 A.M., the injection last evening produced a complete collapse of the abdominal wall; patient slept nearly all night,—is very comfortable this morning; pulse 106; no pain; no tenderness, except at seat of wound. 5 P.M., improving steadily. Thursday, 6th, 8.30 A.M., pulse 100; no unpleasant symptoms; wound around the pedicle discharging freely; takes considerable nourishment. Friday, 7th, same as above. Saturday, 8th, removed the remaining needle which transfixed the pedicle,—ligature firm. Sunday, 9th, wound looking well; patient looking and feeling very well,—appetite good.

Case 2. I was called, Nov. 5th, 1862, to see Miss P., aged 20 years, at Eleroy, Illinois, suffering with ovarian dropsy. There had been nothing remarkable in the condition of her health, although of rather spare and fragile form, until February last, when she first observed a tumor, almost the size of an orange, in the right iliac region. Her menses ceased to make their regular appearance about the same time. The tumor had grown quite rapidly, and she had suffered several weeks severely from pressure. At the time of examination, she seemed larger than most women at full term of pregnancy. The patient was examined by, and in consultation with, Drs. L. A. Mease, B. J. Buckley, F. W. Hance, E. C. De Puy, and John Charlton, of Freeport, Dr. R. Hayes, of Lena, and Dr. J. A. Darling, of Eleroy. It was unanimously de-

cided that the tumor was multilocular; and as it had grown so
rapidly, and the patient had begun to suffer from its great size—
she could not long survive if not relieved—that extirpation was
the only means of cure advisable, and that owing to the probabilities of a large portion being solid, and the existence of adhesions,
the chances of success were less than ordinary. The conclusions
of the consultation being submitted to the patient, with a heroic
determination that I think had much to do with her recovery, she
begged us to give her what she considered the only chance of escape from a lingering and sure death. With the assistance of the
above-named gentlemen, the operation was performed in the following manner : After anæsthesia was induced by chloroform, an
incision in the linea alba, midway between the umbilicus and symphysis pubis, about two inches long, exposed the tumor and evacuated several pints of peritoneal effusion. Upon introducing the
finger to survey the tumor, some slight adhesions were torn through.
A large trocar was next plunged into one of the presenting cysts ;
as no fluid flowed out of the canula, it was withwrawn. Attached
to it was a thick glutinous semifluid, that was so tenacious as to
admit of being drawn into a string two feet long. It was evident
that the contents of these sacs could not be thus evacuated. The
external incision was enlarged until it was about five inches in
length; the abdominal walls were pressed closely against the tumor,
a free incision made into the cyst, and the contents, almost as
thick and dark as tar, pressed out. The same procedure was repeated upon several sacs until the size of the tumor was considerably decreased. Upon drawing the partially collapsed tumor forward, and examining its sides, firm and extensive adhesions were
discovered in every-direction in which the examination was pushed.
Much of them gave way under the fingers by using considerable
force ; there were, however, five bands of fibrine, from two to three
fingers wide; two of them were far around toward the spine,
which, on account of their firmness, had to be separated by the
ecraseur. The external wound was again enlarged upward until
it was about nine inches in length, and the tumor lifted out of the
abdominal cavity. After passing a double hempen ligature through
the centre of the pedicle, and securing it by tying each side firmly
as possible, the chain of the ecraseur was passed through it close
to the tumor, and above the ligature. Owing to the careful atten-

tion of the gentlemen present assisting, very little of the contents
of the sacs, and probably no blood, found their way into the abdominal cavity. The external wound was now closed by four pins
and several silver sutures; the stump was transfixed and retained
in the wound, with its surface even with the external surface, by
the pin nearest the pubic extremity of the cut. I should have
before stated that the great omentum, which lay on the upper part
of the tumor, was adherent throughout the whole extent of contact; but these adhesions were so feeble, that they gave way under
pretty smart force exerted by the fingers for that purpose.

Very little blood was lost; and the patient bore this terrible
operation without any appearance of shock or depression whatever.
The time occupied in completing the operation and dressing was
forty-five minutes. A compress wet with water was placed over
the wound, and secured by a broad flannel binder. The tumor,
with its contents, weighed thirty pounds. After witnessing the
extensive adhesions, peculiarity of contents of the tumor, &c., all
present joined in expressing the opinion that recovery was hardly
to be thought of as a possibility.

It will be noticed that, in these two cases, although as great care
as practicable, on the account of the difficulties of the case, was observed in avoiding extravasation or effusion in the peritoneal sac,
no effort by sponging or wiping among the intestines, was made
to remove any substance that did escape. Such was the case,
also, in the first operation I ever performed; and I cannot but express the conviction, that the amount of ovarian fluid should be
very considerable, or acrid in quantity, to justify the rough operation of sponging it out. The notes of the case, after the operation, were kept and forwarded to me by Dr. J. A. Darling, of
Eleroy, and I have not altered them, believing them to be a faithful exhibit of the progress towards a cure. Although somewhat
lengthy, they show but one interesting circumstance, which is, that
"our patient" recovered from the effects of the operation without
a single bad symptom.

Operation finished Nov. 5th, 12.30 P.M.; pulse feeble, and 120
per minute; countenance pale; nausea and vomiting from effects
of chloroform; complains of pain in back; took teaspoonful of
tinct. opii. 2.30 P.M. nausea continues; less pain in back; lies
on side, disposed to doze; pulse 120, good volume; natural color

returning to face; respiration 40 per minute. 5 P.M., pulse 112, full and soft; resting well; respiration 40 per minute; complains of pain in back; some thirst; took teaspoonful of tinct. opii, which was immediately rejected. 7 P.M., pulse 104; nausea; skin cool, soft and natural; ¼ gr. morphia, which was immediately rejected; complains of pain in back; slight desire to urinate, which was relieved by catheter; has slept some last two hours. 8 P.M., pulse 120; more urgent nausea, otherwise comfortable; ¼ gr. morphia, immediately rejected. 8.45 P.M., ¼ gr. morphia, retained. 10 P.M., ¼ gr. morphia; some restlessness. Nov. 6th, 2 A.M. ¼ gr. morphia; dozing. 5 P.M., has slept some; ¼ gr. morphia; pulse 125; desire to urinate, relieved by catheter. 8 A.M., pulse 118; skin cool, soft and natural; ¼ gr. morphia; some nausea. 10 A.M., ⅓ gr. morphia. 11 A.M., pulse 120; inclined to doze. 1 P.M., skin moist; breathing good; pulse 120, and soft. 1.15 P.M., removed bandage, wound looking healthy; no pain or tenderness; desire to urinate relieved by catheter; pulse 108; 1 gr. opii (Tilden's preparation); skin soft and cool; no nourishment taken. 3 P.M. took a little crust coffee. 3.45 P.M., 1 gr. pil. opii, with crust coffee for drink; inclined to doze. 5 P.M., pulse 120; complains of back, otherwise comfortable. 5.30 P.M., 1 gr. pil. opii; complains of occasional shooting pains in abdomen; pulse 129. 7.15 P.M., 1 gr. pil. opii, rejected. 7.30 P.M. 1 gr. pil. opii, retained; inclined to doze. 8.45 P.M. 1 gr. pil. opii. 10 A.M., relieved bladder by use of catheter. Nov. 7th, 6.30 A.M., have given through the night 1 gr. pil. opii every hour and a quarter; patient has rested well; skin cool; respiration natural; has no pain except in back.

Nov. 9th.—Our patient, thus far, is exceeding our most sanguine hopes. I will continue notes as taken from my book:

Nov. 7th, 3 P.M., pulse 120; tongue somewhat dry and red at tip, with thirst; no nausea. 5 P.M., saw patient with Dr. Charlton; commenced the use of ess. beef, which relishes well; removed bandage, wound looking well; have decreased the opii to 1 gr. every four hours; tongue dry and slightly coated. Nov. 8th, 3 A.M., patient resting well; tongue more natural; pulse 135. 7 A.M., pulse 130. 10 A.M., removed bandage, wound looking good, with very slight suppuration; pulse 130; commenced the use of Tilden's f. ext. veratrum viride, 2 min. every two hours, continuing 1 gr. opii every four hours ; patient cheerful and happy ; continues use of beef ess. and crust coffee ; tongue moist. Nov. 9th, 8 A.M., have continued above treatment through the night ; patient rested well ; slight sweating when sleeping ; complains of some flatus ; gave 3 gr. carb. soda ; pulse 110, with all other symptoms favorable ; have discontinued use of verat. viride. I would here state that I have emptied the bladder regularly. 5 P.M., saw patient with Drs. Buckley and Charlton ; removed bandage, wound looking healthy ; slight suppuration, healing mostly by first intention ; pulse 120 ; tongue moist ; skin cool. Nov. 10th, 7 A.M., patient has rested well through the night ; pulse 120 ; tongue slightly dry ; has no pain ; have given through the night opii, as usual, with verat. viride, every four hours ; patient feels well and cheerful ; thinks she has grounds for hope that she will recover ; there is but slight distension of abdomen.

Nov. 12th.—Our patient is prospering finely ; thus far, everything looks favorable for a recovery. Her pulse this morning is 84 ; tongue moist and clean ; the only disagreeable symptom is wind in the bowels. I moved them yesterday, and shall give another injection this morning. I have continued treatment with opii and verat. viride, the same as at first, also anise-seed tea. I am giving her all the nourishment she will take in a liquid form.

Nov. 14th.—Our patient is still doing well, has no fever, nor any unpleasant symptoms. I have moved the bowels: I used simply an injection of soapsuds with a little turpentine. She begins to eat toast and some roast potato, with a good supply of beef essence. The wound is looking well ; the superior one-half is entirely healed. I have removed one pin and two sutures ; shall remove one pin to-day. I have continued treatment same as formerly.

Nov. 17th.—Our patient is gaining as fast as could be expected. The wound is healing gradually, the superior one-half is entirely closed, the other is suppurating some. I have removed two of the pins and the three sutures, the others I shall allow to remain for a few days. I am treating her now with opium and quinine, 1 gr. each, every four hours. Her pulse is about 110 per minute, soft ; tongue clean and good ; skin soft and moist ; and all appearances favorable. We have strong hopes of a recovery ; I am giving her plenty of nourishment.

Nov. 19th.—Our patient is gaining as fast as we could hope for; she begins to have an appetite for food. I am continuing treatment with opium and quinine; the wound is healing slowly. I have not removed the two lower needles, yet the stump looks well. I have moved her bowels with injections about every other day; they have moved once without any medicine whatever. I see nothing to hinder a favorable termination.

Nov. 22d.—Dear Sir: Your favor of 19th was duly received. I will first answer your questions: 1st. There has not been any hemorrhage whatever from the stump or wound. 2d. There has not been distension of the abdomen at any time of any account; she only complained one day of flatus.

She is now getting along as well as could be asked for: has a good appetite, feels cheerful, and says she wants to sit up. I have removed three of the pins; I thought it was best to allow the lower one to remain for a few days, yet the stump is about on a level with the abdomen. I am giving her quinine, with reduced doses of opium.

February 14th.—Both patients have completely recovered from the operation, and are in good health.

Case 3. Mrs. R——, aged 31 years, is the mother of four children, her last being six months old. Immediately after getting up from confinement with the youngest, she perceived a tumor in the abdomen, which rose above the umbilicus. Since first discovered, it had grown quite rapidly, until now, April 10th, 1863, it reaches to the ensiform cartilage, fills the abdomen to great distension, and causes her much suffering. Upon examination the tumor was pronounced ovarian multilocular dropsy. She desired to be relieved of it in any way that might be deemed best. After consultation with Prof. N. S. Davis, ovariotomy was determined upon, and the 15th instant appointed for the time. The patient was prepared by taking tinct. ferri chl. gtt. xxv, three times a day, good, full diet, and a soluble condition of the bowels preserved by an injection once a day. At the time appointed, the operation was performed in presence and by the assistance of Drs. Davis, Shumway, Bartlett, Cheeney, and four medical students. The chloroform, although very carefully administered, caused so great prostration and difficulty of breathing that much apprehension was felt for her safety.

There was one large cyst, from which about ten pounds of thick serum was removed by the trocar canula, after a small opening through the linea alba had been effected in the usual way; but the other cysts, which were very numerous, were small, and filled with such thick fluid that it would not flow through the largest instrument. The opening was enlarged, in consequence of this fact, to about nine inches in length, when the tumor was turned out of its bed of intestines and removed from the abdominal cavity. Some adhesions, which were not very firm, were overcome by disruption, one only requiring the ecraseur; but what embarrassed the operator most was the great fragility of the walls of the small cysts. These were so tender that the manipulations necessary to remove the mass caused rupture of some of them, and the consequent effusion of their contents into the peritoneal cavity in spite of all care to avoid it. The pedicle was thick and very vascular, hence it was thought best to tie it with a double ligature, as in the two last cases. The wound was dressed with silver pins as usual. The operation, which occupied about forty minutes, was completed at $3\frac{1}{2}$ o'clock P.M. The temperature of the room during the operation was about seventy degrees, and the atmosphere rendered moist by evaporation from boiling water. The shock of the operation, or the effect of the chloroform, left her quite depressed. The pulse was low and weak; the surface pale and unnaturally cool. Dr. Bartlett remained with her. At 4 o'clock and after, he gave her a teaspoonful of brandy in some water every half hour. He gave her also at that hour morph. sul., gr. $\frac{1}{8}$th, as she commenced complaining of pain in the location of the wound. The morphia was administered every half hour also until the pain was relieved, which occurred at 5.30 P.M. The brandy was continued all night every half hour, and the morphia every three hours in same dose as before mentioned. Her extremities, which were too cool, were wrapped in warm, dry flannel. At 1 A.M., 16th, the urine was drawn through the catheter. 8 o'clock A.M., pulse 85; surface natural, tongue moist and clean, spirits good, and the patient in every way comfortable, except occasional nausea.

The bladder was again evacuated, the brandy discontinued, and one tablespoonful of beef essence every two hours. At 1.45 P.M., pulse same in number, but stronger and fuller; some serum oozing from the wound; the skin moist, but a little too warm; some pain in the back and wound; tongue somewhat red on the tip and edges, and coated over the central part white; some thirst, which has never been urgent. Take ice and small quantity of ice-water, continue beef essence, and take ⅛th grain sul. morph. often as necessary for pain.

8.15 P.M., patient restless; countenance anxious; pulse 120, firm, but not full; skin moist, but too warm; pain in the back; abdomen somewhat distended. The patient complains of urgent thirst, and expresses doubt about recovery. Morphia sul., gr. ¼ every four hours; use the catheter every eight hours; give ice *ad libitum;* beef essence as before.

8 A.M. 17th, tongue moist, red at edges and tip, white fur in centre; pulse 120; respiration 32 per minute; still thirsty and restless, with pain, requiring the morphia as above directed. The abdomen is more distended and tympanitic. The wound continues to discharge serum, which is becoming somewhat fœtid. Continue treatment, with the addition of brandy, as much as the patient can be induced to take with ice-water.

11.30 P.M., tongue dry; pulse 140; respiration 32 per minute; cheeks and extremities cool; body hot and moist, particularly the abdomen; pain in back more considerable; abdomen greatly distended; the patient complains of great weakness, approaching to syncope; treatment continued. I forgot to mention that soon as the abdomen began to increase in size, it was covered with linseed-meal poultices, which were kept moist and warm up to the time of her death. The patient continued to sink; the abdomen to increase in size until it was enormously distended. She expired on the 18th, at 5 o'clock P.M. We could not procure a post-mortem examination, but I think there is no doubt that peritoneal inflammation was the cause of death.

CHAPTER XXIX.

DISEASES OF THE MAMMÆ.

INFLAMMATION attacks the mammæ of infants, children, and youth of both sexes, and of women childless or senile; but in the present article, I desire to confine myself to the disease as manifested in the pregnant, puerperal, and lactiferous conditions of females, the conditions in which the mammæ are functionally active, or preparing for the discharge of their duty.

Inflammation may invade the tissue in and about the breasts, as

1st. The involucra.

(*a*) The skin and integuments, areolar glands, follicles, &c.

(*b*) The suspensory fascia covering and containing the whole breast, and its intra- and sub-glandular processes and laminæ.

2dly. The lymphatic glands, superficial and deepseated;

Or the structures entering more intimately and essentially into the formation of these, as

3dly. The nipple and milk-ducts contained within it, constituting the eliminatory apparatus.

4thly. The sub-areolar expansion of the milk-tubes, called by Sir Astley Cooper, milk-reservoirs.

These reservoirs actually occupy nearly the whole front part of the breast, immediately beneath the integuments and fascia, and lie above the gland in all parts of the breast except the margin, where the hard substance of the gland may be felt.

5thly. The lacteal gland and cellular tissue, which pervades every part of it, and which is the medium of connection between the lobes, lobules, tubes, vessels, and nerves of its substance.

Although very few cases of mammary inflammation occur in which the disease is confined to one structure, and many in which several are simultaneously invaded, yet I think an intelligent anatomical division will conduce to clearer views on the subject. I shall, therefore, base what I have to write about mammary inflammation upon the foregoing consideration with respect to its seat.

It might be supposed that the integuments, or involucra of the breast, were as liable to disease at one time as at another; and hence, at the time when the various processes connected with generation are passing, should enjoy their usual exemption from disease, but observation proves the contrary.

It may not be expected that I shall dwell at any length upon the eruptive or specific diseases which may attack the breast, for they may occur at any time; nor erysipelatous and rheumatic affections, which more frequently than is generally believed attack the structures. Rheumatism of the fascia of the breasts I think I have witnessed repeatedly; it is manifested by the usual characteristics as in other parts. Almost the only sort of inflammation to which the integuments are subject, that can particularly interest us now, is phlegmonous. Of course, the real seat of the inflammation, or, at least, its beginning, is in the areolar tissue beneath the skin. Generally, it is circumscribed and single in locality; often there are several simultaneous or successive foci; less frequently it is quite diffuse, involving a large surface, causing great deformity and damage to the organ, and attended with serious constitutional disturbance.

Phlegmonous subcutaneous inflammation in the breast is attended with the symptoms which usually accompany it elsewhere, —pain, heat, redness, swelling, hardness, tenderness in the early stages, varying in intensity with the extent and acuteness of the affection. We may generally diagnosticate this from inflammation in other tissues of the breast by isolation. There is usually no trouble in the secreting, eliminating, or containing apparatus of the breast. The functions of the whole organ are properly discharged. The inflammation is one generally of inconvenience instead of damage. It is superficial, and we may ordinarily get below it by manipulation, so that we can assure ourselves it is outside the mamma. Most frequently the areola is the seat of this kind of inflammation. There can be no question, however, but that the deep cellular tissue is as often the subject of inflammation as any other of the deep structures; and, indeed, some good pathologists think it is the seat of disease, when we suppose the gland to be the part affected. However this may be, inflammation of any of the deep tissues generally brings this into the morbid mass. Inflammation of the superficial areolar tissues occasionally involves the reservoirs or glands by contiguity. I have but little doubt that the diffuse intromammary suppuration which we see sometimes take place, and produce such prodigious quantities of pus, often melts down the connective areolar tissue distributed between the lobes, lobules, and tubes of the organ, without always at least attacking the more essential structures. I cannot hope, however, nor do I design attempting, to distinguish between deepseated cellular and glandular inflammation. The distinction, if made, so far as I can see, would not lead to any practical result. Chronic superficial cellular inflammation does not often occur, except as it becomes chronic by a long-continued succession of small abscesses. It is possible, also, that the chronic sequela of cellular inflammation, as exhibited in hard tumors, may be of this character in some instances. When this is the case, we should expect to find the hardness not so defined, but shaded off into other parts, somewhat regular in outline, and not sharp, irregular, and lobulated.

Inflammation of the Nipple.—This may be accompanied with abrasions, fissures, or ulceration. Abrasion is most frequently seen on the apex of the nipple, and is the condition in which the delicate epidermis is removed by action of the child's organs in sucking, leaving the dermis naked, bleeding, and raw. It may, however, be observed on any part of the nipple. Not unfrequently, these abrasions are increased in depth by ulceration, until a greater or less portion of the nipple may be destroyed. Cracks or fissures, likewise, often affect the nipple. These cracks are located either on the top, sides, or at the base of the organ. The apex of the organ, sometimes, is so deeply fissured, as to lay it open to the bottom of this projection, and leave it split in halves; but usually it is much less extensive, and it simply lays open the top of the nipple to the depth of the skin. The worst fissures that occur on the nipple, however, generally more or less completely encircle the base of the organ. To such an extent are fissures of the base carried by ulceration, sometimes, as completely to amputate this little projection. Abrasions and fissures lead almost invariably to ulceration; and we may consider these as the first stage, so to speak, of ulceration.

This ulceration, of course, resulting as it usually does from abrasions and fissures, occupies the place which I have assigned to them.

The symptoms which accompany these three conditions of the skin of the nipple do not differ each from the other, and without inspection we would not probably be able to distinguish between them. There is great pain upon handling the part, or when the child sucks; indeed, it is so very severe as to render it entirely intolerable to the patient, and cause her to resist every request, or even command, to nurse the child. When the child is put to the breast, in addition to the pain, they bleed so as to disorder the milk, and sometimes sicken the child, and cause it to vomit up the contents of the stomach. The extent to which ulceration may proceed, under the irritating influence of nursing, is sometimes very great.

I remember an instance in a patient affected with stomatitis materna, where the nipple was completely destroyed, and the place where the nipple had been, excavated below the surface before ulceration was arrested. Every experienced physician must have seen cases where the nipple was cleft, cut off, or very badly mutilated. Ulceration has its origin in many cases, also, in small phlegmonous inflammations of the cellular tissue of the nipple. It not unfrequently happens that small pimples arise, suppurate, burst, and, on account of the constant irritation of nursing, remain open, and pass into a state of ulceration which is often very obstinate.

Small ulcerations occasionally occur in the same way on the areola, but not with anything like the frequency of those of the nipple. Neither are they so painful as when situated on the nipple. The parts being less firm, the swelling does not so greatly press upon and distress the surrounding parts. Such diseases of the areola get well much easier than those upon the nipple, because they are less disturbed in that place by the child when sucking.

Inflammation of the Lymphatic Glands of the Mamma.—It is important, in a diagnostic point of view, to bear in mind the frequency of inflammation of these glands. As in other parts of the body, so in the breasts, they inflame in consequence of the passage of acrid or unhealthy lymph through them derived from inflamed tissues. Ulcerations and abrasions of the nipple and areola are frequently followed or accompanied by the inflammation of these bodies. No doubt enlargement by deposit, leading to chronic inflammation, may also sometimes occur independent of inflammation. The indolent tumors over the gland and near its margin, on the inner, outer, and upper circumferences, are in frequent instances chronically inflamed lymphatic glands.

The *symptoms* of inflammation of these glands do not differ in the acute form from those attendant upon superficial phlegmon. All the distinguishing circumstances of inflammation are experienced. They are, probably, more circumscribed than ordinary; the margin is more defined, and does not shade off into the healthy tissues, but appears, as it were, encysted. This is the case, however, only at first, as the inflammation often spreads to surrounding tissues, when the difference cannot be clearly made out.

As the inflammation subsides, hardness is left for a much longer time than in phlegmon of the integuments. The acute symptoms merge into chronic, and hardness, tenderness, and, in many instances, discoloration, last a considerable time. Suppuration does not occur so quickly as in phlegmon, and resolution much oftener. To make out a diagnosis, we should remember the more common seat of the two. They are ordinarily both (phlegmon and inflammation of the lymphatic glands) small in size, usually not larger than an English walnut; but phlegmon occurs about the areola, while the other is usually located over the gland, and near its margin. The phlegmon may occur in any direction from the nipple with reference to circumference, but lymphatic inflammation is situated at the inner or outer upper edge of the mamma. In scrofulous or broken-down patients a chronic condition of inflammation is likely to take possession of these glands, or they may be filled with albuminous accretions and undergo indolent changes, which might lead the inexperienced to fear malignant disease. I have a patient who has a deep lymphatic tumor in the breast on the axillary margin, who assures me it has been in the same condition for six years. This tumor is hard, round, regular, a little flattish, freely movable, and resembles lymphatic enlargements at the clavicle and groin in the same patient.

They may be usually distinguished from malignant tumors when indolent and not tender from inflammation (for malignant tumors are not sensitive in the beginning), by being more rotundly irregular, without the sharp outline generally characterizing malignant disease. If they are livid, they are also tender; if they involve the skin, they are tender to the touch, and the skin is inflamed. Neither of these conditions obtains in malignant cases. The malignant tumor may be livid and almost insensible. It may involve the skin, attach itself to it, and not inflame it.

The lymphatic tumor is hard alike all over; if softer in one part, that part is the centre. The malignant is harder in the centre until nearly ready to ulcerate. When the lymphatic tumor has ulcerated, the cavity is regular, and red or pale about the edges, and secretes pus. The malignant ulcer is ragged and exceedingly irregular; in fact, sharp irregularities of edge and cavity mark peculiarly malignant ulcerations; the edges are livid, not red nor pale, and ichor instead of pus is produced. In the ulcerated lymphatic there is no smell ordinarily, certainly none but the smell which may arise from uncleanliness. A malignant ulcer will smell in spite of us, and the smell is peculiar, and when once noticed will be recognized without difficulty again. Lymphatic glands may be inflamed singly or in numbers, several being the subjects of inflammation at the same time, or only one. As I have before intimated, the disease may be chronic or acute—indolent or active.

Milk Abscess.—Passing to the deeper structures of the breast, we encounter inflammation of the containing portion of the mamma, the expanded milk-tubes, the milk-reservoirs. There are from fifteen to twenty-five of these expanded tubes, holding from one to two drachms each in the natural condition. They are separate and distinct, each tube representing a lobe of the gland. One or more of these may inflame, ulcerate, and discharge the milk, mixed with a greater or less quantity of pus. Inflammation, followed by ulceration and discharge of pus and milk of these reservoirs, is alone what should be called milk abscess. Abscesses from this part of the breast do not occur singly as a general thing. Several are going on at the same time, one arriving at the ulcerated stage after another, so that we have a succession, each abscess involving one tube, and sometimes, but not often, more. They are seated under the anterior surface of the breast, mostly within an inch of the areola, and sometimes under it. In some persons the reservoirs are large, extend a considerable distance in every direction from the areola, and overlay the gland almost to the margin of the mamma. Milk abscess need not necessarily be near the central portion of the organ, although they generally are not far from the areola. They usually proceed somewhat slowly, taking longer to arrive at the suppurative stage than in superficial phlegmon. Swelling and tenderness are felt near the areola; it increases steadily until an apex is observed in the tumor; the integuments are thinned, fluctuation is observed, and rupture follows. This process requires a very different length of time under different circumstances. If the milk is secreted rapidly, the tube is distended faster; if secretion is scanty, the advance is slower. The inflammation depends upon distension of the reservoirs by milk which cannot find its way out of the milk-tubes. Retention of milk is caused by several different circumstances, which I shall have occasion to mention after awhile. I wish now to be understood as saying that it is the essential cause of the inflammation in this form of disease. The milk is secreted, but not eliminated from the reservoir affected. It acts as an irritant by its great accumulation, until inflammation is the result. The secreting capacity of the organ is not necessarily disturbed, and the excretion of the milk may be ready and easy through all of the tubes whose reservoirs are not affected, and we may think it is being evacuated entirely while it is retained in one or more reservoirs by the stoppage of the nipple-tubes. When evacuation, either spontaneously or by the lancet, is effected, pus and milk flow in moderate quantities at first. The pus gradually diminishes, the milk becomes more pure, until a milk-fistula occurs, which lasts a greater or less time. Should the eliminative tube become open, and allow the milk to flow from the affected reservoir through the nipple, the adventitious opening may entirely heal, and the integrity of the part be restored; but, as is most frequently the case, the fistula remains open until the breast ceases to secrete, all the milk produced by the lobe whence the reservoir is supplied flowing out at the place.

Sometimes, again, after breaking and discharging, it suddenly heals up, distension recurs, and the process of ulceration and discharge is repeated.

The sympathetic symptoms are not generally so great as in some other varieties of mammary inflammations. Fever does not

31

run so high, aching of the head, limbs, &c., do not distress the patient so much. Yet they sometimes are quite considerable, and require alleviation by appropriate remedies. The damage done to the breast by inflammation attacking these parts is not so great as results from glandular inflammation generally, though I have known instances in which nearly all the reservoirs were destroyed, and the breast henceforth remained useless. One of the worst features of the case is derived from the persistent repetition of abscesses, wearing out the patience of the medical attendant and the powers of endurance of the patient. It is always complicated by disease or deficiency of the nipple. Besides this ulceration or phlegmonous inflammation of the milk-reservoirs, there is another form, in which blood and pus are discharged through the nipple-tubes, the passage from them being free. Very few experienced physicians but have seen this discharge of pus, blood, and mucus from the milk-tubes, with tenderness and some tumefaction under the areola. It is generally considered to be an abscess discharging in this way, but it is ordinary inflammation of the lining membrane of the milk-reservoirs discharging its products through the nipple. Abscesses occurring as the effect of over-distension of the reservoirs do not give origin to those deep, ungovernable sinuses that sometimes trouble us in glandular inflammation; and while there is often milk-fistula following them, these close as soon as the secretion ceases, and we have no further trouble.

Several times in my life I have met with these abscesses during pregnancy, in which the accumulation of pus and milk was very great, so that when they are opened, many ounces of pus and imperfectly formed milk were discharged. Several months since I was called in consultation in a case, in which the disease had begun three months before labor; and when I saw the patient, the child was two months old, and large collections of pus and milk existed, pent up in the reservoirs, with impermeable tubes in both breasts: and while some of the reservoirs contained and their tubes discharged milk upon nursing, half of them were the subjects of purulent inflammation. Generally, the inflammation which causes the evacuation of the milk and pus checks the secretion of milk, and the patient recovers before the time for labor. This is fortunate when it occurs. According to my observation, this is the most common of mammary abscesses; indeed, I think, by a large majority.

Glandular Abscess of the Mammæ.—This is the most grave of acute inflammations of the breasts occurring during lactation. I am not aware of ever having seen an instance of mastitis proper, unless caused by violence, in any other than nursing women. When the inflammation takes place early in nursing, it usually comes on about the third or fourth day. Mastitis cannot, in the first few hours, be distinguished from the intense congestion which occurs at the time the secretion of the milk is first produced. In either case the woman is seized with a severe chill, in which it is not uncommon for her to shake and chatter as in violent ague. In the course of an hour or sometimes longer, sometimes in a few minutes, the chill gives place to a violent reaction; a high fever, pain in the head, limbs, back, and often abdomen, annoy the patient. All the phenomena of severe inflammatory fever occur. When the congestion subsides into a copious effusion of milk in the cells of the gland, the fever declines, a copious perspiration appears over the whole surface, and comfort succeeds great uneasiness and sometimes alarm.

When, however, the gland is not completely relieved by secretion, this transition from a state of febrile reaction is imperfect, and the patient is left with more or less of the symptoms of fever.

Simultaneous with these general symptoms there is pain, tumefaction, tension, heat, and tenderness of the mamma. If the secretion is established, the breast, as the sweating stage advances, becomes soft, cool, and less sensitive, until it is entirely comfortable. On the other hand, if inflammation is to succeed congestion, some part of the organ is left in a hard, tender condition. A hard lump, of greater or less size, continues to occupy some deep portion of the breast. Tenderness, tumefaction, heat, and redness increase, until inflammation is permanently fixed. Without early, energetic, and appropriate treatment, the woman will lose part of the mammary gland by destructive suppuration.

In the beginning of glandular inflammation, if the part be attentively examined, the shape and position of the lump will enable us to determine the seat. It will be either deep in the central portion of the breast, or in the marginal region. The tumid part is irregularly lobular; depressions and elevations may be observed, nodular, not sharp ridges. Very soon after the inflammation begins, particularly should it be advancing, this nodular feel is merged in diffuse hardness of the surrounding parts, until the whole tumor may become smooth and irregularly defined. Inflammation, hardness, and tenderness, increase for a few days, when the centre becomes slightly soft at first, growing more so, until distinct fluctuation is perceived.

At this time we find a soft fluctuating locality completely margined by hardness all round. This, then, will be the feeling of a mammary abscess, whether acute or chronic. Glandular abscess differs from milk abscess, by being at first much deeper, having a covering of integuments, &c., half an inch or more in thickness, while milk abscess, though quite hard, seems to be immediately beneath the integuments. When fluctuation is *first* perceptible in milk abscess, it is shallow; in mastitis it is deep, and makes its way slowly to the surface. When pus arrives at the surface, and ulcerates through, or is evacuated by the lancet, its flow is much more difficult, and the evacuation less complete; relief is not so sudden and perfect. Extensive destruction takes place, both in the internal portions of the organ and in the integuments; and so tortuous and irregular are the tracks of transit, in some instances, toward the skin, and the pus finds its way out with so much difficulty, that the sinuses are sometimes extremely difficult to heal. This state of things may last for many weeks and even months. We not unfrequently find cases in which these sinuses are numerous, tortuous, and lengthy, so as almost to riddle the internal structure of the organ, and discharge large quantities of pus, thus draining the system of the woman, inducing hectic, exhaustion, and, in extreme cases, death.

Often, instead of beginning at the time of puerperal congestion of the mamma, mastitis shows itself late in lactation. When occurring at such times, it may spring up suddenly, inducing all the general phenomena above described, in a greater or less degree of intensity; or it may be slowly established, and not bring the system into so decided sympathy and perturbation. Yet in the latter case, as the inflammation becomes more completely established, fever is pretty certain to be manifested, its intensity being greater or less, according to the extent of tissue involved, the rapidity with which it advances, and the susceptibility of the patient.

The first thing noticed, perhaps, is what the woman would call cake in the breast, of moderate, yet decided tenderness. This consists in inflammation in one or more lobes of the mammary glands. It gets worse; the swelling becomes greater, the tenderness more considerable; instead of the well-defined nodular tumor, the swelling is more diffuse; other parts are involved, redness in the skin is observed, sympathetic fever sets in, and then it passes through the different grades above mentioned in the acute variety, with less intensity.

In glandular inflammation, milk is suppressed more or less perfectly, owing to the amount of tissue involved.

Causes of Mammary Inflammation.—As I have intimated, the pregnant, puerperal, and suckling conditions of women may be regarded as predispositions to mammary abscess. Women are much more liable to them when in these conditions than at any other time. Hence it would not be improper to say, that these states of the system are predisposing causes of mastitis and its associate inflammation. The physiological congestion preceding and accompanying the commencement of lactation very frequently is carried too far, and merges into pathological congestion, and this again into inflammation. When inflammation arises from this cause, it will almost invariably be mastitis, or glandular inflammation. This sort of congestion may occur later, but usually it is in the puerperal condition. Another sort of congestion, which often runs into inflammation of the glands, is brought about by sexual intercourse in very excitable nursing women. I think I have known several instances of this kind. Other passions, as anger, may be succeeded by like results. Vascular excitement from stimulants will endanger the breasts in puerperal women, also. External causes may give origin to similar sorts of inflammation, as bruises from blows, tight lacing, stays of whalebone, &c. These last are productive of a good many cases. Not unfrequently our patient gets up well from the effects of labor, and the first time she dresses to go out, pinches her excitable gland with lace-strings, or punches it with the end of a piece of whalebone during the whole of her round of fashionable calls, and comes home with the breast excited to inflammation. Cold, acting partially upon the person, as the feet, the breasts themselves, or even upon the general surface, repels the blood to the already blood-loaded gland, producing congestion as the first step of inflammation. Other external causes operate upon the nipple and surface of the breast, irritate the skin, or destroy its integrity, &c. The child often sucks off the epidermis, and by thus abrading the nipple, ulceration is brought about.

Allowing milk or saliva to remain in contact with the delicate skin of the nipple, or areola, long enough to undergo decomposition, too often is the cause of ulceration, more especially when the saliva of the child is rendered poisonous by the existence of aphthous incrustation upon the tongue, gums, and roof of the mouth. The cracks so often found upon the nipples, I think, are almost invariably produced by the habit of allowing the fluids deposited upon the delicate skin to slowly evaporate, and thus carry off, or otherwise neutralize, the sebaceous unction of these parts, which is intended to keep the cuticle pliant and soft.

There is a class of causes which I am disposed to call pathological, very prolific of grave mammary diseases. One affection may act in producing another. Thus, ulceration of the nipple prevents proper efforts to draw the milk from the reservoirs; they become distended to a degree that causes inflammation; or the ulceration on the top of the nipple, by the swelling it causes in the intertubular issue, lessens the diameter of the tubes, or entirely closes up their mouths, so that milk cannot find its way out, or be drawn; accumulation results, and inflammation follows. Cracks, of course, will do the same; or, again, the inflammation originating on the nipple, may creep down the lining membrane of the milk-tubes into the reservoirs, or even farther, through the ramification of the radicles of these ducts, to the substance of the gland itself. In either of these localities, suppurative inflammation may arise, and proceed through all its most aggravated forms. Contiguity of inflamed parts may awaken inflammation in other parts. Integumentary inflammation may extend to the reservoirs or glands, by spreading from one tissue to another. There can be but little doubt that acute, and, in most cases, chronic inflammation of the lymphatic glands, is generally secondary to inflammation and ulceration of the nipple and areola. It would probably be too strong an assertion to make, to say that inflammation of the lymphatic glands always has its origin in this way; for in cases of strong predisposition to this disease—and there are numerous instances of that kind—it would probably arise without much cause of excitement. Certainly, I cannot be mistaken in supposing that I have seen several such cases.

Anatomical causes of inflammation of the breast exist to a great extent. They are sometimes congenital and hereditary; but I think, for the most part, brought about by improper dressing. The flat, undeveloped, or retarded nipple, is one form of anatomical peculiarity which prevents the perfect performance of suckling, as is represented in Fig. 27. The retention of milk will lead to milk abscess. Nursing is often impracticable in this breast. Fig. 28 represents a breast with a very broad but extremely short nipple, entirely too large for a child's mouth, and so short as to

Fig. 27.

Interior View.

add to the difficulty of prehension. Fig. 29 represents a breast with scarcely a trace of the peculiar, warty, tissue-like nipple, simply pouched slightly where the nipple ought to be; or, a very small nipple, where the milk-tubes seem to be bound in such a contracted bundle as not to allow free egress to the milk, is represented in Fig. 30. These four specimens of nipples, which we often meet with, are almost impracticable. The first and third quite so; and the second and fourth so difficult that we are generally driven to the necessity of abandoning it after the best directed efforts to make the breast available. The danger to breasts furnished with such nipples is, that the milk will not be properly evacuated, and that milk abscess will result. In Fig. 31 we have

Fig. 28. Fig. 29. Fig. 30.

a nipple large enough to be easily apprehended and drawn by the child, but it is too constricted at the base. The milk-tubes, upon entering it, turn too acute an angle. A little swelling of the sub-areolar tissue from retention of the milk will stop them entirely up, so that the milk will not pass out. In order the better to illus-

Fig. 31.

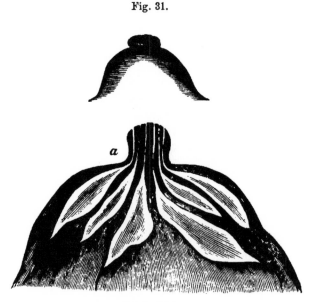

Interior View.

trate what I mean, I add a sectional view of this kind of breast and nipple. At *a*, the milk-reservoirs may be seen contracting at the nipple, forming the milk-tubes, which turn abruptly upward

and even a little outward. This will be made still plainer by
giving what I call a model breast and nipple, Fig. 32. It speaks
for itself. The nipple is slightly conical, the base being larger than
the apex. I add also a sectional view of this breast. As will be
seen, the milk-tubes are free from pressure everywhere. Their
entrance into the nipple is by a slight curve instead of an angular
turn. The milk will flow spontaneously from this kind of breast,
and there can be no accumulation in the reservoirs. In nipples
represented by Fig. 31, one danger is that milk, saliva, and mucus
will collect in the groove around the base, decompose, and thus
induce mammillitis with its attendants and consequences. This
could not well occur in the case of Fig. 32. There is no lodging-
place; the nipple would be wiped clean of all these accumulations
by the mouth, and return of the breast inside the clothing of the
mother. The shape of the mamma may predispose it to disease,
but not in so striking a manner. The more conical a breast the

Fig. 32.

Interior View.

better. A flat sessile mamma is more likely to inflame. Although
the above-mentioned varieties of nipples are not the only ones predisposing to mammary abscess, yet by drawing the attention of

the profession to the subject thus distinctly, it is believed there will be no difficulty in recognizing adverse anatomical peculiarities whenever they do occur.

It might be appropriate to examine into the cause of these anatomical differences in the shape of this interesting organ, but the length of the subject will not allow me to indulge in this direction. Like all other formations, the nipple would doubtless differ under the same circumstances in different persons naturally, but I think there is no doubt much of the deficiency is produced by tight lacing, and the pressure made directly upon the nipple for a series of years during its development.

More regard to dressing, as well as education, is bestowed upon fitting the young lady to get married, than to perform her functions properly after getting married.

Treatment.—I can better give my views of the treatment of the affections above described, by observing the same general division with reference to the application of the processes of cure. Inflammation of the nipple will come up in this order of the arrangement for consideration first. Our means of cure for mammillitis should be arranged under three different heads, as follows: prophylactic, palliative, and curative. The first have for their object the preparation of the nipple for the trials through which it has to pass at the time of nursing. As has been seen, the causes operating upon it produce abrasions or chaps, and their action is greatly facilitated by the natural and acquired tenderness of the structure, particularly the epidermis and skin. The prophylactic means to be used are such as harden these. As elsewhere, so in the nipple, the skin becomes tough, and the epidermic scales abundant and adherent, upon exposure to air and friction. The contrary condition will obtain—tenderness, &c.,—from pressure and covering with impermeable or great thickness of dress. In this condition it is protected by extraneous covering, and hence does not furnish its own proper defence. The epidermis will be thin and light, and the skin tender. The nipple, therefore, should be covered lightly during pregnancy and nursing. The thinner and more permeable the covering the better. It should be of such a character as freely to admit the air. At the same time it should be subjected pretty constantly to moderately rough friction.

An excellent dressing for the nipple for the last two months is a rough, coarse sponge, so cut as to cover the areola, and surround and cover loosely, but touch, every part of the nipple. Over this there should be but one thin texture of raiment, so as to allow of the evaporation of fluids as fast as secreted, and the free admission of atmospheric air. In cold weather, when going out, the breast, of course, would be covered by all the clothing that is used for the protection of other portions of the person. It is a great mistake to cover these important organs,—important on account of their usefulness instead of their beauty,—so thickly as they usually are; they bear exposure with great impunity. When we wish to harden the nipples, we should bear in mind the circumstances which harden our hands, and make use of them; we should equally avoid the circumstances that soften our hands. When a lady wishes to soften and whiten her hands, she wears kid gloves, and does not allow them to touch hard substances. In a like manner she may soften her nipples if she wishes to do so. To occasionally moisten them with water, and to allow it to evaporate slowly on exposure to air is a good expedient to harden them. Friction, with a dry towel or the fingers, will assist in this process. It is a matter of great question whether the various washes used to harden the nipples are not injurious instead of beneficial. They generally exert a chemical as well as physiological effect, while this last is all that is desired. During lactation the same exposure to air and lightness of covering should be observed, and, after nursing, the nipple should be wiped clean and dry before being returned under the clothing. This is a rule that should never be neglected. Those who have observed the effect of allowing the udder of a cow to dry spontaneously after the calf is taken from her, will understand the importance of attending to this matter. It will be all the better to use a little glycerine or very fine olive oil after they are dried each time, particularly if we have reason to apprehend danger of chaps or cracks. Such prophylactic measures will very generally enable us to avoid the occurrence of distressing chaps or cracks. When, however, the nipple becomes inflamed, these are not sufficient to satisfy the demands of the case, and we must resort to palliative and curative measures; and first of the palliative. As the nipple must be used in order to preserve the functions of the breast, and as every time the child sucks the healing processes that have begun must be more or less interrupted, it becomes important to procure such means as will preserve the breast from the effect of these interruptions as much as possible. The chaps and abrasions that occur and give rise to inflammation and ulceration, may be located anywhere upon the nipple, at its summit, sides, or base; and when the child nurses, the tongue and labia embrace it so closely that none of these places escape. The artificial means used to palliate the effect of sucking intervene between the mouth of the child and the nipple, and should be selected with special reference to each case. The shield of ivory or britannia answers very well when properly managed. They are made in the form of a conical hat, having a rim, a crown cavity, with a draught-tube rising out of the top for the milk to pass through. Now, having in mind that these three parts must vary in length and size for different-shaped nipples, and cases in which the locality of the abrasions or chaps are different, we will have no trouble in making a profitable selection. The rim should be large enough to cover the areola, the crown or nipple-cavity large enough to pass over the nipple, merely touching it on the sides. These things should be observed in all cases. The depth of the nipple-cavity is a matter of the greatest importance. If the abrasions or chaps are on the summit of the nipple, it should be so deep that when drawn the top of the organ will not touch, or else it will cause pain. There should be no pressure on the top. But if the cracks or abrasions are on the sides or at the base of the nipple, then the cavity of the shield must be shallow, so that the top of the nipple touches its bottom in such a manner as to prevent stretching the organ, and bring the pressure on the top altogether. In this latter case, the bottom of the cavity should be as smooth as possible, and correspond in shape to the summit of the nipple, in order to prevent unequal pressure. The shield, of proper shape, size, &c., will afford great relief to the patient, and prevent very much the disturbance to the healing nipple. It is not a matter of indifference either what material we use as an envelope for the shield. Gum-elastic or cow teats are always clumsy, and easily become foul or hard, and sometimes taste in spite of our best efforts. A soft linen rag, properly adjusted over the draught-tube, is better and cleaner than any other envelope. It has the advantage of being cheap and always at hand, and so abundant that it may be replaced by a new one after each operation of sucking.

But a very ingenious contrivance is mentioned by M. Legroux, which I will describe:

$$
\begin{array}{lll}
\text{R. Collodion,} & \ldots & \text{ppts. xxx.} \\
\text{Ol. ricini,} & \ldots & \text{`` ss.} \\
\text{Ol. terebinth,} & \ldots & \text{`` jss.} \\
\text{Mix.} &
\end{array}
$$

This is a fluid mixture, which is quite adhesive, and dries less quickly than collodion. It is applied upon the areola with a brush, so as to encircle, but not touch, the nipple, the width of an inch. While yet soft, the nipple is covered by goldbeater's skin, and pressed well down around it upon the mixture. The skin adheres to the adhesive material, and thus forms a perfect, smooth, and pliant covering to the nipple. All that remains to finish it, is to prick several holes through the goldbeater's skin with a needle to let the milk through. This has the advantage of not changing the shape, size, and feel of the nipple to the mouth of the child, so that it sucks more readily than it would an artificial nipple made with a common shield. But while this is the case, it allows the pressure of the lips upon the nipple at every point, and only partially relieves the mother from the pain.

In most cases I would rely more upon the judicious selection and management of a shield than this contrivance, ingenious and neat as it is. This may be imitated by other adhesive mixtures and tissues. Before sucking, the goldbeater's skin must be moistened with a little sugar and milk. Much of the suffering under nursing, while the nipple is raw from chaps, abrasions, or ulcerations, may be avoided by being drawn by the mouth of an adult, so shaping the vacuity, produced for the purpose of drawing, as not to touch the sore part. If the lips are so placed around the nipple as to press upon the areola and not touch the nipple more than very gently,—and I am sure this is practicable by any intelligent adult who will make a persevering trial,—the draught can be accomplished with comparatively little pain. Violent action should not be used; a gentle but constant pressure with the lips on the areola, with persevering but very gentle draught, will usually suffice, and powerful suction is sure to aggravate the cause of the retention of milk. I have often sat down, and by encircling the nipple with my fingers without touching it, and pressing upon the areola, caused the milk to flow freely, when, with great difficulty, it could be drawn out. In thinking upon this subject, we should remember that it is the pressure of the atmosphere upon the outside of the breast, combined with the elasticity of the integuments and coats of the milk-reservoirs, that urges the milk forward through the nipple into the vacuum caused by excluding it from around the top of the nipple. The vacuum will not be necessary, if the pressure can be made with sufficient firmness without injury of the part. Why may not some ingenious individual invent a milking apparatus of gum-elastic, that would act by pressing upon the areola and front of the breast without causing a vacuum on the nipple? This would often save a great deal of trouble and suffering to our patients. In thus viewing and treating the subject, we would push the milk out instead of, as we, upon a superficial look at the matter, suppose, pull it out.

The above palliative means do not enable us to avoid the causes of inflammation of the nipple; but by their use we may render the operation of them less mischievous, which is often sufficient, in favorable cases, to effect a cure. In considering the curative remedies for sore nipples, I must protest against the simplicity with which we use the term, and think of, sore nipples. We speak and think of it as though there was no variety of sore nipples. The same treatment is not applicable to abrasions, that is, to chaps or cracks, nor to ulceration, nor to all the conditions of ulceration. Nature tries to cure cuticular abrasions by an effusion upon the naked surface of a viscid albuminous layer, thus defending the delicate tissue from contact with atmospheric air, or other irritating substances; and if this is allowed to remain undisturbed, it will as it falls from the place, leave a well-formed delicate cuticle. And I think the nearer we imitate nature in this respect, the more good we will do. We may use starch or mucilage to cover the abrasions, but any astringent or stimulant application is inadmissible. Abrasions, however, do not last long without becoming ulcerated, and the treatment may be different. When there are numerous fine chaps, covering a large surface of the nipple, or when single, if very shallow, the treatment for abrasions will usually answer every purpose. Ointments of a mild, unirritating, or even a soothing quality, are probably more applicable than in abrasions. The following is a very good one:

R. Cerat. alb., . ℥ij.
Ol. amyg. dulc., . ʒj.
Mel. desp., . . ℥ss.
Mix. Dissolve with gentle heat, and add Bals. Canad., ʒijss.

This should be applied every time after nursing. When the cracks are deep, it is indispensable to quick cure that they should be closed up, and kept so until complete adhesion of their sides takes place. This may usually be done with great facility in the following manner, viz.: press the nipple in such a way as to close the crack, and while thus holding it, apply a thick layer of collodion over the surface. We should apply the layer thickly, and have it extend some distance in every direction, so that it will keep the crack together. The collodion is not easily sucked off by the child, and if the nipple-shield be used, it need not be disturbed at all until completely healed. We should watch the coat of collodion, and remove it when it seems to be becoming deficient by violence of nursing. In most cases this covering, if kept up unbroken for a week, will suffice to complete a cure, if suppuration is not going on in the chapped place. If this is the case, and the surface becomes an ulcerated one, it will fill up by granulation alone, and falls into the category of ulcerations. In this part of the body, ulceration does not differ from the conditions it assumes in other places; and it cannot be expected that I should dwell upon every variety that may occur. General principles must guide us here as elsewhere. There are two conditions, however, one of which is apt to obtain a prominence and give character to this ulcer,—acute and chronic; in either of these conditions, the ulcer may be exceedingly irritable to the touch, and painful; and in the latter, indolent and atonic. The acute variety is apt to be attended with considerable heat, tumefaction, color, and tenderness. These conditions should be removed by depletion, as by leeches,—one or two will generally do,—cold emollient poultices, large enough for the nipple alone, and removed as often as they become warm. Or we may envelope the nipple in a thin layer of thick mucilage, covered by oil-silk, so as neatly to fit the organ, kept cold by ice applied in a small bladder or India-rubber bag, or we may wrap the ice in oil-silk.   In whatever, envelope it is used, it should not extend beyond the inflamed part, and should be separated from it by a thin layer of cotton-wool, or something of that kind.

When such remedies are not necessary, because of the non-existence of these symptoms, we should content ourselves, in the very early stages of ulceration, with similar mucilaginous and bland ointment applications as in abrasions; but as the process goes on, and the acute symptoms entirely subside, astringents become useful, and these will vary in character and strength according to indications of atony and flabbiness, &c.   Alum and tannin are excellent applications at first, but will have very little effect after the ulceration has continued for any great length of time.   Sulphate of zinc and borax will come next in respect to time.   One scruple of tannin to one ounce of rose-water, five grains of alum, the same quantity of sulph. zinc, are all good in the earliest stages of ulceration of the nipples.   When the more acute symptoms have subsided, the following formulæ are often very useful:

R. Glycerine,   .   .   .   ℈ij.
      Sodæ subboras,   .   .   ℥ss.
      Aquæ rosæ,   .   .   .   ℥jss.
Mix.   Use as a wash each time after sucking.

Or,

R. Sodæ subboras,   .   .   ℈ij.
      Cretæ prep., .   .   .   ℥j.
      Spts. vini,
      Aquæ rosæ, āā,   .   ℥iij.
          Mix and dissolve.

This last may be used when the ulcer is becoming somewhat indolent.   Tinct. kino, tinct. nutgalls, and, in fact, almost every astringent, has been used in these ulcers.   In chronic ulcers, still stronger astringents or stimulants will become necessary in conjunction with other remedies.   A skilful use of the sulph. cupri and nit. argent will do a great deal to heal up and shorten the course of these chronic ulcers.   The nitrate has done the most good in my hands.   It should be applied in substance to the surface of the ulcer, and never be used oftener than once in eight days, when a second application becomes necessary.   Between times, the ulcer may be dressed with some of the milder astringents, alum or tannin, for instance, in solution.   In the irritable variety, some narcotic extract should be made into ointment,—belladonna, hyoscyamus, opium, &c.   An excellent expedient, and one that will often entirely change the character of these ulcers, is to anæsthetize the part with ice, as is directed to be done on a part before the performance of an operation.

We are very apt, after we begin to use curative measures, to neglect the palliative.   This is a great mistake, for they can have but little good influence while the causes are allowed to act with all the power that is necessary to produce the disease.   We cannot attach too much importance to the measures of palliation.

Treatment of Inflammation of Lymphatic Glands.—The causes of lymphatic inflammation should receive our attention first, as the abraded or ulcerated nipples, inflamed areola or integuments of the breast, or, when chronic, the constitutional condition in addition to the local excitements.   When acute they will require, in addition, the antiphlogistic measures adopted in other inflammations,—leeches, cooling lotions, fomentations, cathartics, &c.; when chronic, alteratives, iodine, tonics, liniments, irritants, &c., which will be adapted by every physician, according to his own judgment, to the peculiarities of his case.   If we are accurate in our diagnosis, and separate this affection from those of the deeper-seated structures, there will be no great difficulty in adjusting the treatment of it.

The treatment of milk abscess is one, however, of greater importance, because of its frequent occurrence and destructive effects. The remedies naturally range themselves into preventive and curative.   The preventive has reference to the management of the anatomical and pathological conditions of the nipple, which prevents the free elimination of the milk.   Of the latter, I have written quite as extensively as the limits of this work will allow. Can we change the anatomical deficiencies or depraved shape of the nipple, of congenital or acquired origin?   It is a matter of the utmost importance to the health and happiness of the patient, that this question should be decided promptly and properly.   Much will depend whether our attention was drawn to the case early in pregnancy, or not until the time of labor, or even afterwards, as to the probability of success in many cases.   In other cases, we can decide the nipple to be impracticable at the first sight, at whatever time we examine it; and I would insist upon the im-

32

propriety of compelling a woman to pass through the terrible pain and exhaustion which attend these cases, where the nipple, for instance, is entirely wanting, and prehension impossible. If our attention be not drawn to the nipple until after labor, and the functions of the breast are required, we ought not to hesitate to decide against nursing, or attempting it; and so far as I am concerned, individually, I would advise against the endeavors to use the breast represented by Fig. 27, if I was aware of its condition at the beginning of pregnancy. Fortunately, this deficiency is rare. When there is an approximation to this, but not complete absence or depression of nipple (the breasts shaped like Figs. 28 and 29), much may be done toward rendering them useful, provided our efforts are judicious, and sufficiently prolonged. They should be commenced as soon as pregnancy is known to have taken place; and if in the state of society it were practicable, the prospect of success would be much better, could we have the management of our patients as soon as menstruation began. If mothers were well instructed in this matter, and would carefully attend to it, the probability is, that almost no cases of anatomical unfitness for nursing would present themselves.

Flat or undeveloped nipples (Figs. 28, 29, and 30), if not observed by the practitioner until after parturition, will be almost certain to give him trouble, and he will be scarcely able to prevent extensive milk abscesses. The first and most important principle, is to take perpendicular pressure entirely off the top of the nipple, and this would probably be sufficient to prevent the difficulty, if complete. This little projection, on account of the fashions of female dress, is kept constantly pressed back into the soft yielding mammary tissues, until it becomes hopelessly imbedded into them. Now what we want is to counteract and remedy the effect of this mischievous habit. A number of devices have been resorted to for the purpose of starting the nipple forward from its imbedded condition. They have for their object, as a general thing, the production of counter pressure around the nipples, upon the areola and central portion of the breast, in such manner as to press the central tissues beneath the nipple, and thus cause it to protrude. If this object can be effected by such gentle means, continued for a sufficient length of time before the birth of the child as to make it a permanent state of this organ, the treatment will be effective. The misfortune is, we can seldom get the important desideratum, time, and we are under the necessity of beginning our treatment often too late to effect anything. When called upon to remodel a nipple before or during pregnancy, we may make use of a shield of stiff silver, or iron wire, large enough to embrace and actually fit the anterior surface of the breast, with a cap-like projection from its centre, into which the nipple may project. There may be some soft substance, very thin cotton or wool, to protect the surface from the wire placed immediately beneath it. This should be worn for months under the dress, and receive all the pressure from it, and distribute it over the front of the mamma, and protect the nipple from any pressure. Such a shield is far better than ivory, wood, India-rubber, or any other impermeable substance, as it does not interfere with the transpiratory functions of the skin, or the secretion of the areolar follicles and glandulæ.

When we are not called upon to treat these rudimentary nipples until the time of or after parturition, such treatment will not avail. The effect must be brought about more promptly, on account of the necessity for immediate use. In many cases the nipple can be made available by temporarily inducing its erection by simple titillation with the finger, moving it gently around it, and then immediately applying the child. An excellent way of erecting the nipple, when there is considerable depression, is to place a thick layer of collodion around it on the areola. When this dries and contracts, the nipple will be elevated quite prominently; the child should then be placed to the breast, and allowed to nurse.

When the nipple is protruded in some of these ways, the milk may usually be drawn, so as to more or less completely empty the reservoirs. This will prevent milk abscess very generally. When inflammation of the reservoirs has fairly begun, it will be exceedingly difficult to prevent suppuration. The curative means consist in thoroughly evacuating and keeping empty this set of vessels. Several modes of doing this have been recommended, such as drawing with a glass tube shaped like a pipe. Various shapes of breast-tubes and pumps are in use, but I must object to all of these. It is a very easy matter to injure the delicate tissue of the breast by the hard rim of these instruments, and I think the accident often happens.

A puppy is often brought into requisition for this purpose, but is rough, and sometimes irritates the nipple and even sucks the skin off it. The only proper thing for drawing the milk is the mouth, and when these reservoirs are inflamed, it should be the mouth of an adult, who can vary the pressure or force to suit the tenderness of the part. Another very useful class of measures are those intended to suppress the secretion of the milk, and thus relieve the reservoirs from the distension. The narcotic substances taken internally, or applied externally to the breast, do a great deal towards stopping the secretion of the milk. Opium in large doses, so as to keep the patient very thoroughly under its influence, aids very much in arresting the secretion of milk. Applied externally in ointments, so as to produce a decided impression upon the system, it has a similar effect; but belladonna seems to have acquired most renown for this purpose. Mr. Richard Marley reported forty cases to the Obstetrical Society of London treated successfully by inunction of belladonna. Dr. Tanner corroborates Mr. Marley's conclusions by his own experience. American physicians testify in favor of belladonna by furnishing to medical periodicals a great many favorable cases within the last two or three years. It should be remembered that many of the extracts sold in the shops, if not entirely inert, are at least much below the standard in strength. Our opinion of the efficacy of these, of course, will vary from this circumstance, and hence, doubtless, the discrepancy in the testimony of different observers in regard to the use of belladonna for the purpose of suppressing the secretion of milk. The inunction of ointment made with the extract should be carried to a sufficient extent to produce some of its characteristic effects upon the system. Its use should be as profuse as the system will well bear. Cold, as a local application in cases of milk abscess, has several good effects. It anæsthetizes the part, rendering the patient more comfortable; it decreases the secretion, constringes the reservoirs of milk, and allays excitement in the capillary circulation. In supplying cold to the breast, the temperature should be about forty or forty-five degrees, and kept as steady at that as possible. Water of that temperature might be kept running through an India-rubber bag enveloping the organ. A bladder partly filled with ice and water, with a piece of flannel between it and the skin, would also do very well. When we do not desire to promote secretion of milk, cold may be used. I do not believe there is any danger from it while its application is confined to the part affected, and its bad effects are usually produced by wetting the clothing, or allowing it to get applied to other parts of the person. I cannot express with sufficient force the evil effects which the prejudices of a former age in medicine have fastened upon the minds of at least a part of the public in the practice of keeping the breasts wrapped in thick layers of cotton or lamb's wool. It is promotive of the secretion of milk by drawing blood to the gland, and thus keeps up the state of things we desire to avoid. For internal treatment, a saline cathartic every other day, and two grains of iodide of potassium every four hours, may be relied upon as materially aiding the other treatment. In this affection, antiphlogistic treatment is merely auxiliary, and should not be pushed to an extent usually considered necessary in other inflammatory affections. In this case, over-distension is the cause of the inflammation, and its removal in the early stages is generally sufficient to cure.

Treatment.—Acute inflammation of the glands of the breast, when it occurs as the effect of congestion immediately preceding the secretion of milk, is apt to be very extensive, sometimes involving the whole of the gland, and will require energetic treatment. For the first few hours we should try warm fomentations with the hope of establishing the secretion. This probably would be unavailing if actual inflammation had begun; but we cannot always determine the point when this intense congestion passes into inflammation, and hence we are justified, I think, in making the effort. If the patient is robust, and the fomentations fail wholly or partially to bring relief, a decided venesection will often turn the balance in favor of resolution. When we bleed, the object should be to produce a decided impression, and in order to do this the patient should be in a sitting posture, and the blood allowed to run until the pulse is affected and syncope approaches. I have so much faith in verat. viride in combating inflammation, that I begin its use immediately after venesection, and if the patient is strong, give it in six-drop doses every four hours until the pulse is brought down to sixty in the minute, and then, by administering it in decreased doses, keep it as nearly at that as possible. One grain of calomel, with a quarter of a grain of sulph. morph., may be given occasionally, if the pain is urgent, every four or six hours. This kind of promptitude and energy of treatment will frequently arrest the inflammation and bring about resolution. And when we remember the amount of suffering and damage it may prevent, nothing should deter us from urging our patient to accept the treatment. Should this not be sufficient, it is an important question whether depletion can be carried further. One good full general bleeding, if followed by veratrum, will be sufficient generally, but sometimes it will be expedient to use leeches, and produce a general alterative mercurial influence. A lotion made of one part of sulph. ether to two parts of alcohol, will be a good soothing adjunct after the inflammation becomes permanent. If the inflammation begins later, the extent of disease is apt to be less, and may be confined to one lobule, or, at most, a part of the gland only. In this case, a brisk cathartic of calomel, aided by some saline, leeches to the part, followed by cold lotions, tinct. verat. viride, or solution of tart. ant., given at sufficient intervals in proper quantities, will afford us efficient treatment. If this treatment is begun early we may expect much good from it. It has always been an interesting question with me, after the inflammation has existed for a length of time, and we cannot avoid the formation of pus, whether we should abandon antiphlogistic means and resort to warm poultices and fomentations, to promote suppuration. I think that this is not justifiable in many instances. The probability is that if we continue our general and local antiphlogistic treatment until suppuration is clearly evident, we may limit the extent of that termination, lead to resolution in a larger part of the gland than would otherwise take place, and thus save much of the glandular tissue. When the *whole gland* is inflamed, there is no necessity, in fact, I think it injurious, to institute and continue strenuous efforts to draw the breast. There is little or no secretion, and when a part of the gland only is inflamed and milk is produced by the rest of it, it is questionable whether anything but the most moderate means for this purpose are admissible. Retained milk is not the cause of inflammation in this case as in milk abscess. Very frequently glandular inflammation is complicated with inflammation of the reservoirs. Then we must combine our treatment to suit the case, local and general antiphlogistic, with means to arrest the secretion and empty the reservoirs of the milk already contained in them. Chronic inflammation of the gland will be cured by much the same treatment successful in other glandular inflammations of this grade,—leeches, mercurials, iodine, and vegetable alteratives, perseveringly administered internally and locally applied. Much reliance can be placed upon well-regulated and graduated pressure, with adhesive straps, pressing the part diseased against the ribs, or collodion encasing the breast thoroughly. When suppuration has taken place, what are the indications to be relied upon to justify us in evacuating it? There can be no doubt, I think, that the earlier the matter is let out the better, for several reasons. The cavity becomes larger by allowing it to remain; it burrows through the surrounding tissues. The longer it remains the greater the amount and duration of the irritative fever that accompanies its retention. But, notwithstanding the desirableness of getting rid of the pus, we should hesitate to cut through the uncondensed tissue to any extent. In cases where the inflammation and suppuration are deep in the gland, it is desirable to wait until the pressure from within has lasted long enough and in a sufficient degree to cause the condensation of the tissue. Otherwise, it will require a very large opening to allow a free discharge. I think we should not lance the part until fluctuation is quite evident, and the pus has made its way to the fascia or integuments. It is never desirable to cut through any part of the uninjured gland or milk-ducts, and, altogether, I should feel more inclined to allow it to approach the integuments very closely before cutting.

In the case of milk abscess, the earlier the opening is made the better. As soon as it is evident that suppuration is inevitable, the opening may be made. The smaller the opening to allow the escape the better. Should the disease still exist that caused the retention, the opening should be preserved. Often the evacuation of one or two reservoirs will suffice, and the rest will continue to discharge through the nipple. The effect of suppuration and evacuation of a milk-reservoir is often to entirely destroy its cavity, but in other instances it continues to discharge through the artificial opening, and a milk-fistula remains. This may be closed by an occasional application of the nitrate of silver in pencil. Worse than these are the tortuous lacunæ, that sometimes result from the deep glandular abscess of the breast. They are generally difficult to cure. · Injection of iodine is the remedy most relied upon for these troublesome sequences to suppuration. The most effective way to inject is to insert a soft flexible cathæter, if possible, to the bottom of the twisted canal, and throw the injection through it, so as to apply it without dilution to the bottom of the pus fistula. I think this important when practicable, because it favors the shallowing instead of the narrowing of the cavity.

Of course it is never advisable to slit up these obstinate puriferous ducts in the breast, as it sometimes is in other parts of the body, because of the amount of tissue that might be damaged which it is desirable to save.

Chronic Inflammation of the Breast.—Inflammation of the breast, instead of pursuing this definite and decidedly phlogistic course, proceeds with such gentleness, and so slowly, as to render its diagnosis obscure; and there is no doubt but that we meet with instances in which a correct diagnosis, at the time we first see the case, is quite impossible. These instances are not numerous; and a careful examination of all the conditions enables us to come to positive and correct conclusions.

Symptoms.—The inflammation begins, in some instances, with an obvious degree of pain, giving the patient inconvenience that is almost constantly experienced, or the pain and soreness are noticed only at intervals. In other cases there is no pain whatever. The first indication of the disease is the discovery of tumefaction in some accidental way. When examined, there is no discoloration at all in many cases, in some again but slight. If the tumid part is examined by the test of touch, the patient experiences but slight inconvenience from considerable pressure. And in order to discover tenderness, pressure must be made in every direction, when, in some particular place, or by pressing in some particular direction, the patient will complain of tenderness, generally deep-seated. I have occasionally been able to elicit complaints of tenderness, by pressing behind the tumor, or attempting to lift it up from the chest. The tenderness is generally deep-seated and obscure, but I believe I have not met with a case in which there was an entire absence of it when thus carefully tested.

When the tumidity is first discovered, it may be very small, not larger than a hazel-nut, and never attain to any great size; or it may increase very much, so as to involve a large part of the breast.

It may also commence by implicating a large part or the whole of the organ. The cases I have seen have nearly all been large.

It may require only a few months for the disease to run its course to suppuration and exulceration; but sometimes several years are consumed in these slow steps, as many as threė, four, or even five. In connection with the deep inflammation, we sometimes witness inflammation of the skin and integuments. It occasionally proceeds to discoloration and considerable ulceration, and must be regarded as a complication, and not a part of the deep inflammation, unless we consider them both the results of a general scrofulous diathesis.

It occurs almost invariably during the childbearing period, and is more frequently discovered during the time of nursing. The patients are usually persons of feeble, broken-down, or impaired constitutions, with a thin, unhealthy state of the blood, and, many times, of tuberculous tendencies. Sometimes, however, we meet with it in robust persons, to all appearance sound and healthy. In these patients we generally have more of the inflammatory symptoms. From the fact that this chronic inflammatory tumor makes its appearance generally in weak, unhealthy patients, it is sometimes called scrofulous tumor of the breast.

Diagnosis.—It occurs in the nursing,—not in the aged, as does cancer, nor in the sterile and single, like the adenoid tumor. The patients are under forty years of age. It is always tender, if properly examined, by lifting up, shaking, or pressing in every conceivable manner. This is not the case with cancer, the adenoid, or encysted tumors. Sometimes the parts are discolored in cancer and chronic inflammation. In cancer, tenderness and discoloration do not correspond: they do in inflammation. If there is much discoloration in the inflammatory tumor, there is considerable tenderness; but there is not much tenderness, however purple and dark the skin may become, in cancer. The inflammation commences softening in the centre, and the soft part is surrounded by hardness completely, showing hard inflammatory walls. Cancer softens irregularly, and generally contemporaneously with exulceration. There is, at most, very doubtful fluctuation in cancer. In chronic inflammation there becomes evident fluctuation for a considerable time before exulceration. The adenoid tumor does not soften except under the process of inflammation. The diagnosis may be rendered demonstrative by the exploring needle in the discovery of pus.

Prognosis.—The prognosis is always favorable. The patients will get well spontaneously; or may be more readily cured, provided there is not fatal complications connected with it.

Treatment.—In robust patients, the treatment should be antiphlogistic and alterative: massa hydrarg., grs. x, every fourth night, to be followed in the morning by a saline laxative. Anodynes at night, if the patient is in much pain, are also admissible.

In patients whose constitutions are impaired by long suffering, tonics should be added to a gentle alterative medication. The bowels should be kept regular by the above course of alteratives. If the stomach is debilitated, the bitters, as infusion of gentian, cinchona, serpentaria, quassia, as they may agree with the patient; to which, if acidity be troublesome, some form of alkali, as carb. soda or potassa, may be added. Iron should not be omitted, unless some peculiarity forbids it. Cod-liver oil not unfrequently does better for these cases than any other medicine. It is not often retained without combining with a stomachic or stimulant, and should always be given after meals, as it will do more harm by the nausea it produces than good.

The local treatment should be conducted with a view to promote the discharge of the pus. Poultices of bread and water, or bread and milk, linseed meal, or other emollient, will suffice for this, if persevered in sufficiently. Soon as fluctuation is evident, the pus should be evacuated by puncture, and its evacuation be encouraged by poultices.

Mastodynia, Mastalgia, Neuralgia of the Mammæ.—This is a true neuralgic affection of the breast, unattended by permanent structural disease of the parts. Sometimes both mammæ are affected simultaneously, but very often the pain is confined to one; and singular as it may seem, the left side is most frequently attacked. When both mammæ suffer, one is pretty sure to be more painful than the other. Sterile married persons, or single women, are the most common subjects of this affection; and it generally occurs between the ages of sixteen and forty. The menstrual period exercises a perceptible influence upon it, making it much worse; and, in many instances, only at this time is the pain experienced. Pregnancy sometimes either originates or aggravates the pain.

The general health of patients in whom mastalgia is observed, is usually feeble; they are often nervous, and often broken down in constitution. They are subject to menstrual irregularities of some kind, displacements, inflammation, or some other diseased condition of the uterus; and there is no doubt in my mind that neuralgia of the mammæ is secondary, and that it originates in uterine or ovarian disease. It is certainly possible that this painful affection may be primary or idiopathic in some instances, but such must be very rare, and I have never seen any such case.

Symptoms.—The pain varies in intensity, in different instances, from mere inconvenience to severe suffering. It is sometimes confined to the mammæ, but often it radiates to the shoulder, axilla, down the side, into the chest, up the side of the neck, and down the arm. The character of the pain is not always the same. It will sometimes be a sense of painful tenderness, but generally the pain is neuralgic. Sometimes the breasts are tumid, and not unlike in appearance the post-partum breast; sometimes there is a distinct lump, which seems to be the point whence the pains originate, and occasionally the axillary glands are tender, enlarged, and the seat of the same kind of pain as the breast. In a large number of instances, the mammæ undergo no apparent change.

With all this there may be neuralgia elsewhere, as in the face, head, back, extremities, &c. The individual attacks of mastodynia last ordinarily several days, sometimes a week or more; and the attacks recur for months, and even years, with an obstinacy difficult to overcome.

Prognosis.—Hence the prognosis, so far as ready relief is concerned, is not very good; but on the other hand, the disease never endangers the life of the patient. In this connection, however, it should not be forgotten that it is occasionally accompanied with fatal disease of the uterus, as cancer, for instance, and although not in itself fatal, is indicative of mortal mischief somewhere else.

Treatment.—The treatment is palliative and curative; and although this last is not as effective as we might wish, in many instances, yet in the majority of cases a cure may be effected. During the paroxysm the various anodynes used, locally and generally, may be made to mitigate the suffering very materially. The belladonna plaster, or a plaster of equal parts of ext. belladonna and powdered camphor gum, may be applied as a good local anodyne. Ointments containing chloroform, opium, and narcotic extracts, are useful in allaying the suffering. If these are not sufficient, internally the same remedies are more effectual. The tinct. of gelsemin, or the resinoid, may be used with the hope of relief. The resinoid prepared by Keith, in quarter of grain doses, may be given every four hours until vertigo is experienced; or four grains ext. hyoscyamus, with an equal quantity of gum camphor, given every four or five hours, will often give relief. Should these fail, opium in substance will pretty certainly bring the required ease. As a general thing, the pains of mastodynia are not so excruciating as to make it necessary to resort to active anodyne medication, and it should be avoided if we can.

The curative treatment is constitutional and local. The constitutional should be conducted with a view to the correction of any constitutional depravity. The patient is usually weak, the stomach, liver, and bowels, need correction, and sometimes the skin and kidneys do not discharge their functions in a healthy manner. Small doses of hydrargyri, in the shape of blue pill, calomel, or corrosive chloride, should be given every three or four days. The bowels ought to be kept free by Rochelle salts, aloes, or rhubarb, owing to the state of the secretions. If the passages are dry and hard, the salts will do best; but if the bowels are simply slow, a combination of aloes and rhubarb will do better. The tonic course should be a mixture of vegetable bitters and iron. Infusion of quassia, or gentian, or the compound tinct. of cinchona, immediately after eating, followed by tinct. of iron two hours after eating, three times a day, will very generally agree well. Sometimes with the bitter infusion, instead of iron, ten or twelve drops of the liquor potassa will do better. If the urine is highly acid, this alkali is decidedly indicated. The iron may be given in pill if the tincture should be offensive. We not unfrequently find it impossible to administer ferruginous preparations to these patients, on account of their causing headache, or an increase of the nervous symptoms. This will not often be the case, however, if the bowels are kept sufficiently free, and the secretions in good condition. We should try it in very small doses, and when it fails us entirely, we have an excellent substitute in arsenic. Two or three drops of Fowler's solution, given some hours after the bitter infusion, or tincture, will often produce a very happy effect. If we can succeed in bracing up the system in this way, we may sometimes relieve the sufferer entirely; but almost always the beneficial effect of this treatment is only temporary, and the patient is not cured until the state of the uterus upon which the mammary suffering depends is removed, and then it entirely and permanently disappears. I have seen this occur quite often.

It will here be necessary only to counsel the student to make a thorough examination of the condition of the genital organs, and when anything wrong exists to rectify it. In proportion as the accomplishment of this last part is complete, will be the relief of the neuralgia of the breasts. The local treatment, as well as the mode of examination, will be found in another part of this work.

Agalactia—Absence of the Secretion of Milk.—This affection may be met with in two different forms, viz., when the milk fails to make its appearance after confinement, or ceases to be secreted during the ordinary term of lactation.

The local conditions which cause agalactia are deficiency of the formation of the gland, inflammation, or injury of it by accident. Many general circumstances will produce the same effect, as fever of various kinds of long duration, depressing passions, want, exposure to cold, &c.

The suppression of the secretion during lactation may also take place as the result of pregnancy. It may be further stated that any disease, accident, or other cause, producing a powerful deleterious general influence, will occasionally have the same effect. The suppression of the secretion of milk, so far as I am aware, does not produce disastrous effects of any kind upon important organs; and I am persuaded the consentaneous evils, as tuberculosis, &c., which are said to be the result, are more causative than sequential.

Agalactia is generally an obstinate difficulty. Of course it is irremediable in cases where the gland is deficient. And great perseverance is often required to induce a good state of lactation in persons who have commenced childbearing at very late periods in life.

Treatment.—Well-directed and persevering efforts at suction, with the child's or other person's mouth, is one of the most effective as well as entirely natural stimuli to the flow of milk. The child should be applied at intervals of four or five hours for a number of days. In the intervals of nursing the breasts should be oiled and fomented with warm water, or poulticed with some warm emollient substance. A thick pancake applied warm, and re-placed with another soon as it becomes cool, is an excellent means of promoting the secretion of milk. They should be persevered in for one or two weeks, or until the secretion is established. We are recommended by almost every medical work on the subject to use the leaves of the castor-oil plant (Ricinus communis) wilted in hot water, and applied warm over the breasts, changing. them as they become cool, or to foment the breasts perseveringly with a strong decoction of them. It is a popular belief that they have a specific effect in causing or promoting the secretion of milk.

Electricity is also recommended as an excellent stimulus to the secretion of the mammæ, as well as to that of almost all the rest of the organs. The continuous current of a galvanic battery of weak power, or the gentlest effects of the electro-magnetic appa-ratus in ordinary use, may be tried. The current should be so gentle that it is barely perceptible to the patient, and persevered in for several hours. The internal treatment is not very impor-tant, unless there is some very serious general disorder. Then, of course, the treatment will be governed by the indications presented in the case. What, perhaps, ought to have been mentioned in the treatment, is the removal of any of the causing conditions stated as producing this affection. Potations of some savory fluids in considerable quantities, when the secretion is once established, will aid in promoting its abundance. Chocolate, milk, and, if the patient is weak, lager beer, porter, ale, of the malt liquor kind. Soups and fluid diet generally may be used to the same end.

Galactorrhœa—Unusual quantity of Milk, or long continuance of it.—Sometimes when the patient is nursing the flow is immod-erately great, and she is very much weakened by the excess. In other instances, the flow may be unusually great, and after the child ceases to nurse continue for a great length of time; or, al-though the flow is not excessive, it does not cease when the child is weaned. In all of these varieties of abnormal secretion the general health of the patient will suffer in a greater or less degree.

The conditions of the system upon which galactorrhœa depends are various. They all, however, seem to affect the blood to a greater or less extent. General debility is one of them, diseases of the uterus, indigestion, perhaps of a character to produce more than an ordinary amount of glucose.

The prognosis is favorable so far as the ultimate termination is concerned, but the disease lasts for a great length of time sometimes in spite of the most judicious management. It has been known to last for five years, as we are informed by Montgomery.

Treatment.—The main object is to remove the causing condition when practicable. If the uterus is diseased, displaced, or otherwise in abnormal condition, it should receive special attention. Debility, so common an attendant, should be remedied when possible by suitable tonics, alteratives, nutritious diet, proper exercise, &c. The preparations of iron, as the tinct. ferri chl., gtt. xx, three times a day, after eating, syrup of the iodide, in like quantities and at the same intervals, may be mentioned as especially adapted to the purpose. The cod-liver oil is also a good nutrient, taken in tablespoonful doses, combined with some alcoholic stimulant, soon after eating. As different cases will require some variation owing to their peculiarities, and almost all practitioners have their own experience in the use of the roborant regimen and course of medicine, a good deal must be left to a discriminating judgment.

But, in addition to the rational course suggested by the symptoms and circumstances of the patient and attack, several remedies are recommended for their special effects in lessening or suppressing the lacteal secretion. Dr. Scanzoni calls them ischogalactics. The one that seems to be a favorite with the profession is the iodide of potassium. It may be given in three-grain doses every four hours in solution. Although I have repeatedly administered this medicine with a view to its effects in suppressing the secretion of milk, my opinion of its efficacy is not positive. In some instances its effects were apparently moderately favorable; in others I could see no effect whatever. I do not know whether it is entitled to a place as a remedy in this respect or not.

With belladonna, however, the case is very different. It, too, will entirely fail in instances where we may expect it to succeed; but in the majority of cases we may rely upon its efficacy, not immediately to suppress the secretion, but to aid rational treatment very materially. I have also seen decidedly good effects from sage (salvia). There is often an appetite for some such aromatic substance, and when this is the case, I think it does more good for the patient to eat it in considerable quantities. We may, however, use a strong infusion of it as a drink and fomentation to the breast. The fomentation should be cold, as it has a much better effect when of that temperature. Ice has also a very decided ischogalactic effect. We may use it freely for some hours at a time. The mode I have almost invariably adopted is to put the ice in a bladder, partly fill it, and then lay it upon the breast, with thin flannel between. If the sense of cold is oppressive or disagreeable in any way, the breast may be kept covered with two thicknesses of flannel. If general chilliness occurs, we ought not to continue it. Although I have tried it in this way in quite a number of cases, I have not seen it cause glandular inflammation in any. Indeed, I think there is too much dread of cold to the mamma, and that upon more extended trial the prejudice will be dispelled.

Cancer of the Breast.—This most malignant affection appears in the breast in the form of scirrhus more frequently than in any other, although all the different varieties have been observed in this organ. Persons beyond or about the end of the childbearing age are most subject to it. Although persons above forty are much more frequently attacked, instances of a younger age are not unfrequent; and even in quite young girls it sometimes occurs. Such cases almost always are of the medullary variety. A no less noted fact is, that childbearing women are the subjects generally selected; the single and sterile, although not exempt absolutely, suffer much less from it. All parts of the breast are occasionally the points of inception: none so often as the gland, and the tissues contiguous to it. We may say, then, that it commences—1st, in the gland; 2dly, in the tissues surrounding; 3dly, the nipple; and, 4thly, the skin and integuments. When it attacks the gland, a portion of one lobe may suffer first, or the whole gland. When the nipple and skin are the places of beginning, the form is apt, but not always, to be cancroid, while the scirrhous and medullary form are found in the gland.

Such a strong tendency exists in all forms of this affection to implicate contiguous tissues, that the point of beginning is soon lost in the surrounding parts.

Symptoms.—At the very beginning, the patient may not experience much pain in the part affected; in a short time, however, the peculiar pain is experienced,—a sharp darting paroxysm, lasting only long enough to create an excruciating perception, and then subside into complete ease. These pains are in the part, and seem to extend in some direction from them. At first they may not occur oftener than three or four times in twenty-four hours; but as the disease advances they become much more frequent, until they cause sleepless nights and suffering days.

These are the pains of scirrhus more particularly, but they occur likewise in the medullary, less frequently and later. I do not know that they are experienced in the cancroid variety. As the disease advances, a distressing weight and distension are complained of; and as soon as ulceration is fairly established, new kinds of pain may be looked for as the result of the exposed conditions of the deeper structure, the excoriations and inflammation from the discharge.

A wide variety of suffering from scirrhus will be observed, and no doubt is caused, by the different susceptibility of persons. Some have almost no pain, certainly complain of very little, while others suffer excruciatingly almost from the commencement.

When the attention of the patient is first directed to it, we may find a small or large lump deeply seated, or near the surface, toward the margin, or in the centre, in the case of scirrhus; generally in the beginning movable, but very soon it becomes attached to the neighboring parts above, below, or at the sides, while it may be made to glide over parts in every other direction. As the case advances, the attachment becomes general, and the part affected fixed among the neighboring tissues. As it approaches the surface, the skin becomes closely adherent to it, and changes its color gradually until it becomes deep purple. Unevenness and nodulation are seen, and very soon the skin begins to ulcerate, and before long a foul, irregular, rough-edged fetid excavation occupies the outer part of the tumor. From the beginning there is a strong inelastic hardness to the sense of touch, as distinctive when attentively observed as almost any other symptom. This feature is not lost until ulceration is completely established. Another sense

33

is very forcibly impressed by a cancer in the ulcerative stage, viz., the smell. An ichor, formed of acrid serum, blood, pus, and minute sloughs, yields a stench sufficient to render the patient loathsome to herself and others. The *medullary* form is never so very hard, but from the beginning often feels soft and so yielding to the touch, as to suggest the presence of fluid. The sense of fluctuation is always obscure enough to leave a doubt upon the mind of the observer. As it approaches the surface, the softness becomes more marked; points of prominence, generally more than one, sometimes several, are observed, pressing the skin upward, which becomes purple, and soon gives way. The disease rapidly bursts out into a bloody fungus, which springs above the surface, becoming detached in sloughs, and sending forth an excoriating fetid discharge, not unlike, but generally more copious, than the scirrhous variety. When the skin is attacked, the point first affected is likely to become more than ordinarily dense, and attached to the parts below, so as to give it a pinched appearance; discoloration sooner or later occurs, and finally ulceration and rapid destruction of the parts. The nipple, if attacked, increases in size, becomes purple, cracks open, and disappears in sloughing ulceration. Very soon after, the parts immediately around the part first affected become implicated, the absorbents take up the malignant matter, and carrying it to the axillary glands, these become centres of disease, which slowly advances in the march of disintegration. The glands inside the chest become implicated in the same way, and at the same time. Distant parts in various portions of the system begin to degenerate, until not unfrequently the patient is affected in a multitude of parts, and the cancer becomes a general disease.

Whether these separate and distinct deposits are the effect of transmission and lodgements of cancerous matter, is more than our science has determined; some pathologists think so, while others believe that it is primarily a blood disease, and that the first as well as the last deposits or growths are the results of it.

The constitutional symptoms vary with the stage and extent of the disease, and the constitutional condition of the patient. My experience, of course, will not compare with many who have written upon the subject; but I am led to believe, from observation, that a large majority of patients are robust persons when the cancerous growth is discovered. In some of my cases this is a remarkable feature; the patients seemed to enjoy excellent health, and suffered from no sort of inconvenience, much less cachexia. The fact of the breast being on a very exposed part of the person, a part which frequently comes under the eye and hand of the patient, would lead to an early discovery of disease in it, than would take place when it is situated in the internal organs. This will account for the fact, that recognized cases of cancer of the breast are not always, from the beginning, connected with a cachectic condition of the system; while the cachexia is manifested before we discover the cancer in the uterus or other internal organs.

I have been induced to infer from authors, that cancerous cachexia was a precedent circumstance to the development of local carcinoma; while my own observation has driven me to the belief that it is but a symptom of the advanced condition of that affection.

The peculiar straw-colored translucency of the surface in cancerous patients, does not, as a general thing, appear obvious until the growth is large enough, and the deposits numerous enough to influence hæmatosis, as a like amount of disease of a benignant nature would do. It sometimes is induced by the exhausting influence of cancerous discharges, the watchful restlessness, and disturbances of digestion and nutrition. It is merely the result of blood depravity and poverty, and may be classed in the same category with anæmia. We may be prepared, therefore, to find cancerous growths without cachexia, but we also see cancer commencing in broken-down constitutions; but I think we ought not to attach undue importance to this fact.

Entertaining this view of the connection between constitutional and local symptoms, it is not necessary to describe with minuteness the establishment and progress of the cachectic declivity down which the patient glides to her final home. The symptoms are much the same as cachectic suffering from any protracted disease.

Diagnosis.—The diagnosis of cancer of the breast is not generally difficult, as its characteristics, for the most part, are strongly marked; but sometimes there is great doubt whether, in a given case, the patient is laboring under malignant or benign disease. I need not say that the distinction is a subject of very great importance. Tumors of other character than cancerous may so much resemble this, and the converse, that it may be well to mention and contrast the differences in this connection.

Malignant growths are rapid; benign, slow. Malignant growths occur in patients over thirty-five generally, and most frequently in childbearing women; while benign tumors grow in persons who are single or sterile, and under thirty-five years of age.

Carcinoma is painful, but not tender; benignant tumors are neither painful nor tender. The former contract adhesions readily and early, while the latter are always moved with facility. Malignant tumors are often irregular and knotty, but not lobular; while the adenoid tumor has a regular lobulated feel, that by care may be almost always verified. In cancer, the nipple is very soon retracted; while in the non-malignant, retraction does not take place, or at least very seldom. Chronic inflammation begins during lactation, in persons under forty, is usually, if not always tender, and has none of the lancinating pains of carcinoma, is not so stony hard as scirrhus, before suppuration, nor so soft as the encephaloid variety, and is generally more extensive than cancer in the very beginning, and more regularly developed.

The diagnosis of cancer after ulceration is not difficult. The sharp irregular projecting edges, the uneven surface, the very large stony granulations, and the foul ichorous discharge, with the fetid, intolerable smell, are unmistakable characteristics. The fungous or encephaloid, in its open condition, is not similar to anything else with which I am acquainted. It grows with almost explosive energy; in a few months from the beginning, the part is very large. The surface, uneven, nodulated, and purple, feels firm in some places, in others soft enough to contain fluid. The skin soon gives way, and large, red, bleeding, suppurating, sloughing fungi shoot through it, and shed upon the surrounding parts an acrid, excoriating, fetid ichor, that irritates everything it touches. The fungus spreads and sloughs, and bleeds and smells, as nothing else on a living being can. There is no danger of confounding it with anything beside.

When any doubts exist as to the diagnosis between this variety of cancer and chronic inflammation in the suppurative stage of the one, and the subcutaneous growth of the other, the exploring needle will demonstrate the difference.

Prognosis.—The almost invariable fatal termination of carcinoma, under every variety of management as yet within the possibility of our profession, paralyzes every hope of the patient when she learns the nature of her disease, and discourages resort to radical measures, only in exceptional cases. It is fatal with treatment, and without it. Its duration, compared with other chronic diseases, is short. This is particularly the case with the encephaloid variety. The younger the patient, the more rapidly it marches through its different stages to dissolution. From nine months to three years are the limits, or nearly so, of its duration. One reason why it terminates more rapidly in young persons is, that scirrhus does not often occur in the young, while the fungous variety occurs much more frequently.

Causes.—Very little is known of the causes of cancer. It is believed to be hereditary in certain families, and perhaps this influence has as much effect in bringing it about as any other; but I think the veil thus used to cover our ignorance should be lifted, by the manly acknowledgment of our want of knowledge. Instances occur that are attributed by the patient to a blow or bruise; but this is more likely a coincidence than a cause. In fact, I think we know nothing of the etiology of cancer, although many circumstances under which it occurs are well determined.

It is to be hoped that, at no very distant day, the numerous and effective means of research so successfully brought to bear upon other obscure subjects, will result in a definite settlement of the question, What are the causes of cancer, and what the cure? It is not too much to believe that such will be the case.

Treatment.—Notwithstanding the hopelessness of treatment in cancer, so far as the general result is concerned, much may be done to palliate the sufferings through which our patient must inevitably pass, and some hope is yet indulged that life may be somewhat protracted. It is not my purpose to argue the propriety or impropriety of pursuing a given course, but succinctly to give my convictions in the several items of treatment I shall present. In the early stages there is not much indication for general or constitutional remedies, as the health of the patient is not much, if at all, deteriorated; but, as its ravages make inroads upon the function of nutrition, nutritious diet and bracing regimen, with medicinal tonics, will do a great deal towards postponing the fatal prostration. Of the tonics, the preparations of iron are, perhaps, the best, and yet we often increase their efficacy by the co-ordinate employment of the bitters. The latter are used to influence the stomach under digestion; the former to promote hæmatosis. There is no specific for cancer; nothing that will contend with its eccentric cell-formations and destroy them. We must promote the healthy discharge of the important functions, and preserve as far as possible the integrity of the vital organs. To do this best we must watch their aberration, and treat them upon general principles. After the disease has advanced to a certain degree, the all-absorbing symptoms are connected with the nervous system; the patient suffers excruciating pain, is annoyed by sleepless nights, nervousness, and depression of spirits. A well-directed use of anodynes will afford the patient great relief from all these symptoms, and ease her to the grave. And the physician should remember that the profession and humanity demand of him all the ingenuity and skill he is capable of exercising to fulfil these objects in the completest manner. Opium is, of course, the great solace of the patient; and while it should not be doled out so niggardly as to fall short of great relief, it ought not to be so lavishly given as to arrest digestion, or destroy its usefulness by overdoses. With some patients the different preparations of alcohol are equally effective in the relief of pain. In such cases an alternation of them with opium will preserve the patient in a better state of general health than either alone. The anodyne may be made to do some good by local application, particularly subcutaneous injections of morphia. To sprinkle the powder of morphia upon the ulcerated surface, or wash the part and keep in contact with it a watery solution of the aqueous extract of opium in lint, saturated with the solution, will be often effective.

There is, perhaps, no radical cure for cancer of any kind; and yet we are sometimes justifiable in taking what in other cases are usually regarded as radical measures. I have seen some cases of cancer of the breast amputated with what seemed to me to be the prolongation of the patient's life, and hence I do not hesitate, in a case where the development is slow, where there is no discoloration of the skin, and when the maxillary or other lymphatic ganglions, so far as we may be able to determine, are not implicated, and the patient is in good general health, to amputate the diseased gland. When any of the visible parts are affected, the skin, the ganglions, or other tissues, amputation will not even prolong life.

It may be very properly said that the case I describe as suitable for amputation is the kind of case that the patient will live longest under, and the worst to determine the worth of that operation,— and that is true. I may be also wrong in thus directing amputation, but they are certainly the only cases in which ablation has seemed to me to be beneficial. As to the mode of removing the breast, I am decidedly in favor of the knife, but must freely acknowledge that my experience is very limited in the use of caustics, never having seen, I believe, but two thus removed.

CHAPTER XXX.

PUERPERAL CONVULSIONS.

PRESSURE of the gravid uterus within the abdomen seems to be the originating cause of this kind of convulsions, for although there may be cases of uræmic convulsions arising to some extent in the same state of the blood and nervous system as in Bright's disease, and perhaps others, the chain of pathological changes originates as the effect of some other causes; while in the pregnant condition they are so unfrequently operative as scarcely to be worth serious consideration.

As the uterus increases in size within the pelvis, it presses upon the rectum and bladder, especially the first, preventing the free and easy evacuation of the fæces, thus giving rise to such a loaded state of the alimentary canal as to cause it to occupy a larger space in the abdomen than usual. This pressure upon the rectum is what perhaps causes the general gaseous distension of the abdomen in the first few weeks of pregnancy. As it rises higher and occupies more room, the veins passing through and along the brim of the pelvis are pressed upon, so that the labia, perineum, and lower extremities, become unusually filled with venous blood; and after pregnancy is further advanced, this stagnation goes so far that serum is pressed out of the capillary extremities of the veins into the areolar tissue, causing œdematous tumefaction of the parts. The legs and labia swell quite largely in many cases. The veins themselves become greatly distended, their calibre is increased, and sometimes their walls give way, large varices are formed, and in rare instances rupture takes place. The blood thus long detained and so slowly returned into the general circulation must undergo excessive venusation. The venous changes in the blood must be excessive compared to what they would be if that fluid flowed with its wonted rapidity, and probably deteriorated in quality by the detention, either destroying or impairing some of its ordinary elements, or impregnating it with deleterious materials.

Increasing still more in size, the uterus comes to press upon the iliac and aortic arteries, preventing the blood from reaching the lower extremities in the usual quantities, and by the backward pressure thus made upon the column of blood causes more of it to be sent to the upper part of the body and head. This upward distribution of the blood is augmented by the pressure exerted upon the small arteries and capillaries in the abdomen, excluding it to a considerable degree from this cavity when distension becomes great. From these two conditions arises a true hyperæmia of the head, and consequent increased general excitability. Whether the spinal cord partakes of this state, I am not aware, but it most likely does. The diaphragm is pushed up further into the chest than usual, the cavity of the thorax is thus very materially diminished, and consequently contains less blood than ordinary. After looking at these effects of pressure upon the systemic and pulmonary circulation, no one can be surprised at the hyperæmic excitability so frequently observed in the last weeks of pregnancy. But pressure upon the abdominal organs not only causes unequal distribution of the blood, but deteriorates its composition. The excretory and secretory capacity of the organs is impaired. The mucous crypts and intestinal glands do not produce their full supply of secretion, nor is the watery exudation from the mucous membrane as great as usual; the fæces become dry as a consequence, and pressure upon the sigmoid flexure of the colon and the rectum adds all that is necessary to bring about constipation more or less obstinate. The liver and pancreas are doubtless likewise prevented in the same way from pouring their stimulating fluids into the alimentary canal as plentifully as common.

Another and more deleterious concatenation of circumstances begins in the pressure upon the emulgent veins and substance of the kidneys. Pressure upon the emulgent veins retards the return of blood from the kidneys to the general circulation, the capillaries are over-distended with blood until some of the serum of that fluid transudes their sides and appears in the urine, and when the urine is properly tested the albumen thus effused is detected in it. This exudation is in itself a matter of minor importance, otherwise than as an indication and evidence of embarrassment in the excretory function of the kidneys. The passive or venous hyperæmia which exists when albumen is discoverable in the urine, is sufficient to prevent the excretion of urea. *Generally*, therefore, the urine that contains albumen in considerable quantities is deficient in urea. The azotic elements of this excretion are retained in the blood, and may be detected by chemical reagents, according to some observers, in the form of carb. of ammonia. We are, I think, yet hardly warranted, however, in deciding the precise form they assume after failing to find their way out of the blood, but ample observation attests their deleterious effects upon the nervous centres, both animal and vital. The palpitation of the heart and susceptibility of the stomach to irritant influences, show how the great sympathetic plexuses are affected by it; while the abnormal sensations and movements, the neuralgic pains, convulsions, &c., demonstrate the deleterious action exerted upon the brain and spinal cord. Although convulsions occur in all kinds of patients suffering from albuminuria, yet their frequency and urgency are beyond all comparison greater in puerperal patients. This, no doubt, arises from the fact that in ordinary Bright's disease, the albuminuria of children in eruptive diseases, &c., &c., there is not that hyperæmic state of the nervous centres caused by pressure upon the large arteries, and abdominal and thoracic organs, that there is in puerperal women. The coincidence of uræmia and cerebral hyperæmia constitutes the peculiarity of the latter class of patients, both exalting the excitability of the nervous centres; the hyperæmia furnishing an unusually high nutrition to the nerve-cells, thereby making the changes in them more easy and rapid, the uræmia by a direct stimulating influence upon them by contact. This conjectural explanation may not be correct; the main facts nevertheless of cerebral hyperæmia, uræmia, and puerperal convulsions, probably, almost if not quite always, go together.

This is too short a sketch of the effects of pressure of the gravid uterus to contain everything relating to it. The remarks are intended more as suggestive than as full explanations. I cannot forbear, however, in this connection, from alluding to the theory of anæmia as contrasted with the plethora of pregnancy. I think the weight of authority is in favor of the idea that *most* women—certainly not all—are rendered somewhat anæmic by pregnancy. Considering the great pressure upon the abdominal organs, how it would prevent perfect chylification and lacteal absorption, we at least have something of an explanation of the manner of its occurrence. It should be borne in mind, in connection with the above explanation of the predisposing conditions—for they are usually only predisposing circumstances, many patients having them all without being convulsed—that the pressure is effective in the production of them only in patients whose abdominal muscles are rigid and comparatively unyielding, not permitting the uterus to distend them much, but keeping it pressed tightly against the posterior wall of the abdomen. The large vessels are, thus pressed against the spinal column. Primiparæ present this state of the abdominal muscles more frequently than any other kinds of patients, and the older the primipara, generally the more rigid the muscles are. And we find that convulsions occur much more frequently—eighty per cent. of the cases—in patients pregnant with the first child. The pressure of the uterus in the abdomen is greatest during pregnancy about the end of the eighth month, and fore part of the ninth. I have seen more cases of convulsions about this time than any other. It will be recollected, that during the ninth month the organ settles down lower into the pelvis, presses more upon the veins of the lower extremities, and less upon the great arteries and abdominal organs. The convulsions may take place as early as the sixth month. I knew one fatal case at this period of pregnancy. They frequently occur during labor, and after it less often. The conditions above may be present in sufficient intensity to cause convulsions, though I think there is almost always present some exciting cause proper. They may be regarded as predisposing in their effects usually, and speaking in the language of Marshall Hall, we may call them centric, because they affect the nervous centres directly. Most other causes are excentric, and affect the nervous centres indirectly, and excite them to the production of convulsive movements in the muscles. Perhaps the most common of these last is labor, the pains of which, operating in a reflex manner upon the brain and spinal nervous centres, wrought up to an unusual susceptibility by the hyperæmia and uræmia, caused by uterine pressure, are sufficient to induce convulsions. Gastric, intestinal, and cervical irritation have a similar influence. In the same way the pressure of the fœtal head upon the cervix uteri, vaginal walls, sacral plexus of nerves, perineum, external organs, &c., &c. Any circumstance that causes unusual pain or local nervous excitement may, by reflex effects upon the spinal centres, corresponding to the

affected locality, set up a chain of phenomena that will result in a
convulsive paroxysm.   The emotions or passions, fright, the effect
of bright light, loud noises, all these, and many others, are not
unfrequently sufficient exciting causes.   I think that but very few
persons are so susceptible by reason of the delicacy or susceptibility
of their natural organization as to be thrown into convulsions by
any of the causes here enumerated as exciting, and believe they
must be preceded by some such morbific influence as very much
exalts the irritability of the general organism.   I cannot, therefore, subscribe to the doctrine of the sufficiency of these reflex
causes alone.

There can hardly be said to be any distinctive morbid anatomy
of puerperal convulsions.   The fatal conditions are often evanescent or inscrutable.   In some instances there is much cerebral
congestion, and *very rarely* sanguineous extravasation in the brain.
More frequently there is serous effusion between the membranes
or in the sinuses.   Still more common is œdema of the areolar or
interstitial tissue of the brain observed.   The lungs are pretty
constantly filled with œdematous appearances, and the air-cells
and small bronchi gorged with tough mucus; sometimes great
sanguineous congestion colors their structure extensively, while
there is sometimes subserous emphysema.   Probably as often as
any other way the lungs and brain betray no evidence of violent
action within them.   I think it ought to be well understood,
that these anatomical changes are the effects and not the causes
of the convulsions.   They are doubtless the causes of death in
very many, if not all the cases in which they are observed, but I
know of no well-informed pathologist of the present day who believes them to be primary links in the chain of morbid states in
the body.   As I shall have occasion to explain hereafter, the convulsive paroxysms overwhelm these vital organs, by projecting the
blood into them with great force, and preventing its return from
them through the veins.

The kidneys are more constantly and obviously affected, perhaps, than any other organ, and yet sometimes when uræmia has
been an indubitable fact, they exhibit scarcely any appearance
of disease.   Ordinarily they are merely injected with an unusual
amount of blood, the venous capillaries are more distended, and
the cortical substance mottled less frequently than is, serous, fibrin-

ous, or even sanguineous effusion in the areolar tissue. It is not often the case that actual degeneration takes place; very generally there is nothing more than vascular turgescence, which readily passes off when recovery takes place, and disappears as an effect of the post-mortem distribution of the blood. The uriniferous tubules sometimes contain albuminous casts and clots.

Symptoms.—The first thing that attracts our attention in pregnant women, and leads us to suspect uræmia, is œdema. When this symptom is confined to the lower extremities it has not much significance, merely indicating pressure upon the venous trunks leading through the pelvis from below; but when, with or without this, there is dropsical puffiness of the hands, face, arms, cellular tissue over the chest, hips, back, or any other portion of the upper parts of the person, we ought at once to suspect that the peculiar thin condition of the blood is caused by a loss of albumen. The œdema of uræmia, or rather albuminuria, in pregnancy, is not often either general or excessive. The hands become swollen more than ordinary, or the face. Sometimes the extent of the œdema is small, and it makes its appearance in one place for a short time, subsides, and appears in another. We may, for instance, find cases where, a part of the day, one *side* of the face swells up, and another part of the day one arm, and the face is relieved. With this symptom the patient is ordinarily unusually nervous. She has disturbed and dreamy sleep, startings and twitchings, neuralgic pains, headache, &c. There is often, too, indigestion, pain in the stomach, vomiting, and more or less febrile disturbance, pain in the loins, mental depression, and sometimes mental hallucinations. The bowels are constipated, the skin torpid, and the secretion of urine scanty and high-colored. The tongue is loaded with a whitish-yellow coating, the saliva abundant and offensive, causing almost a constant bad taste in the mouth, and disgust for food. Although these symptoms are pretty uniformly present, and sufficiently prominent to excite the attention of the patient and physician, yet there are instances in which they are so slight as not to attract notice. I think the œdema will scarcely ever fail to show itself. Since my attention has been turned to the fact, I have not known convulsions to occur without it. The measure of its extent and duration has but little to do with the intensity and duration of the attack. It often occurs without convulsions, and it should be remembered that the quantity of effusion is only a measure of the extent of attenuation and not intoxication of the blood. Other causes may so deteriorate the process of hæmatosis as to thin the blood very materially, while the function of the kidneys is not much if any disturbed, when this symptom will be greater in proportion than the results. All this is intended to show that although an invaluable, it is not an infallible symptom, indicating the presence of uræmia.

If the urine is tested in these cases, it will be found holding albumen in solution, and the relative quantity of albumen will indicate with some degree of accuracy the imminence of danger of an attack. An error in this respect may be very readily committed by examining the urine voided at different times in the day. The first in the morning is apt to be the richest in quantity. The more fluid the patient takes, the more dilute the solution. Perhaps, upon the whole, the best way is to save the whole amount discharged for twenty-four hours and test portions of this as a whole. How long these premonitory symptoms precede the convulsive attack depends upon a great variety of circumstances, many of which are entirely inappreciable, and vary in duration from a few days to several weeks.

The paroxysm is sometimes preceded, for some hours or but a few moments, by more marked and obvious phenomena. One patient whom I attended saw a bright light for several minutes before the first paroxysm, and described its peculiarities. Blindness is often complained of. Double vision, half vision, intolerance of light, deafness, great sensitiveness to sound, and various versatile delusions, not unfrequently precede the convulsions. Unusual motions of one extremity, or numbness of it, a rolling or inability to direct the movement of the eyes, and occasionally an aura not unlike epileptic warning, begins in some part, and spreads towards the head or epigastrium. At other times the paroxysm bursts upon the victim with a suddenness as unexpected as alarming to the attendants. In whatever manner it may be initiated, the first general condition is that of great muscular tension; the head is slowly drawn back, and often to one side, the eyes and mouth are opened widely, the hands are clenched and drawn closely and forcibly up against the chest, the legs are stretched straight downwards or slightly backwards, the feet are extended until almost upon a line with the limbs, the muscles of the abdomen, chest, and back, become rigid, and generally extend the body somewhat backwards. This general rigid condition of the muscles is maintained for a few seconds, during which time the tongue is protruded beyond the teeth, sometimes projected very much, the eyes are closed, and the jaws brought together suddenly, wounding the tongue; the muscles of the face and extremities suddenly relax, and as suddenly contract so repeatedly, that convulsive contortions of the face present every phase of ludicrous grimace, while the limbs and back heave and throw the body about in every direction, and cause it to writhe and contort in every conceivable manner. One remarkable feature distinguishing this and other epileptic convulsions from most other forms of irregular muscular movements is, that the motion is symmetrical and synchronous in the two halves of the body. The two arms are moved in the same manner at the same time; the two legs, the two sides of the face, keep time and measure with each other. As the paroxysm draws to a close, the period of relaxation becomes longer, and the length of duration of the contraction shorter, until the relaxation is complete and universal, when the whole body and limbs assume a posture of helpless and unconscious repose. Although this relaxation in most instances is confined to the voluntary muscles, sometimes the sphincters relax, and the urine and fæces escape from their receptacles without the knowledge of the patient. During the early part of the paroxysm, the air is expelled from the lungs with great force until the chest is as nearly as possible empty of it; and during the whole of the convulsion, there is but very little air admitted, sometimes none at all. At the conclusion, and with general relaxation, the rigidity of the thoracic and abdominal muscles gives way, and allows the air to enter into the lungs rapidly and in large quantities. The large amount of air thus admitted during the deep inspiration is expelled with a hissing noise through the teeth and mouth partly or wholly closed, sending the saliva, generally colored with blood, which flows from the wounded tongue, copiously over the person of the patient and the bed. This sibilant and deep respiration continues for a time, and gradually gives place to more calm, and, after a while, natural breathing. A circumstance of great interest to the observer is the change of color noticed in the face, so obvious during the paroxysm. At the commencemynt of the paroxysm, the face is natural in color, or perhaps pale; during the fit it becomes turgid with blood, and red in color; soon the redness becomes dusky, afterwards crimson, and finally very dark purple. The color is greatest at the time of the solution of the paroxysm, and passes slowly off after the respiration becomes good. I need not inform the student that this change in color depends upon the highly carbonized state of the blood circulating in the capillaries of the skin; and we have but to reflect that the blood circulating in the nervous centres is in the same condition, to understand the cause of the solution of the paroxysm. The blood surcharged with carbon, circulating in the brain and spinal cord, induces a true carbonic anæsthesia, hence the entire relaxation of the muscles, and the coma that succeeds the paroxysm; when the blood is decarbonized and re-oxygenated, the coma subsides, the irritability returns, and the paroxysm, after a greater or less length of time, is repeated. Marshall Hall tells us that the glottis is closed as the first step in all these epileptiform convulsions, and looks upon it as the initiatory if not the causing condition of them. Whether this is the case or not, the state of the thoracic and abdominal muscles renders perfect respiration impracticable, and the cessation of respiration ought to be regarded as the method adopted by nature of resolving and shortening the paroxysms, and serves as a basis on which to found rational practice. The state of the pulse cannot be ascertained during a paroxysm, but after its subsidence it is generally slow. The duration of a paroxysm is from half a minute or a few seconds only to several minutes. The more violent and universal the muscular action, the shorter the time of duration.

This description of the phenomena of a convulsive fit is applicable to the first or few first only, for as they recur, some of the symptoms become more intense and somewhat changed in appearance. The color of the face is darker in the first few paroxysms, and the stupor is of very short duration. In fact, after the first paroxysm, it is more like perfect relaxation than stupor, the breathing is seldom stertorous, and hissing and spirting but a short time. After each convulsion, there is more and more marked torpor, until it becomes profound and prolonged coma. The patient does not arouse to consciousness, the eyes become turgid and continue half open, the inferior maxilla falls and leaves the mouth wide open, the tongue is swollen until it seems to fill the mouth and falls back upon the glottis, and the whole face is swollen and injected. The respiratory murmur, at first clear and complete all over the chest, soon becomes masked by the moist râle, and after a time a coarse mucous rhonchus obscures all other sounds, the breathing is more rapid, the pulse accelerated, and the general powers more prostrate, until the respiration becomes gasping, the pulse is very rapid, and the skin covered with copious thin perspiration, and the patient expires; or the coma is more profound, the pulse slower, and the extremities cold, respiration less frequent and perfect, until the patient, after a longer or shorter struggle, expires. In the intervals, at first, there is apt to be restlessness; even where the patient is unconscious, the reflex sensitiveness is quite marked; the patient will move her limbs when touched, the features become distorted when the skin on the face is lightly brushed; but after the convulsions have continued for some hours, this passes off, and there is no reflex sensitiveness perceptible. This *interparoxysmal coma is the result of cerebral congestion*, and is very different from the anæsthetic condition which resolves the paroxysm early in the case. As the case grows worse, the anæsthesia runs into and is complicated with coma; but the attentive observer will be able to see the difference throughout the whole progress of the disease. It is not difficult to understand how the coma and difficult respiration should result from the effects of the recurring paroxysms. The compression of the abdominal and thoracic cavities expels much blood from the capillaries of their tissues, and presses it into the larger venous trunks, and thence into the heart, from which, on account of its accelerated movements, the blood is impelled rapidly and in large quantities into the great arteries; and, as much resistance is made to its entrance into the small arteries of these cavities, there is a larger quantity driven into the brain, which does not return through the veins into the chest so readily as usual, on account of constriction of these by the muscular tension. A temporary congestion of the brain is the effect, and after repeated and prolonged paroxysms the congestion becomes permanent, and great effusion of the serum of blood is produced. The longer the paroxysms and the more frequent their recurrence, the more rapid and deleterious

34

this effect of them, until apoplectic congestion is set up, and so much damage is done to the brain,—and perhaps spinal cord,—that it fails in its functions. The lungs are very powerfully compressed during the convulsion, mucus accumulates in the bronchi, while the want of the ordinary sensitiveness of the mucous surfaces prevents the inconvenience to the respiration from being observed by the reflex nervous centres. This accumulation aids in deteriorating the quality of the blood, and consequently undermining the powers of the system. In all the fatal cases I have observed anything like closely, I have witnessed a distressing degree of dyspnœa from this cause, and could not resist the conviction that the fatal catastrophe was accelerated more by this than any other condition.

The frequency of the paroxysms varies in different cases from a few moments of intervals to many hours, say twelve, and even twenty-four. Sometimes there is marked periodicity in their return,—every half hour, every two hours, &c. &c.

Generally, they are preceded in their return by restlessness and other evidences of pain. After the convulsions have continued for a certain length of time, if there is to be a favorable termination, they cease. Sometimes they cease gradually, the intervals becoming longer and the severity less, until they entirely fail to return; but generally they come to a somewhat sudden termination, the last paroxysm being as bad or even worse than any of its predecessors. However this may be, the patient is left in a state of insensibility which lasts a longer or a shorter time, according to the damage done to the brain. Consciousness usually returns slowly, the movements receiving intelligent direction, and having some purpose in their object; sounds affect the patient, or she may be induced to move by touch or annoyance of any kind, until she opens her eyes and directs them in a semi-sentient manner, and then falls asleep. This gradual awakening becomes complete in ten or twelve hours, or it may be two or three days. I knew one patient to remain unconscious for six days, and yet completely recover. After the recovery is fully effected, the memory of all that occurred during one, and sometimes a number of days prior to the attack, is wholly obliterated. It is a blank in her existence.

Puerperal convulsions are said to be followed quite frequently by some other form of puerperal disease. This is not in accordance with my observation. Indeed, I cannot now recall a case where the patient did not recover without further puerperal accident or trouble.

Uterine symptoms are not uniform; sometimes there are none; for so far as we can judge by examination of the uterus, and observations of the manner of the patient or any complaint, labor in many cases does not commence before the convulsions. Rarely we meet with cases that recover from them and are not delivered for several weeks afterwards. The first case I ever saw, the patient, after having convulsions for twenty-four hours, and ten or twelve in number, recovered perfectly, and in five weeks afterwards was delivered of a fœtus that probably ceased to live at the time she had the convulsions. Where there are no uterine symptoms at first, labor is apt to commence soon after the convulsions become severe, and proceed slowly to a termination, or advance so tardily as not to be complete before a fatal termination of the disease. Again, we often meet with cases where labor is in active progress before the convulsions commence. Where this last is the condition of things, labor is not generally retarded, although occasionally it is brought to a complete stop. The symptoms may show themselves in the second or third stage of labor, or even after the whole process is finished. I knew one patient seized with these convulsions on the third day after delivery. Attacks after labor are more likely to take place very soon—in a few minutes, or a very few hours at most.

The diagnosis is not generally difficult after the affection is completely formed. In a practical point of view, it is quite important to diagnosticate the predisposition to this form of disease, and proper care will generally enable us to perceive the tendency in this direction when well marked. With the nervous and gastric derangement described above, there is œdema of the *upper* portion of the body, upper extremities, or face. While œdema of the legs and feet is more frequently produced by pressure upon the veins coming from the lower extremities, this symptom occurring in the last months or weeks of pregnancy in the upper part of the body is generally the effect of albuminuria. When this form of œdema shows itself, the urine should be subjected to tests for albumen. Whether uræmia is always present when there is albuminuria, and whether uræmia does not sometimes occur without albuminuria, are questions which have not been positively decided, so far as I know; but we can, with sufficient assurance, regard it as an invariable rule, that albuminuria indicates uræmia. If albumen is found in the urine, we should prove by tests that there is or is not a deficiency of urea and uric acid in the secretion. These two facts would show a predisposition to puerperal convulsions. The diagnosis, then, of the predisposition is made out by the presence of œdema of the upper part of the body, face, and limbs, albumen in the urine, and a deficiency of urea and uric acid. After the attack has supervened, the diagnosis, for the most part, lies between epilepsy, apoplexy, hysterical and puerperal convulsions. There may be other affections with which there is a possibility of confounding these convulsions. Epileptic convulsions very rarely occur under puerperal circumstances in epileptic patients, and when they do, the convulsive seizure is not repeated. Apoplexy is ushered in sometimes by convulsions of short duration, to which characteristic profound and prolonged coma succeed. The convulsions are not repeated, the patient does not soon become conscious as after the first paroxysm of puerperal convulsions, and there is almost constantly nausea and vomiting with apoplexy. When coma is of more than very temporary duration, there is congestion of the brain, either as the result of the convulsions or as a primary condition; hence, we have a combination of apoplexy and uræmic convulsions as a not unfrequent thing. I think, however, that apoplexy, uncomplicated with uræmia, is quite uncommon in the puerperal state. I remember to have seen only one instance. As I have endeavored to show, in speaking of symptoms, the supervention of this complication is, perhaps, the most frequent cause of death, and it is what must, if possible, be avoided. Hysterical convulsions are distinguishable from the uræmic by symptoms arising from an opposite state than those of apoplexy, constant consciousness, more or less complete, and want of coma properly speaking. In hysteric convulsions, the color of the face is not so deep, the paroxysm is not terminated by carbonic anæsthesia, which shows the existence of apnœa; there is, in fact, almost no embarrassment to the function of respiration in hysteria, while there constantly is in puerperal convulsions. In hysteric convulsions, the muscular contractions are not completely symmetrical, but one arm will be acting violently and the other firmly fixed, and the legs do not move synchronously like shocks that succeed the first general contraction in uræmic cases; and I hope I may be permitted to say, that in hysteria the mental manifestations are perversities of disposition instead of the real inebriety of the uræmic disease. We must remember that they may be complicated, so that in the beginning we may have hysterical symptoms, but these are soon obscured by those arising from congestion of the brain. Perhaps, as in the premonitory condition, the most important items in a clear diagnosis are albuminuria and œdema. Examinations of the urine should be made when practicable.

Prognosis.—This has reference, 1st, to the safety of the mother, and 2d, that of the child. According to Braun, 30 per cent. of the cases, under ordinary circumstances, prove fatal to the mother. My own observation leads me to a more favorable prognosis, and in the practice of my friends and myself, I think 25 would be a large estimate of the mortality of uræmic puerperal convulsions. The particular circumstances influencing the prognosis are various. One of the most important is the time of the attack; commencing before or in the early steps of labor makes the case more dangerous than if it takes place later, for, according to Braun, the accomplishment of labor causes a cessation of the convulsions in 37 per cent. of the cases, their amelioration in 31 per cent. more, and that about 32 per cent. continue without change. There is less danger in post-partum cases than those commencing during labor. Cases in which the inter-paroxysmal coma is profound and persistent, or the breathing and pulse are very much accelerated, or where the bronchiæ seem filled with mucus, and sibilant râle is heard over a large part, or, as is often the case, the whole of the chest, the prognosis is grave; as is also the case, should the convulsions recur frequently, less than every hour, or continue for a long time. Cerebral, pulmonary, and circulatory deviations are the guides to the prognosis. The nearer the breathing and pulse continue to the natural standard, the more favorable the prognosis; the greater their departure from it in any respect, the greater the danger. This statement needs but very little qualification. The mortality to the *fœtus*, if I can judge from the cases that have come under my own observation, is greater than that to the mother. I think at least half of the children born during puerperal convulsions were dead. The cause of this great mortality among the children is not clearly agreed upon by the profession. Some think the convulsive action of the uterus cramps the circulation, causing asphyxia. The most likely cause, I think, is the uræmia. I have known two well-marked cases of uræmia where convulsions did not take place, but the fœti in both cases were dead-born. In most cases, so soon as the labor has terminated, and the convulsions cease, the albuminuria passes rapidly away, all evidences of uræmic poison disappear, and the patient uninterruptedly convalesces. At other times, several weeks elapse before the excretion returns to its normal state, when the convalescence is interrupted by the occurrence of dropsical accumulations, causing death or tedious recovery, or there may be waning or great nervous prostration and fever. Very rarely the damage done to the brain is so great as to leave some part of the body or limbs partially or wholly paralyzed, or bronchial or pneumonic inflammation may be established sometimes, and not readily subside.

Treatment.—The treatment is very properly and naturally divided into the medical and obstetrical.

Although possibly not the most important, it is most convenient to consider the medical treatment first. The application of treatment is modified by the time when instituted; that is, either as preventive or curative. As I have endeavored to show, we may often determine the presence of a predisposition to puerperal convulsions quite conclusively some days, and even weeks, before the supervention of an attack; and not unfrequently, by proper treatment, entirely avoid the development of them.

The two main indications in the medical treatment are: First, the removal of hyperæmia of the brain and spinal nervous centres; and, second, removal of the uræmia.

To overcome both these morbid conditions is very desirable but sometimes impracticable, yet we can generally entirely get rid of one of them and thus avoid the commencement or continuance of the paroxysms.

We will often be able to accomplish *one* of the indications of a cure by preventive measures, viz., to remove hyperæmia of the nervous centres. The means to accomplish this end are bleeding, the administration of cathartics, diaphoretics, and diuretics. When the patient is plethoric, with a red injected face, a full and slow pulse, bloodletting should be resorted to. We may bleed the patient until these conditions are removed or very much reduced. I hope I may be allowed to remark that when the predisposition is certainly present, we are justified in using, nay the case demands energetic measures, and that we ought not to be deterred from the employment of them for fear of inducing premature delivery, for really this would add to the chances of both the patient and fœtus. And for fear of being misunderstood by the student, I will repeat the main items diagnostic of the predisposition, viz., œdema of the *upper* part of the body, face, or limbs, or all together, albumen in the urine, and the various nervous symptoms I have mentioned. When to these are added those of decided plethora, an active antiphlogistic medication is generally very beneficial. One decided venesection in the sitting posture, until approaching syncope is observed, will sometimes do much good and entirely avert the attack. But active catharsis is also attended with very satisfactory results. The compound powder of jalap is one of the best of cathartics at such times. It should be administered in sufficiently large doses to cause copious watery evacuations. Should the patient not be decidedly plethoric, we may, with propriety, omit the bleeding, but the cathartic should be given. This should in either case be succeeded by the wine of colchicum in decided doses every four or six hours, and at least once a day the patient should have a steam bath to cause free diaphoresis.

At any stage of this treatment nervous suffering should be promptly relieved by opium. The Dover powder is a good form.. Very often such treatment is followed by premature labor, and when such is the case the labor is much less apt to be attended by this dreadful complication. I think this plan is very much more appropriate than an immediate resort to the intentional induction of premature labor without first reducing the cerebral hyperæmia. This cerebral hyperæmia should be held up to us as affording one of the most valuable indications of treatment; very often its removal is sufficient to render a very threatening case safe. The preventive treatment I have recommended is not always followed by premature labor, as this process may not take place at the end of term. Sponge baths of acetic acid, friction with acetic acid, and acid drinks and medicines, may be used with the hope that they may enter the blood and neutralize the ureal accumulations in that fluid.

During a Paroxysm.—The first thing that should be attended to during the paroxysms is to prevent the injury which so frequently results to the tongue. A large cork, or a soft pine stick, should be placed between the molar teeth on one side while the jaws are separated, and thus retained, until the convulsion ceases. We ought not to be betrayed by any excitement into the use of a spoon-handle or other metallic substance, lest damage may be done to the teeth. The clothing should be loosened, so that the neck, body, or limbs, may not be constricted by it. The throat must be made entirely naked when practicable, and we must observe that the violent and irregular movements do not twist the clothing about some part of the person to an injurious degree. The attendants ought not to restrain the movements of the patient too greatly; indeed, perfect freedom to the whole body should be allowed when it does not become clear that it will result in throwing the patient off the bed or bruising her person. It is very harmful, for instance, to take hold of the limbs and try to hold them still. The bed should be large and firm enough to bear the weight of the attendants, if necessary. Plenty of air should be admitted into the room, and the latter should not be too warm. The face may be sprinkled or washed with cold water. I have tried to interrupt the convulsions by dashing the face suddenly with ice-water, but have not seen any effect from it that was desirable.

Inter-paroxysmal Treatment.—It is a matter of the greatest moment to prevent a recurrence of the paroxysms; if this can be done we are almost sure of success. The fewer the convulsions the less the danger, should be an ever-present fact with us. It is the frequent repetition of them that produces the fatal effects upon the brain. By the judicious use of chloroform we may prevent them from returning in a great many instances, if not all of them. Dr. Braun, of Vienna, recommends us to give chloroform as soon as there is any sign of a supervening paroxysm, and continue it until sound sleep is induced; but I am sure that to administer it in this way we shall fail to get all the good effects from its use that are practicable, for in many instances the premonition is so short we will not prevent the convulsion. The only safe way, and the one I now practice and would strongly recommend, is to keep the patient under its influence all the time sufficient to prevent a manifestation of the premonitory symptoms.

This direction may be followed by sitting by her constantly, and administering the anæsthetic about every ten minutes generally, but sometimes it may be necessary to have a small amount mixed with the air inhaled constantly. The intention is to very much lessen, if not entirely subdue, all reflex excitability, and keep it controlled until the cause of the trouble, uterine pressure, can be removed. The effect of the chloroform is to remove, or rather to subdue, the explosive excitability of the nervous centres caused by the circulation through them of the urea, carbonate of ammonia, or whatever other exciting stimulant has resulted from the retention of substances that ought to have been excreted by the kidneys. It is not likely that the chloroform chemically neutralizes the poison in the blood, but that it renders it inoperative by producing a more powerful but antagonistic influence upon the brain and medulla. It is not at all directly curative, it only suspends the action of the mischievous agent until curative means may be made effective in removing the pathological conditions upon which the convulsions depend. And I must insist, in *all* cases of uræmic puerperal convulsions, where there is enough excitability to keep reproducing the paroxysms, chloroform is applicable and demanded, whether attended with a uræmic, paralytic, or even apoplectic state of the brain and spinal cord or not. It is precisely to control this morbid excitability of the nervous system that it is used; and although I have been governed by this principle for several years, I have never seen any unpleasant effects from its use in convulsions. And whenever I had cause to regret anything of my management of this most excellent remedy, it was that I had been too sparing with it, that I had not kept my patient sufficiently under its influence. It would hardly seem necessary to say that it is entirely useless during a paroxysm, except in very rare instances, in which respiration continues, because it cannot be inhaled, as respiration is suspended, and that it is injurious immediately after the paroxysms, while the face is livid with carbonated blood. We wait until the face loses this asphyxiated appearance, and the blood is again charged with its usual amount of oxygen. As soon as this state is brought about we give the chloroform continually, and maintain an anæsthetic condition that will keep the patient entirely free from the convulsions. She should never be allowed to come out of this state until delivered, if this can be done by any allowable means, even should it require twenty-four, thirty-six, or forty-eight hours. After she is delivered she may be treated generally without further use of chloroform, but if the convulsions recur we should resort to it at once, and keep it up until renal and intestinal secretion is fairly established. Keeping the patient thus constantly anæsthetized beyond the convulsion-point, we ought to institute curative measures with as much energy and activity as possible.

Bloodletting is one of the first things to be thought of. The intention in using it is to reduce or remove the cerebral hyperæmia, by lessening the amount of blood and reducing the arterial and cardiac *vis à tergo*. To effect this we must take a large quantity compared to the amount for other conditions of the system. From twenty to forty ounces will be required in one or two operations. The pulse must be affected by it. Sometimes a convulsive paroxysm occurs while the patient is bleeding; this ought to be prevented by chloroform. I have several times bled patients under the influence of chloroform. The anæsthesia should be slight during the bleeding, but sufficient to prevent the paroxysms, and the patient must be more closely watched. She should be supported in a sitting posture while the blood is running, and as soon as the pulse becomes decidedly quicker, we should lay her down and stop the blood, and if necessary repeat it when reaction has resulted. One copious or two moderate bleedings is allowable almost always, and in very plethoric patients two or more copious bleedings. Those old worthies in the profession, Gooch, Armstrong, and others of like stamp, cured many cases of puerperal convulsions by powerful depletory treatment alone. Gooch thought bleeding *the* remedy. As I have endeavored to show, the apoplectic congestion of the brain is an important item, although not, as was supposed by the older authors, the only item. It is probably of secondary importance, but its removal goes a great way towards a cure. I insist that bleeding is one of the very best curative measures, and should be considered in all cases. At the same time that we bleed, or if we do not think best to institute bloodletting, we may, with great propriety, administer an active cathartic. And in doing so we are to remember that promptness is a necessary quality; for the longer the bleeding is delayed, and the longer the cathartic is in operating, the more valuable time is lost.

If we can procure copious and early catharsis without pain, we make a very favorable impression. The medicine I have used most, and still continue to recommend, is croton oil. I usually administer one drop in an emulsion every hour until it operates thoroughly, which, if good, it will do after two or three doses. This treatment will seldom fail to very much modify the severity of the symptoms, and in very many instances completely arrest the convulsions. I know that before the days of chloroform, at a time when the pathology of these cases was not as well understood as now, twenty-five to thirty years ago, many were cured by this treatment; and I am cognizant of two cases in which the convulsions were arrested, and delivery did not take place in one until three and the other five weeks afterwards, each patient carrying a dead child that long, and being delivered without interference or further convulsions. We then bled in both arms in a full stream until there was a close approach to syncope, and sometimes twice, and even thrice, in the same case, and gave the croton oil cathartic at the time we bled. And with all deference and respect for those who have learned to do better without than with bleeding, I must assure the profession that such success as attended our efforts then would not put them to the blush now. I am afraid that the tendency now is to magnify to an undue extent the effects of uræmia, and depreciate the importance of cerebral hyperæmia. And in saying this I feel assured that a proper estimate of uræmia will make it a very dangerous element of the disease, while I think I but express the conviction of the most judicious and thinking portion of the profession, when I say that we have no remedy for it that acts with anything like the promptitude that those do which prove effective in the cerebral congestion. That we may very much modify if not relieve, the latter, in a few hours, nay minutes, while the uræmia cannot be remedied in less than *one or two days*. After proper depletion, or with it, cold to the head, ice in bladders or India-rubber bags, will aid in the same thing, reducing cerebral congestion; and opium administered in sedative doses may so affect the nervous centres as to materially lessen their violence; it operates somewhat in the same manner that chloroform does by diminishing the excitability of these organs. It does not exert any curative power, it only holds the convulsions in abeyance until the uræmia and hyperæmia are removed. Acids undoubtedly act beneficially by attacking the irritating compound circulating in the blood. We must not expect too much from them, however. Acetic and citric acid drinks, vinegar baths, acid compresses over the abdominal surface, back, &c., should be resorted to as much as in the nature of the case is practicable. Then we should resort to diuretics and hydragogues of a particular character. Colchicum, or its preparations, is a powerful diuretic, and has the reputation of inciting the kidneys to the secretion of urea. When given in large doses it proves a hydragogue cathartic and emetic. A teaspoonful of the vinous tincture every hour, until perceptible effects result, and then continued in such doses as the patient can bear without great irritation in the alimentary canal, is the usual mode of using it here. Other diuretics would probably be useful.

A very important part of the treatment is obstetrical. If the views above expressed are true, so long as the uterus maintains its extreme size, the pressure on the large vessels, and the vessels and substance of the kidneys, is continued, the evil-working cause is kept up; and while the subsidence of that organ in the pelvis and the consequent relief to the tension of the abdominal muscles, takes off some of this pressure, and sometimes this extends the critical period, if it does not entirely arrest it, yet it is not generally sufficient, and the patient is not usually safe until it is emptied.

The obstetrical treatment will vary somewhat, owing to the time when the convulsions occur. During pregnancy, any time before term, premature labor is apt to supervene as the effect of the convulsions or the condition of the system upon which they depend. And should this be the case it may be accelerated by means which I shall presently indicate; but if there are no signs of labor, the treatment should be wholly medical, and may be conducted as I have directed above. If labor has commenced and dilatation is slow and difficult, we should dilate the os by artificial means. The only one I have ever resorted to is the finger. By placing one or two fingers in the mouth of the uterus and gently but constantly keeping up pressure upon it from within outwards, we will very much shorten the stage of dilatation. The effort to dilate should not be abated during pain, nor in the intervals, but a constant fatiguing pressure made. So soon as the os is dilated to two inches in diameter, we ought to rupture the membranes and allow the liquor amnii to flow out. This will generally add energy to the contractions and somewhat lessen the size of the uterus. We should not desist from dilatation until the head has emerged from the cervix entirely. If the head is not readily expelled, we must apply the forceps and assist by judicious traction. If labor proceeds, however, with reasonable activity, and we can succeed in controlling the convulsions, we must abstain from operative procedure entirely. An indispensable condition in all cases where it is necessary to accelerate delivery by any means whatever, is to keep the patient deeply under the influence of chloroform, so much as is necessary to preclude the danger of additional irritation thus caused, bringing convulsions. To dilate, rupture the membranes, and use the forceps, should be the extent of our interference, unless when the feet, knees, or breech presents. Where either of these parts can be seized, we may aid very materially by traction upon it. Craniotomy is not justifiable until we can demonstrate the death of the fœtus, unless indicated by circumstances independent of the uræmia, such as disparity of size between the head and pelvis, or ineffectual efforts with the forceps. Neither are we to think of turning. In fact, we must do no more than is absolutely necessary, and do that in such a way as to cause the least possible irritation, and be sure to use chloroform liberally. I say that I have never used anything but my fingers to dilate the uterus, but there is a variety of very efficient means used for this purpose. The compressed sponge is one of the means very effi_cient, quite handy and available. Dr. Barnes's dilators are also very eligible and efficient. These consist of gum elastic air-bags of different sizes. The smallest should be introduced into the cervix empty, by means of the probe, and then inflated through a tube attached to it; in an hour this may be replaced by the next larger, and so on until the mouth is fully dilated. If the compressed sponge is used, a piece large enough to pass easily into the cavity of the cervix may be introduced, and in three hours a larger one if necessary, and so on until the process is completed.

Convulsions commencing after labor is finished, necessitate an examination of the uterine globe externally and internally, by introducing the finger into its mouth, to see that the organ is well contracted, and there are no imprisoned clots that are causing irritation. If contraction is not good and blood coagula accumulate, we may use friction above the pubes and give ergot enough to secure the desired effect. Depletion, cathartics, diuretics, and opium, with chloroform, are to constitute the treatment, as in the other condition above mentioned.

CHAPTER XXXI.

PHLEGMASIA ALBA DOLENS, OR CRURAL PHLEBITIS.

Nature and Anatomy.—Inflammation of the uterine branches of the hypogastric vein, the hypogastric, iliac, and femoral vein, and, in some instances, the popliteal and lower branches of the veins of the lower extremities, seems to be the initiatory step in phlegmasia dolens. There is not merely inflammation of these veins, but they are also filled to a greater or less extent with coagula, of sufficient firmness to obstruct and even entirely interrupt the flow of blood through them. The blood returning from the extremity is thus crowded out of the principal channels, and made to seek the ascending vena cava through more circuitous routes and less commodious vessels, while the arteries corresponding to the occluded veins are acting with more than ordinary frequency, and with their natural capacity. The effect of this condition of the circulatory apparatus of the limb is to amass an unusual quantity of blood in the capillary vessels between the femoral artery and its branches, and the femoral vein and ramifications, that can find its way out with difficulty. The walls of these minute tubes are thus distended to their utmost capacity, and are permeated by the thinner portion of the blood,—the serum,—which penetrates the inter and intra-muscular and subcutaneous cellular tissue in every direction, thus producing enlargement of the limb by the accumulation of the watery elements of the blood. If the obstruction is complete, and the number of veins involved large, the extremity is enormously distended, and becomes hard and shining.

Thus far the swollen limb is œdematous, and before excessive distension has been reached pits upon pressure. In the progress of the case, however, it becomes too firm to pit, and often remains stiff and hard for many months. The hardness is, doubtless, dependent upon coagulation of the fibrine contained in the serum as it escapes from the vessels.   This coagulation in the cellular tissue, within and around the muscles and beneath the skin, is somewhat permanent, and not reabsorbed perhaps entirely, in some cases, in the lifetime of the patient.   She is, therefore, in future, more or less completely deprived of the use of the limb.   There are very few cases in which the efficiency and usefulness of the member are not very much impaired for a long time.   The fibrinous coagulation takes place generally in from four to fifteen days.   When the effusion in the limb is not great, the hardness will not be very considerable; but if excessive, the hardness and stiffness of the limb are correspondingly prominent circumstances.   The veins, when dissected, show signs of decided inflammation in the thickening of their walls, deposition of layers of fibrine, and, after a time, in containing pus near the ends of the coagula.   If the cellular substance is cut into in the first few days, serum will be evacuated to such an extent as to decidedly lessen the size of the limb.   Punctures may thus be made to relieve the painful distension attendant upon the early stages of the affection.   After the fourteenth or fifteenth day, the punctures will still allow serum to escape, but not with the same facility, nor to a like extent, as before, and the limb will retain much of its firmness, however much it may be thus punctured.   And if the tissue is cut into it will be found slightly firmer, and more dense than natural, showing adventitious substance of a consistence greater than fluid.   This, I think, is a fair exposition of the condition of the limb in phlegmasia dolens, and from it most pathologists infer that inflammation of the veins is the beginning of the disease, coagulation in their cavity the second step, effusion of serum in the cellular tissue the third, and coagulation of the fibrine in the cells of the cellular tissue the fourth step.   But some others, as McKinzie, Simpson, and others, think that behind all these is a peculiar condition of the blood that starts all the other phenomena into existence.   Again, while Dr. Robert Lee, who has done more to elucidate this disease than any other one author, believes that the inflammation of the veins is the cause of the coagulation of the blood in them, Virchow thinks that the coagula are the cause of the inflammation.   The probability is, that while all the blood may not be diseased in the very beginnning, the absorption of the poison of decomposing substances in the uterus may so affect that part of it in the uterine branches of the hypogastric vein as to start up inflammation in their walls, and, possibly, directly affect the blood in them, so as to promote the coagulability of it, and in that way begin the chain of phenomena. In the cases I have observed, offensiveness of the lochia was a very noticeable circumstance. The remarks I have made and shall make in reference to the character, cause, and treatment, are intended more particularly to apply to puerperal phlegmasia dolens, but I think they will also be, in the main, appropriate to the disease as it occurs under other circumstances and in the opposite sex. For we shall find, by investigating the subject extensively, that it is in nowise confined to the puerperal condition, the lower extremity, nor the female sex.

Etiology.—Phlegmasia dolens occurs most frequently in women recently delivered,—the necessary inference from which is that there is something in this condition which predisposes to it. In what post-partum circumstance the predisposition consists, it would be difficult to say, but if we adopt the theory that the venous inflammation begins in the uterine veins and spreads to others, the unusually enlarged state of the vessels, and greater accumulation of blood in and about the uterus, would afford sufficient reason for the susceptibility to morbific agencies. It will be rendered more likely that this is the rational mode of accounting for the predisposition in such cases, by the fact that an exciting or congestive influence exerted by cancer in the uterus, rectum, or vagina, the inflammation caused by the pressure, but more particularly the inflammation or disintegration, of uterine tumors, injuries of the pelvic organs in males or females, stricture of the urethra or wounding of this canal by the catheter, cancer of the rectum of the male or the male genital organs, or, in rare instances, of deranged menstruation, all act as predisposing causes of phlegmasia dolens. It is difficult to understand why these conditions should predispose to crural phlebitis, unless it is through the congestion they produce in the small veins which empty into the iliac or femoral vein. Disease, as suppurative inflammation in the axilla, or pressure upon the veins in that part, sometimes predisposes to phlebitis, and perhaps excites it,—that leads to phlegmasia dolens of the arm. It would seem that injuries done to the uterus by turning, the use of instruments, or the rough management of the os in the first stage of labor, act as exciting causes. The absorp-

35

tion of putrid matter from the uterus by the veins excites inflammation in the walls of the veins. The putrid matter may be retention of a portion or the whole of the placenta, which occurs sometimes in cases of abortion, retained membranes, or even blood at full term, &c. Outside causes, acting in conjunction with the internal predisposition, must also be enumerated as efficient exciting causes, as exposure to cold, too much exercise, too early resumption of the erect posture and ordinary business duties. The mind, doubtless, in some instances, has a powerful influence in inducing phlebitis of the extremity. Other causes not enumerated here may probably be influencing the patient to the assumption of this disease also.

Symptoms and Progress.—Crural phlebitis may attack the lying-in woman any time from the fourth day after delivery to the end of the month or even later. I have a patient under treatment now, who was taken ill with phlegmasia dolens on the fortieth day from the birth of her child. The most common time is within the first fourteen days. The onset is generally rather sudden, accompanied with a chill, and succeeded by fever; but often its commencement is not marked by any prominent symptom, and three or four days elapse before there is sufficient positiveness in its manifestations to enable us to discover the nature of the attack. The fever not unfrequently is the first symptom; at others the pain precedes it; or again in some instances, the two begin consentaneously. When fully formed, the extremity is swollen, hot, painful, and in some parts very tender to the touch. The swelling begins at various parts of the limb in different cases. It sometimes begins in the upper part of the inside of the thigh, just below the groin; at others, it commences in the calf of the leg; and again in the popliteal region, and still, in other cases, in the foot. The tumefaction sometimes remains partial, not extending beyond the point of commencement.

The case above referred to, is confined to the leg below the knee, with a small place on the inside of the thigh. Generally, however,—and this is particularly true of the cases occurring within the first ten days after delivery,—the swelling spreads from the places where it first shows itself all over the limb. The degrees of distension are not constant, being sometimes so slight as to be merely easily recognized, while at others, the limb is increased to double its ordinary size. When the swelling is slight, the limb is soft; but if the tumefaction is great, there is a peculiar hardness that prevents indentation without considerable force. In most cases in the beginning, pressure with the finger will pit the surface as in other kinds of œdema; but as before remarked, pitting cannot be produced by pressure with the finger when the swelling is very great, and in all cases the enlarged part becomes harder after a few days. The swollen limb is not red, or otherwise discolored, it usually retaining its natural hue; but in cases of excessive tumefaction, the skin becomes very white and shining. This white and shining aspect of the limb encouraged the use of, if it did not give origin to, the term "milk-leg," applied to it in ancient times. The swelling, when general, is equable, and seems to be governed in shape by the natural contour of the extremity. When, however, the swelling is partial, this is not the case; and occasionally one part of the limb is tumefied and subsides, and some other part goes through the process, to be succeeded in these phenomena by still another part. The tumidity begins to subside, in the majority of instances, after twelve or fourteen days, and the subsidence is complete, or nearly so, at the end of from forty to ninety days; but in some cases the limb is not entirely relieved from swelling for several months, and in rare instances, the limb is not restored to its natural shape and size after the lapse of a long lifetime, it remaining swollen and stiff, and the veins enlarged. The effect upon the motion of the limb is various, also. The patient, in slight cases, can move the limb with great freedom, while in others, the least movement produces the most intense suffering. The suffering caused by movement seems to depend more on the pressure exercised upon the veins by the contracting muscles, than any tenderness in any other tissue in the limb; and, indeed, it will be found that any part not traversed by vein trunks, may be handled with impunity, but the moment the veins are pressed upon, there is suffering commensurate with the intensity of the inflammation in them. The pain is slight in less severe cases, but in the more intense the pain is often very severe, requiring the use of anodynes to secure rest at night. The pain begins most frequently, perhaps, in the groin, where in most of the cases there is hardness and tumefaction before elsewhere apparent. In others, and not unfrequently, it is first felt in the popliteal space, and in still others in the ankle or foot. In slight cases, the pain is confined in one locality not unfrequently; but we often meet with instances where the pain extends from one end of the extremity to the other. It will be observed generally, too, that the pain is in the locality of the inflamed veins, extending in lines corresponding with the course traversed by them. The veins may be usually traced throughout much of their course by their uncommon hardness; appearing like large round cords, which may be rolled under the finger, as well as by the pain and tenderness. The temperature of the affected limb is generally higher than natural, but in slight instances it is not changed, and, perhaps, even lower than ordinary.

After the acute symptoms have subsided, the limb not unfrequently remains large, stiff, and hard, impeding locomotion for a long time, and forever impairing the usefulness of the member.

With these local symptoms we generally observe general ones of a febrile character, commencing, as before remarked, sometimes before and sometimes after the supervention of the local symptoms. The temperature of the whole surface is increased, the pulse is more frequent than natural, sometimes hard and bounding, but generally less distinctly changed in any other respect than in frequency. The pulse is sometimes very frequent, varying from one hundred and twenty to one hundred and sixty in the minute. This is in a severe form of the affection, occurring soon after confinement; in less intense attacks the circulation is not so very much excited, and occasionally scarcely at all. The tongue is coated white at first, afterwards more dirty, and finally brown in some cases. Pain in the back, aching of the head, bones, of the extremities, &c., are the accompanying symptoms. The nervous system sometimes partakes of the general morbid condition, when there is restlessness, sleeplessness, and delirium. The stomach enters into the circle of sympathetic derangement, so that we have loss of appetite, even nausea and vomiting, while the bowels are constipated, or affected with diarrhœa. In some cases, the lax condition of the bowels is a prominent and very troublesome symptom. The lochia are not generally suppressed, although sometimes they are scanty, and at others, the disease does not make its appearance until after they have ceased to flow. This discharge is usually fetid, or otherwise offensive, and occasionally so acrid as to produce abrasions on the external parts.

The secretion of milk is almost uniformly influenced; in bad cases it entirely ceases, while in very mild instances it is only rendered more than ordinarily scanty.

Although phlegmasia dolens is usually confined to one limb, it has been occasionally observed in both at the same time. It very seldom invades the two limbs simultaneously,—first, almost running its course in one, and then attacking the other, proceeding through its various steps to resolution or otherwise.

Diagnosis.—It will not be necessary to draw a differential diagnosis, as phlegmasia dolens is so distinctive in its history, relations, symptoms, and physical appearance. When it occurs at other times than the puerperal state, and in the upper extremities, there may, for a time, be more reasons for doubt than usual. The œdema, confined to one extremity, with pain and soreness in the course of its principal veins, and the enlargement and hardened condition of them, feeling like cords under the finger, are sufficiently distinctive.

Prognosis.—There are two items in the prognosis,—1st, as to the recovery of the general health; and, 2dly, the restoration of the usefulness of the limb involved in the disease.

In a large majority of the cases we meet with in practice, the tendency is to resolution of the inflammation, the recovery from the general depressing influences of the acute attack, and the removal of the effusion from the textures of the limb,—in short, thorough recovery.

The length of time in which these different conditions are removed, varies greatly in different cases. Most frequently, perhaps, the disease arrives at its acme in from fourteen to twenty-one days, when the pain, fever and swelling begin to subside: the pain first, the fever second, and lastly the swelling. In the course of from four to six weeks, the patient will be able to be up, and engage in some of her household duties; but several months will elapse before the hardness, swelling, and stiffness of the leg, have entirely disappeared. This is favorable. Many instances are observed in which the patients are confined to their rooms, if not to their beds, for many months, and the limb is not restored for years, and sometimes never, but remains a misshaped mass of fibrinous deposit, and deficient and impervious veins, rendering the patients cripples for all time to come in spite of the very best management.

Treatment.—In robust patients, remembering that the disease is inflammatory, we should resort to antiphlogistic remedies early in the disease. If the inflammation in the veins can be checked or moderated so as to prevent their occlusion, the deposit in the limb may be prevented. I think I have seen the disease thus cut short in more than one instance by a tolerably energetic use of the antiphlogistic means. It will do but little if any good after the swelling in the lower part of the limb is fairly begun, to deplete. The damage is already done, and we can do nothing more than palliate the symptoms, and remove the effects of the disease. Attention to the beginning of the case will often enable us to detect the swelling in the groin, sometimes in the pelvis, and always in the upper and inner part of the thigh, before œdema of the lower part has begun. This is the time when we may derive invaluable service from energetic treatment. If the patient is quite strong, the pain severe, and the fever high, venesection to a degree approaching syncope is, beyond all odds, the best step with which to begin the treatment; and if the pain and tenderness of that part are not very much relieved in eighteen or twenty hours, from fifteen to thirty leeches should be applied over the most painful part. Or if the patient is delicate, or the pain, swelling and fever less intense, the leeches may suffice without the venesection. The bleeding is to prevent, not to cure, the phlegmasia dolens, and is not to be thought of if there is much increase of size in the limb. We should, therefore, learn to detect the first pathological link in the chain of the phenomena,—inflammation of the veins. I think the dispute about the propriety of bleeding, either locally or generally, has arisen and is perpetuated by forgetting or rather ignoring the fact that the swelling of the limb is not the inflammation but the consequence of it, and that not it, but the inflammation, can be benefited by the antiphlogistic treatment. Soon after, or even before bleeding or leeching, a dose of calomel and jalap,—ten grains of the former and fifteen of the latter,—should be given. After this has operated, and even before, we may begin to administer sedatives, diuretics, and diaphoretics. Two drops of tinct. verat. virid. every hour until the pulse is brought down to sixty or seventy per minute, will be of great service. Much good may be done also by giving ten grains of the nitrate of potash every four or five hours. This treatment should be diligently and promptly applied until the disease is arrested, or the leg becomes distended from effusion, and then it must be abandoned for an alterative and anodyne course; or, as is almost always the case, we are not called until the swelling and œdema have become quite marked. The fever, if high, denoted by quick, sharp, or strong pulse, heat of surface, &c., after the limb has commenced to enlarge, is a reason for a moderate antiphlogistic course of treatment. In making the change or adapting the anodyne and alterative course, we must bear in mind that they are merely palliative. The alterative treatment should consist, in the first place, of gently but hardly perceptibly inducing the specific effects of mercury, and afterwards administering the iodide of potassium in decided doses. I usually give five or six grains in solution every four or five hours. The anodynes ought to be administered as sparingly as possible to enable the patient to rest with any comfort. At first, Dover's powder at bedtime, or in the daytime, when the sufferings are very great. Opium in some form, or its alkaloids, is the only kind of anodyne that may be depended upon. The bowels ought to be kept in a soluble state by laxatives. Cream of tartar as a drink, or citrate of magnesia in wineglassful doses every four or five hours, will generally answer. After the lapse of sufficient time to produce debility, tonics must be added to other general treatment. Quinia, in good, liberal doses, say two grains every four or six hours, some wine or malt liquors, and generous diet, as the gravity and chronicity of the attack seem to demand, will be indispensable. I am partial to the tinct. ferri. chl. It may be given in doses of twenty drops every six hours, and, in many cases, is a very valuable tonic and deobstruent. Various local applications to the limb afford the patient comfort, if they do not more; but some of them, in certain stages, are productive of decided good. A bran poultice, applied so as to cover the whole limb, acts, in some instances, very soothingly, and, what is nearly as good, is warm, dry bran, in a flannel bag, large enough to completely envelop the extremity. Cotton batting, in sheets large enough to cover the whole of the limb, is used by some practitioners. This last may be covered again with oil-silk, so as to keep the part dry and of equable temperature. Dry flannel, in sheets, wrapped several times around the limb, or in the form of a roller, commencing at the toes and extending to the upper part of the thigh, will occasionally afford a grateful support. Flaxseed-meal poultices, corn-meal mush poultices, camomile flowers in bags, wet in some instances and in others dry, hops in the same way, &c., are all applicable and often beneficial. The above array of local applications will enable the practitioner to select suitable ones for his patient, and it will generally be found that some of them will agree well in some cases and afford no relief in others; and again, some of them will be praised by the patient in the early part of the same case, and replaced by others before the end of the treatment. The same may be said of the different kind of liniments, oils, and ointments, so frequently resorted to. Soap liniment, camphorated oil, oil and laudanum, chloroform and lin-seed oil in equal quantities, tincture of aconite, camphor ointment, mercurial ointment, ointment with opium, belladonda, hyoscyamus, and other anodynes, may all be employed for the purpose of sooth-ing and quieting the pain in the limb. After the acute symptoms have subsided, and the stiffness of the leg remains, combined or not with soreness and pain, the character of the treatment must be changed. The tonics and iodine alteratives may be given inter-nally. Iron, tinct. cinchona, and, in fact, any of the bitter tinc-tures, would be very applicable to support the strength of the patient. Travelling to the seaside, a residence at some of the popular watering-places, or other equally complete change in the residence and circumstances of the patient, may be made to con-tribute very materially to the general health of the patient, and hence promote the absorption of the deposit in the limb. Much good may be done by local means to aid in restoring the usefulness of the member. If there are enlarged and varicose veins, the silk stocking, or the flannel roller, applied as far as the knee, will be of great service. The appliances have a good effect upon the deposit, and assist in reducing the size of the leg, and making it more supple and active, and also frictions, dry or with liniment, with a flesh-brush, coarse flannel, or linen towel, repeated as often as can be borne without causing much soreness. Great patience will be necessary to derive the full effect of any of these remedial means. After the tenderness and pain are all gone, electricity may be tried to promote absorption. A matter of the utmost importance is exercise. In the beginning of the disease, of course, perfect quiet is indispensable to the subsidence of the inflammation in the veins. But after the pain and soreness are so completely gone from the limb that it gives no direct pain to step with it, we should urge the patient to a moderate degree of exercise. Exercise, to a moderate degree, on foot, will promote the general health, and most powerfully aid other means in rendering the muscles useful. After the phlebitis has entirely disappeared, there is no reason for complete rest of the limb, and I think absorption will be very greatly promoted by the use of the muscles.

The treatment of phlegmasia dolens, occurring under other circumstances, although to some extent modified by the locality of the affection and the circumstances surrounding it, in the main will be the same as in the puerperal state.

CHAPTER XXXII.

PUERPERAL FEVER.

In order to have an intelligent idea of puerperal fever, we must remember that the term has been applied by different authors and practitioners to almost every form of fever and inflammation incident to the puerperal condition; and that while the patient may be attacked with a great variety of diseases soon after delivery, not necessarily in any way connected with the state of the system at that time, such diseases are almost invariably, if not always, modified to a greater or less extent by it. The confused use of this term has misled the inexperienced, and been the cause of grave error in teaching on the part of the older members of the profession. It is plain to the intelligent reader, therefore, that we ought entirely to discard the term puerperal fever, or use it only in a generic sense; and in this last case, affix to it a specific term, to define the meaning in particular instances. My own preference would be in favor of dropping it, believing that the most common forms, if not nearly all of them, are inflammation, and its consequences, of particular viscera, or collection of viscera. I desire to be understood, therefore, as using this term in a general sense; and I will try to so define the subjects treated of under it, as not to be misapprehended. I think I can express my own views better by describing toxæmia as it occurs in the puerperal state, sporadically; metritis, as we observe it sporadically; and metro-peritonitis, in sporadic and epidemic forms; and I think almost every practitioner of experience has observed and recognized these different forms of puerperal diseases.

It will be seen by this statement of the arrangement, forced upon me by my own observation and reading, that I consider the diseases mentioned above identical, essentially, in this and all other conditions of the system, but very greatly modified by the puerperal state in some instances, and in others, the superadded condition of epidemic influence.

In justification of the prominence I give to the difference between the sporadic and epidemic conditions of puerperal fevers and inflammations, I would appeal to medical men of extensive observation, as well as my own convictions and observations. What a difference in the mortality of accidental cases of inflammation of the uterus, and such as occur during the prevalence of a devastating epidemic! All must have observed it. In the one class of cases recovery is almost universal, while death is the rule in the others; the intensity and grade of action being pretty uniform and moderate in the sporadic or accidental variety, while in the epidemic form they are overwhelmingly destructive, and all this, too, in spite of the most judicious management in both instances. These differences, no doubt, depend on the impression produced upon the general system prior to the actual attack of disease, and for which I know no better expression than "epidemic influence." I think no better opportunities are offered than occasionally fall under the observation of rural practitioners, where there are no hospitals to serve as points of origin, the population sparsely situated, and imperfect ventilation impossible on account of the open and exposed character of their buildings.

A severe, fatal, and almost universal epidemic of puerperal metro-peritonitis, prevailed under my observation in the woods, amongst inhabitants of almost strictly primitive habits. The patients were attended at their own homes by midwives and physicians, both having patients with the disease. In that neighborhood, the majority of families invaded were from one to three miles separated from their neighbors, living in houses without glazing, or other hindrance to a perfect ventilation. Under these circumstances, death occurred in some instances in less than forty-eight hours from the attack, and at least sixty per cent. of the patients died.

Those who have seen epidemics prevail in frontier rural districts, cannot resist the idea of an all-pervading influence affecting the whole community,—predisposing to diseases that are determined in their nature by circumstances attaching to the individual patient, the prevalence of which cannot be accounted for from the narrowed opportunities of contagion or infection alone, but which may, nevertheless, be rendered more unerring in their onslaught by these two additional, efficient, and co-operating causes.

This all-pervading, incomprehensible, subtle, and deadly influence, is what I mean by "epidemic influence." It is manifest in numerous attacks, in limited districts, of a peculiar incurable form of pneumonia, decimating neighborhoods in sparsely settled countries noted for their general salubrity; or, under similar circumstances, erysipelas becomes malignant and deadly in its nature.

The observation of intelligent practitioners as to the incursion of destructive epidemics in the healthy districts of the comparatively wild regions of the northwestern portion of this continent, where marsh miasm cannot, by any possibility, be dragged into the account, may yet exercise a wholesome check upon the vagaries of hospital attendants. We have the same forms, the same grades, the same deadly epidemics, in the woods and prairies of Indiana and Illinois, that are described by the Parisian and London practitioners, or any of the other Eastern writers, as hospital in their origin and propagation.

I desire to make a record of this fact, because the circumstances that will admit of a repetition of the observations are rapidly giving place to an entirely different state of things, particularly with reference to the houses, habits, and diet of the people. These remarks are based upon observations extending back for thirty years, to the days of poverty in this country, when the mode of living more nearly approached the natural than they do now. A number of different kinds of epidemics have prevailed under my observation in that time, and among them, one of puerperal metro-peritonitis, which was well described by Gordon, Hay, Lee, Armstrong, Gooch, Velpeau, Baudelocque, and others. This epidemic of puerperal fever occurred in the years 1838, '39, and '40, in a neighborhood noted for its good health. I may mention in passing, that this epidemic was most successfully treated according to the instructions given by Gordon and Hay.

So far as I am able to form a judgment with reference to that epidemic, and others I have since encountered, there were no obvious causes for the spread of the disease at that time; no apparent causes that were not present in as full perfection when the epidemic did not prevail as when it did. We are hence compelled to fall back upon some general explanatory term that may account for the effect which alone we perceive. Epidemic influence not only causes a difference in frequency of attacks of disease, but it impresses on it a modification or peculiarity in each instance.

Toxæmic puerperal fever is often attended with phlebitic inflammation, but not always. As a sporadic affection, it is generally mild, recovery being the rule; and when epidemic, is probably the least dangerous of any other variety of acute inflammation or fever in the puerperal state. In some epidemics it is mild, in others exceedingly fatal. This is, doubtless, the form of puerperal fever described by Dr. Butler as prevailing in Derbyshire, in 1765 to 1775, and alluded to by Dr. Gooch, and which was cured by a gentle cathartic of rhubarb and cordial, administered every day until the stools became natural.

The attack takes place from four to ten days after labor, and in rare cases later. It is generally ushered in by a chill of greater or less intensity, sometimes amounting to severe rigors, but generally confined to a sense of chilliness. The chilly feeling recurs frequently throughout the fever, and sometimes with so much regularity as to induce the suspicion of periodicity; at others quite irregularly, but once or oftener in twenty-four hours. The skin is more than naturally warm, and for a good part of the time bathed in perspiration; sometimes the perspiration is very copious, and occurs at night, or whenever the patient falls into sleep. The tongue, at first, is nearly coated with light white fur, but is otherwise healthy; it afterwards becomes dry and red, and coated with a dark layer, or is clean, red, chapped, and dry. There is generally thirst in the beginning, which may continue, but more frequently subsides, and leaves the patient in a few days. The appetite is usually poor, if not entirely absent. The pulse ordinarily is rapid, ranging from one hundred and twenty to two hundred in the minute, but soft and compressible, each stroke sometimes giving the sensation of a thrill, instead of the solid shock of sthenia. In the milder forms it seldom goes above one hundred and thirty, and is more frequently above one hundred and twenty in the minute. The nervous system is almost always, comparatively, seriously affected; the patient is wakeful, apprehensive, and despondent. When sleep does come, it is generally disturbed and dreamy, and the patient often wakes unrefreshed, and believing that she has not slept. She often talks and starts in her sleep, the tendons twitch, and she becomes decidedly delirious, but when aroused she is conscious, answers questions, and talks intelligently. In other instances not so common, she is stupid, and sleeps a great deal, mutters and moans, and works the muscles of the face with grimaces. The stomach is not generally affected much; there is not often nausea or vomiting, or other disagreeable symptoms connected with it, but the bowels are generally very much deranged. Diarrhœa is one of the very common symptoms. The stools are thin, dark-colored, and fetid. The diarrhœa is not generally, but sometimes, a very early symptom, coming on after other symptoms of the fever have been in existence for some time; at others, it begins first. The abdomen is generally somewhat, though not excessively, distended with gas; the tympany is hardly perceptible in many other instances. In severe and fatal cases, toward the last, the intestines become considerably distended. There is not ordinarily much pain; sometimes there is severe pain attending the beginning or progress of these cases, and generally, there is but slight sense of soreness over the uterine region and ilia. The pain is sometimes situated in other than the abdominal region, as the shoulders, arms, head, and elsewhere, and seem to be neuralgic in character. The lochia and milk continue to flow usually for several days, but the secretion of milk diminishes at first, and progressively becomes less until it entirely ceases; the lochia however, is not often much affected in quantity, but in quality is fetid and very offensive, at once suggesting the idea of putrefaction. It is also so acrid, sometimes, as to excoriate the parts over which it flows.

During the progress of this fever, even when comparatively mild, there often supervenes intense inflammation in some organ distant from the pelvis. One of the lungs, for instance, becomes the seat of general inflammation, and the advance is oftentimes so rapid that it is completely engorged in twenty-four hours. No air penetrates further than the large bronchi; vesicular sounds of all sorts give place to bronchial respirations and bronchophony. Sometimes in four or five days these phenomena pass completely off and the same process is set up in the opposite side, or some other part of the system suffers from a similar sudden invasion of inflammation. But not unfrequently the pneumonia passes into the stage of real hepatization, and the organ if not permanently spoiled is a long time in recovering from the consequence of this severe attack. All the large joints, the knee, shoulder, and elbow, and some of the smaller, are liable to these sudden and severe attacks of inflammation. It may likewise in them prove to be evanescent, or permanent and destructive. Even the muscular system, such as those of the thigh, leg, arm, back, and chest, are also complicated. Large bags of pus may thus be formed. I have a case in my care now where the left thigh has just been relieved by incision of a large collection of pus.

The duration of this toxæmic form of puerperal fever is often protracted, continuing for many days, and even weeks. Three, six, or even ten weeks, may elapse before it subsides entirely, and the declination is so slow, as to be hardly perceptible from one day to another. Sometimes its progress is extremely and disastrously rapid, overwhelming the powers of life in a few hours.

Diagnosis.—The diagnosis is not difficult, I think. The diarrhœa, want of pain, and tenderness of the abdomen, copious perspiration, &c., mark the difference between it and the inflammation proper.

Complications.—We sometimes meet with cases in which, with these symptoms, there is unmistakable evidence of inflammation of the uterus, as evidenced by pain and tenderness over the uterus, an enlargement of that organ, suppression of the lochia, &c. And sometimes we also have signs of peritoneal inflammation, which I shall describe when I come to speak of it.

Prognosis.—This will depend almost wholly on the general epidemic constitution. There can be but little doubt that it occasionally, though rarely, appears in a very fatal form *epidemically;* but usually, the epidemics of this kind of puerperal disease are mild, and in the large majority of instances they terminate favorably, and require very little treatment. As I have seen it *sporadically,* it scarcely ever proves fatal. When it does terminate unfavorably, it does so generally by exhaustion, as the effect of its long duration and profuse cutaneous and intestinal discharges.

Cause.—From the offensive character of the lochial discharge I have been inclined to believe that it almost invariably arises from absorption of the products of decomposition. Indeed, I think I have seen some instances traceable to retention of pieces of placenta or membranes until decomposition takes place. From inefficient uterine contractions, accumulations of blood in the uterus may result, and remain until they become putrid, and thus furnish the material for absorption and poison. So far as my observation is sufficient upon which to base an opinion, I would say that this is the manner of organization in sporadic cases. Where it prevails as an epidemic, there is doubtless a general predisposing cause operating upon the whole community, rendering the exciting cause I have mentioned efficient and operative when applied in a slight degree.

Nature and Morbid Anatomy.—As I have before intimated, I think this form of puerperal disease is a true toxæmia or septicæmia. We know so little about acute diseases of the blood that I have not sufficient data upon which to base a theory of this affection that will admit of minute application. Whether in the blood is set up a series of changes that result in an alteration of its composition to such a degree as to induce these morbid phenomena, or whether the functions are disturbed thus gravely by the circulation of the small amount of poison admitted in this fluid in its original state, or while undergoing further chemical changes, are questions we may fairly conclude as unanswerable in the present state of science. I am inclined to believe the latter to be true in a general way. But, in cases of overwhelming epidemics, may we not suppose very plausibly that the predisposition consists in such a state of the circulating fluid as will permit of a rapid decomposition in it, or septicæmia? However this may be, the main difference between the mild sporadic form and the rapidly fatal epidemic variety would seem to be the preservation of the integrity of the composition of the blood in the one, and the sudden reduction of it to a state but little better than putrilage in the other.

Recorded facts attest very positively that instances of this form of fever are not unfrequently observed which run a very rapid fatal course, and yet present no traces of disease in the solid organs after death. More frequently, however, the fatal cases are attended with post-mortem evidences of metritis, phlebitis, peritonitis, or all these combined. Hence we cannot regard this fever as possessed of a morbid anatomy in the proper sense of the term. The inflammations of the pelvic and abdominal viscera I regard as complications, instead of essential portions of a whole. I cannot refrain from remarking in this connection that, in my opinion, the combination of this toxæmia with the dangerous inflammations of the uterus and peritoneum, has formed the true nature of some of the most appallingly fatal epidemics on record, and that it is for the want of the philosophic contemplation of this mixed or complicated variety that has given origin to so much acrimonious and, in many instances, unprofitable debates as to the nature of puerperal fever. Of course, toxæmia is always attended with an asthenic grade of febrile reaction, and the more decided this item in any given case, complicated or simple, the more rapid the declension of the powers of life. Inflammations, attended with an intense toxæmia, are, as a consequence, influenced by it, and must fall under the general head of asthenia. I do not wish to anticipate what I shall have to say of this complication under the head of puerperal peritonitis, and will only add that, unattended by toxæmia, the inflammations of the puerperal state are not essentially different from the same disease occurring in the same organs at other times.

Treatment.—From what I have said as to the nature of this form of fever, the student will very properly infer that the treatment is alterative and supporting. We should endeavor to stop the process of poisoning, and support the system until the fever has spent its force. The probability is that the first item may be furthered by remedies that enter the blood, and there influence the elimination, neutralization, or destruction of the poison in that fluid. The first thing, however, that presents itself for correction is the state of the contents of the genital canal. The uterus and vagina should be kept clean as possible, and no substances allowed to remain long enough in them to undergo decomposition. To this end, copious vaginal injections are indispensable. It will be well to make these injections of some soapsuds. The patient may be placed on a bed-pan, and half a gallon of suds, made of fine toilet soap, passed through the vagina by means of one of those perpetual syringes. This may be done every eight or twelve hours, as the foulness of the discharges may require. A teaspoonful of carbonate of soda to a quart of water is highly detergent, and may be advantageously used. We are also to attend to the contents of the uterus, not by injecting water in its cavity, but the organ should be examined with the finger, so as to be sure that the mouth is not occluded with a clot of blood or membrane. The finger may very properly be introduced into the mouth of the uterus to encourage the discharge of imprisoned fluids. Whether in certain instances it might not be advisable to throw water into the cavity of the uterus, my observation does not enable me to determine. The organ may be pressed and agitated somewhat to aid in voiding its contents. This is for the purpose of preventing further ingress of poison into the blood. We may aid nature somewhat in the excretion of it. To assist in its elimination we may encourage a soluble condition, when necessary, of the bowels, as being favorable to the early resolution of the case; and we should not interfere to restrain the diarrhœa, unless it is clearly exhausting the powers of the patient. It will not be inappropriate, when the evacuations from the bowels are not very copious but very offensive, to administer from four to ten grains of rhubarb and the same quantity of carbonate of soda, to promote the excretion and neutralization of the offensive substances in the stomach and bowels. Hyd. cum creta, in from three to five-grain doses once a day, will have a like effect. The skin should be thoroughly cleansed with tepid soap and water twice a day, and then sponged afterwards with a weak solution of hydrochloric acid each time. The sulphites of lime and soda are now administered to arrest septicæmia, or neutralize the poison circulating in the blood; and although my experience in their use in this form is limited, I am disposed to favor their use as probably promotive of that end. We may give the sulphite of lime in doses of ʒj, mixed with water, three or four times in twenty-four hours. It is, I think, the pleasanter of the two, and perhaps as beneficial. I have been very much in the habit of giving the chlorinated tincture of iron—ten drops every two or four hours in plenty of water—for the double purpose of exerting a tonic and antiseptic influence. When there is great depression, we should endeavor to derive all the benefit possible from tonics. Sul. quinia is an excellent tonic in such cases, and may be given with the tincture of iron. Two grains every two or four hours is a suitable quantity as a general thing. The iron and quinia may and ought to be persevered in when there is much prostration. If there is very great prostration, to these may be added alcoholic stimulants, of strength and in quantities indicated by the grade and intensity of the symptoms. Wine will do in the milder forms, while brandy, variously mixed with milk, water, broths, &c., will be necessary in great copiousness in the very severe forms. Of course, these medicinal tonics and stimulants are but temporary in their effects, and it will be necessary to give nourishment to produce permanent results. Probably the tonics do good by promoting the digestion and sanguification of the nourishment taken; for this toxæmia is the very state of the system in which the patient may eat and digest but not be nourished. The great function of sanguification being in abeyance, brandy and milk may be mixed as stimulant and nourishment. Beef essence, beef tea, mutton soup, chicken tea or soup, and, in fact, any of the animal broths, may be given in such quantities as the patient can take, are always in place, and should all be mixed with stimulants, to promote the changes necessary for nutrition.

In addition to this treatment, we may often do much to prevent exhaustion and ameliorate the sufferings of our patient, by the use of astringents and anodynes. When the bowels are moved excessively, opium and tannin combined will be very efficient in restraining them, and thus closing the exhausting outlets. A mixture I have used with much advantage in correcting the character of the intestinal secretions, allaying the irritation of the alimentary canal, and soothing the nervous system, is made as follows, viz.:

R. Tinct. opii deod.,
   Magnesia sul.,
   Acid sul. arom., āā .   .   .   .   .   .   ℥ij.
   Aqua, ...   ℥ij.
     Mix and dissolve.

A teaspoonful every four hours or often as necessary to answer these purposes. The opium may be increased in the mixture to meet the indications in different cases. Sometimes, when there is no substance to the stools, ℥ss. tinct. opii deod., and ℥jss. tinct. rhei, mixed, and given in ℥j doses, produce very favorable effects.

The indications for astringents ought to be plain before they should be energetically used. We ought merely to moderate the diarrhœa when there is reason for interfering, desiring by such interference, to prevent exhaustion. Opium given alone, or combined with stimulants, is necessary and useful in most cases. The patient very often passes sleepless nights from nervous watchfulness, and if there is no idiosyncratic objections to opium, may generally be quieted into grateful slumber by it. When opium cannot be borne, grs. v of ext. hyoscyamus may be occasionally substituted. I have recently, in some cases, found much good to result from the administration of bromide of potassium in scruple doses at bedtime, and repeated in four or six hours; or bromide of ammonia in two-grain doses every four or six hours. This last may be given in some kind of syrup. The valerianate of ammonia in syrup, given in six or eight-grain doses every four or six hours, is an admirable anodyne stimulant when the patient is very nervous and wakeful. This treatment is, of course, adapted to comparatively mild cases of toxæmic puerperal fever, and would not be sufficiently energetic for those cases of overwhelming severity attended with extensive inflammation, and I must refer the reader to puerperal metro-peritonitis for treatment in such instances.

Puerperal Inflammation of the Uterus—Puerperal Metritis.— Inflammation of the substance of the uterus is probably of all puerperal affections the most frequent. It commences generally with a chill of greater or less intensity, usually not very severe, which is succeeded by febrile excitement. When reaction is established, the bones and head ache, the surface is hot and generally dry, the pulse accelerated and somewhat firm. The pulse is not often very rapid, seldom exceeding 110 or 120 in the minute. The patient complains of pain, pretty severe, and sometimes excruciating, in the hypogastrium, or one of the iliac regions; and, upon examination of the lower part of the abdomen, a well-defined tumor is perceived, formed of the uterus, which is exceedingly tender to the touch. The tongue is white and moist; sometimes, though not always, there is nausea, and the bowels are generally constipated. When the patient has a stool, it often gives her considerable pain. There is also, frequently, an urgent desire to urinate, which is sometimes attended with pain. The pain in the region of the uterus is generally dull and aching in its character, but sometimes there are exacerbations of greater severity. The lochia, in this form of metritis, is always scanty, if not entirely suppressed. I think it is almost always entirely suppressed. The secretion of milk is not generally much affected at first, but if the disease continues for a number of days it generally becomes very scanty, and finally fails. The time of the attack varies from the third or fourth day to the fourteenth. It sometimes seems to result from the first efforts to sit up, or resume ordinary avocations or habits, continues with some intensity for from one to three weeks, and terminates in resolution or chronic metritis, which runs on indefinitely. I have never observed a fatal case of this kind of uncomplicated sporadic uterine inflammation.

Diagnosis.—This is not difficut, and will, for the most part, be sufficiently clear from the ordinary history and symptoms of the case; but, if doubt exists, it may be removed by physical examination. By introducing two fingers in the vagina, and tenderly surveying the pelvis, pressing upon the sides of the pelvis high up towards the ovaria, upon either side, the bladder, rectum, and uterus, we will find the inflamed organ. It will be still more plain if, with the other hand, we press upon the uterus above the pubis, and thus include it between the two hands. If the uterus is the organ inflamed, there will be pain when we exert pressure with either hand in this position.

Causes.—The cause is generally traceable to an unusually difficult labor, in which the uterus has been extraordinarily excited for a long time, torn, bruised, or otherwise damaged,—the application of cold, improper or too early exertion, stimulating drinks, excessive passion, &c. In this sporadic form of acute inflammation of the uterus after delivery, none of the effects of contagion, epidemic influence, want of ventilation, or other deleterious influence acting upon the blood, operate as a cause; on the contrary, the causes all seem to directly influence the uterus itself.

The *prognosis* is favorable; the disease, as I have before remarked, generally and even spontaneously terminating in resolution. It may be protracted for a considerable time by inefficient treatment or imprudent conduct, and run into the chronic form, but probably never directly proving fatal.

Treatment.—The treatment of ordinary sporadic metritis is simple and generally successful. It is, in the main, antiphlogistic and alterative. Most practitioners, I think, following the teachings of the present times, fall short of efficient energy when they act in the right direction. I am free to say that I have more frequently erred in the omission of the use of antiphlogistic remedies than any other way, and that I have been more gratified with the effect of them than any other measures in these cases. If the patient is robust, we omit one of the most useful means of cure if we do not bleed enough to produce a decided impression early in the course of the case. And my experience is decidedly in favor of venesection to a degree closely approaching syncope. It almost invariably breaks the force of the disease if done within twenty-four hours after the attack, and paves the way for complete success in the use of other less energetic measures. If the patient is not sufficiently robust to bear such a decided bleeding, we may cup over the sacrum, so as to draw ten or fifteen ounces of blood, or place over the hypogastrium twenty leeches, and allow the bites to bleed as long as they will afterward. Decided sanguineous depletion is the most important remedy in the beginning of the disease, and it should be measured by its effects upon the patient at the time it is used. After the first forty-eight hours, the beneficial effect of bleeding will be less marked, but still local depletion, with cups or leeches, will be useful, and ought not to be omitted. A good active cathartic should follow the bloodletting almost immediately. Hyd. mit. chl., grs. iv, to be followed up by the citrate of magnesia, sul. magnesia, or other agreeable and active saline cathartics, in sufficient quantity to act briskly, may be used to great advantage. If the pulse should be quick and frequent after these means, we may very properly administer of tinct. verat. virid., gtt. iv, every three or four hours, until its sedative effects are positive. At night, and even at other times, it will be very proper to quiet the nerves of the patient and relieve the pain with a good, liberal dose of opium. If the case should last without very decided relief four or six days, our treatment should be shaped so as to induce a scarcely perceptible mercurial influence upon the system. One grain of submur. hyd., twice a day, will usually do this in a very short time and gentle manner. During the treatment, the bowels ought to be kept in a soluble condition by saline cathartics. After the specific sedative effects of the verat. virid. are produced, we ought to keep it up by giving half as much as at first. We will be able to derive much good from emollient applications in the early part of the disease. Among these, the best of them, I think, are corn or Indian-meal mush poultices, the water compress,—made of several folds of napkin or towel wet in tepid water, and covered with a dry bandage,—fomentations of hot vinegar and water, &c. In the advanced states, counter-irritation over the hypogastrium is very serviceable, provided the patient is not too sensitive to such measures. When the mercurial alterative is not deemed applicable, the iodide of potassium, in eight or ten grains, three times a day, may be sometimes profitably substituted. It will hardly be necessary to remind the student that the more perfectly quiet the patient can be kept during the treatment, the more complete and speedy the cure. It may not be amiss, also, to say, in this connection, that an inflammation in the acute form is more easily cured than after it has become chronic, and that our efforts should be continued until it is completely removed—not merely mitigated.

It would not be inappropriate in this place to describe puerperal perimetritis, but instead of doing so, I will merely refer the reader to the description of this affection in another part of this book, under the general title of Perimetritis.

Puerperal Metro-peritonitis—Causes and Nature.—Puerperal metro-peritonitis consists in extensive and generally overwhelming inflammation of the uterus, peritoneum, and other pelvic and abdominal viscera, and occurs sporadically and epidemically. We are to regard a case as sporadic, when it is the only one in a hospital, or given district,—an isolated case. The epidemic variety is represented by a number of cases in a hospital or neighborhood taking place together, or at short intervals from each other. There are differences between cases occurring sporadically and epidemically, so marked, that a casual observer will easily distinguish them. It is important that the distinction be made, for various reasons, which will appear as I advance in the consideration of the subject; and I beg the reader to bear in mind the difference between sporadic and epidemic forms of metro-peritonitis, as essential to an intelligent treatment of any given case.

The causes of sporadic cases are generally obvious and accidental. Damage done in producing miscarriages, the effects of drastic or perturbating medicines used for the same purpose, excitement from strong passions, as anger, hatred, or the effects of depression, of melancholy, mental anxiety, and alcoholic stimulants, occasionally act as causes of puerperal metro-peritonitis. Incautious exposure to cold during or immediately after labor, severe labor, protracted too long, also, may be considered as causes of the sporadic form. Generally, the disease may be traced to some of these obvious accidental circumstances. It is almost, if not always, sthenic in character or grade of vital action. The only exception to this is when the system of the patient has been predisposed by debilitating influences to an asthenic condition. There is nothing in its etiology to give it an asthenic phase.

As an epidemic, the above causes seem to have almost no influence in propagating it, and no precaution in the circumstances of the patient or labor will insure immunity from an attack. The disease is caused, continued, and impressed with its peculiarities, by an epidemic influence pervading whole communities, preparing the system of its victims during pregnancy for the disastrous attack that follows delivery. According to the preparatory effects of the epidemic influence, will be the grade of vital action. In some epidemics it will doubtless be sthenic, and the disease of a tonic character, requiring for its cure antiphlogistic treatment; in others, it will be asthenic in a high degree, and the vitality is almost annihilated by the predisposition and force of the attack, at once prostrating the patient beyond all hope.

Epidemic influence, in this view, is not only the predisposing, but the qualifying cause; and it will be seen further that the qualifying impression may be very different in different epidemics. Each epidemic must be, therefore, studied separately and comparatively, in order to understand and treat it intelligently. It is only where this strong predisposition exists, that contagion can be reckoned as a cause, if at all, of puerperal peritonitis. The same may be said of the etiological effects of unclean hands. When there is no epidemic influence, the disease will not be carried from one patient to another, by practitioners, as it occasionally appears to be done.

I do not argue against contagion as an occasional cause, but I very much doubt whether it is ever more than co-operative in its effects.

It would seem that other forms of disease often prevail in these epidemics contemporaneously with puerperal metro-peritonitis. Drs. Gordon, Armstrong, Hay, Lee, and others, observed in seasons of epidemics of puerperal fever, the prevalence of erysipelas, putrid angina, scarlatina, and typhus, with an unusual tendency to hospital gangrene and erysipelas. This will strengthen the idea of a general epidemic proclivity to a certain character of disease. The causative effects of epidemic influence are greater sometimes than others. We sometimes meet with epidemics in which not more than five per cent. of parturient patients are affected by it, and, perhaps, even a less proportion than this; while again fifty, seventy-five, and even a larger ratio, are unfortunate. I think the virulence of attack corresponds to some extent with the frequency of occurrence. Where the cases are few the intensity is less than when there is an almost universal predisposition.

The beginning of an epidemic is often marked by the exceeding violence of the cases observed; while at the decline, cases are very much milder and less fatal.

Bearing in mind the difference in the *intensity* and *nature* of the epidemic impressions produced by each season, set of surroundings and localities, we will be prepared to understand and estimate the numerous and contradictory statements of experiences and opinions of the many able observers and writers on this perplexing subject. The modifying influence of this great predisposing cause is apparent in the dissection of persons dead of the disease. In sporadic cases, the inflammatory effusions are plastic, and resemble in appearance those of peritoneal inflammation occurring at other times and under different circumstances; in some epidemics, such is likewise the case, but in others, the exosmotic products are aplastic, or exhibiting strong chemical qualities, and are grumous, offensive, and sometimes almost putrilaginous.

The appearance of the products of inflammation has led to acrimonious disputes as to the nature of the disease, one party contending that it is a peculiar fever, while the other could see nothing in it but inflammation; the first considering the pelvic and abdominal appearances one of the conditions of the fever, secondary in importance, in fact, of no importance at all, and not always present; while the second believe the inflammation, the disease, and the phenomena of the fever, of no farther importance than is usually attached to symptoms. It is probably impossible to decide which of these is right, or whether they are not both wrong for taking extreme positions. It will be inferred from what I have stated that I consider it a true inflammation, but that the inflammation partakes of a sthenic or asthenic form, according to the effects of the epidemic influence at different times exerted.

While upon the subject of difference of opinions as to the nature of epidemic metro-peritonitis, it may not be inappropriate to state that the success or failure of a particular sort of treatment is supposed to decide the question as to whether the appearances are the result of inflammation or fever. Some argue, and the argument is generally accepted by the other party, that if it is inflammation, depletion and the energetic pursuance of the antiphlogistic treatment will cure it; and as this is not always the case, it must be a peculiar fever. Just as though the old notions of an invariable methodical array of remedies were always applicable and beneficial in all inflammations, when the truth is, we have all learned that the evacuant mode of treating it is not always or even generally the best. The pneumonia, for which we were once in the habit of depleting so copiously, is still pneumonia, although we have ceased to treat it as formerly. Questions of sober seriousness to the practitioner of long experience, who has outlived the old regime, is, whether he was right when he let blood so copiously twenty-five years ago, or whether there has not been a great modification of the human constitution, or of the vital grade of the disease, from some peculiar influence. The man of wisdom will never dismiss these questions, but continue watchfully and carefully to investigate them while his responsibilities last, not only with reference to pneumonia, but puerperal fever, and every other form of disease, and hold himself as open to convictions in favor of another change of treatment when philosophical observation leads the way, as he has shown himself to desert his old plan, when proven inapplicable to disease as it now exists. The fetters of habit and association bind many a practitioner to error at the expense of valuable lives; and the only way to avoid this danger, is to make every important case a test of the truth or falsity of our doctrine, formed from the observation of former ones.

I cannot here avoid the declaration, that there is something in the symptoms, anatomy, and treatment, of all the epidemics of motro-peritonitis on record, that completely and convincingly stamps the disease under consideration as inflammation, in the light of a liberal and intelligent understanding of that pathological condition. But I am also ready to admit, in some epidemics, and, indeed, urge, the coexisting, all-pervading depravity of the fluids (so pertinaciously presented as evidence that the disease is a malignant fever, and not an inflammation), and that this state of the blood—with corresponding depravity of innervation—is just what makes the difference between different epidemics.

The observer will find the disease sthenic in some epidemics, and asthenic in others, but always inflammatory; and the welfare of his patients and his own reputation will depend largely upon an accurate determination of all questions in this respect.

Morbid Anatomy.—Upon opening the abdomen, obvious signs of extensive and terrible inflammation present themselves. The intestinal canal is generally very greatly distended with gaseous accumulation, and the peritoneal cavity contains much serum, sometimes colored with blood, sometimes mixed with pus in various quantities, with coagula of fibrine, &c. At other times the serum is fetid, acrid, greasy, and almost a putrilage. The fluid contents of the abdomen, in patients dying of puerperal peritonitis, has the reputation of being highly poisonous in its effects upon the dissector, and instances of death from dissection-wounds in these cases are numerous. It is also regarded as imparting to the hands an infection that is difficult to get rid of.

The peritoneum is usually lightly colored upon its visceral layer, as well as its extension upon the abdominal walls. But in rare cases, where the symptoms, during life, indicated it, the evidences of inflammation are said to be entirely wanting, or at most only to exist in a very slight degree, a little bloody serum being all that is present.

In addition to the redness and effusion of serum, we mostly find also fibrous patches or layers of false membrane covering the peritoneal surfaces. These patches on a red base give the membrane a mottled appearance. The fibrinous effusion is occasionally so copious and so firm as to agglutinate the intestinal convolutions, so as to mass them together in a great lump of inflammation, or it glues them to the uterus, bladder, and sides of the abdomen. In some cases, the fibrine, is destitute of adhesive qualities, or possesses it in a very slight degree, and it may be scraped off the surface to which it adheres, and is not much more tenacious in consistence than cheese. When the uterus is cut into, it may be found healthy below the peritoneal lining, presenting in every way the normal appearance; but generally there are strong manifestations of the effects of inflammation. The fibrous tissue is softened sometimes to such a degree that it may be broken down by the finger, and it may be infiltrated with pus. Sometimes the pus is collected into small abscesses in various parts of it.

The lining membrane, even when the fibrous substance is inflamed, may not show much signs of disease, but generally it is vascular and softened, sometimes ulcerated, and in more intense forms of the disease it is occasionally gangrenous in patches of large size.

In some cases there are collections of pus in the sides of the pelvis around the uterus, and in the ovarian regions, these organs not unfrequently being involved, and in some rare cases softened or dissolved into pus.   Abscesses, or collections of purulent fluid, may also form in the omental duplicatures of the peritoneum, or even in the mesentery.   The veins of the uterus, in this tissue, the spermatic, hypogastric, iliac, and even the vena cava, have been found in a state of inflammation. The lining membrane of the veins is deeply colored, apparently stained with blood, and the cavities are filled with fibrinous concretions, so as partially or wholly to occlude them, and to obstruct the current of blood.   In these concretions, and about them, pus is more or less copiously formed. Dr. R. Lee has met with cases wherein this phlebitic inflammation is the main, if not the only, anatomical evidence left by the disease, the uterus and the peritoneum being otherwise free from any sign of disease.   More frequently these were more or less combined, the whole of the abdominal and pelvic disease being overwhelmed in one grand phlogistic conflagration.

Besides all this localized inflammation, there are sometimes remoter ones, apparently as destructive to the parts attacked, and as certainly fatal in their effects.   The lungs are overwhelmed with sudden inflammation, rapidly leading to gangrene ; the liver suppurates ; large collections of pus take place in the joints, or in the muscular interstices, on the back, sides of the chest, in the nates or limbs.

As before hinted, it is sometimes a remarkable fact, that in the most rapidly fatal cases, no lesions are left behind to point out the seat of overwhelming disease.   Life is probably extinguished by the powerful and extensive impression of the morbific cause before the signs can be fixed upon the viscera by vascular action.

There are no less marked changes in the fluids than the solids of the body.   The blood, in some patients, shows signs of inflammation, as pointed out by the authors of the past half century. The coagulum is firm, cupped, contracted, and covered with a thick, strong, buffy, or fibrinous coat, after cooling, and when first drawn is florid and bright in color. In other cases, and particularly in some epidemics, the blood presents a very different sort of appearance. When drawn, it is dark-colored, and after cooling, there are none of the appearances described; the clot, if formed, is dark-colored, flat, large, and loose in texture, so that it easily breaks to pieces. There is no buffy covering to it, and much dark-colored blood settles to the bottom of the vessel. In still another variety of cases, the blood does not coagulate; it is in a condition termed the dissolved state. The serum and globules do not separate, except slightly, and that by the latter settling down to the bottom of the cup.

These different appearances were formerly, and even now are by many members of the profession, regarded as signs of the sthenic or asthenic forms of inflammation. If the blood separates readily into serum and clot, and the last contracts, and is strongly cupped on the top, and is yellow or buff-colored, the indication is to bleed, because the inflammation is sthenic, and may be cured by it. But if the clot is large, loose, soft, and flat, without the buffy covering; and, *à fortiori*, if the blood does not coagulate at all, the inflammation is asthenic, and bloodletting contraindicated. These old-time notions are too little regarded nowadays, and without sufficient reason are discarded.

Symptoms and Mode of Attack.—Although, as in every other disease, there is some variety in the mode of attack of metro-peritonitis, it is generally pretty uniform. Most frequently, without any premonition, and with the usual health for the circumstances, the patient begins suddenly to experience pain in the hypogastric region, attended with tenderness upon pressure or movement of the body. Then ensues a sense of chilliness, a mere coldness, or amounting to rigors, which are sometimes very severe. The attack dates ordinarily within forty-eight hours from the time of delivery, but several days may elapse before its occurrence. An attack may also occur before confinement. These variations as to time are not very frequent compared to the whole number of cases. The mode may vary occasionally as well as the time. The fever comes on gradually before or soon after delivery, so gradually as not to be marked in the beginning by any remarkable symptoms. Instead of rigors, we may have nausea and vomiting as the first symptoms in an attack. We sometimes have premonitory indisposition for several days, instead of the sudden unlooked-for attack. After an attack has merged into a fully formed case, we have fever, pain, tenderness, and enlargement of the abdomen. The tenderness and pain, in the earlier part of the attack, are confined to the region of the uterus, which feels larger than it ought at that time. The pain, tenderness, and enlargement, in a few hours, extend over the whole abdomen. The distension in mild cases requires two and even four days to become great, but in the severe forms it arrives at its maximum in a few hours. When we examine the abdomen with reference to the distension, we find it to be tympanitic, caused by an accumulation of gas in the intestines. In very rare instances the pain and enlargement, one or both, are absent, or so slight as to escape observation. The fever has some peculiarities that at once attract our attention. It is not generally attended with a great amount of heat, but the rapidity of the pulse is remarkable. It seldom is less frequent than 120 to the minute, and very commonly it is much more frequent, running up to 150, 160, and even 200 strokes. This excessive frequency of the pulse is the rule in the disease. There is not so much pain in the bones or head as there usually is in most cases of excessive arterial reaction. The pulse varies very much in other properties besides that of frequency. It is sometimes small and weak,—I think it is seldom very full,—at others, it is hard and sharp, and possesses a peculiar thrill. The peculiarities of the pulse of puerperal peritonitis are rapidity and smallness; besides these it varies as it does in other diseases.

The tongue is at first coated white, and is moist, but it is apt to become dry and red, or brown, and even black. The teeth are usually covered with a similar coating, which becomes dark sordes, not unlike typhus, in some cases; in others, there is no remarkable change in the appearance of the tongue. The stomach ordinarily sympathizes with the general disorder. There is want of appetite almost invariably, often nausea, and sometimes, in fact, not unfrequently, vomiting. It is singular, however, that we occasionally meet with cases where the stomach in every respect retains its comfortable state, if not its complete functional action. In other instances, while there is no nausea, vomiting, or gastric discomfort, the organ seems to lose all power to respond to or even recognize the presence of anything taken in it,—medicines of all kinds, as well as nourishing ingesta, lying in it inert and useless; the vitality, as well as sensibility, being entirely lost. When there is vomiting, the change in the character of the ejections mark the progress toward an unfavorable issue. The ejections, being at first only the ingesta, become acrid, green, mucous, bloody, grumous, &c., until the coffee-grounds emesis of other malignant forms of disease is imitated. The bowels are often not at all disturbed, particularly at first. They are generally torpid, but sometimes there is diarrhœa in the beginning, which remains a more or less obstinate and urgent symptom during the whole course of the disease. When there is diarrhœa, the discharges are apt to advance to a very disagreeable fetor. A loaded saburral condition of the alimentary canal is also sometimes observed at the time, or after the attack. The most remarkable condition of the bowels, as before remarked, is the great accumulation of gas, which distends them painfully. The head is not remarkably affected in the early part of the course of the disease as a general thing, but is very apt to become affected later.

Headache, delirium, and restlessness, however, sometimes begin with the attack, and even usher it in. But delirium is more common toward the close of fatal cases when the general powers of the system are failing. The delirium is usually a mild, constant muttering, instead of the boisterous and violent kind. Again, there is occasionally melancholy and suicidal tendency. The skin is seldom dry and hot throughout the course of the disease; it is much more frequently moist, and in some cases it is bathed in a copious watery perspiration, that is almost constantly present; at other times it occurs only at night. The urine is generally scanty and high-colored, sometimes bloody, often thick and muddy with sediment and mucus. The milk, if it has made its appearance, becomes entirely suppressed, and the breasts become flabby, as though the milk was absorbed. The lochia is not often suppressed; it frequently continues the same in quantity and quality with its normal condition under ordinary circumstances at the same period after delivery. Occasionally it is rendered scanty and even suppressed; more frequently it is fetid and offensive.

After these symptoms are established, they usually increase in intensity. The pulse becomes so frequent as to be scarcely nu-

merable and weaker; the abdomen is greatly swollen, being larger
than before confinement, very tender and sonorous to the slightest
tap of the finger; the mind wanders, and at length becomes completely clouded; the patient picks at her person in bed; the extremities first become cool and then cold; the respiration is hurried on account of the great abdominal distension; and in two to
four days the patient dies. If, however, the disease is less overwhelming, the symptoms continue severe, keeping the patient for
a number of days in a state of great suffering and doubt, until
they gradually begin to decline, and a tedious convalescence ensues; or the force of the constitution breaks down, and the patient
is exhausted and worn out, and dies after suffering from one to
many weeks. Sometimes, after the first five or six days, the fever
assumes something of a hectic character; remissions and exacerbations of a very distinct nature occur once or twice in twenty-
four hours; some particular part of the abdominal walls becomes
very tender and distinctly swollen; a livid patch appears; the
place finally ruptures, and a discharge of pus follows after several
weeks of almost mortal fever, and the disease gradually subsides.
At other times the eruption of pus takes place into the vagina or
rectum, and it is voided, to the great relief of the sufferer. These
large abscesses sometimes continue to discharge until the patient
is exhausted, from the great amount of pus evacuated, and the
long time required for the closure of the sac. The iliac regions
are the most common seat of these purulent collections.

Diagnosis.—It is only in the early stages that there is danger
of mistaking any other disease for puerperal metro-peritonitis and
the converse; but then, as this is the time when a correct diagnosis is of most importance, we should, if possible, have clear
ideas with reference to it. The four main items in the diagnosis
are pain, tenderness, fever, and tympanitic distension; and when
present all together, may be regarded as sufficient under the circumstances to decide the nature of the case. In the very beginning, however, some of these symptoms are either entirely absent
or imperfectly developed, but there is almost invariably sufficient
of them and agreement among them, together with the history, to
be quite conclusive. And one part of the history is the prevalence
of the disease. It is true this is not always available, but it very
often is. It is distinguishable from after-pains by the absence in

them of tenderness and rapid pulse and distension, and then the after-pains are intermittent, wholly or nearly. In peritoneal inflammation the pain may be exacerbating but not entirely paroxysmal; in the after-pain it occasionally entirely ceases. In milk fever there is not any marked pain and tenderness in the hypogastric and iliac regions, but a swollen and tender condition of the mammæ, and a greater amount of heat, headache, boneache, &c. In fact, the *fever* is highly exaggerated in intensity compared to the amount of local trouble, and this is in the breast. Puerperal peritonitis seldom begins with great fever properly; the pulse is very quick, but the surface is not correspondingly hot and florid, nor are the pains in the limbs apt to be great, but the pain in the back is ordinarily quite severe.

Another condition that will occasionally perplex the inexperienced practitioner, is the great tympanic distension sometimes observed to occur a few hours after a severe protracted labor. The first case I met with was in an unmarried woman, aged thirty-five, confined with her first child. She had been in labor over twenty-four hours, when I was called to operate with forceps after the pains had almost entirely ceased, the extremities cold, and pulse beating at 120, from exhaustion. In ten hours after delivery, the abdomen was enlarged to the size it was before the confinement, from gas in the intestines. I remarked that I could see the convolutions of the colon. There was no pain or tenderness, and the pulse had dropped down to 100 in the minute. This patient recovered very kindly with nothing but stupes of warm camphor. The distension in this case was too early and excessive, unattended with very quick pulse, or pain. I have met with several similar cases that terminated equally well.

Prognosis.—This is essentially unfavorable. Statistics fearfully verify this assertion, and yet there are none that I regard as sufficiently reliable on which to base an average of mortality. But if we examine the subject somewhat in detail, we find that many circumstances qualify the prognosis. The same influence that exerts itself, and is recognizable in stamping the character of the disease, must be taken into consideration in forming our opinion as to the termination of any given case. The sporadic form is generally milder than the epidemic variety, and it will be, probably, more profitable to consider them somewhat apart, bearing in

mind that what is said of the import of the difference in intensity
of the symptoms in the sporadic, is quite as applicable to the epidemic disease. As a general rule, the earlier a patient is attacked
after labor, the more rapidly disastrous the course of the disease.
It follows almost necessarily, that when the onset dates anterior
to delivery, the attack is severe; and I cannot now call to mind
an instance of this kind under my own observation that has not
proved fatal in a very short time. The symptom upon which,
above all others, our reliance is justifiable for correct prognosis, is
the rapidity of the pulse. When the pulse does not range above
120 strokes in the minute, we hope, with much reason, for recovery;
on the contrary, a pulse that ranges above 140 in the minute for
a great part of the time, indicates great danger; and the more
rapid the pulse, the more hazardous the case. We habitually,
also, rely upon the pulse as the best indication of the effects of our
remedies. Next to the pulse, the disturbance in the nervous system affords us the clearest insight into the ravages of the disease.
Delirium, jactitation, restlessness, or aimless change of position,
or desire to be removed from one bed, or one room, to another,
indicate great danger. The sympathies of the stomach are justly
regarded as indicative of certain tendencies. Nausea and vomiting, particularly the latter, and the more so if the ejections are
vitiated, instead of the superabundant natural secretions, are serious complications, as well as indications of evil. Diarrhœa supervening in the early part of the attack, or before the beginning
of the other symptoms, is not equally as unfavorable a symptom,
but renders the case more dangerous than if it did not exist. It
is sometimes regarded as critically favorable, when it does not appear until some days after the disease is established. In either
case, our judgment must be influenced by the character of the de-
jections. The more vitiated, fetid, or otherwise morbid, the worse
the prognosis. There is a class of rapidly fatal cases in which, so
far as the heat of the body and extremities is concerned, reaction
is imperfect. The feet or hands, or both feet and hands, remain
below the natural temperature in these, and are generally moist,
and sometimes bathed in a profuse perspiration. I ought to have
mentioned that entire indifference to the child, or absence of af-
fectionate manifestations, is also unfavorable. Copious secretion

of urine, continuance of the lacteal secretion, and warm natural perspirations, may be regarded as favorable signs.

The first thing to be remarked with reference to epidemics is, that they are, as a rule, more fatal than the sporadic disease. Each epidemic must be judged apart, some of them being much more fatal than others. Another remark is, that the beginning of an epidemic is almost invariably attended with a greater number of fatal cases proportionately. This is so true, that we often find ourselves flattered with the opinion that we have discovered a course of treatment that will avail us in future, because of our greater success; while we also indulge in regrets that we had not fallen on the happy course of treatment earlier in the prevalence of the scourge, when, in truth, our greater number of cures result from the change in the intensity of the disease, instead of our own improvement. The fearful mortality which sometimes attends this disease when it prevails epidemically, may be realized by reading that in those occurring in Paris in 1746, in Edinburgh in 1773, and in Vienna in 1795, all the patients attacked died. This is too dreadful an account of the effects of this disease to represent the prognosis of puerperal peritonitis now. So far as I can judge, from my facilities for comparing recently recorded facts and my own limited experience, I should feel that I had pretty closely approximated the results in epidemic puerperal peritonitis, when I set down its average mortality at sixty-six and two-thirds per cent. of cases. Even this is scarcely paralleled by any other disease.

Treatment.—A very important part of the treatment of puerperal peritonitis is prophylactic, or preventive; and as epidemic influence constitutes the principal and efficient cause in many instances, it should be our object, 1st, to avoid it; and, 2dly, when we cannot do this, as nearly as possible to counteract it. Probably the only way to fulfil the requisitions of the first proposition is, to remove our patient entirely beyond the infected district. We can, I think, more effectually and certainly accomplish this with pregnant women, than patients predisposed to attacks of any other form of epidemic disease, for we know precisely when to expect an attack of puerperal peritonitis, and, fortunately, we can know this for weeks, nay, even months, beforehand. The woman will not be attacked until she is delivered. When any other epidemic prevails in a community, everybody is liable to it at all times; every sacrifice ought, therefore, to be made to send our patient, several weeks before the expected confinement, entirely beyond the region endangered, and allow her to remain there until the circumstances which render her susceptible have passed away.

There is no doubt but this prevention would neutralize the morbid tendency that ends in the death of many patients. If a hospital is the field of devastation, it should be shut against such patients, of course; and if the disease is scourging a city, or county, the lying-in woman should leave her home, with all its dangers, for some distant but healthy neighborhood. This is almost always practicable.

It may be said that the fleeing patient will transport to her place of retreat the dangers she endeavors to shun, and serve as a focus of a new epidemic. There is no certainty of a pregnant woman being capable of an epidemic morbific impression, at least as a general rule; and if she were, several weeks' residence away from the causes of such impression would enable her to free herself from it. Nor is it probable that she would impart such impression to a person outside of the field where the general cause of the disease exists. When we cannot comply with this prophylactic measure, we must next protect our patient as effectually as may be from the action of the epidemic influence. Every hygienic measure within our control should be attended to. Good ventilation is, perhaps, of all hygienic means, the most effective. The room should not only be large and airy, but the air must be changed frequently, and the temperature should be low,—the rooms of lying-in women are generally too hot,—nor should the patient be kept too warm. A cheerful quiet should be maintained in the presence of the patient. Absolute isolation from friends is not desirable, but the patient should not see any person in whose presence she feels the least restraint. She may, and ought to, see and converse with her own family,—such members of it as are old enough to behave themselves sedately. No excitement is allowable, produced by gossip or improper reading.

When we have reason to fear an attack of peritonitis, our suspicions should find no utterance save in such proper care and attention as the patient needs, and will not lead to apprehension. A nourishing, but bland and unstimulating diet, will be most appropriate. In fact, the surroundings of the patient should be the best hygienic circumstances that she can command. While there can be no question of the propriety of hygienic prophylactics, there is some doubt as to the necessity or benefit of drugs. Can we give any medicine that may render the attack less certain? I am persuaded that the strong cathartic—calomel and jalap—recommended and administered by Hay and Gordon, should be administered with much caution. Where diarrhœa is the tendency, as is not unfrequently the case, I think they may do harm; and, as Mr. Hay says, some of the worst cases he ever had commenced after the operation of the cathartics. Yet when constipation has preceded delivery, we may, with great propriety, give a gentle laxative of calomel, grs. iv, to be followed by a saline, if it does not operate in ten hours. I am not aware that opium has been recommended as a preventive; there is probably more promise in it than any other medicine as a preventive measure. Experience, however, and not theory, must be allowed to determine this question. I need say but little about the preventive effects of emetics; I think they should not be thought of. The main co-operating causes, contagion or infection, should be guarded against in every possible way. Friends or nurses, who have been in the presence of the disease, should be sedulously excluded from the entire premises,—not merely from the room of the patient, but from the house, for several days before and after the accouchement. During epidemics, the medical man should observe every precaution to avoid spreading the disease among susceptible patients. He should be in the room of the patient no more than is necessary to do his duty towards her. If he has been exposed to the disease, he ought not to risk contaminating the atmosphere of the room unnecessarily with his breath; and during the time he occupies it ventilation should be as free as possible. And when attending upon puerperal peritonitis, he ought to avoid much contact with the person and bedding of the patient, for fear of making himself the nucleus of contagion. When he is in such attendance, or after having been engaged in treating erysipelas, gangrene, &c., he should always wash his hands, as recommended by Prof. Meigs, in warm water, immediately before approaching his parturient patient. It will add to the safety, also, if he wash his hands in a

solution of chloride of lime, as recommended by Semelmeir. Au-
topsies by accoucheurs in active business should be done by proxy.

Curative Treatment.—I hope I will be excused for emphasizing
the fact that there are two circumstances indispensable to the success of any remedial course in this disease. The first is, that the
treatment must be instituted as soon as possible after the attack,
when practicable contemporaneously with it; and the second, that
an *immediate* and *powerful impression* must be produced by it.
If either of these circumstances is wanting, the objects of treatment will not be accomplished. In order to understand the grounds
upon which this pointed statement rests, we must remember that
in a few hours all the damages resulting from the overwhelming
power of the disease are done, and that after the lapse of these few
hours our only office is to aid the system in the *repair of them*, and
the curative means, so appropriate and effective in the early part,
become prejudicial. Let me make myself understood. The control of the excessive excitement of the first twenty-four hours by
powerful means prevents the damage that would otherwise occur.
Should these means be used after the damages are sustained, and
the powers of the patient very much exhausted, their only effect
would be to increase the prostration. Then the treatment must be
palliative and sustaining, upholding the powers of the patient
until convalescence results from the subsidence, and not interruption of the course of the disease. Whatever is effected in the way
of a cure must be done generally in the first twelve hours after the
attack. There are but very few remedies sufficiently prompt and
powerful in their action to accomplish these results, probably but
one,—venesection. Medicines or medicated appliances require to
be administered several hours before their impression is complete,
with the exception alone of an emetic. And I have no doubt, from
the testimony now on record, that on account of its prompt and
powerful effect *at the very beginning* of the attack, an emetic often
entirely interrupts the disease; but I am equally certain that in
six hours after the establishing of the inflammation, its effects are
scarcely if at all useful. It may be regarded, however, before this
time as useful; as injurious after it. I think, as is here said, that
we may regard bleeding and emetics as the only remedies capable
of interrupting the disease. The advocates for the use of these
remedies contend that they stop the advance of the inflammation.

The opponents to venesection do not claim that their course of treatment will break up the disease; indeed, they generally contend that it cannot be done by any means. Their main objects are to moderate the violence of it, and *conduct* it through a course, when it terminates in recovery, to a favorable end. And what is curious, there is no attempt to collect statistics of epidemics of metro-peritonitis, so clearly described as to leave no doubts of their character, treated in any other way but by venesection, so discouraging are the results of every other kind of treatment. There are general statements of them, but no well-arranged tables of statistics that I have seen. Perhaps I ought to except the records of Prof. Alonzo Clark and Fordyce Barker, of New York City, in favor of opium and verat. virid. In making this statement, I freely confess that my opportunities for extensive literary research in this direction are limited. As I have very little confidence in statistics made up by collecting cases from all sources, I do not wish to lay much stress upon this fact, but it may, to a certain extent, indicate the inefficiency of treatment not interruptive in its character. I am not fully convinced, but think it probable, that epidemics of this disease prevail of so strongly an asthenic grade that depletion is not applicable at any time, even in the inception of the disease, but I am fully convinced that in this class of cases there must be less hope than any other of doing good by any course of treatment whatever. Dr. Gooch attempts to draw a distinction between sthenic and asthenic epidemics, and after speaking of the symptomatic indications, gives us the very judicious advice to determine the question by treatment, making general bleeding the test. If bleeding did manifest good, it is to determine our course in favor of vigorous antiphlogistics; if not, this must be abandoned for an opposite course of treatment. This is probably the best advice that can be given, provided we meet with the opportunity of bleeding at the beginning, and judging of its results only in such cases. Otherwise, for reasons above stated, the test would not be fair. I think the most difference of opinion as to the nature and treatment of this disease arises from the mode of using remedies for the cure of it. We must remember that the time, quantity, mode of drawing blood, and position of the patient, are everything in the application of venesection. To be a test of the nature of a disease, it must be used at first before the affection has resulted in its consequences; with the patient in the sitting posture, the stream must be as large as the vein will allow, and until there is an approach to syncope, and as soon as reaction occurs, repeat it. This is the way we must proceed when we determine to bleed for metro-peritonitis; and I should advise the student to pursue this course if he is called in less than twelve hours from the beginning of the attack; to abstain from it altogether if it is later than twelve hours, and not to expect with much confidence to cure his patient if the twelfth hour has elapsed, but to build strong hopes upon an immediate beginning. When it is remembered that so early an application of remedies is not frequently practicable, it will be seen that venesection does not often have a perfectly fair trial. When an epidemic is raging, and the people alarmed, they are more likely to summon the physician early, and this duty ought to be emphatically enjoined when we can do so. I have stated the time for effective efforts to break up the progress of the disease to be twelve hours after the beginning, but there are doubtless cases in which bleeding might be effective after the lapse of double that time, and to a certain extent we may judge of such cases by the pulse, temperature, strength of the patient, and epidemic constitution of the season. On the contrary, there certainly are cases in which fatal damage is done in much less than twelve hours. Of these we may likewise form an opinion from the same conditions. Should we be called in time and. bleed, we may immediately afterwards use other remedies to further the patient toward convalescence. If there is no diarrhœa, I should be in favor of an active cathartic. Hyd. mit. chl., grs. x, and pul. jalapa, grs. xv, mix, and take at once. If this does not operate in six hours, administer enough sul. mag. to produce brisk catharsis. A linseed mush poultice, large enough to cover the whole abdomen, should be immediately applied; also renewed as often as every four hours. At the same time, or immediately after the cathartic is taken, we ought to give the patient ten drops of tinct. verat. virid. This last may be repeated in doses of four drops every hour, until the pulse is reduced in frequency to at least below the normal numbers. So soon as this reduction of the pulse is effected, we must lessen the dose of the tincture to two drops, or even one, every hour, increasing the quantity as much as necessary whenever the action of the heart and arteries is increased. All this cannot be done without personal supervision from the attendant or well-educated medical assistants, instructed with special reference to the case in hand. After the cathartic has operated thoroughly, it will be well to resort to the free use of opium, combined with small doses of calomel. Three grains of the former with one of the latter every four hours is a good combination. When diarrhœa exists at the beginning, we may omit the cathartic, and at once begin the use of calomel and opium. These two may be given until the specific effect of mercury to a slight but obvious degree has been brought about, when, unless the disease is rapidly subsiding, the calomel should be withdrawn, and the opium continued. If the patient is much better, the opium may be given in diminished doses, and soon entirely omitted. With reference to the use of opium in peritonitis, I shall have more to say further on, and I will merely add now, that these doses of opium are more likely to be too small than too large.

If we are not called soon enough to be justified in the venesection, we may, with great propriety, institute the same treatment with that exception. After we are convinced from sufficient careful observation and trial that bleeding is not profitable in the epidemic with which we are engaged, I should strongly advise an emetic at the very instant of the appearance of the first symptoms. The emetic may be left in the hands of an intelligent monthly nurse, after instructing her with reference to the symptoms. Əij of pul. ipecac may be put in a teacupful of warm water, and half of it administered at once. If emesis does not take place in thirty minutes, the other half should be given. Good ipecac in these doses will pretty certainly act, and it is a grateful consideration that we have a remedy so prompt and powerful in action. Its impression is perfect in thirty or forty minutes from the time it is given. I do not partake of the fears expressed by some physicians as to the effects of succussion upon the abdominal organs, and would not hesitate to give the emetic at once. So soon as the effects of it have pretty well passed off, we may institute the course of treatment above recommended to be used after it.

Dr. Fordyce Barker, of New York, has very ably and judiciously dwelt upon the mode of administering and great efficacy of verat. viride in puerperal fever, in a discussion before the New York Academy of Medical Science. It will pay the student to carefully read and weigh his suggestions. Alas! no matter how sound in his views, prompt and energetic in action the practitioner may be, he seldom has an opportunity to effectually try the remedial course I have recommended. He is called after the stage of sthenic activity is passed, and is confronted by conditions of great damage to the tissues and vital powers. In such cases he is not to be idle; much may yet be done, at least sometimes. These are the circumstances in which the stimulating, supporting, and soothing plan of treatment, so highly recommended but generally unsuccessful as the *treatment for puerperal fever*, is applicable. I need not specify at any great length the mode and means of conducting this course. The degree of prostration, nervous and vascular, will readily suggest to the intelligent practitioner all that is necessary. Quinia, wine, brandy, carb. ammon., camphor, capsicum, cantharides, opium, and turpentine, under certain circumstances, are indicated and beneficial. The terrible rapidity and overwhelming power of the disease require corresponding promptness, energy, and skill, successfully to combat it after it has passed—if I may be allowed the expression—victoriously through the stage of activity, and laid prostrate every conservative energy of the system. We must administer these remedies in such way as to have the greatest possible effect, and persistently maintain it until there are indications of a return to activity and integrity of function. We should, in addition to the medicinal support thus given to the patient, furnish her with as much nutrition as possible. Essences of beef, strong broths, milk punch, egg-nog, wine-whey, &c., will be the chief articles of this sort, combining as they do stimulants and nutrition. In concluding what I have to say on the treatment of puerperal peritonitis, I think I shall add much value to the subject by quoting a summary given by Prof. Alonzo Clark, of New York, to Dr. Keating, the editor of "Ramsbotham's System of Obstetrics," of his observations on the use of opium in this disease. It commences on page 534 of the last edition of that valuable work.

"Puerperal peritonitis has frequently visited the lying-in wards of Bellevue Hospital during the last twenty years, and the recovery of those who have been attacked by it, up to the winter of 1851–52, was the exception rather than the rule. Having acquired

great confidence in the efficacy of large doses of opium in simple
peritonitis, when the puerperal form of the disease made its appearance, in December, 1851, I resolved to try its virtues in this
more formidable affection.

"My first attempt was a complete failure. The house physician did not fully comprehend my views; and though the patient
took three grains of opium for the first dose, the prescriptions for
the succeeding twenty-four hours were no more unsuccessful than
inefficient.

"Three other women were attacked in rapid succession, two
within twenty-four hours of the first,—severally, two, ten, and
three days after labor. The following case will give a good idea
of the treatment of all. It will illustrate at once the efficacy and
the dangers of administering opium in heroic doses. It will be
seen that the medicine was given at first somewhat timidly, afterwards more boldly. We did not know our ground, and every new
case must be an experiment, so far as the susceptibility to the
effects of opium is concerned.

"CASE.—Anna N——, aged 20 years, married, was delivered
of her first child October 1st, 1851. The labor was natural, and
of usual duration. Convalescence was not attended by any untoward symptom, except a slight diarrhœa. She was, however,
of delicate constitution, having an hereditary tendency to phthisis,
and had been under treatment for cough the previous five months.
On the tenth day after confinement, she was seized with a chill at
10 o'clock A. M. The chill lasted an hour. The abdomen was
markedly tympanitic, extremely tender, and the seat of lancinating pains, which caused sobbing; pulse 120; respirations thoracic;
tongue clean. Ordered solution of sulph. morph. (gr. xvi to f℥j
of water) ℥x every hour. She slept most of the day and succeeding night quietly, but not profoundly.

"2d day, 10 o'clock. Tympanitis increased; tenderness somewhat diminished; extreme thirst; occasional vomiting of a bright
green fluid; bowels quiet; lochia had ceased previous to attack;
pulse 120. Other symptoms as before. It was obvious that not
much impression had been made on the disease in the first twenty-
four hours, and the solution was given ℥x every half hour. This
produced deep sleep, and by evening the pulse had fallen to 90.

"3d day. Tympanitis and thirst continue; retention of urine;

intellect rather obtuse. Other symptoms as before. The morph.
sol. was now increased to ℥xxx every hour. These doses were
continued from noon till 6 or 7 o'clock in the evening, when the
signs of narcotism became rather alarming. She slept heavily;
was roused with difficulty; would not reply to questions; pupil
contracted almost to a point; pulse 68, and respiration 7 in the
minute. The opiate suspended, and tablespoonful-doses of strong
coffee administered. The narcotism continued for six or seven
hours, and gradually subsided.

"4th day. Tympanitis diminished; tenderness markedly less;
features composed; intellect rather dull; pulse 78; tongue dry
in the centre; green vomiting occasionally; slight hysterical symptoms; respiration 15. Morph. sol. resumed, ℥x every half hour.
Narcotism recurred at about the same hour as last evening, in
which the sleep was very profound, and the respiration only 5 in
the minute. Treatment as on the previous evening. Towards
morning these unpleasant effects of the medicine passed off.

"5th day. The disease was subdued. The tongue was dry and
slightly brown, but the patient declares that she feels quite well;
tympanitis has nearly disappeared; no pain or tenderness; pulse
80, and respiration 15; some appetite. She complains of nothing
but thirst, and inconvenience from retention of urine (the latter
relieved by catheter, as before). At 11 A. M., morph. sol. resumed,
℥x every two hours. At night, wakefulness, and pain in the abdomen, and the prescription was changed for opium, gr. iij every
hour, under the influence of which she slept quietly.   .

"6th day. Sleeps most of the time; easily roused, but answers
questions slowly, and with an effort; tongue moist and not furred;
tenderness, that recurred last night, less; pulse 80; still vomits
green fluid. Treatment continued.

"7th day. Symptoms all improved; opium, gr. j, every hour.

"8th day. Abdomen soft, and free from pain or tenderness;
thirst slight; she says, 'I feel so much better.' She was decidedly
convalescent. From this time the opium was continued in gradually diminished doses. The bowels were not opened till the thirteenth day; a dose of calomel, followed by castor oil, having been
previously given. After this, all medical treatment was discontinued, and she was discharged in a few days in her usual health.

"The relief experienced after the narcotism of the third day had passed off, was very striking, and inspired us with confidence in the curative virtues of the medicine. Still more did that of the fourth day. But these occurrences also taught us how short is the distance between such extinction of disease and the extinction of life. It will, however, be borne in mind that we were feeling our way through thick darkness, that experience had not informed us what degree of narcotism was necessary to subdue the inflammation, or to what extent it could be carried with safety to life.

"Two other cases were under treatment at the same time. Neither of these were affected to anything like the same extent by the opium, yet both took a larger quantity of the drug, and one very much larger doses. The first took gr. ij, then gr. j, the hour; on the second day, one-third of a grain of sulph. morph. (in solution) every half hour; the fourth day, opium gr. iv every hour; fifth and sixth days, the same, or part of the time half a grain of sulph. morph. every half hour; making, for the first day, 25 grains of opium; for the second and third, each 8 grains of sulph. morph.; for the fourth, 96 grains of opium; for the fifth and sixth, each 24 grains of sulph. morph. No narcotism in the least degree alarming was induced by these doses, but the symptoms of the disease have markedly diminished, and by the seventh day the hourly doses were gradually reduced.

"The other took what appeared to me, then, surprising doses, and yet had no deep narcotism. She was seized on the 13th. She took at first ⅓ gr. sulph. morphia, and in the evening of the same day, gr. j every hour, or about 20 gr. sulph. morph. in all. The second day, 1⅓ gr. an hour, or gr. xxxij; and this dose, or gr. vj of opium, was continued for several days. The quantity was even increased in this case to .gr. xij of opium the hour, for many successive hours, without marked narcotism.

"As I have said, then, these recovered. When it was apparent, on the 10th and 11th, that puerperal fever, as it was called, had attacked three patients in one ward of ten beds, during twenty-four hours, no time was lost in removing the well from the sick, and in taking every possible precaution to prevent the spread of the disease. Except the case that began on the 13th, no other occurred in the Institution till the 3d of December. The symptoms in this case were so masked that the nature of the affection was not recognized till the third day, and then the disease had made such progress that all medication was believed to be hopeless. The opium treatment was faithfully but unsuccessfully tried.

"Precisely the same words may be used to sum up the sixth case, which occurred on the 17th of the same month.

"The seventh case, December 26th, recovered.

"The eighth, December 29th, died; and after death it was found that the peritonitis was inconsiderable, but that the principal disease was purulent metritis.

"The ninth, January 4th, 1852, recovered.

"The tenth, January 9th, recovered.

"The eleventh, January 16th, died. Post-mortem examination revealed but slight evidences of peritonitis; but, as in case eight, the chief disease was found to be purulent metritis.

"The twelfth case, March 25th, was also one of purulent metritis, with slight peritonitis. The treatment had no influence on the progress of the leading disease. Patient died.

"The thirteenth case, March 29th, recovered.

"The fourteenth, occurring also March 29th, recovered.

"Of these fourteen cases, it will be seen that six died and eight recovered. The result stated in this way will hardly appear a triumph to any but those engaged in hospital practice. Yet it is a great improvement on the results of treatment in the preceding years at Bellevue. I have known in one season thirty women attacked, of whom only one recovered; and in general, the recoveries have not been more than one in five. But when these fourteen cases are examined, it will be found that the first was not treated with large doses of opium, and should be thrown out of the account entirely; that on the fifth and sixth the treatment was not commenced till the third day (when, I may add, Dr. Foster, my colleague, agreed with me that all treatment was hopeless); that the eighth, eleventh, and twelfth, were cases of purulent metritis, and died more from pyæmia than from the ordinary effects of inflammation. We have eight cases left, in whom a leading element of the disease was peritonitis, but not without the symptoms of metritis in some,—as the examination of the recorded cases in my possession will show, and in whom the treatment was commenced on the first day,—none of whom died.

"These were all the cases that occurred in the hospital from October to April. But during this period I saw seven other cases in consultation with Drs. Gilman, Rockwill, Higgins, McLaury, Smith, and others, six of whom recovered, and the seventh died of uterine hemorrhage while convalescent from the puerperal peritonitis.

"In reviewing the cases of which an outline is given above, I think the following conclusions are justifiable:

"1st. When a prominent element in 'puerperal fever' is peritonitis, the treatment with large doses of opium is more successful than any other that has yet been proposed.'

"2d. To be successful, this treatment must be commenced early, and the patient must be brought under its influence as rapidly as the susceptibility of the system can be ascertained by trial.

"3d. The quantity of opium required to produce a safe but desirable degree of narcotism varies greatly in different cases, so that it is necessary to begin with doses that cannot do mischief, and increase every two hours till the influence of the opiate is sufficiently decided.

"4th. Every dose, during at least the whole tentative period, should be administered by the physician himself, or by some person on whose knowledge of the effects of opium and whose watchfulness and discretion he can rely. Some young physicians are too bold, and endanger the life of the patient; others are too timid, and do not control the disease.

"5th. The opium treatment alone will not cure 'puerperal fever' when its leading element is purulent metritis, though there is reason to believe that it will control and even prevent the peritonitis which generally accompanies it. This conclusion has been confirmed by recent observations.

"6th. The tolerance of opium in some cases of puerperal peritonitis almost surpasses belief. Yet in private practice I have not found more than half or two-thirds of a grain of sulph. morph., every two hours, necessary, and have generally begun with less, except for the first dose.

"7th. The influence of the opium should be kept up till the pain and tenderness subside, the tympanitis diminishes in some degree, and the pulse falls below 100; then, with the concurrence of other symptoms, it should be gradually diminished, and at length discontinued.

"A few remarks and statements may be needed to make some of these conclusions intelligible.

"The usual effects of opium given in efficient doses for the cure of this disease are, a disposition to sleep, but not profoundly; a contracted pupil; perspiration, often profuse; sometimes a red, blotchy eruption; diminished frequency of the respiration; subsidence of pain and tenderness; slight suffusion of the eyes; and, after a variable time, reduced frequency of pulse. Of these effects, three have been chiefly regarded as criteria by which each particular dose is to be governed. If, when a dose is due, the sleep is profound (the amount of sleep is of little importance if the patient is easily roused from it), there is reason to hesitate; if the respiration has already been reduced to twelve in the minute, and is *very* irregular and sighing, the dose should be diminished or wholly withheld; yet so long as the tenderness continues, it is desirable to urge the opiate, but, of course, always within the limits of safety.

"The respiration appears to be the most certain indication of danger. I have not generally aimed to reduce it below 12 the minute. Yet in almost every case it has fallen, once or twice in the course of the treatment, as low as 7, and sometimes to 5. In no instance, however, has the narcotism, taken as a whole, been so profound as in the case detailed above. No instance of fatal narcotism has yet occurred under my observation, nor among the many cases reported to me by others.

"Regarding the tolerance of opiates in some of these cases,—at the risk of being charged with rashness and trifling with human life,—I will make some extracts from Case 7. The treatment was commenced at 10 A.M. on the 26th of December,—two grains of opium hourly. At 2 P.M., no change in symptoms; dose increased to gr. iv. At 3, gr. iv. At 4, gr. v. At 5, gr. v. At 6, gr. viij. At 8, gr. x. At 9, gr. xij. At 11, sol. morph. sulph. (16 gr. to f℥j) ℥jss. At 12, ℥j. At 1½ A.M. (respiration 6), 0. At 6 A.M. (respiration 12), opium gr. xij. At 10, sol. ℥j. At 12 M., opium gr. xij. At 1½ P.M., sol. ℥ij. At 2½, ℥ij. At 3½, opium gr. xxiv. At 5, gr. xij. At 6½, sol. ℥ijss. At 7½, ℥ij. At 9, opium gr. xiv. At 10, gr. xvj. At 11, gr. xviij. 28th, at 1 A.M., sol. ℥ijss. At 2, ℥iv. At 3½, opium gr. xx. At 4, sol. ℥ijss. At 5, ℥iij. At 6, ℥iijss. At 6½, opium gr. x. At 7, sol. ℥iijss. At 8, opium

gr. xxij.   At 9½,.sol. ʒix.   At 10, ʒiij.   At 11½, ʒiij.   At 12, 0.
Thus, this woman took in the first 26 hours of her treatment opium
gr. lxviij, and sulph. morph. gr. vij; or, counting one grain of
sulph. morph. as four grains of opium, one hundred and six (106)
grains of opium.   In the second 24 hours, she took opium gr.
cxlviij and sulph. morph. gr. lxxxj, or opium four hundred and
seventy-two (472) grains.   On the third day, she took 236 grains.
On the fourth, 120 grains.   On the fifth, 54 grains.   On the sixth,
22 grains.   On the seventh day, 8 grains; after which the treatment was wholly suspended.   This woman was not addicted to
drinking, and after her recovery she assured me repeatedly that
she did not know opium by sight, and had never taken it or any
of its preparations, unless it had been prescribed by a physician.
This is, perhaps, 'horrible dosing,' and only justifiable as an experiment on a desperate disease.   Yet this woman is alive to tell
her story, as are several others who took surprising quantities of
this drug.   But later observations have shown that the tenth to
the twentieth part of this maximum is efficient in controlling the
disease.   So this case is referred to, not for imitation, but because,
with similar cases, it is a medical curiosity, and may, perhaps,
open some new therapeutical views.

"The results of the opium treatment in the hands of my professional friends in this city have not been uniformly successful.
This was to have been expected.   When the path to success is so
narrow and so little trodden, though beset with dangers on both
sides, it is unavoidable that many will lose it.   But I believe I am
authorized in saying that those who have seen most of this mode
of medication are most attached to it.   It is not to be expected
that in a disease so dangerous as the one under consideration any
plan can be uniformly successful, even with advantages of accurate
diagnosis and early treatment; but when it is remembered that
the diagnosis between purulent metritis and puerperal peritonitis
is not always easy, and that this medication is successful in proportion to its early adoption, we may probably find reason for its
failure in other hands as well as in my own.

"By way of illustrating the vigilance and discretion which must
be exercised in the administration of each successive dose of the
opiate in this mode of treatment, I will add that it could never
have been fairly tested by me without the zealous, intelligent,

and untiring assistance of the house-physicians of the Hospital.
They visited the patients every hour by night as well as by day,
and every dose of the medicine, from the first case to the last, was
given by them, and proportioned to the hourly exigencies. Dr.
Stephen Smith, of this city, now surgeon in the same Hospital,
had the immediate charge of the first three successful cases. It
was his judgment in carrying out the details that gave me confidence in the new treatment. Others, then of the house staff, to
whom my acknowledgments are also due, are Dr. Brodie, of S. C.;
Dr. Frederick Nash, of this city; Dr. Charles H. Rawson, and
Dr. Moneypenny."

I think I shall also do a favor to my readers by quoting from
Prof. Fordyce Barker the report of a case which he rendered to
the New York Academy of Medicine in discussion of puerperal
fever, October 7th, 1857. It is intended to show his mode of administering verat. viride.

"Kate Short, aged 23 years, fell in labor in full term at 2
o'clock P. M., February 25th, and was delivered of a healthy child
at $8\frac{1}{2}$ o'clock on the morning of the 26th. Nothing unusual occurred in her labor, except that the second stage was somewhat
prolonged. Placenta came away in due time, and was not followed
by hemorrhage. First pregnancy.

"February 28th, at 8 A. M., she was seized with a very severe
chill, followed by increased frequency of pulse, and pain over
hypogastric region, extending as high up as the umbilicus. This
pain was very much increased by taking a full inspiration, or by
the application of pressure. Tympanitis very considerable; the
discharge abundant and very offensive; pulse 140; respiration 24.

"At 1 o'clock P. M., Dr. Barker saw her, and recommended that
she should be transferred to the Fever Wards, and put on the use
of the tinctura veratri viridis.

"At 2 o'clock P. M., after having been removed to the Fever
Wards, her pulse was 140; respirations 24; pain over hypogastric region intense; tympanitis very considerable; discharge abundant and very offensive; no mammary secretion. Dr. Barker requested that she should be seen hourly by one of the house staff,
and that her condition, as to the state of the pulse, respiration,

and other symptoms, and the dose of the veratrum viride given, should be recorded at each visit. The following is the record thus kept:

	Hour.	Pulse.	Resp.	Drops.	
Feb'y 28th.	2 P.M.	140	24	10	
	3	127	22	10	
	5	140	22	10	
	6	132	12	10	
	7	120	20	10	
	8	80	20	9	Bowels moved once.
	9	75	16		Vomited a greenish-colored fluid. Bowels loose.
	10	66	16	4	Vomiting ceased. Bowels moved once.
	11	65	22	7	
	12	58	13	2	
March 1st.	1 A.M.	64	52	6	Respiration very irregular. Inclined to sleep.
	2	58	25	2	Sleeping.
	3	59	21		Hiccough and headache.
	4	60	18	1	Hiccough still continues.
	5	66	20		Severe headache. Vomited a greenish-colored fluid
	6	66	21		Headache severe, and very restless. Vomited several times within last hour. Hiccough.
	7	58	20		Vomited once since last visit. Vertigo and headache.
	8	52	28		Sleeping.
	9	60	19		
	10	68	21	1	Slight hiccough.
	11	70	23	2	
	12	80	28	3	Tenderness over abdomen, marked. Tympanitis somewhat diminished. Discharge dark, bloody, and very offensive.
	1 P.M.	80	20	4	Visit of Prof. Barker.
	2	92	24	8	
	3	76	24	8	Face flushed.
	4	76	28	9	Sleeping.
	5	68	28	8	Sleeping.
	6	66	28	8	
	7	68	26	6	Slight hiccough. Bowels moved once.
	8	66	18		Vomited a greenish-colored fluid.
	9	68	24		Vomited once since last visit.
	10	60	28		Sleeping.
	11	64	28		Still sleeping.
	12	66	28	2	Sleeping still.
March 2d.	1 A.M.	56	32		
	2	70	24	3	Complains of pain in left thigh. There is slight swelling, and along its internal surface, over the course of the veins and lymphatics, the tenderness is so great that she can scarcely bear the lightest touch. Tenderness over abdomen still continues. Slight tympanitis. Discharge abundant, dark, bloody, and very offensive. No mammary secretion.
	3	76	24	4	
	4	65	20	3	Sleeping.
	5	78	22	8	
	6	68	22	4	
	8	64	24	4	
	9	72	24	6	
	10	64	28	2	Bowels moved once.

	Hour.	Pulse.	Resp.	Drops.	
March 2d.	11 A.M.	72	28	6	
	12	70	24	5	
	1 P.M.	64	24	3	
	2	60	20		
	3	64	24		
	6	68	28	3	
	7	72	28	5	
	9	80	28	6	Face flushed.
	10	80	26	6	
	11	80	28	8	
	12	80	28	10	Sleeping.
March 3d.	1 A.M.	80	29		Vaginal discharge now ceases to be offensive. No mammary secretion. Tympanitis still remains. Tenderness over abdomen still continues, though not so well marked. Tenderness and swelling in left thigh still continues.
	2	78	28	10	
	3	80	28	8	Slight hiccough.
	4	72	20	4	
	5	68	28		Vomited a greenish-colored fluid. Headache. Hiccough. Bowels moved twice.
	6	64	24		
	8	60	24		
	10	68	24	5	
	11	70	24	3	
	12	76	28	6	
	1 P.M.	80	28	6	
	2	80	22	8	
	3	76	30	4	
	4	76	26	5	
	5	72	32	4	Sleeping.
	7	64	32	2	
	8	72	28	5	
	9	68	30	4	
	10	68	28	3	
	11	72	28	5	
	12	70	30	7	Sleeping.
March 4th.	1 A.M.	72	32	8	Tenderness over abdomen not so intense. Slight tympanitis. Vaginal discharge now appears to be natural. Tenderness and swelling on internal surface of left thigh now seems to be diminishing. No mammary secretion.
	2	70	.30		
	3	64	28	2	
	4	64	28	3	
	5	60	24	2	
	6	60	28	2	
	7	60	28	2	Bowels moved twice.
	8	58	28		
	9	60	28		
	10	56	28	2	
	11	64	32	3	
	12	72	24	4	
	1 P.M.	78	32	6	
	2	80	28	8	
	3	80	24	8	
	4	80	30	8	
	5	80	28	8	Sleeping.
	6	60	32		

	Hour.	Pulse.	Resp.	Drops.
March 4th.	7 P.M.	64	24	6
	8	60	24	2
	9	60	28	2
	10	60	24	2
	11	60	26	
	12	58	24	
March 5th.	1 A.M.	60	22	3

She now says she feels much better. Her countenance looks much brighter, and she appears to be much improved in every respect. The tenderness which has been so intense over the abdomen, now is scarcely noticeable. Tympanitis very slight. Discharge very scanty, but normal. No mammary secretion. The swelling and tenderness on the internal surface of the thigh, in the course of the veins and lymphatics, has now disappeared altogether.

	2	68	26	4

Sleeping.

	3	60	22	2
	4			
	5			
	6	70	30	6
	7	64	24	4
	8	76	24	6
	9	76	24	6
	10	72	28	6
	11	64	24	3
	12	68	24	6
	1 P.M.	64	28	5
	2			
	3	56	28	
	4			
	5	64	24	5
	6			
	7			
	8	68	26	4
	9			
	10	72	24	4
March 6th.	8 A.M.	70	24	6

Feels well; improvement marked. No tenderness on pressure over abdomen. No tympanitis. Discharge still scanty, but normal. Slight mammary secretion.

	11	76	24	4
	12			
	1 P.M.	72	24	
	5	78	28	8
	6			
	7	76	26	
	8			
	9			
	10	72	24	4
March 7th.	9 A.M.	76	24	

She says she feels well and hearty. No tenderness over abdomen. No tympanitis. Vaginal discharge healthy. No tenderness or swelling in left femoral region. Appetite good. Bowels regular.

March 8th.	10 A.M.	76	24	

Continues to improve very fast.

"From this time she continued to improve, and in a short time was discharged as well and hearty as she ever was.

"Now here is a case occurring in a hospital, at the time of an epidemic, presenting a combination of symptoms which all familiar with the disease would pronounce truly alarming. By the verat. virid. the pulse was brought down from 140 to 60 per minute, and it was never permitted to rise above 80. The quantity administered varied according to the condition of the patient, two, three, or four drops being frequently sufficient to control the vascular excitement. No other medicine was used. In many other puerperal cases, I have seen equally striking results. I will briefly mention one which I saw, in consultation with Dr. Sayre, the tenth day after confinement. She was a primipara, and her convalescence seemed perfectly normal, until the sixth day, when she began exhibit some appearance of mental disturbance. She was especially anxious in regard to her religious condition. Gradually a high state of nervous excitement was developed, with insomnia, and when seen by myself, she had been decidedly maniacal for more than twenty-four hours. Her respiration was short and hurried, her pulse very rapid, her countenance anxious and frightened; she was incessantly talking and starting with apprehension, from the slightest movement in the room. No physical exploration could be obtained, but there were no local symptoms indicating pelvic trouble. She sat up in bed, and moved from one part to another with great rapidity. The verat. virid. was now given, and by its influence the pulse was brought down below 70 per minute, the respiration became slower, the mind tranquil, and she was enabled to sleep. I am informed by Dr. Sayre that in the course of a few days there was developed, in the pelvic cavity, an extensive abscess, which pointed externally, near the sacrum. Her convalescence was somewhat prolonged, but she eventually recovered."

I quote from these two eminent teachers with the hope to enable them to impart to my readers not only the information but the spirit so apparent in them. The great quantities of these medicines administered, and the profound effects produced by them, may be considered by some as recklessness, and, in many acute diseases, such imputation would be just, but in puerperal peritonitis anything short of this would be culpable dereliction.

CHAPTER XXXIII.

STOMATITIS MATERNA—NURSING SORE MOUTH.

THIS, as its name implies, is a disease peculiar to those who are, or who are about to be, mothers, and is attended with painful inflammation of some portion of the lining membrane of the mouth.

Although inflammation of the mouth is a symptom considered necessary to the full development of the disease, it must be regarded only as a symptom attending a general condition of the whole system, or, at least, of some one of its elementary constituents, perhaps the blood, which, by its own peculiar modification, implicates the solid parts in an action which they would not otherwise take upon themselves.

This view of its pathological seat, it is believed, is the only one which will enable us satisfactorily to account for many of the phenomena presented, both in respect to the time of occurrence, and the particular solid tissue affected.

What this modification of the condition of the blood may be, we can only conjecture; as, in the present state of science, the investigations, which have extended only to the physical and chemical qualities of this fluid, do not afford the means of ascertaining with any exactitude many of the most important changes which occur in it.

What are its vital conditions under varied circumstances, or indeed, in any case, perhaps, is entirely beyond the reach of our philosophy; and, it may be, will ever elude the imperfect means of research attained by the ingenuity of man.

If the pathological condition of the system is that of anæmia resulting from pregnancy and lactation, there must be some peculiarity about it, judging from its effects, differing from anæmia arising from other causes.

Several cases have occurred in my practice in women who were pregnant with their first child, which continued throughout the remaining time of gestation and during lactation. I remarked that all the patients in whom these cases occurred were very young, of scrofulous diathesis, weakly, and labored under most of the symptoms of anæmia resulting from other exhausting influences, such as pallor, languor, shrunken veins, &c.

I think that without other influences, pregnancy and lactation are not sufficient, and hence I believe we must look for extrinsic causes; and these are, probably, endemic and epidemic. By the former, I mean such morbific agents as are operating in the immediate locality of the patient; for, so far as I can learn, it is not of very general prevalence.

By the epidemic influence, I mean the extensive change which has taken place in the general cast of the diseases of the West, especially along the course of the large rivers, from the ordinary endemic bilious fever, and other miasmatic diseases, to the typhoidal, or continuous type, attended for the most part with affections of the mucous membranes, particularly of the alimentary canal.

There are three different varieties of this disease. The first includes the most simple variety, so far as local symptoms are concerned. It is characterized by a superficial and often diffused inflammation of the mucous membrane of the mouth, which may be confined to a small part, as the lips, or the end of the tongue; or it may spread throughout the whole cavity of the mouth and fauces.

The parts, upon examination, are found of a scarlet-red color, and dry; but as a general thing not much, if at all, swollen. This appearance may be of transient duration, lasting, probably, only for a few hours, more generally, however, for several days, when in a great many instances it completely subsides, leaving the patient to all appearance quite well, with the exception of a little debility.

In some cases the subsidence is not so complete, and amounts to only a very considerable remission of the soreness and distress.

After an uncertain length of time—in slight cases longer, and shorter in severe ones—the inflammation returns, and runs a similar course to the former paroxysm.

The paroxysms usually commence suddenly, with a sense of burning or scalding in some part of the mouth,—oftener, perhaps, on the end of the tongue than elsewhere,—which rapidly spreads, involving the parts continuously, until the whole mouth feels as if it had been scalded, and the acts of mastication and deglutition are intolerably painful. The subsidence of the pain and suffering is as sudden and gratifying as its onset was unexpectedly afflicting.

The second variety seems to engraft itself, as it were, upon the diffuse and superficial inflammation of the first. In addition to the above appearances described, a crop of vesicles are scattered over the whole or a part of the inflamed surface. These are often so clear and transparent, that without attention they may be overlooked; sometimes, again, they have an aphthous appearance, and are quite obvious. The duration of this eruption is about eight or ten days; but it often lasts much longer, and then consists of successive crops.

Although the symptoms subside sometimes as completely as in the former variety, the respite from suffering is commonly shorter and less complete.

In the progress of a case, it is not unusual for the appearances described as the two varieties to alternate with each other.

In the third variety, the whole force of the paroxysm is concentrated upon a small part of the surface, always, in my experience, upon the tongue, either upon its side or inferior surface. I have seen it begin with a fissure gradually leading to an ulcer, from a hardened tubercle, from the bursting of a vesicle, or simply from an inflamed point. However it may commence, a rapid ulceration destroys the substance of the tongue, until a ragged notched line half completes its amputation. Suddenly it ceases, the cavity granulates, fills up, and heals, but the organ is left distorted. The patient flatters herself, upon the cessation of each paroxysm, that some newly applied remedy is an all-sufficient sanative against the ills which she knows by experience are in reserve for her. But, unfortunately, with a returning paroxysm, she finds her suffering unmitigated. Notwithstanding this fearful ulcerative process, this variety is less dangerous and affects the constitution more mildly than either of the other varieties.

A very important consideration in connection with the local manifestation of the disease is, that the first two varieties are migratory, travelling from the mouth along the surface of the mucous membrane to all the neighboring cavities, down through the pharynx and œsophagus to the stomach; and thence through the whole extent of the alimentary canal, frequently finding permanent lodgement in some section of the extensive tube, and destroying the patient by originating chronic gastritis, duodenitis, ileitis, &c., or passing through the larynx, trachea, and into the bronchia. And if it does not, by establishing inflammation in some portion of these tubes, exhaust the patient (by originating chronic gastritis or duodenitis), it may awaken into existence the more fearful affections of the substance of the lungs. It has also followed the nasal passages into the different cavities of the skull, or maxillary antrum, and there induced permanent inflammation. At other times it travels through the Eustachian tube to the tympanum, and thence to the mastoid cells. And I have seen one case where permanent deafness of one ear and exfoliation of bone from the mastoid process occurred. The most common course for it to take, is into the alimentary canal and lungs. It is very prone to fasten fatal disease upon the lungs, when it is allowed to run on for any considerable length of time.

The date of the commencement of the above local symptoms is various. Hitherto, I believe it has been considered that they date from some time during the term of lactation, especially where the subject is young, and with the first child, but that they might reappear during subsequent pregnancies; and that they never are present in pregnancies, unless the patient has been subject to the disease during some previous term of lactation.

I am not certain that this is the general rule; but it is unquestionably the fact, that the woman will be more likely to experience such trouble after having once labored under the malady.

Accompanying the above array of symptoms are those of a general character. Perhaps, of all others, disorder of the digestive organs is the most prominent, as well as first in importance. Difficulty of deglutition, indigestion, and diarrhœa, form a part of these. All of these symptoms, like the local, are more or less paroxysmal,—the diarrhœa particularly. For some time it will harass the patient, exhaust the resources of her system, and then disappear, and allow her to recruit strength, to be prostrated again by its return. Indigestion is, probably, more or less constantly present.

Difficulty of mastication and deglutition vary, of course, with the local symptoms. Emaciation is also often considerable, and generally keeps pace with the digestive disorders.

Many other general symptoms might be enumerated, but as they are not peculiar or so important, they will not be noticed.

Treatment.—The first step in the cure of any disease, should be the removal of the cause when practicable. In cases where more than one cause contributes to the production of disease, the removal of one of them may so far interrupt the chain of impressions as to accomplish a cure.

Occasionally, this is found to be the case in the disease under consideration. The patient, by a change of her residence for the balance of the time of lactation, may get quite well. This, however, is not generally the case; and in all instances where the objections are not too weighty, the child should be weaned, or transferred to another nurse. I have seen so many cases of unfortunate terminations, and regard the condition of my patients so uncertain while laboring under this ubiquitous inflammation, in which, without any warning, some vital organ is involved, and it becomes the seat of destructive organic alteration, that I deem it a matter of great importance to take *immediately* the most effectual course within my knowledge to place the patient under the most favorable circumstances.

After an experience of many years' duration, I cannot feel quiet while my patient, if at all seriously affected, continues to nurse her child. I am thoroughly convinced that, in many instances which I have known to be followed by fatal secondary diseases, had the connection between the sore mouth and them been properly appreciated, and the causes of it understood, the mother's life would not have been sacrificed in a useless attempt at nursing; and I am well assured that, with the best management in grave cases, there is much more likelihood of the patient becoming worse than better while she nurses her child. I will also say that, in some cases, weaning will not of itself cure, although this is the general rule.

The most obvious and urgent indication for treatment will be found in the emaciation and debility. We must meet them with tonics and proper diet. Animal food, in as liberal quantities as the enfeebled digestive apparatus will allow, must be given. Milk, eggs, and mutton, I have found to agree best.

First upon the list of medicines may be placed cod-liver oil. It must be persevered in during the whole term of lactation, and as long after as any traces of the disease exist. The dose should be regulated by the capacity of the stomach, as much being given as can be retained. This remedy has the advantage of any other alterative and tonic in its soothing effects upon the irritated bowels when this difficulty exists.

This is a circumstance of the utmost importance, as we are often interrupted in the use of remedies by diarrhœa.

When there is no diarrhœa, or between the paroxysms of it, the most useful tonic, next to the oil, is the carb. of iron, prepared and used according to the following formula:

> R. Carb. potass.,
>   Sulph. ferri, āā, ... ʒjss.
>   Gum acacia mucil.,   .   .   . ℥iv.

Pulverize the potass. and dissolve in the mucilage; then pulverize, and add the sulph. of iron; mix well in an earthen mortar. Dose: half an ounce three times a day, gradually increasing to as much as the stomach will bear.

For persons very much reduced, of lax habit, brandy or wine, taken during meals, will sometimes do good. Very often they both disagree, when we may substitute ale, porter, or beer. Vegetable tonics and aromatics are useful in certain cases, where the iron disagrees with the stomach.

The diarrhœa, so exhausting to the patient and perplexing to the physician, should claim our attention also. Indeed, while this symptom exists, our efforts to restore the wasted energies of the system, and remove the unnatural condition of the blood, from an inability to introduce effective tonics, will be futile.

Entire quietude, while the diarrhœa is in active existence, in the horizontal position, should be strictly enjoined. Morphia, combined with acetate of lead, will often control it very completely. We may replace the latter remedy by sulphate of copper or nitrate of silver. This latter, when given with solid opium, often answers an excellent purpose in quieting the irritation of the bowels, and acting also as a tonic. Astringent injections and suppositories may also be used to advantage. Indeed, in some instances all our resources will be vainly tried to relieve this symptom.

A great variety of local remedies have been put in requisition, and all have enjoyed more or less reputation. I have used with benefit a solution of sulphate of copper, of different degrees of strength, the vinous tincture of Hydrastis Canadensis, a weak solution of nitric acid, sulphate of alum, borax, and sulphate of zinc; some agreeing in one case, and some in another. The only way I have ever been able, with certainty, to adapt the local remedies to different cases, was by trying them.

INDEX.